BRIDGE MANAGEMENT

Inspection, Maintenance, Assessment and Repair

This volume consists of papers presented at the First International Conference on Bridge Management, held at the University of Surrey, Guildford, UK, from 28th to 30th March 1990.

ORGANISING COMMITTEE

INTERNATIONAL TECHNICAL COMMITTEE

SPONSORS

Institution of Civil Engineers
1, Great George Street, London, SW1P 3AA

Institution of Structural Engineers
11, Upper Belgrave Street, London, SW1X 8BH

Department of Transport
2, Marsham Street, London, SW1P 3EB

BRIDGE MANAGEMENT

Inspection, Maintenance, Assessment and Repair

Edited by

J. E. HARDING, G. A. R. PARKE

and

M. J. RYALL

Civil Engineering Department, University of Surrey,
Guildford, Surrey, UK

Springer-Science+Business Media, B.V.

First edition 1990
Reprinted 1994

Ursprünglich erschienen bei Chapman & Hall 1990
Softcover reprint of the hardcover 1st edition 1990

ISBN 978-0-419-16050-2 ISBN 978-1-4899-7232-3 (eBook)
DOI 10.1007/978-1-4899-7232-3

A catalogue record for this book is available from the British Library

Library of Congress Cataloging-in-Publication Data available

Preface

The last 10 to 15 years have witnessed a growing awareness amongst bridge engineers around the world of the need for a properly formulated strategy of bridge management to ensure that the existing stock of road and railway bridges remains in service for as long as possible. Bridges represent a large capital outlay in any road or rail network, and are important not only because of their location, but by virtue of the cost implications if their capacity is impaired or if they fail outright.

Proper management is essential if this capital investment is to be protected and begins from the moment the bridge is first conceived until the moment it is replaced. In the past, such management has not been properly perceived and many bridges were constructed with little or no thought given to their inevitable deterioration due to ageing, accidental damage, and overstressing from increased weights and volume of traffic.

Bridge inspection of the last decade has revealed an alarming rate of deterioration to the world's stock of bridges, and the fields of maintenance, repair and strength assessment are growing in importance as the volume of traffic on our roads increases and heavier vehicles are sanctioned by government authorities. This is particularly so on secondary road systems where many older bridges are located and on primary roads which have attracted a density of commercial vehicles greatly in excess of that envisaged in the original design.

The management task is a daunting one and very often engineers have worked in isolation to try and provide an adequate programme of Inspection and Repair within strict financial limitations and with little idea of how their counterparts are tackling the problem elsewhere.

The aim of the *First International Conference on Bridge Management*, held at the University of Surrey during March 1990, was to improve communication by providing a focus for the presentation and discussion of the various management tasks facing bridge engineers throughout the world and to help them in the formulation of a sound Bridge Management Strategy at both national and local levels. We believe that this aim was realized and the papers presented provide a wide pool of knowledge from which engineers can now draw in order to develop a management system which will work best within the framework of their own particular constraints.

The volume contains 67 papers emanating from more than 25 countries, covering the fields of Inspection, Protection, Structural Assessment and Evaluation, Maintenance, Repair, Strengthening, Rehabilitation and Performance Monitoring as well as important papers on Bridge Management Systems currently operating in several countries.

The book provides a very useful reference manual for engineers working with Transportation Departments, Consultants, Contractors, Regional Councils and Local Authorities, who are interested in developing anything from the latest 'state of the art' expert system for Bridge Management on a cost effective basis, to the sharp end tasks of ensuring that bridges are adequately protected against the vagaries of aggressive environments.

The editors and conference o.ganisers all come from the University of Surrey which has had a varied interest in bridge design, construction and performance spread over many years. A Master's Degree course in Bridge Engineering has operated for the last 22 years. The course covers the main areas of analysis, design and construction by means of lectures, tutorials and laboratory sessions. The course structure has recently been remodelled to include design study modules where bridge designs are covered in-depth to latest codes of practice; and elements covering the major Bridge Management tasks have also been introduced.

Research in Bridge Engineering at the University has concentrated on the areas of material behaviour and structural assessment. In particular, a novel non-destructive method of determining the residual stresses in pre-stressed concrete bridges was developed here. The method is currently being developed to increase its application to other types of structure and materials such as reinforced concrete and masonry. Research is also in progress into the viability of composite bridge design using concrete and timber.

Also in preparation by the editors is a *Manual of Bridge Engineering* which will be of general use to bridge engineers worldwide, and should provide in one volume a mass of information which at present is very widely spread and often difficult to find. It will include comprehensive chapters on Construction, Analysis, Design, Management and Planning.

Finally, we would wish to offer sincere thanks to our sponsors who provided great encouragement and help with the conference; to our Technical Committee for their patience and diligence in poring over the many abstracts which were submitted and the comments and advice which they tendered on each; and last, but by no means least, to our Conference Secretary, Mrs E. Ryan who managed to cope admirably with the typing and administration of the numerous conference documents.

Contents

Preface v

Management Systems

1. The Centenary of the Forth Rail Bridge (1890–1990) 3
D. BECKETT

2. Bridge Management—An Overview 17
K. SRISKANDAN

3. Implementation of Bridge Management and Maintenance Systems (BMMS) in Europe and the Far East 29
A. B. SORENSEN and F. BERTHELSEN

4. Management of the Bridge Stock of a UK County for the 1990s . 39
J. PALMER and G. COGSWELL

5. Bridge Management within the Swedish National Road Administration 51
L. LINDBLADH

6. Local Agency Experience with the Utilization of Bridge Management Systems in Finland and the United States 63
A. R. MARSHALL and M.-K. SÖDERQVIST

7. The Pennsylvania Bridge Management System 75
R. M. MCCLURE and G. L. HOFFMAN

8. Data Information System for Structures: DISK 89
M. EL-MARASY

9. Optimization of Bridge Maintenance Appropriations with the Help of a Management System—Development of a Bridge Management System in Finland 101
A. Kähkönen and A. R. Marshall

10. Highway Bridge Management 113
J. W. S. Maxwell

11. Bridge Management in Cyprus 121
P. H. May and S. Vrahimis

Maintenance Strategies

12. Bridge Rehabilitation: Department of Transport's Fifteen-Year Strategy 135
D. A. Holland and P. H. Dawe

13. Comparative Maintenance Costs of Different Bridge Types . . . 145
D. Lee

14. Programmed Maintenance of Motorway Bridges: Italian Experience in the use of 'Expert Systems' 155
G. Camomilla, A. Dragotti, G. Nebbia and M. Romagnolo

15. Engineering Management of the Tamar Bridge 173
W. I. Halse and R. L. C. Stephens

16. Modelling and Predicting Bridge Repair and Maintenance Costs . 187
M. Bouabaz and R. M. W. Horner

17. Bridge Operation and Maintenance Costs 199
H. Ingvarsson

18. Clifton Suspension Bridge: An Historic Monument that Earns its Keep 205
D. Mitchell-Baker and S. Cullimore

19. A Systematic Approach to Future Maintenance 215
A. van der Toorn and A. W. F. Reij

20. Management of Bridgeworks Maintenance in the UK 223
N. J. Smith

Protection

21. Crack Bridging by Surface Treatments to Concrete 233
J. G. Keer and B. H. Le Page

22. Keeping Water Out of Concrete—The Key to Durability . . . 243
M. B. Leeming

23. Reinforced Concrete Bridge Protection in Northern Ireland . . 259
F. R. Montgomery and A. McC. Murray

24. Rebar Corrosion—FBECR: The Fight to Cure the Problem . . 267
J. A. Read

Inspection and Monitoring

25. Experiences with the First Generation of Prestressed Concrete Bridges in Germany 287
B. Göhler

26. Movable Bridge Machinery Inspection and Rehabilitation . . 295
C. Birnstiel

27. Application of Radar and Thermography to Bridge Deck Condition Surveys 305
D. G. Manning and T. Masliwec

28. Inspection Based Reliability Updating for Fatigue of Steel Bridges 319
A. G. Tallin and M. Cesare

29. Diagnostic Dynamic Testing of Bridges on Brenner Motorway . 327
R. Flesch and K. Kernbichler

30. Experience with the Management of Cable Stayed Bridges in Korea 339
H. Wenzel

31. Performance Monitoring of Glued Segmental Box Girder Bridges . 349
P. Waldron, M. Ramezankhani and B. Barr

32. Remote Computer-Aided Bridge Performance Monitoring . . 361
T. D. Sloan, J. Kirkpatrick and A. Thompson

33. Inspection and Repair of some Highway Bridges in Italy . . 373
M. P. Petrangeli

34. Inspection and Strength Evaluation of Concrete Highway Bridges in Czechoslovakia 383
K. Dahinter

35. Prestressing with Fibre Composite Materials and Monitoring of Bridges with Sensors 395
R. Wolff and H.-J. Miesseler

Assessment and Evaluation

36. Bridge Capacity Assessment and Control of Posting, Permit and Legal Vehicle Loads 405
F. MOSES

37. The Use of Reliability Analysis in the Assessment of Existing Bridges 417
C. MIDDLETON and A. LOW

38. Strength Assessment Methods for Concrete Bridges 429
P. A. JACKSON and R. J. COPE

39. Assessment of Stresses in Post-Tensioned Concrete Bridges . . . 439
C. L. BROOKES, S. H. BUCHNER and S. MEHRKAR-ASL

40. Assessment of Prestressed Bridge Beams 447
D. CULLINGTON

41. Fatigue Assessment of Orthotropic Steel Bridge Decks 459
C. BEALES and J. R. CUNINGHAME

42. Assessment and Rehabilitation of Suspension Bridges 475
P. G. BUCKLAND

43. Structural Assessment of a Bridge with Transversal Cracks . . . 489
C. ABDUNUR and J.-L. DUCHÊNE

44. Reliability Analysis Applied to Deteriorating Bridge Structures . 501
J. G. M. WOOD, R. A. JOHNSON and C. ELLINAS

45. Computer-Aided Sketching of Load Paths: An Approach to the Analysis of Multi-Span Arch Bridges 515
W. J. HARVEY and F. W. SMITH

46. The Assessment of Masonry Arch Bridges—The Effects of Defects 523
C. MELBOURNE

47. Theoretical and Experimental Investigations on Railway Bridges Dating from 1856 to 1895 533
F. MANG and Ö. BUCAK

48. Structural and Material Damage to Concrete Highway Bridge Decks in Saudi Arabia 549
M. Y. AL-MANDIL, A. K. AZAD, M. H. BALUCH, A. M. SHARIF and D. PEARSON-KIRK

49. Traffic Load Simulation Programme 563
D. LEBEK

50. Canada's Advanced National Standard on Bridge Evaluation . 575
P. G. BUCKLAND

51. Serviceability Assessment of Masonry Arch Bridges Using Vibration Tests 585
A. J. PRETLOVE and J. C. A. ELLICK

52. Assessing the Dynamic Properties of Existing Bridge Structures by Hammer Testing 595
J. R. MAGUIRE

53. Serviceability Performance of a Steel Highway Bridge 607
I. ROSENTHAL and M. ITZKOVITCH

54. Monitoring of Traffic Induced Strain in the Steel Reinforcement of a Concrete Bridge Deck 619
J. CAIRNS

Repair and Rehabilitation

55. Cracks in Steel Orthotropic Decks 633
P. MEHUE

56. An Analysis of the Behaviour of Reinforced Concrete Beams Following Deterioration and Repair 643
J. CAIRNS

57. An Investigation into the Effectiveness of Silane for Reducing Corrosion Activity in a Chloride-Contaminated Reinforced Concrete Bridge Structure 655
G. P. HAMMERSLEY, M. J. DILL and J. J. DARBY

58. Bridge Strengthening Using Load Relieving Techniques 667
B. PRITCHARD

59. The Integrated Construction and Conversion of Single and Multiple Span Bridges 677
M. P. BURKE JR

60. Inspection and Rehabilitation of Steel Trusses for Highway Bridges 695
A. G. LICHTENSTEIN

61. The Renovation of a Victorian Swing Bridge 705
B. SIMPSON and M. F. BLYTH

62. Cost-Effective Strategies in Bridge Management 715
R. S. REEL and C. MURUGANANDAN

63. Tension Drop in Cable-Band Bolts on Suspension Bridges . . . 725
Y. Kagawa and A. Fukushi

64. The Design of a Flexible Surface Mix for Use at Bridge Expansion Joints 737
A. R. Woodside and W. D. H. Woodward

65. The Repair of a Composite Concrete–Steel Bridge 747
P. H. Besem, M. Wouters and C. Warnon

66. Hydrodemolition—A Modern Technique of Concrete Removal in Bridge Repair 765
R. Medeot

67. Aluminium Extrusion Bridge Rehabilitation System 777
L. Svensson and L. Pettersson

Index of Contributors 785

Subject Index 787

MANAGEMENT SYSTEMS

1

The Centenary of the Forth Rail Bridge (1890–1990)

DERRICK BECKETT
School of Civil Engineering, Faculty of the Built Environment, Thames Polytechnic, Oakfield Lane, Dartford, Kent DA1 2SZ, UK

ABSTRACT

The Forth Rail Bridge, opened on 4 March 1890, celebrates its centenary at a time when bridge engineers are facing the need to formulate a strategy of bridge management to ensure that the existing stock of road and rail bridges remains in service for as long as possible. The bridge has over 54 000 t of steel in its superstructure with a surface area for painting of over 6 300 000 ft^2. This paper describes aspects of the design and construction of the bridge and the current maintenance programme.

HISTORICAL BACKGROUND

In the period that Thomas Telford was working on major projects to improve transportation in Scotland—the Caledonian Canal and a network of over 1000 miles of roads—there were two proposals to provide an alternative to ferry boats crossing the Forth between South and North Queensferry. The first, in 1805, was for a double tunnel—one for comers and one for goers. Outline details were produced but nothing further came of the scheme.

In 1818 James Anderson, a civil engineer and surveyor, produced a design for a three-span suspension bridge with a main span of 2000 ft, at a location close to that of the present rail bridge. It appears that the design required about 2500 t of iron and Westhofen[1] suggests that this quantity of iron, if distributed over the total length, would have given the structure a very light and slender appearance, so light indeed that on a dull day it would hardly have been visible, and after a heavy gale probably no longer to be seen on a clear day either. The project was fortunately abandoned and the same applied to a scheme drawn up by Thomas Bouch some 40 years later to cross the Forth 6 miles to the west of South Queensferry via a series of 500-ft spans with a total length of 2·5 miles.

In 1873 the Forth Bridge Company was formed for the purpose of constructing a suspension bridge with two spans of 1600 ft, again to the design of Thomas Bouch. Work on the structure was started, but on the night of 28 December 1879 the bridge over the River Tay, completed by Bouch some 18 months previously, collapsed during a severe gale. This shook public confidence in Sir Thomas Bouch's design and the suspension bridge project was abandoned.

The Bridge Company instructed their engineers, Messrs Barlow, Harrison and Fowler, to examine various forms of construction for both a bridge and tunnel crossing. The tunnel solution was not practical and cantilever construction was wisely preferred to the suspension form. The contract was let on 21 December 1882.

DESIGN

At the time of construction of the Forth Bridge, the world's largest span was John Roebling's Brooklyn suspension bridge (1595 ft), completed in 1883. Nineteenth-century engineers were aware of the problems of maintaining the stability of suspension bridges under heavy moving loads (the test load for the Forth Bridge was two 900-t trains) and thus Fowler & Baker chose the cantilever form of construction with a main span of 1710 ft. It remains the second largest of its type in the world and, apart from James B. Eads' St Louis bridge (the arch ribs were built out in cantilever), completed in 1873, it was the first major structure to use steel as a replacement for wrought and cast iron. Over 54 000 t were used in the superstructure.

The principle of the cantilever form of construction was clearly demonstrated by the 'human cantilever' devised by Benjamin Baker (see Fig. 1). . . . Two men sitting on chairs extend their arms and support the same by grasping sticks which are butted against the chairs. There are thus two complete piers, as represented by the outline drawing above their heads. The centre girder is represented by a stick suspended or slung from the two inner hands of the men, while the anchorage provided by the counterpoise in the cantilever end piers is represented here by a pile of bricks at each end. When a load is put on the central girder by a person sitting on it, the men's arms and the anchorage ropes come into tension, and the men's bodies from the shoulders downwards and the sticks come into compression. The chairs are representative of the circular granite piers. Imagine the chairs are a third of a

FIG. 1. Benjamin Baker devised this 'human cantilever', which elegantly illustrates the structural principle of the bridge.

mile apart and the men's heads as high as the cross of St Paul's, their arms represented by huge lattice steel girders and the sticks by tubes 12 ft in diameter at the base, and a very good notion of the structure is obtained. The man supported by the two human cantilevers was Kaichi Watanbe, a Japanese student of Fowler & Baker who was invited to participate in the model to remind audiences of the Far Eastern origins of cantilever construction.

The principal dimensions of the bridge in elevation are shown in Fig. 2 and the view looking along the bridge from the South Queensferry shore (Fig. 3) clearly indicates the 1 in 7·5 batter of the central towers which gives the structure its lateral stability. The tubular columns to the central towers are spaced 120 ft apart at the bottom and 33 ft at the top. Extensive experimental work led Baker to adopt, with some exceptions, a tubular form for the compression members and open lattice girders for the tension members (see Fig. 4).

In order to achieve balance for the dead load of the outer Queensferry and Fife cantilevers, it was necessary to load the ends at the junction with the approach viaducts to compensate for half the weight of the central girder, plus the effects of train loading. The central or Inchgarvie cantilever is balanced for dead load and the out of balance loading due to trains passing over a central girder is allowed for by making the base of the central tower

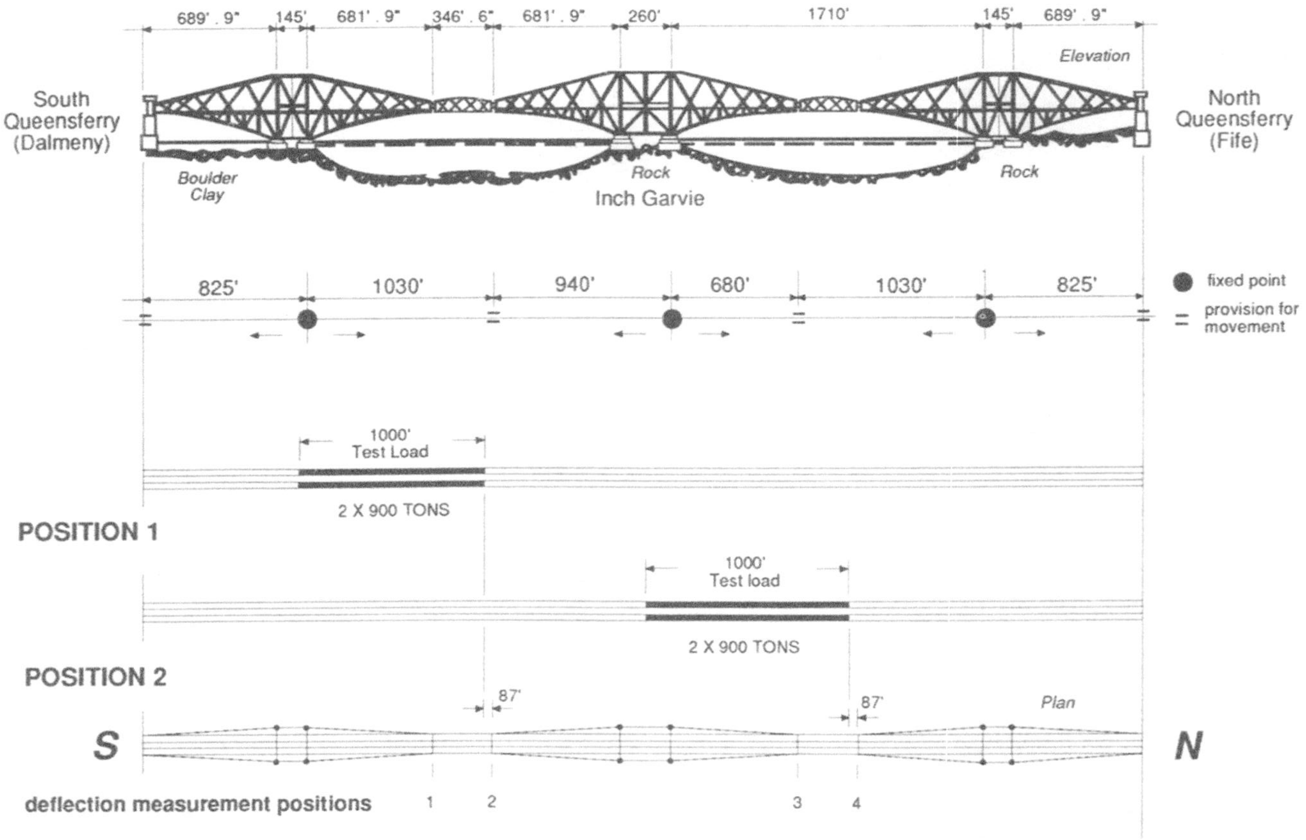

FIG. 2. General arrangement of the bridge superstructure showing the fixed points and test load positions.

FIG. 3. A view from the South Queensferry shore shows the batter (1 in 7·5) of the central towers.

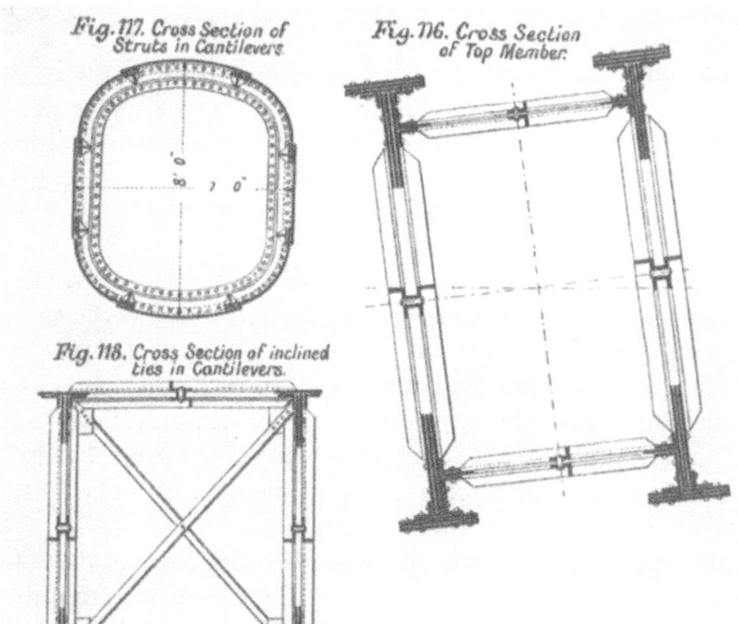

FIG. 4. Details of the cantilevers, tubular construction for the struts and lattice girders for the ties.

260 ft long in contrast to 145 ft for the Queensferry and Fife central towers. Thus uplift is avoided under the worst out of balance loading conditions.

The loads from the central girders and the internal viaduct carrying the permanent way are transmitted via the cantilevers into the central towers. At the base of each central tower there are four circular masonry piers. A skewback is provided over each pier to transmit loads from five tubular and five lattice members into the foundations.

CONSTRUCTION

Space permits only a brief outline of the construction of the bridge, which is described in great detail by Westhofen.[1] The logistics of the construction work listed at the end of the paper are staggering, even by contemporary standards. About 60 acres were used for workshops, storage areas, etc., on the south side of the river and a 2000-ft jetty was built out to the site of the Queensferry piers. Additional temporary works were required on the Fife shore. Construction of the foundations for the approach viaduct piers and the central towers supporting the cantilever structure required extensive work under water. Ordinary cofferdam work was used whenever possible, but for two of the Inchgarvie and South Queensferry circular piers it was necessary to construct pneumatic caissons. These were built on the shore and towed into position. Each circular pier is 70 ft in diameter at the base, reducing to 49 ft at the top. From low water point upwards the piers have a granite facing to the central mass of Arbroath stone set in cement. The piers were built to a height of about 18 ft above high water level and at this stage erection of the superstructure could begin.

The superstructure was built to a height of about 100 ft above high water level and then a lifting platform was constructed to allow the towers to be built to their full height. A start was then made on the erection of the bottom members of the cantilevers. As soon as the 'vertical' columns had reached their full height, work on the top members of the cantilevers was started. In the meantime, the approach viaduct piers were being raised. The 168-ft span double-lattice girders were erected to a convenient level and then lifted at the same time as the building of the masonry piers.

The final stage in the construction was the connection of the 350-ft span central girders to the cantilevers. From Fig. 2 it can be seen that the central girders were fixed at the Queensferry and Fife ends with provision for expansion at the other ends. Since the connection of each half of the central girders was to be made at midspan, both ends were temporarily fixed to the cantilevers until they were joined at the centre of the span. The expansion ends could then be released. A half bay of the central girder was constructed on temporary girders attached to the bottom members of the cantilevers and then moved back and connected to the cantilever. Work on the girders was then continued using cranes which had moved downhill from the top members of the cantilever arms. The connection of the two halves of the central girders depended on the surrounding air temperature. The lengths of the bottom booms were fixed so as to leave a gap of 4 in between the ends at a temperature of 60°F.

On 10 October 1889, when the temperature was about 55°F, alignment was achieved with the west boom, but the gap on the east boom was still about 0·25 in. A quantity of waste was soaked in naphtha and placed in the bottom booms for about 60 ft each side of the gap and set on fire. Alignment was achieved and the bolts inserted. The top booms were subsequently connected and the final stage in the construction was to rivet up all the connections which were temporarily bolted. Over 6 500 000 rivets were required in the superstructure, over half of which were fixed by means of hydraulic rivetting machines.

PROVISION FOR MOVEMENT

Observations at the time of construction of the bridge[1] indicated that expansion or contraction amounted to about $\frac{1}{160}$th of an inch for each degree temperature change for every 100 ft of girder length. This is equivalent to a coefficient of expansion of 0·0000052/°F. The various 'influences' to which the structure is exposed are: (1) expansion and contraction by changes of temperature acting in the direction of the longitudinal axis of the bridge, and to some extent also transversely upon the circular masonry piers; (2) influence of the sun's rays to one side or the other of the structure; and (3) wind pressure acting at right angles or nearly so to the centre line of the bridge. The fixed points on the bridge were established as (a) the southeast circular pier of the Fife cantilever, (b) the northeast circular pier of the Inchgarvie cantilever, and (c) the northeast circular pier of the Queensferry cantilever. The connection of the superstructure to the remaining piers was detailed to allow sliding movement in a longitudinal and transverse direction. The fixed points are shown in Fig. 2, and the expansion/contraction lengths are:

(1) 825 ft from the fixed point on the southeast circular pier of the Fife cantilever to the south end of the north approach viaduct (3·6 in for 70°F temperature change).

(2) 1030 ft from the fixed point on the southeast circular pier of the Fife cantilever to the south end of the north central girder and 680 ft from

the northeast circular pier of the Inchgarvie cantilever to the south end of the north central girder. The two movements overlap, giving a total length of 1710 ft (7·5 in for 70°F temperature change).

(3) 940 ft from the northeast circular pier of the Inchgarvie cantilever to the north end of the south central girder and 1030 ft from the northeast pier of the Queensferry cantilever to the north end of the south central girder. This gives a total length of 1970 ft (8·6 in for 70°F temperature change).

(4) The movement from the northeast circular pier of the Queensferry cantilever to the north end of the south approach viaduct is 825 ft (3·6 in for 70°F temperature change).

Thus expansion joints were provided at the south end of the north central girder, the north end of the south central girder, the north end of the Queensferry approach viaduct and the south end of the Fife approach viaduct. Expansion joints are also provided in the approach viaduct girders every second span.

TEST LOADING

The specification called for the contractor to ... test the strength of the girders as directed by the engineer, and if the girders fail to bear satisfactorily such tests as shall have been required, the work so failing shall be rejected and replaced at the cost of the contractor... This onerous task was commenced on 21 January 1890. Two trains, in parallel, entered the bridge from the south end. Each train comprised two 72-t locomotives at the head followed by 50 wagons, each weighing 13·5 t, and a single rear locomotive of 72 t. Each train weighed 900 t in total with a total length of about 1000 ft.

The trains were moved forward until the front locomotives were three-quarters through the central girder connecting the Queensferry north cantilever and the Inchgarvie south cantilever (see Fig. 2, test load position

TABLE 1
Observed vertical deflections for test load positions 1 and 2 (see Fig. 2)

Test load position	Vertical deflection (in) downward at location			
	1	2	3	4
1	5	$1\frac{3}{8}$		
2			$6\frac{7}{8}$	$2\frac{1}{2}$

1). This was considered to be the most unfavourable load position for the Queensferry north cantilever. The deflections recorded are shown in Table 1.

The train was moved to load position 2 (see Fig. 2) and the deflections recorded are shown in Table 1. The tests (as was the whole of the construction period) were monitored by two Board of Trade engineers, Major Marindan and Major-General Hutchinson. The observed deflections were well within the calculated values.

THE WORKFORCE

The construction work at its peak involved 4600 men, and working on a 360-ft high superstructure in adverse weather conditions, below sea level and/or in compressed air is a hazardous business and the construction of the bridge cost 57 lives. Surprisingly, not one death can be attributed to working in compressed air.[2] Benjamin Baker stated ... it was impossible to carry out a gigantic work without paying for it not merely in money, but in men's lives ... He also refers to the Hawes Inn ... the Hawes Inn flourishes too well for being in the middle of our works, its attractions prove irresistible for a large proportion of our 3000 workmen. The accident ward adjoins the pretty garden with hawthorns and many dead and injured men have been carried there, who would have escaped had it not been for the whisky of the Hawes Inn ...

The workforce was in part drawn from the shipyards and engineering works in the surrounding area and large numbers flocked in from England and Ireland. A French engineer, L. Coiseau, who had extensive experience of sinking foundations by means of compressed air, assisted in the substructure works. This attracted Italian, French, Belgian, Austrian and German workers who had experience of deep foundation work.

Housing and lodgings were provided at North and South Queensferry, and special trains ran from Edinburgh, Dunfermline and Leith accommodating both the day and night shift. Electric light was used for night work, but there were frequent failures and productivity was not high. A paddle steamer ran between South Queensferry and Fife, and there was a summer service between South Queensferry and Leith calling both ways at Inchgarvie, where accommodation was provided for 90 men during the sinking of the caissons.

Westhofen[1] refers to ... black sheep who were mere birds of passage, who arrived on the tramp, worked for a week or two, and passed on again to other parts, making little use of their hands except for lifting the Saturday pay packet and wiping their mouths clean at the pothouse ...

A sick and accident club was set up in the summer of 1883 and membership was compulsory with a contribution of 1 hour's pay per week, the maximum contribution being 8 pence per week. Free clothing was provided for men working on the foundations and those on the superstructure with warm clothing at a nominal charge. Shelters and dining areas were erected on the superstructure, and wages were paid by the contractor until any injured man returned to work.

There were several strikes instigated by the organising committees of various trades unions, and an outbreak of smallpox. The *Hougomont*, originally used to transport and store cement, was converted to a hospital ship and towed to an isolated position.

MAINTENANCE

It is a tribute to the designers, contractors and subsequent maintenance teams that a century after completion of the bridge no major structural repairs have been necessary. Some bracings and angle-plates, corroded by steam and smoke, have been replaced. It was estimated that the total internal (tubular members) and external surface area to be painted is about 6 300 000 ft^2 (145 acres) and the original procedure was as follows.

All the steel components on passing through the fabrication shops or yards were scraped with steel scrapers and wire brushes and then coated with boiled linseed oil applied hot. Prior to, or immediately after erection, they received two coats of red lead paint. This was followed by two further coats (iron oxide)—the first a primary coat of dark chocolate brown and the second a finishing coat of bright Indian or Persian red. The paint suppliers were as follows: at Fife, Craig & Rose's; at Inchgarvie, Calley's Torbay; at Queensferry, Carson's; and for central girders, Wollaston's Torbay. Internally, the tubular members received one coat of red and two coats of white lead paint.

One problem with the batter given to the sides of the structure is that the slope of the lattice girder flanges will allow rain water to collect (see Fig. 4). In locations where rain water is unable to drain, an asphalt–concrete compound was applied to allow water to run off by gravity. Alternatively, holes were drilled and the asphalt was laid to falls to allow water to drain to the holes.

Access hatches were provided to all the tubular members. Over 35 000 gallons of paint oils and 250 t of paint were required during the erection of the bridge. Access to many parts of the structure has always presented a problem, and the Health and Safety at Work Act has stopped the practice of using bosuns' chairs and rope ladders.

Currently maintenance work is controlled from an office close to

Dalmeny Station and the Forth Bridge supervisor, James Sinclair, is responsible for a maintenance team of about 40 men[2]—16 painters, 4 riggers, 2 platers, 3 welders, 2 woodworkers, 2 drillers, 2 boatmen, 1 bridge examiner's labourer, 2 watchmen and 6 permanent way men. Considering the risks of working at heights up to 360 ft above water level, the safety record has been excellent. The last fall was in 1970 and the last fatality Mr Sinclair's predecessor, who was unaccountably dragged along by a train when conducting visitors along the walkway.

The author visited the bridge at the end of May 1989 and soon appreciated that the climate can change from a gentle breeze on Dalmeny Station platform to strong winds on the piers at South and North Queensferry. Conditions would be worse on the bridge itself and, for safety, hand-held anemometers are used to check wind speeds. Above a speed of about 40 mph painting is stopped or not started. It is unusual to get more than 100 days a year for painting exposed surfaces.

Repainting is currently taking place on the south approach viaduct and a temporary materials hoist has been erected at the junction of the four-span masonry viaduct and the first steel lattice girder. Shotblasting down to base metal is being carried out behind plastic screens. A five-coat paint treatment is currently specified and spray application has now superseded the brush. A primer coat is applied within 4 h of reaching base metal, and for health reasons the use of red lead has been banned. This is followed by two undercoats, a coat of micaceous iron oxide and a final coat of the famous Forth Bridge Red. It is hoped that this specification will lead to a 15-year cycle by 1996. Scotrail is well advanced with improved means of access to the main spans, including suspended platforms to replace ropes, blocks and tackle, and a trolley system over the top of the bridge.

Another ongoing maintenance item is the replacement of the longitudinal timbers in trough girders which form an integral part of the internal viaduct. The original design with teak longitudinal sleepers has subsequently been modified. The timbers are now 28-ft long oak 'logs' and there are about 590 on the bridge. Replacement is carried out on Sunday when trains are worked over a single line.

The annual cost of maintaining the bridge, about £600 000 (wages and materials), compares very favourably with other major bridges, including the Forth Road Bridge, which employs 90 men to run and maintain it.[2] Speed limits on the bridge are 50 mph for high-speed trains and DMUs, 40 mph for standard diesel-hauled passenger trains and 30 mph for freight trains. There is no doubt that the bridge will meet the BS 5400: Part 1 requirement of a useful life of 120 years, with reasonable maintenance, and serve the Scotrail network for decades to come. However, it is of interest to reflect on the fact that the 1983 Review of Railway Finances[3] (Serpell

Report) included an option containing only 1630 route miles with no line north of Edinburgh. Thomas Telford, John Fowler, Benjamin Baker, William Arrol and all other pioneers of transportation in Scotland would have greeted this proposal with incredulity.

LOGISTICS

Total Length

South approach viaduct (Dalmeny)—4 masonry arches 300 ft, 10 steel spans 1680 ft.

Three main cantilevers—South Queensferry–Inchgarvie–Fife—5340 ft.

North approach viaduct—3 masonry arches 126 ft, 5 steel spans 840 ft.

Total length of bridge = 8295 ft.

Heights

Rail level from high water mark	158 ft
Clear headway for shipping	150 ft
Top of main cantilever towers from high water mark	361 ft
Top of main cantilever towers to lowest foundation	450 ft

Materials

Steel	54 160 t from the Welsh Landore Works, the Scottish Steel Company, and Dalzells Iron and Steel Works, Motherwell.
Rivets	6 500 000 (4270 t) made by the Clyde Rivet Company.
Granite	27 400 yd^3 from Aberdeen and Cornwall.
Ordinary stone	48 400 yd^3 quarried locally.
Concrete	64 300 yd^3 local aggregate. The Portland cement was manufactured on the Medway, transported by sea and stored in an old hulk, the *Hougomont*, with a capacity of 1200 t.
Paint	Internal and external surface area 6 316 200 ft^2 (145 acres) paint supplied by Craig & Rose's, Calley's and Wollaston's Torbay, and Carson's. Craig & Rose's still supply paint for the superstructure.

ENGINEERS AND CONTRACTORS

The engineers to the Forth Bridge Railway Company were Sir John Fowler, Bart, KCMG, CE, and Sir Benjamin Baker, KCB, KCMG, CE, and the contractors were Sir Thomas S. Tancred, Bart, Mr W. Arrol, Mr T. H. Falkiner and Mr J. Phillips. On the contractors' staff was Mr W. Westhofen, who wrote the authoritative account of the bridge's construction[1] which appeared in *Engineering* on 28 February 1890.

ACKNOWLEDGEMENTS

The assistance of the British Steel Corporation, the Director of Thames Polytechnic, the staff of Scotrail and Craig & Rose's in the preparation of this paper is gratefully acknowledged.

REFERENCES

1. Westhofen, W., The Forth Bridge. *Engineering*, 28 February 1890.
2. Grant, W. D. F. and Dargue, L. B., The Forth Bridge: its history, construction and maintenance. *Proc. Instn Civ. Engrs* (November 1985).
3. Serpell, Sir David, The Review of Railway Finances and Supplementary Volume. HMSO, London, 1983.

2

Bridge Management—An Overview

K. Sriskandan

Bridge Division, Mott MacDonald Ltd,
St Anne House, 20/26 Wellesley Road, Croydon CR9 2UL, UK

ABSTRACT

Bridge management is a comparatively new concept. This paper attempts to define the term and reviews the actions that constitute bridge management during the various stages in the life of a bridge. It comments on some of the issues and makes recommendations for the future.

INTRODUCTION

Bridge management is a term covering all the actions that need to be carried out to ensure that the bridge remains fit for its purpose throughout its design life without the need for excessive maintenance. Generally management has been considered as beginning only when a bridge has been built and is brought into service. However, much can go wrong with a bridge as a result of actions and or decisions taken at the concept, design and construction stages. It is essential therefore that bridge management is brought into play right at the very beginning.

In its wider sense, fitness for purpose includes safety, serviceability and durability. This paper will review actions taken to ensure fitness for purpose and comment on improvements that can be made.

CONCEPTUAL STAGE

Fortunately thousands of bridges have been built successfully in the past and therefore errors in concept are rare if one follows the well-trodden paths.

Errors in concept are more likely when producing a new or innovative design. Such cases need to be properly reviewed and tests undertaken where necessary to verify the adequacy of the concept.

DESIGN STAGE

It is standard practice in this country for the principles of design, including the concept, to be approved, at least from the safety and serviceability viewpoint, and for the design to be then certified and independently checked in appropriate cases. A similar form of review of the design is carried out in many countries. This tends to minimise errors in concept and the design of the structure. These checks are generally concerned with the safety and serviceability of the structure and may not fully cover all aspects of durability.

Until a few years ago it was considered that, provided steel in bridges was painted regularly and concrete was of good quality providing adequate cover to the reinforcement, there was no need for any further maintenance to modern bridges. However, recent experience all over the world has shown this to be a fallacy, and that durability and future inspection and maintenance should also be considered at the design stage. This includes the following:

(a) Design details that give rise to structural or durability problems.
(b) Design details that cause difficulties for inspection.
(c) Provision of access for inspection.
(d) Aspects of durability that are not covered by 'deemed to satisfy' rules in design codes.
(e) Use of appropriate materials—whole life cost versus initial cost.
(f) Design for maintenance.

Typical examples of each of the above are:

(a) Welding details that give rise to fatigue cracks, parapet base plates that are fixed in depressions in the concrete plinth, etc.
(b) Lack of vertical and lateral space between superstructure and substructure:
 (i) to inspect bearings, and
 (ii) to inspect (and paint) ends of steel beams and also to inspect the underside of expansion joints.
(c) Inspection and maintenance gantries for large bridges and provision of walkways, etc., or runway beams to provide access; manholes from boxes to top of very tall piers to inspect bearings.

(d) Specifying strength, water/cement ratio and cover to reinforcement is not sufficient to ensure durable concrete. Improvements in permeability and resistance to freeze–thaw cycles can be achieved by suitable design of the mix. Designers should specify the mix and associated curing required.
(e) Apart from cement replacement materials, it is also possible to use polymers, etc., to improve durability of concrete, and protective coatings on reinforcing steel and concrete to further enhance durability. The value for money of this extra expenditure should be tested by considering whole life cost of the structure, taking account of future maintenance and traffic delay costs, etc., in carrying out the maintenance.
(f) It is now accepted that many parts of bridges from bearings, expansion joints, waterproofing, surfacing, parapets, etc., to hangers, cable stays and external prestressing tendons all have to be replaced, some more than once during the lifetime of the structure. Account should be taken of this at the design stage so that replacements can be carried out with ease and with minimum delay to traffic.

CONSTRUCTION STAGE

It is necessary to ensure that materials and components manufactured on and off the site, and in fact the whole of the construction process, is carried out in accordance with the specification.

The concept of quality assurance (QA), which originated in the manufacturing industries, is being introduced into the construction industry. Quality assurance is not worth doing unless it also includes quality control, which has traditionally, in the UK, been exercised by an independent party, such as the engineer. It is therefore essential that responsibilities are clearly defined if quality assurance is called for within a contract.

The extent of the control that can be exercised will depend on the type of contract. In a normal three-part arrangement, aspects of durability which are not specified in codes, etc., can be introduced into the design and specifications following discussions between the client and the engineer. However, in cases where the design is also competitive, such as in a design and build contract, performance specifications may not be adequate for durability requirements. All aspects affecting durability will have to be very closely specified in the design brief.

Going one step further to the finance, design, build and operate cases, the client will need to be satisfied that the bridge is maintained adequately during the concession period and will be in good condition when it is handed over at the end of it. Sufficient and suitable clauses need to be written into the contract to ensure this.

Whatever the type of contract, every structure will need to be maintained from the time it is brought into service. Invariably the parties concerned with maintenance are different to those who carried out the design and/or construction, even if in some cases they are from the same organisation. It is necessary, therefore, for the maintaining parties to have a full set of the 'as-built' drawings and maintenance schedules which will indicate frequency of inspection and maintenance for each part of the structure. It should highlight areas where something has gone wrong during construction and which therefore require special monitoring. In large structures individual parts should be referenced, as should accesses for inspection.

IN-SERVICE STAGE

Once a bridge is completed and brought into service, it not only starts to carry traffic but is also exposed to the environment. It is subject to wind, rain and temperature changes, and also to chemical and (in some cases) biological attack. In time deterioration may occur and/or the bridge may have to carry heavier traffic loading than it was originally designed for.

Bridge management, in order to cope with these problems, not on one but thousands or even tens or hundreds of thousands of bridges, cannot be a haphazard reaction to something that happens. It must be a systematic consideration of all the problems to ensure best value for the money spent.

The first requirement is information about all the bridges on the network in the form of a *bridge inventory*. Most of the developed countries and states/counties have an inventory of their bridges, some of which are more complete than others. However, the information stored varies from the very basic to many items of information which are peripheral to the bridge structure itself. The aim should be to plan what information is required for use and store only that, and not each and every item which might be thought could become useful in the future.

The next requirement is to know the state of the bridges at any given time and this can only be obtained by bridge inspection.

Inspection and Testing

Until about 20 years ago there was no systematic inspection of bridges. So much so that in the UK inspection was treated as part of bridge maintenance. Following the OECD Report on Bridge Inspection,[1] most countries have adopted a strategy based on a hierarchy of bridge inspection as follows:

(a) General inspection which is a visual examination at not more than 2-yearly intervals but without the use of special access equipment.
(b) Principal inspections which are a detailed examination of all exposed parts of the structure from within touching distance, generally at not more than 6-yearly intervals.
(c) Special inspections are again a close examination but of specific parts of the structure following the findings of another inspection or some special loading event such as passage of abnormal loads, floods, heavy wind or mining.

In the UK, the Department of Transport has published its requirements for the inspection and records of its bridges[2] and also an Inspection Guide.[3]

Up to now it has been considered that special investigations including non-destructive testing need to be carried out only as part of special inspections. However, recent experience has shown that concrete in structures and the reinforcement and/or prestressing are subject to considerable deterioration due to carbonation, sulphation, chloride ingress from de-icing salts and/or alkali aggregate reaction. It is therefore necessary to undertake some testing as part of the principal inspection. This should include testing dust samples for chloride and sulphate content, half-cell potential measurements, cover metre survey and carbonation tests. Where necessary concrete cores should be obtained for strength tests and also to confirm from thin slices the presence of AAR if any map cracking is observed on the surface.

When inspecting steel bridges it is necessary to check whether corrosion is taking place. In some of the older metal (cast iron, wrought iron or steel) bridges the road surfacing is laid on fill material support by troughing or hogging plates or jack arches. In all cases the top flange is hidden and in the last the web is also enclosed. Therefore the road pavement and part of the fill will need to be excavated for a proper inspection.

On older bridges and even in many modern bridges, particularly the smaller ones, there is no permanent provision of walkways and the like for inspection. The increased recognition of the need for inspection from close quarters has created a demand for different types of access equipment. There are many types of lorry-mounted aerial platforms capable of carrying two to three people up to heights of 60 m. These are generally used to inspect bridge soffits and facias where standing space is available underneath the bridge. Where this is difficult, such as over water or over railways, over and under platforms with a reach of up to 14 m and more are available.

Inspection reports are generally written down on standard forms and then transferred to a computer data bank. It should be possible to produce hand-held data capture devices which would facilitate this process.

Maintenance Management Systems

The bridge inspection will identify the areas where maintenance is required for all the bridges in the network, but what it cannot do is to identify the most cost-effective remedy in each case and then rank them in order of priority.

There are many bridge management systems in operation, but few claim to carry out explicit economic analysis to decide the best maintenance operation. In most management systems the type of maintenance is selected by experience and judgement, and priority between bridges and different maintenance actions is based on rankings for various factors which are assigned subjectively and then added up to give an overall ranking.

However, there are certain routine maintenance operations, such as cleaning of drainage channels, washing down bearing seatings, etc., which are obviously good value for money, especially in bridges where chlorides have not permeated into the concrete. This work should be carried out on all bridges. Frequency and extent of each operation should be obtained from trials and a code of practice drawn up for routine maintenance.

For other maintenance work such as repair of deteriorating concrete, cost/benefit analysis of different maintenance strategies needs to be carried out to determine the optimum treatment. This is best done on a computer and when associated with the full bridge data base it becomes a proper bridge management system.

In order to compare the merits of different maintenance options for a given problem, it is necessary to know the maintenance profile and hence the whole life cost for each option. The whole life cost will be actual costs of maintenance plus the cost of delays and additional operating costs plus the increased costs of accidents, all discounted to present values. The benefits have to be taken as the disbenefits that would accrue if the particular item of maintenance was not done, also discounted to present values.

In order to evaluate the various options it is necessary to know the following:

(a) The maintenance profile for each option, i.e. the selected maintenance treatment, its cost, its effectiveness, subsequent actions and costs, and so on.
(b) Traffic delays and operating costs, and accident costs.

Assessments need to be carried out to determine suitable maintenance treatments. Realistic values for costs, etc., can only be obtained from historic records. It is essential that maintenance histories are stored in the computer, along with the bridge inventory and inspection records, to form the total bridge data base.

The costs related to traffic and accidents should be obtained after carrying out a traffic redistribution analysis given the extent and duration of closures for a given maintenance action. This again is a very complicated computation. The Department of Transport have written a program (QUADRO)[4] which is the ideal tool to use in cases where the traffic-related costs will be high. Some simplification should be possible for smaller schemes.

Structural Deterioration

The degradation of paint and other protective treatment and the subsequent corrosion of steel is the most common form of deterioration in structural steelwork. Inspection combined with regular local repair and total repainting before all the paint films break down will reduce the cost of maintenance and preserve the service life of the structure.

The other common defect is fatigue cracking. Again regular inspection combined with NDT methods are necessary to detect these cracks. It may be possible to arrest the progress of the cracks temporarily by various means, the simplest of which is to drill a hole at the end of the crack. However, in highly stressed areas proper repairs will be necessary. If the detail is repeated many times in a structure, it is well worth while testing the fatigue life of a proposed repair detail in the laboratory before implementing on site.

Cracks due to brittle fracture occur in steels which do not have the necessary notch ductility. Since the much publicised failure of Kings Bridge in Melbourne,[5] designers are well aware of this problem. However, there are many old steel bridges in existence which could well fail by this method in very low temperatures. They should therefore be inspected following a severe cold spell.

The process of corrosion of steel embedded in concrete is well known, and many papers have been written on the subject.[6–8]

Corrosion due to carbonation is generally due to lack of cover and/or high water/cement ratios. Testing for depth of carbonation is quite a simple process; however, if early action is not taken, it may result in general corrosion and subsequent spalling requiring repairs over large areas.

General corrosion due to chloride attack will also lead to expansion of steel followed by spalling, all of which can be seen before the integrity of the structure is affected. However, local pitting corrosion could result in a significant loss of steel area without any outward sign. For this reason it is necessary to carry out some investigations such as testing dust samples for chlorides at various depths and taking half cut measurements to determine whether there is risk of corrosion in the steel. Unfortunately there are no positive non-destructive means of determining whether corrosion is taking place and if so at what rate. Therefore selective examination will be necessary to determine the extent of the corrosion.

The need to repair small areas of concrete has produced a large number of so-called 'repair materials'. Care should be exercised in the selection of repair materials. Wherever possible it is best to use concrete similar to that in the parent structure. Coarse aggregates could be of smaller size than that in the original concrete.

Deterioration due to alkali–silica reaction (ASR) has also been a comparatively recent phenomenon in bridges in the UK. This has created much research, and guidance has been published on the diagnosis,[9] avoidance in new structures[10] and management of structures affected by it.[11] All of this cannot be properly summarised in this overview paper, except to state that ASR is probably not as serious as first thought.

Corrosion of tendons in post-tensioned concrete occurs when salt water gets into ungrouted or improperly grouted ducts. There is a paucity of non-destructive methods to detect such corrosion. Radiography can be used to detect whether ducts are fully grouted. However, this method is expensive and has health hazards and cannot therefore be used for routine investigations. If visual surveys and potential measurements indicate that corrosion may be taking place, the prestressing ducts can be exposed by careful drilling at selected places. This again is not recommended as part of routine inspections.

Assessments

Structural assessments are required if a structure has deteriorated, been subject to accidental damage, and/or is called on to carry a greater loading than it was designed for. The greater loading may be due to a general increase in vehicle weights and/or traffic densities *or* merely for a single event.

In the UK, the recent need for assessments arose from the increase in vehicle weights from 32 to 38 t in the early 1980s and the subsequent increase in European vehicle weights to 40 t. The Department of Transport produces its Standard for Assessment (BD21/84),[12] which was later amended to take in the 40-t lorry.

It is generally agreed that assessments should be done to some limit state method. In its Standards mentioned above, the Department has advocated the use of the relevant parts of BS 5400 for the assessment of concrete and steel structures. The publication of the Department's Standards and the deterioration of concrete bridges have generated a whole series of conferences in the UK and their proceedings[13–15] contain interesting papers on the subject. International organisations such as CEB and the OECD have also produced publications[16,17] on the subject of concrete bridges.

It is now considered that the use of design codes may be too conservative for the assessment of existing structures. Some consider that the serviceability limit state need not be considered at all. If the assessment is required due to increased loading, the serviceability limits which are required for new designs should be maintained for existing structures also. If the increased loading included a significant increase in axle weights, the fatigue life of certain details might be affected. Some assessment should be made of the fatigue life that has already been used up so that an estimate can be made of the time when close inspection of the structure for fatigue damage should commence.

It is also believed that the uncertainty about material strengths and weights in existing structures is reduced. If records exist of the plates and other members of a steel structure, it may be possible to make a better estimate of its characteristic strength. A large number of cores need to be taken from a concrete structure if its characteristic strength is to be updated. Weighing a whole structure by jacking it off its bearings is one means of determining its true weight and hence reducing the partial factor on self weight for assessment.

If upper bound methods and non-linear methods of analysis are used to analyse the load effects, these should be multiplied by a factor greater than 1 (γf_3 in BS 5400). Complex equations for the strength of a member are sometimes simplified in design codes. Where necessary the basic equations could be used in the assessment.

In deteriorated structures, allowance has to be made for the effects of deterioration. In corroded steel structures, the member size should be measured and the assessment carried out on the basis of the reduced section properties. Judgements need to be made about corrosion of steel in reinforced and prestressed concrete structures.

It is also advocated that probability methods be used in the assessment of major structures. In Ref. 16 it is recommended that the safety level for the actual current use of existing structures should be equal to the one generally required for new structures. The economic considerations that apply to the safety level of existing structures are different to the economic considerations that apply for new structures. It is therefore the author's contention that there could be a difference in safety levels for new and existing structures, especially when existing structures are being called on to carry greater traffic loading.

If the assessment shows that the structure is not capable of carrying the full loading, either weight restrictions or lane closures should be introduced to ensure that the total loading that could come on to the bridge does not exceed the assessed carrying capacity. In extreme cases it may be necessary to close the bridge altogether.

Increases in Loading

From time to time there is pressure, on the grounds of efficiency, from the haulage industry and vehicle manufacturers to increase the gross vehicle weight and also axle weights. The effect of this on highways and bridges are considered before increased weights are permitted. Wherever possible axle loads and spacings are adjusted so that the total effect on bridges is not increased. Clearly general increases in vehicle weights need to be managed in this way if bridges are to remain fit for their purpose.

There is, of course, illegal exceedence of vehicle and/or axle loadings Enforcement, using weighbridges at selected spots, can be a deterrent; the resource effort required for total eradication can be quite high. Therefore data should be collected about these overloaded vehicles and allowance made either in the loading or the partial factor on loading.

CONCLUSIONS

1. Bridge management covers the whole period from concept to ultimate demolition or replacement.
2. At concept and design stage consideration should be given not only to safety and serviceability but also to durability, inspectability and ease of maintenance, including the replacement of various parts of the structure.
3. At every stage the whole life cost of the various options should be considered. This should include costs of traffic delays and other similar indirect costs.
4. Some investigations should be carried out during the principal inspection.
5. Bridge maintenance management systems should include an explicit economic analysis of the various maintenance options.
6. Routine maintenance is obviously good value for money. There should be a code of practice for routine maintenance which should give frequency and standards for such maintenance.
7. There should be a code of practice for the assessment of existing structures.
8. Axle weights and spacings should be controlled when gross vehicle weights are increased.

REFERENCES

1. *Road Research: Bridge Inspection.* OECD, Paris, 1976.
2. Trunk Road Management and Maintenance Notice: Trunk Road and Motorway Structures—Records and Inspection (TRMM 2/88). Department of Transport, London, 1988.

3. *Bridge Inspection Guide*. Department of Transport, HMSO, London, 1983.
4. Queues and Delays at Roadworks (QUADRO2). Department of Transport, London, April 1982 (and subsequent revisions).
5. Report of the Royal Commission into the Failure of Kings Bridge. Government Printer, Melbourne, 1963.
6. The Durability of Steel in Concrete. Part 1: Mechanism of Protection and Corrosion (Digest 263); Part 2: Diagnosis and Assessment of Corrosion Cracked Concrete (Digest 264). Building Research Station, Garston, UK, 1982.
7. VASSIE, P. R., Reinforcement corrosion and durability of concrete bridges. *Proc. Instn Civ. Engrs*, Part 1 (August 1984) p. 76.
8. PULLAR-STRECKER, P., *Corrosion Damaged Concrete: Assessment and Repair*. CIRIA, Butterworths, London, 1987.
9. PALMER, D., *The Diagnosis of Alkali–Silica Reaction—Report of a Working Party*. British Cement Association, Slough, 1988.
10. Alkali–silica reaction—Minimising the risk of damage to concrete: guidance notes and model specification clauses. Technical Report 30, Concrete Society, London, October 1987.
11. Structural effects of alkali–silica reaction—Interim technical guidance on appraisal of existing structures. Institution of Structural Engineers, London, December 1988.
12. The assessment of highway bridges and structures. Departmental Standard BD21/84, Department of Transport, London, March 1984.
13. Assessment of reinforced and prestressed concrete bridges. Papers presented at a seminar organised by the Institution of Structural Engineers, September 1988.
14. Concrete bridges—Management, maintenance and renovation. Proceedings of one-day conference, Concrete Society, London, February 1989.
15. Bridge assessment symposium, Leamington Spa. *Construction Marketing*, June 1989.
16. Diagnosis and assessment of concrete structures, Bulletin d'Information No. 192. Committee Euro International du Beton, Lausanne, January 1989.
17. *Road Transport Research—Durability of Concrete Road Bridges*. OECD, Paris, 1989.

3

Implementation of Bridge Management and Maintenance Systems (BMMS) in Europe and the Far East

ANDERS B. SORENSEN and FINN BERTHELSEN

COWIconsult Consulting Engineers & Planners AS,
45, Teknikerbyen, DK-2830 Virum, Denmark

ABSTRACT

This paper describes the DANBRO bridge management and maintenance system (BMMS) and the aspects of implementing the BMMS at bridge authorities.

The system provides a rational and systematic approach to organising and carrying out activities related to bridge management. It consists of:

—*A database system.*
—*Three modules:*
- *The inventory module.*
- *The inspection and bearing capacity module.*
- *The ranking and budgeting module.*

—*Manuals for all activities.*

The implementation of BMMS is illustrated by two case studies: the Danish State Railways (2500 bridges) and the Department of Highways in Thailand (10 000 bridges).

INTRODUCTION

It is the objectives of the BMMS to give the bridge authority a tool that helps to:

—Ensure the safety and network capacity.
—Ensure objective information on all bridges.
—Optimise utilisation of allocated funds.
—Ensure technical–economical feedback.

The BMMS is run on personal computers and minicomputers.

The DANBRO BMMS has been developed in cooperation between the Danish Road Directorate, the Danish State Railways and COWIconsult.

CONCEPT OF THE SYSTEM

Figure 1 shows the main activities in the system.

General

Manuals are available for all the activities shown in Fig. 1. Such manuals are especially needed during the inspections in order to ensure objectivity in the inspector's evaluations. Results from the inspections must obviously be comparable from bridge to bridge if the ranking of structures in need of repair are to be reliable.

Information on a bridge is categorised (administrative, geometry, materials, condition, etc.) and associated to a specific element. The elements are arranged in a hierarchic order as shown in an extract from the BMMS in Fig. 2.

Elements are selected at the highest level for which the information is representative, e.g. conditions can be given for level 1 (the bridge in general) and/or for an element on a lower level as required by varying conditions (e.g. level 4, element, foundation, if deteriorated to an extent different from the

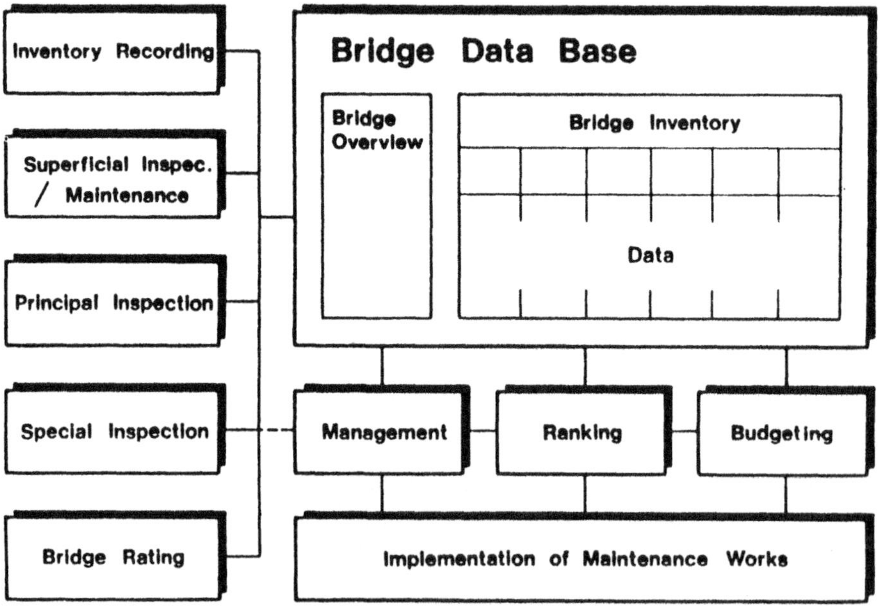

FIG. 1. Main activities covered by the system.

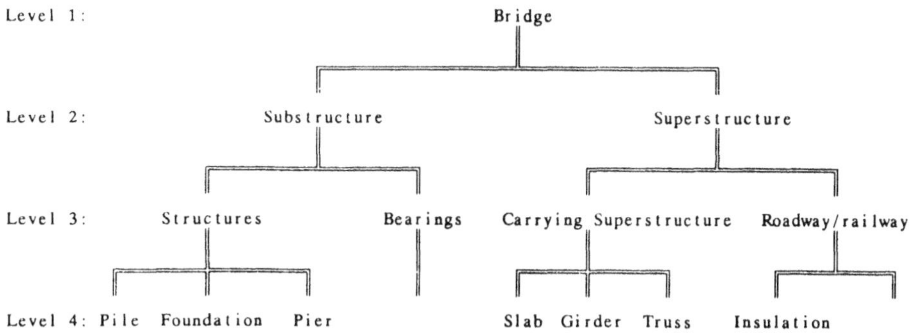

FIG. 2. Hierarchic element structure, extract only.

bridge in general). This ensures the minimum amount of information to be entered and stored in the database.

The System Modules

The inventory module

By means of the bridge information in the database, the inventory module facilitates predefined output forms corresponding to routine enquiries. A typical predefined form is shown in Fig. 3. Further, the user may create individual reports comprising selected information.

A graphical interface is being implemented to the system. The user will on the screen be able to zoom 'down' from a map of all the routes to information on a specific bridge. This facility provides an easy and logical access to the bridge information. The graphical interface is illustrated in Fig. 4.

DEPARTMENT OF HIGHWAYS - BRIDGE OVERVIEW, IN GENERAL - THIS DATE: 31/08/02

Area Code: 610 Nakhon Ratchasima Division Rev. Date: 31-06-01
Route No.: ALL , Control Section No.: ALL , km: -
km: - BK

Route No.	Ctrl Sect No.	Structure No.	PS	km km BK	Name	Pcn	Con/Chan	Superstructure - design - material
2	0702	16	P	350.077	Ban Sida-Phon Huaisokkabuong	1	2506	-slab -reinforced concrete
2	0702	19	P	352.630	Ban Sida-Phon Huaiphailon	1	2506	-slab -reinforced concrete
2	0702	26	P	357.056	Ban Sida-Phon Huaisokkhong	1	2506	-slab -reinforced concrete
2	0702	28	P	357.519	Ban Sida-Phon Huailuk	1	2506	-slab -reinforced concrete

FIG. 3. Bridge overview (example from the BMMS in Thailand). (The Thai year 2531 corresponds to the Gregorian year 1988.)

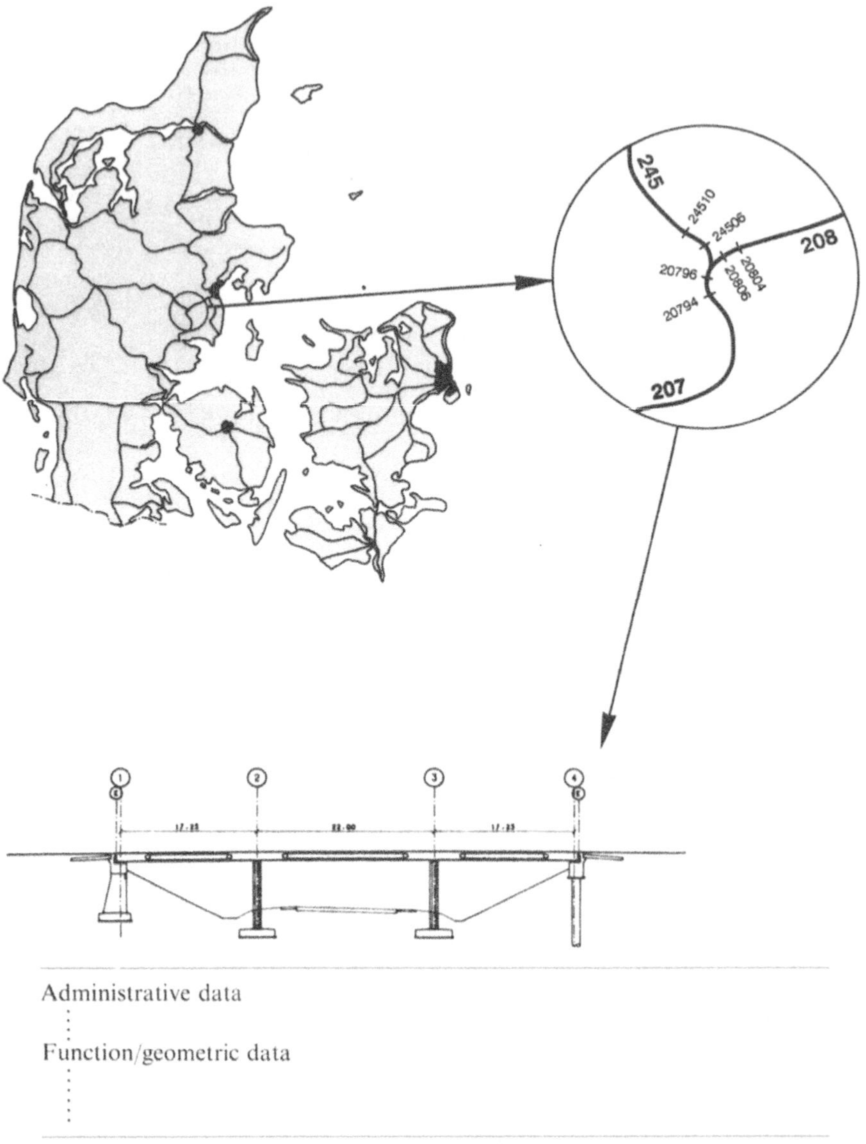

FIG. 4. Zoom from a map of all the routes to information on a specific bridge.

The inspection and bearing capacity module

There are three types of inspections: superficial inspection, principal inspection and special inspection.

—The superficial inspection is carried out by the local road or railway personnel with short intervals. They clean the structural members and repair minor damage in accordance with manuals. The maintenance level for each bridge may depend on the importance of the bridge.

—The principal inspection is carried out by local well-trained engineers, who typically deal with about 500–1000 bridges. Inspection intervals are normally 3 years with a range of 1–6 years depending on the condition of the bridge.

The inspector evaluates the damage and gives a condition mark to selected elements in order to describe the condition of the bridge. Significant damage may be registered if required for documentation. Further, he estimates the remaining life and corresponding repair costs of each element, and recommends the time for the next principal inspection. Finally, it is registered if a special inspection is required.

—The special inspection is a detailed investigation of a bridge structure or parts of it, when the condition or load-bearing capacity is about to reach an unacceptable level. This inspection is either initiated on the basis of the recommendations by the principal inspector or on the basis of the ranking list. Laboratory tests will normally be required. The findings in connection with an inspection will normally lead to alternative maintenance strategies for rehabilitation with corresponding budgets (lifetime and repair costs for elements) (Fig. 5).

The system comprises a bearing capacity programme for rating of the bridges based on the inventory data, including the inspectors' reports on current conditions, and for rating of actual trucks (Fig. 6).

The bridge rating programme calculates the bearing capacity of a bridge and compares the result with the loading from a standard truck. A bridge-bearing capacity class is given as a percentage of this standard truck loading. Material deterioration, as reflected by the condition mark, will often cause lower material strength and thereby lower bearing capacity. The system includes a facility for adjusting the bearing capacity in relation to condition marks.

```
DEPARTMENT OF HIGHWAYS      - SPECIAL INSPECTION REPORT -     THIS DATE: 31/08/02

Route No.:     2 , Control Section No.: 0702 , Structure No.:     19
Stationing: km 352.630 and km           BK

Repair Proposal No.: 1

Superficial Inspection/Maintenance:
        Maintenance level (I minimum, II normal, III high) :     II
        Yearly budget  (1,000 Baht)                        :      5

Repair Works in 1,000 Baht:

Year      Element                          Method          Estimated repair costs
       Code    Name                                              (1,000 Baht)
-----------------------------------------------------------------------------------
2531  BRIDGE  Bridge                       Repair                    850
2535  ABUWAL  Abutment Walls               Strenghtening             250
2541  EXPJOT  Expansion Joints             Exchange                   60
2560  BRIDGE  Bridge                       Exchange                 3000
```

FIG. 5. Special inspection report (example from the BMMS in Thailand).

```
---------------------------------------------------------------------
DEPARTMENT OF HIGHWAYS    - BRIDGE RATING -     THIS DATE:31/08/07
---------------------------------------------------------------------
Route No.:     2 , Control Section No.: 0702 , Structure No.: 16
Stationing: km 350.077 and km          BK
---------------------------------------------------------------------

Rating Document
---------------

Element: 3.10 Carrying Superstructure
Last Principal/Special Inspection: 31-08-02
Element's Condition Mark: 1
Remarks:

Load Case 1:

- Load factor on Dead Load      = 1.10
- Load factor on Standard Truck = 1.25

Inventory Rating Class: 133 %
Operating Rating Class: 104 %

Rating Mark: 0
```

FIG. 6. Results from the bridge rating on a single bridge (example from the BMMS in Thailand).

The vehicle rating programme calculates for an actual vehicle the effects on a set of bridge spans and compares the results with the standard truck's effects on the same bridge spans. The maximum value expresses the actual vehicle's class in terms of the standard truck.

The system thereby facilitates the administration of heavy transports. By a simple comparison of an actual vehicle's class with the bridge class, it is checked whether a given truck or a specific truck type can pass over the bridge.

The ranking and budgeting module

An initial and rough long-term budget calculation is based upon the replacement costs for elements estimated by the principal inspection.

The condition marks and the rating mark of each bridge or its elements form the basis of the ranking of bridges in need of repair. The importance of each element for the function and safety of the bridge and the importance of the route is included in a model that calculates the ranking point for the bridge.

The rank of each bridge (Fig. 7) thus reflects:

—The current condition.
—The actual bearing capacity in relation to the required capacity.
—The importance of each element for the function of the bridge.
—The importance of the route that passes over the bridge.

Alternative maintenance strategies are requested to be prepared by a special inspection of the bridges in the top of the ranking list and of all

DEPARTMENT OF HIGHWAYS - RANKING OF STRUCTURES - REPORT DATE : 31/09/12

Ranking Points have been calculated Year/Month/Day: 31-09-08

Structures are reported for the following unit(s) only:
- Code: 610 Nakhon Ratchasima Division

Route No.	Control Section No.	Structure No.	Km	Km BK	No	Ranking Points Total Pr	Cond. Pc	Bear Cap. Pb	Cond. Mark BRIDGE
219	0500	14	14.961		3	49.2	24.6	24.6	3
2057	0100	27	16.117		10	11.0	12.3	0.0	2
2149	0101	28	11.933		43	5.5	6.1	0.0	2

FIG. 7. Ranking of the bridges in need of repair (example from the BMMS in Thailand).

bridges with a need for immediate maintenance works in order to prepare an accurate short-term budget. Initial selection among alternative maintenance strategies are normally based on net present value, including possible considerations of traffic costs, imposed on users by the work programme (detours, etc.).

The investment schedule for the selected maintenance strategies or inspection estimates from the principal inspection are summarised for all bridges and compared with available budgets (Fig. 8).

When discrepancies between available funds and estimated maintenance costs are found, the user indicates from which period and to which period maintenance costs shall be moved. The system then directs the identification

DEPARTMENT OF HIGHWAYS - BUDGET - REPORT DATE : 31/09/12

The budget has been accepted Year/Month/Day: 31-10-01

Budget Report for the following unit(s) only:
- Code: 610 Nakhon Ratchasima Division

Year / Cost in mio. Baht Route No.	Contol Section No.	2531	2532	2533	2534	2535	2536-2546	2547-2567	2568-2593
2	0702	0.9	0.2	0.3	0.0	0.3	2	4	2
2	0800	0.0	0.0	0.0	0.0	0.0	1	3	6
2	0901	0.0	0.0	0.0	1.0	0.0	0	0	2
23	0101	0.0	0.4	0.0	0.0	0.2	3	0	0
23	0102	0.0	0.0	0.0	0.5	0.0	0	1	4
23	0103	0.0	0.0	2.4	0.3	0.0	1	0	0
202	0500	1.0	0.5	0.0	0.0	0.0	0	0	0
207	0202	0.0	0.7	0.0	0.0	0.0	0	0	1
207	0300	0.0	0.0	0.0	0.0	0.0	0	0	0
219	0500	2.8	0.0	0.5	0.0	1.0	3	2	3

FIG. 8. Final budget for a field division (example from the BMMS in Thailand). (The Thai year 2531 corresponds to the Gregorian year 1988.)

of bridges where alternative maintenance strategies comprises maintenance works in the 'move to' period instead of the previously selected ones in the 'move from' period.

Establishment of the BMMS

The outlines for the system were settled by a working group comprising:

—Representatives of the future users:
- Inspector.
- Coordinator of maintenance works.
- Technical assistant.
- Budget planner.

—System planner, who has long-term experience of bridge design and maintenance projects.

—Database specialist.

The definition of the system was made top-down. The user group defined and described the main items first and detailed later. Preliminary programmes and simple registers were prepared in order to present the ideas:

—Firstly, the users decided the modular activities.

—Secondly, the following selection of information types was settled by the working group as the typical data set:
- Administrative data.
- Function/geometric data.
- Material data.
- Load-bearing data.
- Condition data.
- Maintenance strategy.

—Thirdly, the users decided that all components of the BMMS should be operational immediately after installation and thus be able to perform ranking and budgeting even with limited information. The amount of information that can be stored for old and damaged bridges should, on the other hand, be unrestricted.

The above-mentioned data set was therefore used to describe a bridge element as illustrated earlier in the paper.

—Fourthly, all activities connected to bridge management were described in detail by the system planner and the database specialist. For each of the activities the following were described:
- The contents of the activity.
- Which information is needed prior to the activity.
- Which information is updated or created during the activity.

Implementation of BMMS in the Danish State Railways

The Danish State Railways (DSB) is responsible for the maintenance of approximately 2500 bridges on the overall railway network of Denmark. In the past inspections were carried out on all bridges and the reports were held in large manual archives. The DSB decided to set up a BMMS in 1985 in order to secure an updated overview of their bridges, the proper function of the bridges, and to establish a ranking list of maintenance works and a tool for budgeting of maintenance works. The implementation was completed in 1987.

The following activities were carried out:

—Description of the existing organisation.
—Establishment of new routines and teaching in these.
—Implementation of EDP system and manuals.
—The existing inspection reports were entered into the database.
—A preliminary ranking list was calculated in order to select the bridges where special inspections were required.
—Special inspections, including a set-up of maintenance strategies, were carried out and an accurate short-term budget was prepared.

Implementation in the Department of Highways—Thailand

The Department of Highways (DoH) in Thailand is responsible for the maintenance of approximately 10 000 bridges on the routes that form the superior road network of Thailand. In the past only the seriously damaged bridges were reported—if identified—and subsequent budgets for the repair works were given. This situation caused the DoH to hold a budget reserve each year, and the unforeseen amount varied from year to year. To manage the maintenance budget in the most effective way and to keep the bridges in good condition, the DoH decided to set up a BMMS.

The implementation covers all the modules described in this paper and was initiated in November 1987. From April 1988 to February 1989 a full inventory recording of all the DoH bridges in Thailand took place. Furthermore, a country-wide bridge management and maintenance organisation has been established and the future members have been involved in the design of the system from the very beginning in order to secure its acceptance in the Central Administration as well as in the districts. The majority of the 10 000 bridges have now been inspected. The total implementation is planned to be completed early in 1992.

The following activities are carried out:

—Description of the existing organisation and routines.
—Introduction of a new organisation and new routines.
—Implementation of EDP system and manuals.
—Teaching in new routines in Bangkok and Denmark.

CONCLUSION

It is our experience that an early involvement of the users and well-planned educational sessions are of considerable importance to a successful implementation of a bridge management and maintenance system (BMMS). This will be carried out in order to secure the daily use of the system and avoid a return to 'the old way'.

The implementation both from Denmark and Thailand proves that the BMMS will lead to:

—Establishment of an efficient up-to-date overview of the bridges.
—Consistently updated and objective information on each bridge and easy access to it.
—Systematic evaluation of the condition carried out on the bridges but only when needed.
—Improved basis for budgeting and maintenance planning by objective ranking of the bridges.
—Savings from more reliable and flexible budgeting due to the minimisation of unexpected repair works.
—Savings from a more rational administration of the bridge network with a minimised need for inspections and collection of information.

4

Management of the Bridge Stock of a UK County for the 1990s

John Palmer and Graham Cogswell

County Engineer Department of Surrey County Council, Highway House, 21 Chessington Road, West Ewell, Epsom, Surrey, UK

ABSTRACT

The aim of this paper is to show how Surrey County Council has developed a bridge management system in order to improve control of the stock of structures so that optimum use is made of resources and the overall condition is enhanced.

INTRODUCTION

The bridge office in Surrey has some 2500 bridges or highway structures to manage. These vary from traditional structures such as masonry arches and walls through the whole spectrum to modern post-tensioned structures (see Fig. 1).

Maintenance in the past was the Cinderella of bridge works, often neglected or left until resources were available (or may be just left! If you don't know there is a problem you can't deal with it!). However, ignorance has proved not to be bliss but more like a nightmare.

There is now a growing requirement for local highway authorities to operate on a business footing and target the limited available cash.

The majority of the problems with bridges built over the last few decades or so are material, workmanship or detail related. It is important to record suppliers, types and origins of materials, and details changed on site or during fabrication.

The regular inspection process which gathers information means that the facts have to be assessed and action decided on. It is of vital importance that the right information is collected, stored and used in the correct way.

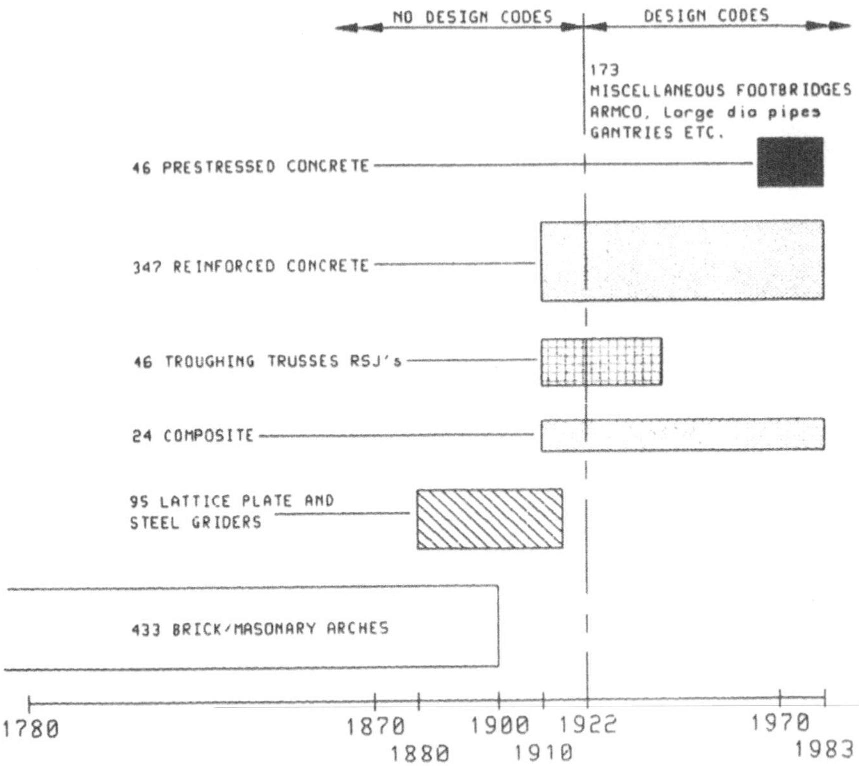

FIG. 1. Bridge deck construction materials and date for county road bridge stock (total number 1172).

A management system is a tool that will allow the best use of resources to ensure that the aim of effective management is achieved cost effectively. It is important also to develop a culture within the organisation that fully utilises all available resources with a business-like approach. The paper shows how this has evolved in Surrey and the future strategy.

FUNDING

An historical look at bridge construction since 1780 in Surrey is shown in Fig. 2.

The prosperity of the Victorian age is reflected in the gradual increase in bridge construction, the growth in canal, railway building and eventually roads showing a peak at the turn of the century. The depression between the wars resulted in a reduction in new bridge works although there were some public works using unemployed miners in the Guildford area during this period.

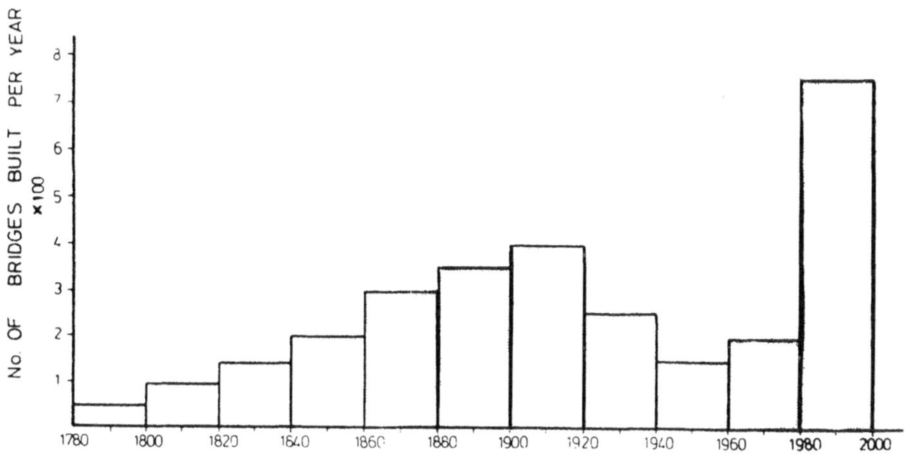

FIG. 2. Distribution of new bridge construction dates.

The bridge construction programme reached a low point in Surrey during the 1940s and 1950s. The national boom in bridge reconstruction of the 1960s and 1970s was not reflected in Surrey until the 1980s.

Assuming a 100–120-year life, then the peak at the turn of the century will be reflected in the next decade's workload.

Since the turn of the century traffic volumes have soared, and larger and heavier lorries are using the roads. This has increased the rate of deterioration of bridges in Surrey.

The capital value of the bridge stock of road bridges only, i.e. the cost to replace them, is estimated in 1989 at:

	Number	*£(m)*
Department of Transport structures	400	300
County structures	1173	250
		550

Assuming an average 120-year life for each structure, then the cost per annum to replace the stock alone is £4·6m, i.e. 0·84% of the capital value.

This does not include any maintenance costs to ensure that the average life is reached, i.e. painting of steel, concrete repairs and other maintenance works. The figures given in the OECD report on bridge maintenance[1] in 1989 gave the average amount spent on routine maintenance in the UK at 0·5% of the replacement value, i.e. £2·75m/annum.

The total estimated cost of replacement and maintenance based on the above is therefore £4·6m + £2·75m = £7·35m/year, i.e. 1·4% of capital value. It must be stressed that this does not allow for deterioration due to increases in traffic loads and weights.

Many of the older structures are exceeding the estimated design life of 120 years, but many of the newer, larger span and more sophisticated bridges, which are more costly to repair, are not. The average age of those requiring restrictions or reconstruction is 77 years, and if the older and inherently stronger arches are ignored this reduces to a life of 63 years. Based on a 77-year life, then the sum required for replacement and maintenance would rise to nearly £10m per year.

CURRENT ALLOCATION IN SURREY

At present schemes in the capital programme (Fig. 3) that replace existing structures, i.e. not new development on 'green field' sites and not including the cost of associated road works, averages £0·5m/annum. The maintenance budgets for 1989/90 is:

		£(*m*)
DTp		0·94
SCC		1·00
		1·94
	+Capital schemes	0·50
		2·44

This represents 0·45% of the capital value of the stock. It compares reasonably well with the OECD figure for the UK in 1981 but well under half of the estimated amount given above to maintain the stock in a reasonable condition.

This broad brush approach deals with average figures and average bridges. It assumes that similar sums have been allocated in the past when in

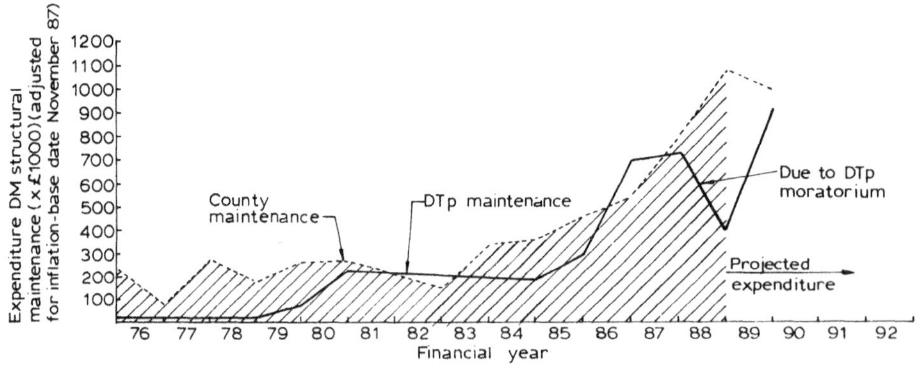

FIG. 3. Expenditure on structures in Surrey (does not include new 'green field' construction).

reality the level is well below current levels let alone required levels of funding. There is therefore a catching-up process to deal with and fund.

This is a very subjective measure of the level of funding required to maintain the stock in good order and is often difficult to use as a persuasive argument when fighting for more funds.

What is required is a management system that records actual cost of maintenance in a form that can be easily interrogated and gives a quantitative measure of the condition of the stock so that the right level of funding can be given to increase or maintain their condition. This is one of the main elements of BRIDGIT.

MANAGEMENT SYSTEM DEVELOPMENT

Prior to 1980 the bridge management system in Surrey was not electronically based. There was a card indexing system and individual bridge files. Maintenance lists and programmes were all manually produced, which was adequate at that time to deal with the relatively small budgets and work involved. As-built drawings were not always properly made and valuable information had been lost, particularly during the various local government reorganisations The inspection cycle was often *ad hoc* or reactive to need or public demand. This changed in 1977 with BE 4/77, which introduced the notion of the principal inspections. Financial control was separate and again largely paper based.

The increase in the number of structures to manage consequent upon the opening of the M25 (Fig. 4) and the necessity for some form of discipline in inspection work led Surrey to develop STREG in 1980.

STREG is a computer management system which stores information on individual structures. It is a batch system on the Unisys mainframe computer which generates standard or *ad hoc* reports as required. Inspection forms are prompted to a predefined programme, completed on site, returned, maintenance work decided, prioritised and programmed. Maintenance lists are then produced. There is some degree of very coarse financial recording. The main financial control is at present carried out using separate computer-based systems on a commitment accounting basis. STREG is very much a system for the 1980s, it is electronically based and has the security that goes with that type of system.

The next development involved converting STREG to BRIDGIT. It is a more advanced management system, giving the engineer a whole kaleidoscope of information to help him in his decisions on maintenance work in the 1990s. The use of data capture devices that will prompt and record inspections electronically on site will cut out the inefficiencies of the current paper-based system although it will work hand in hand with photography, sketches and other methods of recording defects in the inspection process. The system will be interactive and easily accessible to view, change or update the range of information held. This will be extended to include such things as material type, suppliers and stats information, together with location maps linked to the road network. Financial information will be stored in more detail for better future estimating and forecasting. Site instructions for minor works will be printed by the system direct to the maintenance contractor.

A quantitative measurement of the conditions of the individual and total bridge stock will be obtained. The condition of each element of the bridge will be given at the time of inspection on a scale of 1–5, with each classification being well defined. A multiplier is applied to this to obtain the condition factor for each element of the bridge. Each element is given a location factor depending on its structural importance. The bridge itself is given a road factor depending upon class, i.e. motorway, A-class road, etc. These three factors are multiplied together to give a priority rating to that element of the bridge. These are totalled to give a condition factor for the individual bridge and then processed to give the overall stock condition. The key to knowing how the stock is performing is to use a rating system, as from this flows the possibility of making statistical comparisons.

By comparing the overall condition factor on a regular basis, probably annually, then a measure of the effect of maintenance work can be obtained. Cash can be more easily targeted to have the greatest effect. 'BRIDGIT' will give a good objective measure to present to government committees, etc., of the need for extra funds.

Development of 'BRIDGIT' for the 21st century involves a link up with CAD for as-built drawings, drawings of defects, etc., and distance data access either in the inspector's van or home via telephone links. Direct financial management will be within the system and direct feed of traffic, accident and other relevant information to assist easier maintenance decisions from the highway database. It will also be linked directly to digitised OS maps.

INSPECTION STRATEGY

The number of bridges on the register over the past 10 years has increased dramatically with the opening of new motorways and from a concerted effort to find all our bridges (Fig. 4). Unlike other counties, inspections in Surrey are undertaken by technicians throughout the year with support from chartered engineers as required.

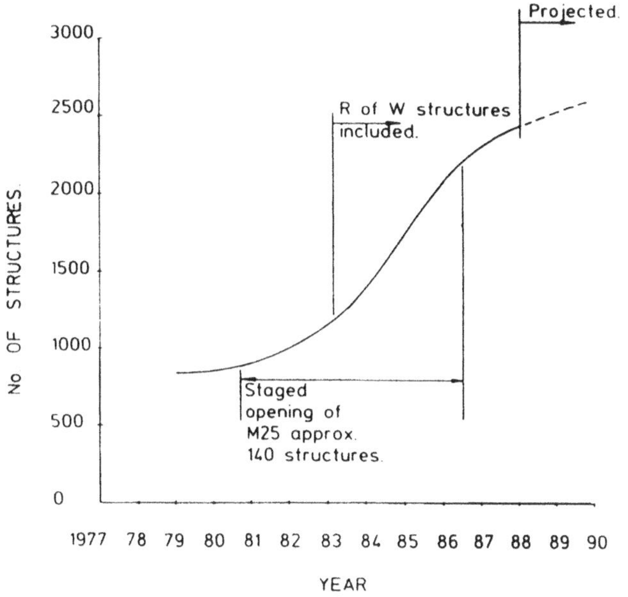

FIG. 4. Structures inspected by Surrey County Council, both county and DTp.

The county is split into four main areas of operation. Motorway and trunk road bridges are the responsibility of the senior bridge inspector. He arranges the programme and makes sure targets and standards are met. The rest of the county is split geographically between three inspectors/clerks of works who report to the senior inspector. The inspectors/clerks of works are responsible for all bridges in their area and making sure that they are adequately inspected and maintained, and records kept.

Principal inspections are undertaken on a 6-yearly cycle with general and superficial inspections in intervening years. The inspection forms are prompted quarterly in an area so that adjacent parishes are done together. This minimises travel costs and time. The individual inspectors plan their work within that quarter. They have to ensure that the work is done to a high standard and target dates met. Within a short time they gain an intimacy and attachment (often sentimental) with individual structures.

The backbone of all the inspections is the *forms sketches, and most importantly photographs*, which can tell more than any amount of written words.

The inspectors, who tend to have a practical background, issue instructions to the maintenance contractor for minor works, i.e. pointing, brickwork repairs, timber deck, repairs to parapets, etc., and supervise the works. This continues the philosophy of keeping them involved and committed.

We are constantly finding substantial structures on rights of way. Students have been employed for the last two summers to walk the rights of way and pick up qualifying structures they find. This programme will be completed in 1989.

The cost effectiveness of the inspection process is constantly appraised. An underbridge unit that gives access to three bridges in a possession (although twice the price) may be more cost effective than a scaffold tower which requires manpower for repeated movement. Annual tenders are let for various categories of inspection plant.

ASSESSMENT STRATEGY

Surrey made funds available in 1986 for assessments to the new Code of Assessment which takes account of the increased weight and numbers of heavy goods vehicles. Eighty-four out of a possible 650 county structures have been done to date with another 40 programmed for 1989. Acting as agent to the DTp, 44 trunk roads and motorway structures are to be assessed in 1989.

This programme presented a great opportunity to thoroughly examine the condition of the bridges. It was decided from the start to combine a thorough inspection and testing programme with the assessment.

This gave an intimate knowledge of the structure. This philosophy was proposed to the DTp in 1986. They were unable to pursue the full testing programme recommended because of shortage of funds. It is interesting to note that the condition report of 200 bridges by the DTp on the condition of bridges recommends just such an approach.[2]

The new Code is conservative in many areas. Structures with no apparent structural distress, only durability problems, are being recommended for heavy restrictions. By using the intimate knowledge gained from this investigation a return to 'basics' has been essential. How is the bridge acting? What load is it actually taking? What happens if it collapses? Is further testing required?

The bridges assessed to date are the older bridges in the county but are of a range of materials (Fig. 5).

While by no means a representative sample because of the way they have been selected (i.e. oldest or known problems first), it is interesting to note that restrictions or recommended works to date are mainly concentrated on metal bridges. The early reinforced concrete bridges fair reasonably well. These are arches or beam and slab constructions with built-in ends, often with no joints and their associated problems.

If there are any lingering doubts then load tests can be carried out. Often a cheap practical method of showing that the bridge is capable of carrying

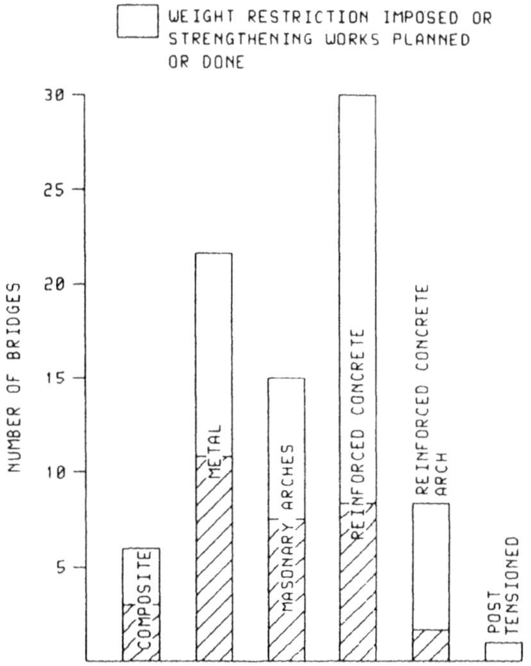

FIG. 5. Bridges assessed to date (under DTp memo BD 21/84).

imposed loads. Two have been done to date in Surrey at a cost of £5000. This is not a thorough scientific examination of the structure but a good practical comparison of actual response to expected response.

The average cost of assessment to date, which includes a principal inspection, is £1100 for county bridges and £3000 for DTp bridges. Extensive testing has also been undertaken, at an average cost of £1200 for county bridges and £2100 for DTp bridges.

Where restrictions have been applied the average costs have been (£/m^2):

(a)	Weight restriction	19
(b)	Strengthening works	1006
(c)	Replacement (bridge works only)	3900

The latter high cost of replacement compared with average new works cost of between £1000 and £2000/m^2 is due to the substantial costs of dealing with statutory undertakers' equipment and maintaining traffic flow.

MAINTENANCE STRATEGY

The thinking in Surrey is for good basic cost-effective schemes which maximise available expertise and resources. The programme is flexible, as is the budget for individual maintenance schemes. Why use a complicated design which can cause future maintenance problems?

A maintenance contractor is appointed annually following competitive tendering. Basically a dayworks contract, the rates are usually very competitive. On average three two-man gangs are used, one for each inspector/clerk of works. These can be increased or decreased to suit requirements and the work given to the contractor varies from minor works to fairly substantial emergency works.

Larger works are let by quotation, using simplified conditions of contract or normal tender by ICE, 5th edition. A 3-year programme is compiled which allows for replacement, strengthening or maintenance works.

Wherever possible existing structures are strengthened, particularly the more traditional types of structures, i.e. Town Bridge and Onslow Bridge, Guildford, and Leatherhead Town Bridge. They are part of our heritage, are often more pleasing and have lasted longer than some more modern forms of construction.

If a new bridge is inevitable then the opportunity to widen or improve is taken. Attention to detailing in new structures is encouraged to minimise the durability problems of the past. A minimum cover of 40 mm is used throughout with better quality concrete. For the future the use of epoxy-coated rebars could make for more durable structures.

OTHER INFLUENCES

The gathering of data via the inspection and assessment process gives the basis for technical decisions for maintenance or strengthening works. Yet more and more of these decisions rely on other influences.

The amount of traffic dictates the type of work. The effects on the local economy of a particularly low weight restriction, for instance, can result in rapid reconstruction or strengthening from below. We are about to let our first contract for minor works (£50 000), where the contractor will be given a bonus for early completion and penalised heavily for being late. A very simple lane rental approach yet vital in the future if traffic is to be kept moving.

The environment is an emotive issue at present and growing in importance. Bridges are invariable landmarks and should be treated accordingly.

Political pressures can be hard to accommodate, yet with well thought-out designs, good communication and justification the right scheme will win through. This is tied up with the selling approach—we want people to buy our schemes. All products need marketing, promoting and selling to often highly articulate, well-informed and well-organised groups.

PRIORITIES

In Surrey we establish priorities based on the following factors:

(a) Need, i.e. condition of bridge.
(b) SCC primary routes.
(c) Volume of traffic.
(d) Cost/benefit (including traffic delay costs and inspection/maintenance costs).
(e) Availability of alternative routes.
(f) Legal requirements.
(g) Environmental impact.

There are a range of tools that can be used at any particular site, as follows:

(1) Monitoring.
(2) Further testing and investigation.
(3) Load tests.
(4) Weight or width restrictions (money has recently been given to trading standards to appoint staff to 'police' these restrictions).
(5) Strengthening, local or major.
(6) Reconstruction to existing standard.
(7) New improved bridge scheme.
(8) Closure.

We try to assess the end of the life of a structure and then include it on a programme for replacement. Clearly it is no use spending vast sums on a structure with limited life. Whole life costing techniques are used to determine the most cost-effective solution for a particular structure for a particular time in its life cycle. The Department of Transport has recognised this necessity in its publication BD 36/88, which gives average maintenance costs. The use of QUADRO for traffic delay cost is also required. Ten years extra life in a structure may give time to construct that bypass.

APPRAISAL

It is important at the end of the exercise to take time to reflect. Has the money been well spent, and the aim achieved? Could it have been done better or more cost effectively? What mistakes were made and can these be shared so that others don't make the same mistakes? Has the targeting exercise been successful?

SUMMARY

Bridge maintenance has been the Cinderella of bridge work in the past, being overridden by the demands of design work for new structures. This boom in new works, sometimes with poor detailing and over-complex structures and materials, has left many problems. Manifestation of these problems, combined with improved inspection techniques and a need to rectify, has meant that maintenance has thrown off its old rags and is now the belle of the ball to be courted. However, unless funds are made available midnight could strike and Cinderella will be in rags once more.

Effective maintenance relies on the gathering of information, an awareness of all the influences and the effects of works so that the right decisions can be made and cash effectively targeted. The management strategy evolving in Surrey is modern and business-like, and will be enhanced considerably once the attractive 'BRIDGIT' is up and running. Book early to secure your copy!

ACKNOWLEDGEMENT

The authors wish to thank Dr J. Bergg, County Engineer of Surrey County Council, for his support over the contents of this report.

REFERENCES

1. OECD Bridge Maintenance. A report prepared by an OECD Road Research Group, 1981.
2. WALLBANK, E. J., *The Performance of Concrete in Bridges.* A survey of 200 highway bridges for the Department of Transport, HMSO, London, April 1989.

5

Bridge Management within the Swedish National Road Administration

LENNART LINDBLADH

Swedish National Road Administration, Head Office, Borlänge, Sweden

ABSTRACT

The Swedish National Road Administration is responsible for a bridge stock comprising some 11 000 bridges, representing a replacement value of some SEK 25 billion at the 1987 price level. During the coming 10-year period the Administration expects to invest about SEK 1 billion per annum in maintenance, repair, reconstruction and replacement. It is of the utmost importance for these allocations to be utilised efficiently in order to be of maximum benefit from a socio-economic viewpoint.

Within the Administration a new aid to systematic administration of the actions required to manage the bridge stock in an optimum manner is currently being developed. Developments hitherto have led to the introduction of an ADP-based information system, Bridge Data, facilitating a number of manual administrative routines during the planning and operating stage. Further developments involve, among other things, an overview and possible computerisation of these manual routines, leading to a complete bridge management system (BMS).

BRIDGE STOCK

The Swedish National Road Administration is responsible for the management of around 11 000 bridges included in the national road network. In this context, a 'bridge' is a structure with a free opening of at least 3·0 m in the longest span. Of these bridges, some 8500 are made of concrete, 1800 of steel and 700 of stone. About 400 of the concrete bridges are made of prestressed concrete.

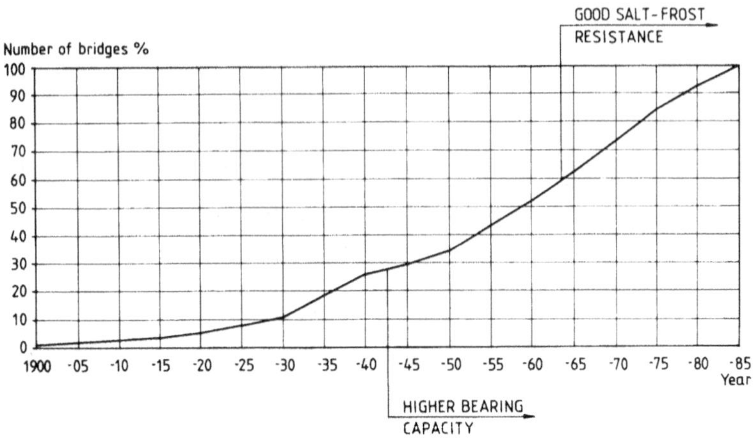

FIG. 1. Bridge age distribution. State and state–municipal bridges.

The age composition reveals that about 3200 bridges were constructed before nationalisation of the road network in 1944 (see Fig. 1). As a rule, these bridges have a relatively low bearing capacity. It can also be stated that some 6600 bridges were constructed before the year 1964, when the requirement of air entrainment in the concrete was introduced, resulting in a higher salt–frost resistance.

The load-carrying capacity of all bridges has been classified in order to assess their trafficability by actual vehicles and vehicle trains. The classification, which specifies the factual bearing capacity with due regard to damage, etc., is expressed as the permissible axle and bogie loading, A/B(t). The road network is normally capable of carrying $A/B = 10/16$ t, a load which can be carried by 97% of the bridge stock.

During 1988 actions have been initiated with the aim of raising the load-carrying capacity standard on the main road network and on the secondary and tertiary road networks in the forest counties. The purpose is primarily to make possible, by reinforcements, reconstruction and new construction, adaptation of the Swedish weight regulations to those ratified by the EC. Roughly 1300 older bridges are affected.

FINANCING

In the current maintenance plan for 1988–92, SEK 170 million per year has been allocated at planning level in the 1987 price level (0·7% of the replacement value) for bridge maintenance. In addition to this sum is SEK 60 million for reinvestments in small bridges with an object cost of less than SEK 3·0 million. Reinvestment is understood in this context to mean

TABLE 1
Annual extent of bridge activities according to current plans

Action (financial)	Number of bridges per year	Cost (SEK m/year)	Cost[a] (%)
Maintenance		170	0·7
Reinvestments			
Durability	60	160	0·6
Bearing capacity	100	320	1·3
Investments	80	250	1·0
Total	240	900	3·6

[a] In relation to replacement value.

reconstruction or new construction solely on account of shortcomings in durability.

Major reinvestments and new investments are evident from the 10-year multi-year plan. According to the plan valid for 1988–97, the annual volume of reinvestment on account of shortcomings in durability is about 30 bridges or SEK 100 million (0·4%). New investments total SEK 250 million (1·0%) or about 80 bridges.

Over and above the regular plans, a special bearing capacity plan has been drawn up for implementation of the aforesaid commitment to load-carrying capacity. From this it is evident that reinvestments will be made in about 100 bridges per year on account of insufficient load-carrying capacity for the future EC-adapted weight regulations.

All in all, the above implies an annual bridge volume during the coming 10-year period as in Table 1.

BRIDGE MANAGEMENT

The term bridge management is used to describe all the activities, both administrative and productive, required to construct, assign to traffic and demolish a bridge (see Fig. 2).

The administrative activities include every stage of economic planning, namely orientation planning, action planning and budgeting, as well as physical planning in the form of preliminary and detailed planning. Also included among the administrative activities is the continuous description of the bridge objects in the form of inspections, condition assessments, classification of load-carrying capacity, etc. Following-up involves an evaluation of the various actions taken from the standpoint of both

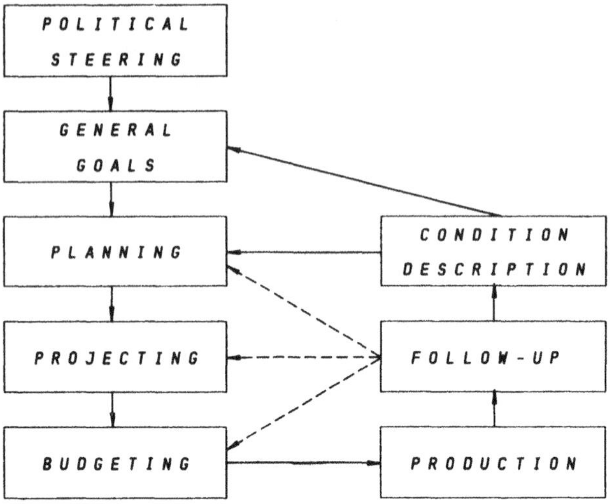

FIG. 2. Bridge management elements.

economy and technique, and forms the basis for formulation and reconsideration respectively of road-keeping goals, strategies, maintenance standards, performance standards, etc.

The allocations placed at the disposal of the Swedish National Road Administration for implementation of its road-keeping assignment are and will in all probability continue to be more or less meagre. In view of this, it is essential for the allocations to be used in such a way that they are of the greatest possible benefit to society at the lowest possible cost. The actions to be carried out must be technically correct, and take place at the correct points in time and on the right objects.

In order for the optimisation to be carried through all the way, an ADP-based bridge management system (BMS), which will make possible systematic administration of requisite actions on the bridges, is currently being developed within the Swedish National Road Administration. The system is being built up around a database, Bridge Data, containing a large quantity of data necessary for the BMS. Compare the description in the section entitled 'Bridge Data'.

In addition to Bridge Data, a fully developed bridge management system, BMS, should include routines for:

—inspections, condition assessment, load-carrying capacity classification;
—selection of planned action, optimisation per shortcoming and bridge;
—prioritisation, optimisation per road network/bridge stock;
—specification of commonly performed maintenance and minor repair tasks;

—economic and technical follow-up;
—reporting; and
—route finding for heavy transports.

Hitherto, developments have been concentrated on designing Bridge Data, which is now largely complete. The ADP system also generates a number of fixed and selected reports.

The inspection routines may be regarded as ready for use with the introduction of new inspection regulations and a system for assessment and documentation of faults and shortcomings.

An aid to the initiation of optimum actions per shortcoming and bridge is also available through the preparation of a standard for assessment of degrees of urgency for requisite actions.

BRIDGE DATA

Bridge Data is included as a subsystem in the road data bank (RDB). The system is stored on a mainframe computer (Sperry) and is built up largely in databases. This will facilitate the transition to base computer storage (VAX), which is expected to take place in 1991.

Bridge Data currently consist of five components, and two new ones are being planned. These components are:

—drawing section;
—administrative section;
—technical section;
—load-carrying capacity section;
—damage section;
—planning section (under planning); and
—projecting section (under planning).

The bridge register, which describes bridges, road tunnels, ferries and jetties along state and state–municipal roads, contains output from Bridge Data.

A brief description of the different sections now follows.

Drawing Section

The drawing section contains information on about 120 000 bridge drawings. Among the stored information, mention may be made of drawing content, format, date of drawing approval, if it has been micro-filmed, and if so, when.

Administrative Section

The administrative section contains information on about 12 000 bridges and about 1500 road tunnels on the state and state–municipal road network (except in the municipalities of Stockholm and Gothenburg). In addition, information is stored on roughly 6000 other bearing structures, including bridges, etc., which have been projected but not yet constructed or bridges which have been reassigned to the private road network or demolished. The administrative data include such things as the bridge name, responsibility for bridge maintenance, year of construction, free distance from underlying road or watercourse, etc.

Technical Section

The technical information relates to, for instance, type of static system, material in superstructure, spans, waterproofing and paving types, support data, mode of foundation, type of bearings and expansion joints, bridge area, etc.

Load-Carrying Capacity Section

The load-carrying capacity figures concern, for instance, permissible axle and bogie loadings, trafficability data for certain heavy types of vehicle, carriageway width, bridge being repaired, etc.

Damage Section

This section contains condition and damage information. The information relates to, for instance, type of inspection, condition class for 15 different structural parts, summary condition class for bridge, type of damage and where it is situated, extent of damage and estimated repair cost, degree of urgency for repair action, etc.

Planning Section

This section will contain in the first instance information that is necessary for economic planning, such as planned and factual action costs, road-user costs, index tables, discount rates, planned action times, etc.

Projecting Section

An expansion of the database with information to facilitate the physical design of actions on the bridges has also been discussed. This section could be used, for instance, for storage of work drawings, specifications, etc., which could be used as source material for implementation of maintenance and minor repair actions.

INSPECTION SYSTEM

Information on the condition of the bridges and any necessary actions comprises the base for the entire management system. The quality of this information is completely decisive for the quality of the result generated by the system, for instance source documentation for plans, transport permits, etc. It is therefore of the utmost importance for regular inspections to be carried out and appropriately documented. To ensure this a new inspection system was introduced within the Swedish National Road Administration in 1987. The rules governing implementation of the inspections are temporary but will become definitive as from 1991.

The new rules will cover the following four types of inspection:

—superficial inspection,
—general inspection,
—principal inspection, and
—special inspection.

With regard to the intervals between different types of inspection, the following will apply in all probability: maximum 6 years between principal inspections and maximum 3 years between a principal inspection and a general inspection and between general inspections respectively. In the normal case, with time intervals of 6 and 3 years respectively, this means alternately principal inspections and general inspections at intervals of 3 years. Superficial inspections will be carried out at least once a year.

The principal inspection comprises a new type of inspection with considerably higher requirements as regards extent and quality in relation to the former full inspection. The purpose of the principal inspection is to detect faults and shortcomings which can affect the function or trafficability of the bridge within a 10-year period. Even such faults and shortcomings which, if they have not been dealt with within this period of time, can lead to higher management costs shall be detectable. All free structural elements both above and in the water that are accessible by means of visual methods shall be inspected. Parts adjoining the bridge, such as road embankments, slopes, cones, fillings and erosion protection, shall be inspected.

The principal inspection shall be carried out visually at 'hand-close distance' and under conditions similar to daylight. Provided that anticipated types of damage are definitely detected, a slightly greater inspection distance, but not more than 3·0 m, may be approved in exceptional cases under extremely good lighting conditions. As a rule, the inspection requirements imply that mobile inspection aids such as bridge lifts, boats and ladders will be required in addition to any fixed devices. Normally divers will be needed for inspection of structural elements in water.

In conjunction with the principal inspections, measurements are also carried out in order to determine

—bottom profiles (foundation in water),
—carbonation depth,
—chloride content,
—plus heights of edge beams, and
—inclinations of supports.

These inspections are performed by personnel with thorough knowledge of the design and mode of function of the bridges and of their durability. The personnel should also be thoroughly familiar with the bridge management system used by the Administration and should be particularly qualified to deal with issues relating to choice of actions from both the technical and the financial angle.

The general inspection is also a new concept. In terms of extent and quality, this type of inspection corresponds roughly to the former full inspection. Special inspections of structural elements are carried out when so considered necessary in order to investigate more closely any faults, shortcomings or other observations found in conjunction with the aforesaid regular inspections. Special inspections are also carried out on the following structural parts or elements, regardless of their condition:

—mechanical and electrical equipment on movable bridges,
—welds in certain steel structures,
—older bridge deck slabs of concrete, and
—foundations in water.

Any faults and shortcomings detected in the course of the inspections shall be assessed and documented in accordance with special standards and lists of codes. The results are registered in the Bridge Data information system.

All faults and shortcomings are described in terms of location, type, cause and action. A fault or shortcoming shall also be assessed on the basis of the functional requirements valid for the structural part in question. On the occasion of the inspection, each structural component is allocated a condition class, which may vary from 0 to 3.

0 Defective function after 10 years (undamaged on the occasion of the inspection).
1 Defective function within 3–10 years.
2 Defective function within 3 years.
3 Defective function on the occasion of the inspection.

Type of damage: corrosion, scaling, spalling
Condition class (CC): the highest CC as below is chosen.

Reinforcement area
The reduction in area (r%) is determined in one section. CC as per table.

Bonding
The reduction in bonding (r%) is determined on a length = the height of the element. CC as per table.

Table

$r >$ %	CC
20	3
15	2
0	1

FIG. 3. Condition standard.

Condition classifications are carried out in accordance with a special standard. An example of such a standard is presented in Fig. 3.

In conjunction with the inspection, a record shall also be made for each fault or shortcoming of information concerning the planned action, including the degree of urgency and the cost. Compare below under the section headed 'Planning—Goals'.

LOAD-CARRYING CAPACITY CLASSIFICATION

All bridges in operation have a load-carrying capacity classification as evident from the Bridge Data information system. This classification reflects the actual bearing capacity of a bridge with due regard to all known faults and shortcomings.

The bearing capacity is expressed as the permissible axle and bogie loading, A/B(t), for a number of real classification vehicles. The classification also includes 14 types of heavy vehicles for which the permissible axle loadings and gross weights are specified.

If a bridge is classified as belonging to condition class 3 in conjunction with an inspection, an investigation is carried out immediately in order to determine the extent to which it may be necessary to reduce the permissible traffic load, i.e. the classification, of the bridge concerned in view of the faults or shortcomings detected. This investigation may vary from a simple static assessment to a regular control calculation. If the investigation results in an indication that the classification needs to be altered this is done immediately in Bridge Data.

PLANNING—GOALS

Prescribed in the economic plan is both the orientation road management should have during a period of 10–15 years and the factual actions necessary in order to attain the road management goals. These actions are evident from 10-year investment plans, multi-year plans and bearing capacity plans, for new investments and reinvestments, as well as a 5-year maintenance plan.

According to the valid orientation decision from 1986, the following standards and actions shall be aspired to for maintenance in the planning of road management:

- —The present technical standard of the road network shall be maintained (status quo).
- —Heavily worn parts of the road network shall be restored to an appropriate technical standard. The lag should largely be recovered by the end of the century.
- —Severely worn bridges should be replaced by new ones when this is appropriate in view of the costs for replacement investment and future maintenance.

When the faults and shortcomings established during the inspections are documented in Bridge Data, the degree of urgency and costs, among other things, shall be specified for the planned remedial actions. The degree of urgency indicates the priority with which a fault or shortcoming shall be remedied in order for the established goals for management of the bridge stock to be attained.

The following degrees of urgency are applied:

3 Action required as soon as possible (on the occasion of the inspection).
2 Action required within 3 years.
1 Action required within 3–10 years.
0 Action required after 10 years.

The aforesaid goals have been concretised in the form of a standard, in which the prerequisites for selection of degree of urgency are specified for individual actions in a number of fault situations. The costs for the most commonly performed actions are evident from an established price list.

During a second planning stage the actions per bridge are optimised. Various action alternatives, all of which satisfy the requirements imposed on the bridge, are studied both technically and economically, making due

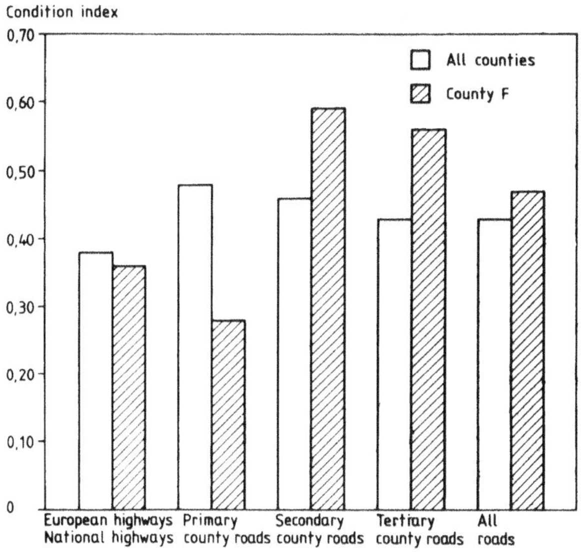

FIG. 4. Functional condition of the bridges. Condition index (mean value of main condition class) per road category. (The main condition class is specified per bridge and refers to weighted mean value of the condition classes of the various structural parts.)

allowance also for the road-user costs. The various alternatives are compared economically with the aid of their current values and a discount rate of 5%. The optimum alternative from a socio-economic viewpoint is selected as a line of action for the bridge and registered in Bridge Data.

In the case of major actions, the finally developed BMS will indicate at least one further alternative in describing the consequences in cases when it is impossible to choose the optimum solution in view of limited funds.

The information registered in Bridge Data as above can, together with other information in the system, be used in different ways as the bridge management process goes on. Such information can, for example, form the basis of prioritisation of objects, drawing-up of new maintenance and multi-year plans, etc. The information can be processed and compiled in accordance with the example in Fig. 4.

6

Local Agency Experience with the Utilization of Bridge Management Systems in Finland and the United States

ALLEN R. MARSHALL

Cambridge Systematics Incorporated, American Twine Building, 222 Third Street, Cambridge, Massachusetts 02142, USA

and

MARJA-KAARINA SÖDERQVIST

Finland Roads and Waterways Administration, Opastinsilta 12, PO Box 33, SF-00521 Helsinki, Finland

ABSTRACT

In recent years significant efforts have been made to collect, review, manage and analyse information on bridge structures in many countries. In many cases these data are organized into computerized bridge management systems which usually include additional analytic and programmatic capabilities beyond mere record-keeping for structures. These systems often are intended to define and establish national bridge infrastructure priorities. In many cases local government agencies have been the primary collection agents for the various data maintained in bridge management systems which include inventory, structural conditions and appraisal information.

The national uses for bridge management systems tend to focus on network level programming and forecasting of resource utilization and have been widely discussed in a variety of different forums and professional publications. This paper discusses the actual uses that bridge management professionals working at the regional or local level have for BMS data bases. The paper is based on interviews conducted with bridge management staff in a local highway agency and an urban transit authority in the US and with district bridge engineers in Finland. The case study discussion is intended to provide BMS system users and designers with a better understanding of the positive and negative consequences of the elaboration of national bridge management systems at the grassroots level.

INTRODUCTION

This paper discusses the essential ways that local agencies utilize their bridge management systems (now and in the future) in two government agencies in the United States and in Finnish RWA districts. Both the US government entities and the Finnish districts have direct responsibility for the collection and organization of BMS data as well as actual construction, rehabilitation and bridge maintenance roles. The first discussion will highlight the uses for BMS systems in the US and the latter section will provide an overview of the Finnish situation.

Public agencies in many countries are utilizing computers in different ways to manage information about their infrastructure, including bridges. A variety of different information is gathered and maintained according to agency information demands, but the basic set of information is usually consistent to a great degree. The computerized information is used in a variety of ways by different divisions of public infrastructure management agencies. In spite of the growth of computerized management information systems, in many cases the data items defined for inclusion in the computerized systems are not totally suited to the information needs of different bureaucratic hierarchy levels and responsibilities. More often than not this mismatch is a result of the system design created by the original bureaucratic division which provided primary impetus and funding for the system development.

US Experience

In the United States most computerized bridge management systems have focused at the most basic level on managing the core set of information first defined by the Federal Highway Administration (FHWA). This set of information has been thoroughly standardized for several years. A recent TRB research report on bridge management systems included a bridge database design that incorporates essentially this fundamental set of information. The initial purpose of the FHWA bridge database organization was to create a national inventory with adequate administrative and structure description information as well as basic condition information which would enable all the states to report the nature and condition of their bridges in a consistent fashion and to ensure that regular bridge inspections are performed in all the states.

In most state governments there is a department of public works that has direct state level responsibility for performing bridge inspections and transferring updated information to the FHWA. In many cases the state agency also develops needs estimates and sets priorities for the repair, rehabilitation and reconstruction of bridges eligible for federal funds within the state. The inspection reports provided to the federal government usually provide one of the basic means of deciding on bridge infrastructure improvement priorities.

It is apparent that more information is required by states to perform actual bridge system management as compared to basic structural inventory and appraisal. The complexity of the most advanced systems in the US (Pennsylvania, North Carolina, etc.) indicates that the information needs defined by the FHWA fall short of the requirements of the states themselves. These systems maintain a great deal of information and perform actual management tasks (for example, overweight permit tracking). The Commonwealth of Massachusetts, for instance, has extended the basic information required by the FHWA to include specific information necessary to perform basic bridge maintenance such as the public works district responsible for the bridge and paint type and area.

This reflects a common federal/state bureaucratic hierarchy problem. States have much different information requirements from those of federal agencies. The result in the area of bridge management is that individual states have been forced to take the lead in developing bridge management systems to address their particular bridge management concerns.

Finland

A somewhat different situation exists in Finland. All management of public bridges is performed by national and regional administrative divisions of the same central government agency, the Roads and Waterways Administration (RWA). The districts and road master subdistricts actually perform construction and rehabilitation while the national administration is responsible for the allocation of public funds for these projects.

In many ways the Finnish bridge system administrative hierarchy is comparable in scale at least to a single large US state, such as New York. The districts and central office have always worked together closely in establishing and carrying out bridge management programs and policy. As a consequence, the types of information and reporting needed are relatively consistent at the different hierarchy levels. Both districts and the national office have collaborated in the development of existing bridge information systems and the new system currently being tested in a prototype version. This has led to the happy consequence that the types of information needed at the district level are a core element of the overall information system specification for the national database.

CASE STUDIES

Metropolitan District Commission—Boston, Massachusetts†

The Metropolitan District Commission is a division of Massachusetts state government with ownership and management of 153 miles of parkways and 157 bridges. From 100 000 to 120 000 vehicles utilize the MDC system of highways on a daily basis since the great majority of the bridges under the aegis of the MDC are located on urban parkways. The MDC bridge stock includes a variety of design types and ages, including some bridges that are exclusively utilized by pedestrian traffic to cross MDC parkways. Many of the bridges are eligible for federal funding for rehabilitation or replacement so there is a strong need within the MDC bridge group to maintain strict compliance with federal inspection standards and to generate regular timely inventory, condition and appraisal reports for federal review. Furthermore, the state mandates a regular inspection of the MDC's bridges so that Massachusetts as a whole will be in compliance with federal inspection guidelines.

As part of the development of a major MDC parkways information system,‡ a simple bridge database system was created to assist the MDC in recording information on their bridges as well as to schedule inspections and perform automatic reporting of federally required information. The system was developed with substantial input by MDC bridge engineers, including specification of information content and reports. The basic organization starts with the federal reporting requirement and extends that data item specification to include several administrative items required by the MDC. The system also includes two additional files of information—a repair project history and an itemization of bridge utilities. These files are intended to help the MDC to track the nature and extent of repairs being performed and to coordinate structural repairs or rehabilitation with utility companies and other joint users of MDC structures such as the local transit authority.

MDC engineers are currently utilizing the bridge inventory for two major purposes: recording and reporting bridge condition based on required periodic inspections, and automatic inspection scheduling and program

† The authors wish to recognize the assistance of Mr David Lenhardt, chief bridge engineer at the MDC, and Mr Dominic Anidi, project manager for the MBTA Bridge Inspection Program, in preparing this and the following sections respectively.

‡ The MDC parkways system, like the others described in this paper, was developed by Cambridge Systematics Incorporated of Cambridge, Massachusetts, USA (617) 354-0167. The system consists of a comprehensive pavement management system and bridge, traffic signal and street light inventories as well as a capital project development system.

tracking. The MDC employs individuals with specific responsibility for performing bridge inspections throughout the jurisdiction. These individuals have received bridge inspector training from the Massachusetts Department of Public Works (MDPW). They utilize reports from the computer database in the field and mark up the data items, mostly condition items, that may have changed in the interim since the previous inspection. These sheets are processed by an assistant engineer at MDC headquarters and the database is updated to reflect current condition of the bridges. No historical record of inspections is maintained by the MDC other than paper records, although a need has been recognized to extend the current computerized condition inventory to track inspection findings over time. No condition items are collected other than those that are federally required but the system does allow an inspector to enter a paragraph of comments in order to extend and amplify the basic inspection condition report. These comments can be viewed or printed at any time.

The inspection scheduling system is quite simple and does not optimize the period of inspection to reflect the condition of different bridges. It allows either a regular 2-year inspection or a more frequent interval to be defined. The type of inspection (regular, in-depth, special) can also be assigned for a bridge and this appears on the report. Two years is the current maximum period between inspections required by the FHWA. At any time the inspection interval can be altered to reflect deterioration or rehabilitation/replacement of bridges. Besides scheduling the inspections, the system can display a completion indicator as well as an estimate of the days or weeks that an inspection is overdue. The MDC is largely satisfied with this simple system but they have planned to add an ability to set a time window for the inspection scheduling. This would enable them to prepare, for example, a 2-week inspection work schedule for the field inspectors at any time.

The subfiles for recording repair projects and utility tracking have been used somewhat less than anticipated. This is due to several factors. First, recording repair projects requires a new procedure to be implemented for providing field reports. Currently the tracking is performed intermittently for outside contractors and almost no in-house/own force repair projects are recorded for subsequent review. Through time the MDC management staff hope to develop a solid record of all repairs performed to their bridges, including nature, extent and cost of the repair actions. They want to utilize this information to help them decide whether a bridge should be replaced or rehabilitated, instead of some other bridge in comparable condition because of its historical repair log.

The MDC has completed an inventory of all utility fixtures on their bridges as a gradual process over the last 3 years. The information on utilities was collected as part of the regular inspections and where the utilities were not readily visible to the eye the information was taken from bridge drawings. The utility type, owner, approximate location and a comment about each utility is maintained in a computer file. This allows the MDC to easily prepare reports on any bridge that shows all co-users, the number of utilities and any special comments related to the various utilities. The reports can be sorted by bridge, utility type or utility owner.

The need for information on utilities is most acute for those bridges undergoing minor repairs or rehabilitation, e.g. shoring where a complete utility survey prior to construction is not likely to be performed. In a total reconstruction or rehabilitation situation a utility survey will always be part of the design/pre-construction work. For minor rehabilitation projects, the MDC can anticipate any utilities that may interfere with the work and proceed accordingly. The comprehensive utility inventory also helps reduce or eliminate potentially problematic situations where the MDC and the utility do not coordinate repair projects. Without an inventory it is conceivable, for example, that a bridge that has undergone deck replacement could have a utility company opening the newly-replaced deck to fix a line or pipe.

Massachusetts Bay Transportation Authority—Boston, Massachusetts

The Massachusetts Bay Transportation Authority (MBTA) has been a leader in developing computerized bridge management systems for transit, railroad and highway bridges. The MBTA has varying levels of responsibility for over 350 bridges, of which approximately 100 are highway bridges over their tracks or property. The MBTA is one of the nation's largest transit properties with a subway and surface trolley system of 62·5 track-miles on five lines and a large commuter rail system of 479 track-miles extending over 50 miles out from Boston to the north, south and west as well as a fleet of over 1153 buses. Daily over 420 000 people ride on MBTA subway lines and 64 000 utilize commuter rail. An additional 435 000 daily riders utilize MBTA bus lines.

Due to expansion of the state's infrastructure management responsibilities under the Executive Office of Transportation and Construction, the MBTA has acquired responsibility for many of the bridges and culverts on the commuter rail lines that were formerly owned and maintained by the Boston & Maine railroad (Guilford Industries) as well as a number of bridges formerly maintained and in some cases owned by other state agencies such as the MDPW and the MDC. In many cases, particularly on the commuter rail system, ownership of the structures is questionable at best and the Authority has found out only recently that it actually owns a number of bridges and should be maintaining them. This clarification of responsibility has also affected the MDC. In fact the MDC has recently determined that it is the agency responsible for three bridges of great importance to the MBTA.

The acquisition of these bridges and the clarification of ownership responsibilities has fomented a management crisis for the MBTA bridge engineers. A number of deficient bridges have come under MBTA control in the last few years with extensive and expensive implications for the Authority. As a result the MBTA Construction Directorate has launched an extensive bridge inspection and reconstruction program which started in earnest only in the last 2 or 3 years. This program is intended to install a regular program of bridge inspections and initiate construction projects to replace or rehabilitate deficient structures.

As part of the bridge management program, the MBTA has developed a computerized database running on a microcomputer to track various types of data related to its bridges. The information is stored in a loosely organized set of computer files tied together by primary bridge identification codes. Each of the files is intended to address specific problems such as tracking bridge inspection contracts or recording construction projects. The differing characteristics of transit and railroad bridges, particularly with respect to operating loadings, as compared to highway bridges has led the Authority to develop new item specifications that are somewhat different from the standard federal data items used for highway bridges. The data files are organized into a menu system that provides an extremely simple means to access the data and print reports but is not comprehensive and does not perform any error checking or house-keeping tasks. No formal database design exists, and the integration of the system is rudimentary. Most use of the database requires a fair degree of sophisticated knowledge of the underlying database software.† As a result the system is utilized almost exclusively in an *ad hoc* fashion.‡

There are a wide variety of reports that can be generated from the system at present. A partial list of available reports is shown in Table 1. It is apparent that several of the reports are in fact quite similar and differ only in the order in which bridge information is presented, e.g. sorted by line, by type of bridge, or some other ordering criteria. These reports are pre-set and can only be displayed in the order shown. The final BMS will allow different sort criteria and querying. As the system is further developed it is anticipated that summary reports will be developed to show yearly construction figures,

† The software used by the MBTA is dBase IV, a product of Ashton-Tate Incorporated. The system operates on a 80386-based microcomputer.

‡ Cambridge Systematics is currently documenting the existing system and will be integrating and extending its capabilities so that users can easily access and update the information and prepare reports. The MBTA project manager is Mr Dominic Anidi, who can be reached at (617) 722-5806 for further information.

Table 1
Typical MBTA bridge management system reports—Prototype system

Transit bridges listed by line
Highway bridges listed by line
Highway bridges that are owned by the MBTA and maintained by the MassDPW
Transit and highway bridges owned entirely or in part by the MBTA
Transit and highway bridges that have questionable ownership or maintenance responsibility
Inspection schedule for transit and highway bridges by month
Bridges that have not been inspected in the last 2 years
Bridges in the MBTA database that have been load rated
Transit bridges ordered by condition indicators
Highway bridges ordered by condition indicators
Rehabilitation program for transit and highway bridges by line by year
Rehabilitation program for railroad bridges by line by year

expenditures and other summary indicators needed by MBTA management to document progress in bridge management within the Authority.

The type of information stored will also be extended. At present no computerized historical records of any type on previous bridge inspections, maintenance or repair are maintained, nor does the MBTA have a comprehensive utility inventory like that of the MDC. The enhancements to the system will address these and other gaps in the information.

In addition to storing more comprehensive information about their bridges, the MBTA is extremely interested in simplifying use of the system at the same time. The current menuing system will be replaced in the near future to allow the project engineers to easily access, update and print out information they need. Capabilities will be provided to sort and browse through the variety of different data items and the system will minimize or eliminate the need for specialized computer knowledge in order to operate the system.

It is also expected that a comprehensive inspection and construction project monitoring system will be created as part of the system. All bridges will be programmed automatically for inspection based on their condition and the legally required inspection frequency as well as an inspection priority indicator that will be developed to determine inspection job order. Another subsystem will track the completion of the programmed inspections. A significant portion of bridge inspections are performed by outside contractors for the MBTA and there is a great need to track contractor invoices and performance as well as the inspection work performed by MBTA's own force.

The MBTA also expects to utilize microcomputers to track bridge construction projects from the design phase through final completion of repairs or structure replacement. The computer will record standard tracking information for each bridge project such as contractor, MBTA project manager/engineer, project scope and budget, milestones and various contractual information, and will particularly emphasize invoice and change order tracking, which are of critical importance to MBTA project managers.

The prototype system currently in use at the MBTA will be replaced with a final version by Fall 1989. A clerical staff person will have primary day-to-day responsibility for the system but several MBTA bridge engineers and outside contractors will be drawing information from the databank.

Finnish Highway Districts—Finland Road and Waterways Administration

The Finnish Roads and Waterways Administration (RWA) is responsible for almost 12 000 bridges on public roads in Finland. The maintenance and inspection responsibilities for the bridges lie with 13 local highway districts. These districts are shown in Fig. 1. Because of varying population and road mileage densities in Finland, the types of bridges and bridge materials in districts is variable and the allocations are shown in Table 2. This table also distinguishes between bridges by their functional road classification. For

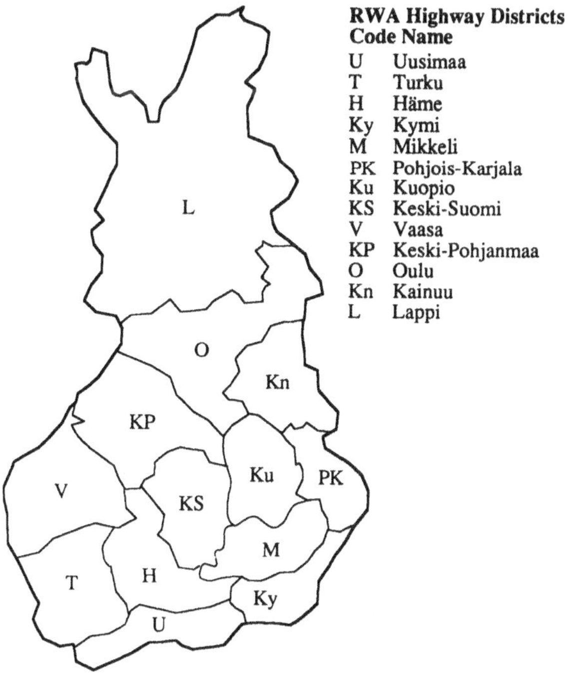

FIG. 1. Finnish Road and Waterways Administration Highway Districts.

TABLE 2

Total bridges and posted bridges, tabulated by road classification and primary construction material[a]

District	Highway bridges				Local roads				All public roads	All bridges
	I	II	III	IV	I	II	III	IV		
U	770	28	11	1	143	30	12	22	1 017	1 131
posted →	2	1			9	10		14	36	36
T	694	51	31	6	224	57	39	122	1 224	1 291
	2	1	1	1	10	5		12	32	33
H	624	20	24	19	110	16	12	96	921	958
	4	1		1	1	4		31	42	44
Ky	322	31	16	93	18	9	19		508	561
	2	3			5	1		6	17	17
M	266	19	2	1	49	18	2	35	392	400
		1				3	1	5	10	10
PK	260	17	3	2	110	17	2	49	460	480
	1				5	3		3	12	13
Ku	385	18	6	12	80	19	2	80	602	619
	2			1	2	2		3	10	10
KS	382	24	10	5	141	24	3	53	642	664
		1			1	2			4	4
V	431	32	5	9	198	92	2	26	795	801
	5	2	2	3	27	1	14		54	54
KP	293	7	3	120	68	30			521	526
	4	2							6	6
O	478	15	3	1	112	60	58		727	736
	1	1	14	4					20	20
Kn	284	13	2	3	83	22	48		455	461
	2	1	3						6	6
L	632	55	3	25	142	55		62	1 037	1 047
				6	4	19		27	86	86
Total	5 821	330	119	84	1 605	496	83	700		
posted →	18	11	3	10	42	95	2	124		
Road type	6 354				2 884				9 301	9 675
posted →	42				263				335	339

Material types: I, concrete; II, steel; III, stone; IV, wood.

[a] TVH Tuotanto-Osasto, *SILLAT 1.1.89* (annual bridge system summary report); TVH Tuotanto-Osasto (Road and Waterways Administration Bridge Management Section), Helsinki, 1989, pp. 1, 19.

each district the number of posted or weight limited bridges is also shown in Table 2. It is apparent that the bridge stock composition is somewhat different in the various districts and that the proportion of 'deficient' bridges also varies. Lappi (Lapland) district, located as it is in a harsh arctic environment that hinders maintenance and accelerates deterioration, shows the greatest absolute number of posted bridges, while Uusimaa, the Helsinki area district, has a very low number of posted bridges. The relatively small number of posted or weight limited bridges reflects the RWA's existing policy to minimize or eliminate postings throughout the system.

Finland has had a database of bridge information for several years which consists largely of administrative and simple descriptive information. The system is currently being revised both to support a national bridge management system as well as to support the information needs of local bridge engineers. (This database and the Finnish bridge management system as used at the national level is discussed in the paper by Kähkönen and Marshall elsewhere in this volume.) The local engineers will be supporting an elaborate new inspection program which will be collecting a great deal of specific condition information on the nation's bridges during the next few years. This information will be used at both the national and local levels. In particular, the local engineers will be able to track specific damage on their bridges as well as the completion of repairs to the damage. Their recommendations for structure repairs, rehabilitation or replacement will be incorporated within the national planning optimization system as well as the local level project analysis system.

Besides tracking damage, the bridge management system used in the local office will enable bridge engineers to completely inventory their bridges, including all structural elements, bridge fixtures such as railing or signage, and any associated bridge utilities and special uses such as bikepaths. Collecting the latter inventory items enables them to anticipate conflicts with other users during any repair or reconstruction projects and to assist in the design of new structures. Once the inventory of associated uses is complete, particularly public utilities, railroads and forest products users, the RWA expects to formalize the arrangements for bridge use by other parties and possibly share operating costs.

Bridge engineers in districts also are required to track their inspection programs. The system will print out inspection reports for all the bridges in the district at any time and a basic report will show the planned schedule for bridge inspections. The inspections are also recorded historically so that an inspector can see the inspection results from a prior bridge visit and also to view the most current inspection. This will help the experienced engineers to highlight any potentially dangerous situations.

A major difference in the new bridge management system is the improved access to information for local district managers. Previously all the information was processed and prepared in the national Road and Waterways Administration Office. Now local engineers will have direct access and control over all information related to the bridges in their district. They will be able to analyse data and prepare their own reports and graphs to display patterns in the condition of their bridges. Up to this point all such reports or special tabulations required the local bridge officials to file a request with the national office. In turn this will free the national RWA officials from the work of preparing these reports for the districts. It is expected that every district in Finland will be equipped with a relatively sophisticated personal computer to operate the bridge management system. In most cases the data will actually be stored on a minicomputer within the district office.

SUMMARY

We have reviewed three different types of bridge management systems ranging from a relatively simple database established to meet federal reporting requirements and manage inspections to a comprehensive nationwide database that has both an inventory and optimization system as major components. A single major finding of this overview is that bridge management systems that are intended to be used at the local level must be designed in close cooperation with bridge management officials. The failure of the federal bridge register to address local needs in the US agencies is apparent and these systems are growing away from the national standard as new local applications for computerized bridge management systems develop. In Finland the close cooperation between the national and district bridge management staff has led to the development of a system that largely satisfies both parties. There is of course still some material being collected by the district engineers that they have little use for in the short run. However, as the network optimization and project development systems that are yet to be developed come on line, these systems that rely heavily on accurate data collection at the local level will have direct utility in assisting local staff to prioritize their bridge projects and effectively utilize the resources available to them.

7

The Pennsylvania Bridge Management System

RICHARD M. MCCLURE

The Pennsylvania State University, 212 Sackett Building, University Park, Pennsylvania 16802, USA

and

GARY L. HOFFMAN

Pennsylvania Department of Transportation, 1009 Transportation and Safety Building, Harrisburg, Pennsylvania 17120, USA

ABSTRACT

A Bridge Management Work Group, as researchers for the Pennsylvania Department of Transportation, developed the engineering concepts and requirements and assisted the system contractor in the development of a total bridge management system (BMS). A contractor working for PennDOT provided system design, development, testing, implementation and training on the use of BMS software.

The general objective for the BMS was to develop a management tool which will enable a systematic determination of present needs for maintenance, rehabilitation and replacement of bridges in Pennsylvania, and to predict future needs using various scenarios, along with a prioritization for maintenance, rehabilitation and replacement, which will provide guidance in the effective use of designated funds.

The software implementation date was 24 December 1986, with full implementation completed 28 February 1987. The department assumed responsibility for the BMS software on 1 March 1987.

INTRODUCTION

Background

A seven-member Bridge Management Task Group was convened in 1983–84 to consider the development of a bridge management system for

Pennsylvania. In their report, entitled 'Pennsylvania Bridge Management Systems', the group unanimously agreed that the development of such a bridge management system is feasible and that it is a very important and urgently needed tool for better management and engineering of the state's large and antiquated system of bridges.[1]

HPR funding was secured for a work group of ten persons to develop the concepts and technical requirements, to pilot test, and to guide statewide implementation of a total bridge management system (BMS). The group worked in Harrisburg approximately 3 days per week. The work group consisted of six department employees and four outside consultants.

The work group prepared a report, entitled 'Engineering Concepts and Requirements for a Bridge Management System', which formed the basis for a request for proposal to develop software for the BMS.[2] McDonnell Douglas Professional Services was selected to provide the development, testing, implementation and training on the use of EDP software. Software work by McDonnell Douglas was performed using state funding.

Needs

The Commonwealth of Pennsylvania currently has approximately 54 500 bridges on the state and local highway systems. These bridges include 22 500 with span lengths of 20 ft or greater and 32 000 having a length between 8 and 20 ft. Approximately 30% of the bridges in the state are classified as structurally deficient or functionally obsolete.[3] The estimated cost to put these deficient or obsolete highway bridges in a minimum acceptable condition is approximately $4·0 billion. The costs to put the system in a first-class condition could approach twice this value.[2]

This information provides a single snapshot of the magnitude of the problem of aging and the accompanying decay of highway bridges in Pennsylvania. To attack the problem, annual funding is generated from federal, state and local sources. The basic management challenge is to make the best use of the available funds in an overall program of maintenance, rehabilitation and replacement.

Objective

The general objective for the BMS was to develop a management tool which will enable a systematic determination of the present and future needs for maintenance, rehabilitation and replacement of bridges in Pennsylvania using various scenarios, along with a prioritization which will provide guidance in the effective use of designated funds.

The specific objectives for a BMS are to develop a system that on demand but at least annually:

—Yields recommendations, with associated cost estimates, for activities required to enable all bridges on public highways and roads to perform their functions in the most cost-effective manner. These activities include various levels of maintenance, various modes of rehabilitation and replacement.
—Predicts present and future needs and associated costs to perform the above activities for all bridges in at least two scenarios, including 'minimum acceptable' and 'desirable'.
—Sets statewide and regional priorities for each of the above activities, based upon functional and physical needs for each highway classification system, and provides a listing of candidate bridges.
—Provides a basis for recommending regional distribution of budgeted funds.

Overall System

The basic parts of the BMS are shown in Fig. 1. The department's computerized structure inventory records system (SIRS) will form the base

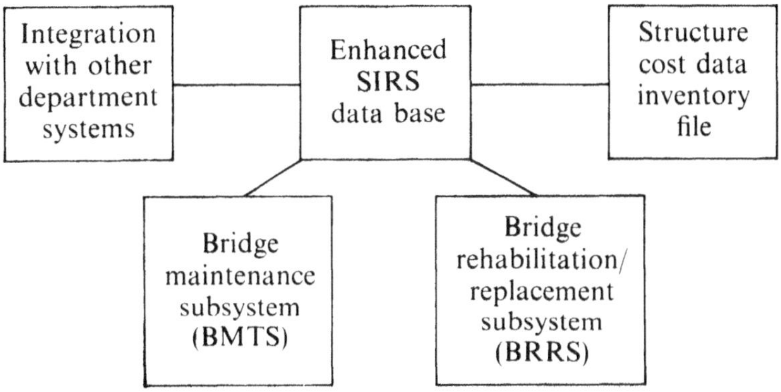

FIG. 1. Flow diagram showing the basic parts of a BMS.

for the development of an overall BMS.[4] Enhancement of the SIRS data base, integration with other data bases and the development of a structure cost data inventory file will form the complete data base. The two main parts of the BMS will consist of the subsystems for maintenance (BMTS) and rehabilitation/replacement (BRRS), as shown in Fig. 1.

DATA BASE

Structural Inventory Records System (SIRS)

The entire BMS project has been structured under the premise that the Pennsylvania system should basically be a derivative of available information. Therefore the department's computerized SIRS will form the base or starting point for the development of the overall BMS.[4] This system includes other data in addition to the 88 data items required by the Federal Surface Transportation Assistance Act of 1978.[3]

Significant modifications in the form of enhancements and additions are also required to satisfy the present and future needs of all users. Since the implementation of a SIRS in 1982, the users have recommended numerous changes or enhancements to improve the utility and quality of the system.[5] These enhancements can be effectively and efficiently added to the system during extensive software modifications needed for the development of the BMS.

Integration with Other Systems

The redundant storage of data elements between SIRS and other department systems, along with the fact that the SIRS does not currently interface with other application systems, indicates that a certain amount of duplicate effort is required to maintain the redundant data. This duplicate effort can lead to data inaccuracies between application systems. Therefore the BMS will be integrated with other PennDOT application systems during the development of the BMS.

Cost Inventory File

Bridge maintenance unit costs for the bridge maintenance activities will be calculated from daily foreman input of labor, materials and equipment. If storage of such data is not provided in the maintenance system, the BMS will be required to provide such storage in the 'structure cost data inventory file' for needs projections and other uses.

New or replacement costs as well as rehabilitation costs, including widening costs, are also needed to predict needs and priorities for rehabilitation/replacement. The 'structure cost data inventory file' will provide storage for these costs which are on the basis of dollars per square foot of bridge deck area. Estimated costs can then be calculated by the system from the square foot costs which are multiplied by the length and width parameters from the SIRS. A bridge manager needs reliable cost data and must be able to track all bridge costs for future use.

MAINTENANCE SUBSYSTEM

Maintenance Inspection

PennDOT has taken steps to expand the inspection effort to more descriptively cover bridge maintenance needs. Seventy-six potential maintenance activities have been identified. SIRS individual bridge data files are being expanded so that the user can include a tabulation of those maintenance needs with each item quantified and prioritized by urgency. This will ensure that routine maintenance needs and their urgencies are adequately documented. A form will be completed by the field inspector giving estimated quantities and priorities for the applicable maintenance activities. A unit price table for the 76 activities is also established. It will be periodically updated based on actual cost experience. The total bridge maintenance needs can be determined from the estimated quantities and unit price values. Maintenance activities completed will also be included in the enhanced SIRS.[6]

The Subsystem

The extent and urgency of overall bridge maintenance needs can be determined and sorting of this work for efficient implementation by either contractor or department forces can be handled automatically by the system. The system will automatically notify the county maintenance manager and monitor the implementation of critical repairs. The work backlog that exists far exceeds that which the department can physically and financially handle. Hence it is important to rationally prioritize this work to apportion limited available funding and manpower, or to relate to non-bridge needs.[6]

Prioritization Method

It is important that guidance be provided to the district and county offices to assist them in selecting the best candidate bridges for maintenance work as well as which activities to perform first. This will help ensure that those deficiencies deemed to be the most critical to the safety features of the bridge, and hence to its users, are brought to the attention of management.

A simple prioritization procedure has been developed.[6] It considers the effect of the most structurally critical maintenance activity needed on the bridge as well as the individual bridge's impact on the road system. The components of the procedure are as follows:

—activity ranking,
—activity urgency,
—bridge criticality, and
—bridge adequacy.

Activity Ranking

The bridge maintenance activities themselves vary in their importance to and effect on the structural integrity of the bridge. Activities such as repairing stringers or repairing abutment underscour would generally be performed on a priority basis while activities like applying protective coatings and constructing abutment slopewalls would tend to be deferred. As a general rule, activities that most directly, immediately and positively impact the continued safety and structural adequacy of the bridge would be performed first while those that have minimal immediate impacts would tend to be performed last.

The activities have been divided into groups from A to E based on their generalized relative importance to the current structural stability of the bridge. Group A has the highest priority and group E the lowest. The activities repair/replace: steel stringers, floorbeams, girders or truss members could be related to existing or potential fatigue damage. If the needs are indeed fatigue related, they are more important and hence would be given a higher deficiency point assignment. This determination can be made by comparing these maintenance activity needs with the type of fatigue prone member that controls the inventory rating. SIRS data provides space for recording the controlling member type as well as the fatigue and load data related to it. If the activity is fatigue related, it will be assigned as group AF and hence given additional deficiency points.

Activity Urgency

The severity of a deficiency can be a reason to increase its priority for repair. The urgency factor for each activity need is coded by the district bridge inspection unit. It yields an informed assessment of how soon the work needs to be completed. As such it is also a measure of the severity of the deficiency. It will judgementally define the promptness of action that is needed for each specific maintenance activity need. The priority code used will range from 0 to 5. Priority code 0 is for a critical safety deficiency where prompt action is required and priority code 5 is for routine nonstructural bridge maintenance which can be delayed.

Bridge Criticality

The importance of a bridge to the road network, as well as the impact of the loss of bridge service to traffic, are other factors that must be considered in deciding the order in which they are to be repaired. It is readily apparent that the road system hierarchy realistically defines importance. That is, if a bridge on the interstate and a bridge on the local access system have similar deficiencies, it is obvious that the interstate highway bridge would be repaired first. However, the impact of a bridge's closure also needs to be weighed. If the detour length is excessive, and hence intolerable, the priority for repair should be raised.

The assessment of the importance of the bridge will be based on the classification of the highway, its ADT and the detour length that would be imposed on traffic if the bridge were to be closed. Multiplying the ADT times the detour length results in a relative measure of this importance.

Bridge Adequacy

The capability of the bridge to safely carry the loads that traverse the route, and to continue to do so, weigh in a manager's decision of whether or not repairs should be implemented. The load capacity rating indicates the current strength of the bridge. It does not give any indication of what can be expected in the future. The condition rating of the most critical component of the bridge can be used to generally assess degradation. By considering both the current load capacity and the lowest condition rating of the structure's components a measure of the inadequacy of the bridge can be obtained.

Deficiency Point Assignment

Most of the data that will be needed to define the above components of the prioritization procedure are already in the SIRS. The only new items are the maintenance activities themselves and their individually assigned urgency ranking. They are important components of the proposed BMS.

Having defined the major parameters that are to be considered, the relative weights to be assigned to them and their elements must be established. To be consistent with the general philosophy of the rehabilitation/replacement prioritization system, a deficiency point concept will be used for the maintenance activity prioritization system. However, it is readily apparent that the factors and methodology being used in each system are quite different. Although it is possible, numerically, for a single bridge to be assigned in excess of 100 deficiency points, the deficiency point assignment will be limited to a maximum of 100. The higher the assignment on a bridge, the higher the priority.

Table 1 summarizes the four major components of the prioritization system, defines the elements in their makeup and indicates the initial or trail weights that have been assigned to each. As the procedure is tested, evaluated and refined the weight assignments could and probably will change.

The maintenance deficiency point assignment for a bridge will be based on the bridge maintenance activity that has the largest sum of deficiency points for activity rank and urgency. The deficiency point assignment and its ranking within the county will be recorded on the bridge maintenance activity needs screen. Hence, when a manager views the subject screen for individual bridges, he has an immediate indication of the relative priority of the most critical repair needs of one bridge compared to another and compared to the worst possible case (100 deficiency points).

TABLE 1
Maintenance deficiency points assignment

Deficiency points	Component	Element	Deficiency point assignment
25	Bridge maintenance activity rank	Group AF	40
	(Note: AF = group A activity that is	A	25
	fatigue prone and controls the	B	20
	inventory rating)	C	15
		D	10
		E	5
25	Activity urgency factor	Code 0	25
		1	20
		2	15
		3	10
		4	5
		5	0
25	Bridge criticality		
	Part A: Interstate		5
	US numbered highway		4
	State highway		3
	County highway		2
	City, Borough St & Twp Rd		1
	Part B: PCN		5
	PCN/coal haul		5
	Agricultural access		3
	Industrial access		3
	Part C: ADT × detour length ≥ 30 000		15
	≥ 15 000 but < 30 000		10
	> 3 000 but < 15 000		5
	< 3 000		0
25	Bridge adequacy		
	Part A: Lowest condition rating < 3		15
	> 3 but ≤ 4		10
	> 4 but ≤ 5		5
	> 5		0
	Part B: Load capacity (inventory rating)		
	(H configuration) ≤ 12 t		10
	(H configuration) > 12 ≤ 20		7
	(ML 80 configuration) > 20 to ≤ 30		4
	(ML 80 configuration) > 30		0

With a deficiency point assignment being stored in the BMS for every bridge, prioritized listings can be easily generated using the particular parameters desired. To facilitate this reporting, friendly programmed report generators with user defined variables are being developed.

REHABILITATION/REPLACEMENT SUBSYSTEM

The Pennsylvania System

The prioritization of bridges for rehabilitation and replacement is based upon the degree to which each bridge is deficient in meeting public needs. Deficiencies are evaluated in three general categories:

—level-of-service capabilities,
—bridge condition, and
—other related characteristics.

These deficiencies are then combined to yield a total deficiency rating (TDR) on a scale which ranges from 0 to 100. Basically, the framework for the determination of the TDR was patterned after parts of the Federal Sufficiency Rating System (FSRS)[3] and parts of the system developed by North Carolina State University for use by the North Carolina DOT (NCS).[7]

Level-of-Service Deficiencies

There are four characteristics included for the level-of-service capabilities: load capacity, clear deck width, vertical clearance for traffic carried by the bridge, and vertical clearance for traffic passing under the bridge. The level-of-service deficiencies are based on comparisons of the actual load capacity, clear deck width and vertical clearances of the bridge with level-of-service criteria which have been developed for the Pennsylvania Bridge Management System.[8] These criteria have been set at three levels: minimum acceptable, minimum design and desirable design. They are primarily dependent upon the functional classification of the highway carried by the bridge, with some additional dependence on volume of traffic. Equations have been developed to calculate load capacity deficiency (LCD), clear deck width deficiency (WD), over clearance deficiency (VCOD) and under clearance deficiency (VCUD).[8]

Bridge Condition Deficiencies

The second category of deficiencies in the Pennsylvania prioritization methodology is based on bridge condition. Bridge condition is included in the Federal Sufficiency Rating System (FSRS),[3] and can comprise a total reduction of up to 64 points in the sufficiency rating. On the other hand, in the North Carolina system (NCS),[7] bridge condition is addressed only indirectly, through the assignment of up to six deficiency points to the estimated remaining life.

In the Pennsylvania system, the evaluation of the bridge condition deficiency (BCD) includes the assessment of the condition of each of the three primary elements of the bridge: superstructure, substructure and bridge deck. The deficiency points for each element, which are directly related to the individual condition ratings, are given by equations.[8]

Finally, the bridge condition deficiency (BCD) is determined as the sum of the conditions deficiencies for the superstructure (SPD), substructure (SBD) and deck (BDD).[8]

Other Deficiencies

Other deficiencies are related to the remaining life, approach roadway alignment and waterway adequacy.

The estimated remaining life entered into the BMS data base is developed by the system as a function of the condition ratings of the superstructure, substructure and bridge deck. The remaining life deficiency (RLD) is then calculated using an equation.[8]

The approach roadway alignment may be the source of additional deficiency points. This deficiency (AAD), which is directly related to the appraisal rating contained in the BMS data base, is also given by an equation.[8]

The adequacy of the waterway is the final characteristic included as a source of deficiency points. This deficiency (WAD) is directly related to the appraisal rating included in the BMS data base and is also given by an equation.[8]

Total Deficiency Rating

Once the deficiency points of the individual categories have been determined, the next step in the Pennsylvania system is the determination of the total deficiency rating (TDR) for each bridge. This is done by summing the deficiency points of the individual categories, applying limiting conditions to several combinations of deficiency points, and applying a factor which reflects the functional classification of the highway carried by the bridge.

The summation of deficiency points is represented by the simple equation

$$\mathrm{TDR} = \phi[\mathrm{LCD} + \mathrm{WD} + \mathrm{VCOD} + \mathrm{VCUD} + \mathrm{BCD} + \mathrm{RLD} + \mathrm{AAD} + \mathrm{WAD}]$$

where BCD = SPD + SBD + BDD.

The last step in the determination of TDE is to apply the factor ϕ, which is

TABLE 2
Functional classification factors

Functional classification	ϕ
Interstate	1·00
Arterial	0·95
Collector	0·85
Local	0·75

dependent upon the functional classification of the highway carried by the bridge. Values of ϕ are given in Table 2.

A tabular representation of the development of the total deficiency rating for bridges is presented in Table 3. Table 3 also shows four combinations of deficiency points which have governing conditions.

Prioritization Listings

After the total deficiency has been established for all bridges, cost information is needed in order to develop the indexes which will be used in the prioritized listings. Initially two costs will be requested: (1) the replacement cost and (2) the cost of rehabilitation. Also requested will be the number of deficiency points removed by the rehabilitation or replacement.

Total deficiency ratings, combined with cost information and other

TABLE 3
Development of TDR for bridges

Deficiency category	Maximum deficiency points in category	Listing conditions (1), (2)	Listing conditions (3)	Listing conditions (4)
LCD	70	70		
			$\sum \leq 80$	
BDD	50			
SPD	50	$\sum \leq 50$		
SBD	50			
WD	15	15	15	
VCOD	15	15	15	$\sum x\phi \geq 100$
RLD	5	5	5	
VCUD	10			
		$\sum \leq 15$	15	
WAD	10			
AAD	10	10	10	
Maximum totals	285	180	140	100

factors, will yield listings of bridge rehabilitation and replacement projects prioritized in order to enable effective management of the bridge system.

RECOMMENDATIONS

The Pennsylvania BMS has been operational since December 1986. The following recommendations reflect knowledge gained through the development and operation of the system which could be valuable to others developing similar systems.

—The system must have administration and top management support.
—Get the users of the system involved early in the development of the system. In Pennsylvania the users were included as members of the bridge management work group during the development of the system. A bridge management coordination has also been set up in the central office and each of the 11 engineering districts to coordinate the operation of the system. Regular meetings are required to discuss problems with the system.
—A commitment must be made to the maintenance and enhancement of the system. At least three full-time system people are needed for the Pennsylvania system.

ACKNOWLEDGEMENTS

The work described in this paper is being funded by the Pennsylvania Department of Transportation and the Federal Highway Administration.

REFERENCES

1. Bridge Management Task Group, Pennsylvania Bridge Management System—proposed strategies to inspect, design, maintain, rehabilitate, replace, prioritize and fund bridge systems. Pennsylvania Department of Transportation, March 1984.
2. Bridge Management Work Group, Engineering concepts and requirements for a bridge management system. Pennsylvania Department of Transportation, January 1985.
3. US Department of Transportation/Federal Highway Administration, Recording and coding guide for the structure inventory and appraisal of the nation's bridges. Washington, DC, January 1979.
4. Pennsylvania Department of Transportation, Structure inventory record system coding manual. Publication Number 100, Harrisburg, PA, June 1982.

5. BRIDGE MANAGEMENT WORK GROUP, The Pennsylvania Bridge Management System. Pennsylvania Department of Transportation, July 1985.
6. ARNER, R. C., KRUEGLER, J. M., MCCLURE, R. M. and PATEL, K. R., Pennsylvania's Bridge Maintenance Management System. Transportation Research Record 1083, 1986.
7. JOHNSTON, D. W. and ZIA, P., A level of service system for bridge evaluation. Transportation Research Board, 63rd Annual Meeting, Washington, DC, January 1984.
8. BRIDGE MANAGEMENT WORK GROUP, The Pennsylvania Bridge Management System. Final Report, Pennsylvania Department of Transportation, February 1987.

8

Data Information System for Structures: DISK

M. El-Marasy

Ministry of Transport and Public Works, Rijkswaterstaat, Bridges Department, Voorburg, The Netherlands

ABSTRACT

The data information system for structures is an automated information system developed by the Dutch Ministry of Transport and Public Works to support the management of bridges and other types of structures. The information stored in the system consists of:

- *Basic administrative and technical data which are almost unchangeable but have to be updated in case of change.*
- *Changeable information which is related to daily management activities of bridges such as inspection and maintenance.*
- *Financial information about maintenance.*

Most of the data have been standardised to meet the requirement for uniform description and evaluation of the deficiencies and the possibility to select and analyse the data.

INTRODUCTION

Bridges are a very important part of the road infrastructure and lack of maintenance can have far-reaching consequences. The developments in science, in building materials and in electronic computation during the last decades have resulted in highly advanced design and construction methods and in the building of very sophisticated bridges. Environmental pollution, the increase in both traffic intensity and heavy transport and the ageing of existing bridge stock have a significant impact on the bearing capacity and life duration of bridges.

The deterioration of the condition of bridges, and the increase in inspection and maintenance costs in view of funding limitations, has in recent years attracted the attention of the highways administration. There is a need for an instrument which can support managers and decision makers in carrying out their work more efficiently and in granting the functionality and safety of bridges in the most reliable and economic way. Therefore it is of great importance to find a model for the management of bridges through which the technical and economic problems related to inspection, maintenance, rehabilitation, strengthening and replacement can be solved. Moreover, it is also required to have an instrument for the assessment of the life duration and the bearing capacity of bridges or bridge components.

The Dutch Ministry of Transport and Public Works has recognised the above-mentioned needs and therefore developed an automated management system, DISK, for all structures under its management. The development of the system has been completed and the implementation has started.

GOALS OF THE DISK SYSTEM

The goals of the system are determined by carrying out an information analysis for the daily activities related to the design, construction, inspection and maintenance of structures under the management of the Ministry of Transport and Public Works. The analyses are carried out by representatives of the different disciplines engaged in the above-mentioned activities and have resulted in the following goals:

- Easily accessible, complete and up-to-date databank.
- Proper planning for the inspection of structures.
- Proper planning for the maintenance of structures and estimated budget requirement for a period of 5 years.
- Contribution to the effort to predict the functional or economic life of the structures.
- Feedback to the designers and other users of the system with relevant information acquired from inspection and maintenance experience.

Types of Structures

The infrastructures consist of various types of structures having different functions:

- Structures for road traffic such as bridges, viaducts, tunnels, etc.
- Structures for navigation and water control such as locks, movable bridges, storm surge barriers, etc.

The above-mentioned structures are built of various materials such as concrete, steel, wood, etc. Moreover, the movable structures include mechanical and electrical installations. All these varieties in form, material and function of the structures justify the development of several information systems. However, because all these structures fall under the management of the same administration it was preferred to integrate all the information systems in the DISK system, which consists of:

- Inventory and administrative information.
- Inspection.
- Maintenance.
- Historical information.
- Special transport.

INVENTORY AND ADMINISTRATIVE INFORMATION

Knowledge of the existing bridge stock is the basic factor for proper management and it will be impossible without it to carry out inspection, maintenance and to plan the required budget and manpower.

The inventory program of the DISK system has facilitated the storage of the bridge stock with the relevant information which supports the users on different managing levels. The stored data are almost unchangeable and consist of:

Identification

- Topographic number related to the topographic map.
- The name of the structure.

Location

- Number of the traffic road or the waterway.
- X and Y coordinates.
- The municipality and province where the structure is located.

Administrative Information

- Names of the departments responsible for design, inspection, maintenance and management.
- Number and location of structure file.
- Dates of construction, modification and demolition of the structure.

Technical Information

- Type, material and dimensions of the structure.
- Relevant members, main members, material, type and manufacturer.

Heavy Transport

- A number of factors for the calculation of bridges for the dispensation of heavy transport.
- Design loading class and actual bearing capacity.
- Allowable passable width and height clearances under and above the bridge.

INSPECTION

Inspection of bridges is a fundamental task for the managers. It is not supposed to be a target but an unavoidable medium to grant the functionality and safety of bridges and necessary activity to supply the decision makers with reliable planning for manpower and budget. The DISK inspection system helps the inspectors by supplying them with several reports and information.

Inspection Planning

The DISK system produces a list of the structures which have to be inspected in a certain period. This selection can be made from the registration of the last inspection date and the required inspection period.

On the report there is valuable information given such as the administrative, expected inspection duration (manpower), the required materials for inspection (crane, boat, etc.) and the possible restrictions for inspection if there are any.

Inspection Drawings

To support the inspectors in doing their work and to facilitate the inspection activity, the data information system introduced the making of inspection drawings. Each structure must have a number of drawings projecting the elevation, side view, bottom and top view and cross-sections of the structure. All structural members which must be inspected are shown on the drawings. The location of the members is defined by two axes.

Each member has its local number. All members which must be inspected are mentioned in a table on each drawing which is adopted as the 'checklist'.

The checklist, the main members to which the members belong, the number of the drawing and the local numbers of the members are stored in the computer.

These inspection drawings are made once and they remain valid as long as the structure is not modified.

The checklists on the drawings are used as guidelines for the inspector.

Deficiencies Report

The inspection of the structures can start according to the plan, the technical instruction for inspection and the available documents such as the drawings, general information, technical reports, preceding inspection report, etc. The results of the inspection and the interim recommendations for required maintenance work are stored in the computer.

The inspection procedure includes the following steps:

- To locate the deficiency of the inspected member according to the method mentioned previously.
- To describe the deficiency according to the standard types of deficiencies.
- To determine the possible reason for the deficiency according to the standard reasons.
- To recommend the necessary maintenance work.
- To assess the reliability of the members with respect to safety and functioning. This is expressed in a numerical code 0–6 (see condition rating).
- To obtain photos of the deficiencies if the inspector considers this necessary.

All this information completed with other general information (identification, location and description of the structure—weather conditions—the inspection directorate—the name of the inspector) forms the interim deficiency report.

Together with the inspection drawings and other documents, the deficiency report forms the inspection report.

Condition Rating

Evidently it is important for the decision maker responsible for maintenance to know the influence of the deficiency on the condition of the member and subsequently on the structure. The condition rating of the member and the structure is expressed in a numerical code for:

- Safety.
- Functioning.

The numerical code of the condition rating is used to get a quick impression of the structure's condition and to use it as an instrument for the determination of the technical maintenance priorities.

There are other factors which may influence the above-mentioned technical priorities, such as the available budget for maintenance and the importance of the structure.

The possible numerical codes for safety and functioning are related to a certain period. The period indicates how long the safety and the functioning of the member of the structure is guaranteed. In other words, within which period the maintenance has to be carried out. The possible numerical codes are as follows:

0 Safety or functioning is guaranteed.
1 There are deficiencies but safety or functioning is guaranteed.
2 Safety or functioning is guaranteed for the next 5 years.
3 Safety or functioning is guaranteed for the next 2 years.
4 Safety or functioning is guaranteed for the next 1 year.
5 Safety or functioning is guaranteed for the next 0·5 year.
6 Safety or functioning is in danger.

It is clear that the choice of one of the above-mentioned codes is a matter of experience and knowledge. In order to be able to make a reliable choice the inspector has to consider the following factors:

- Deterioration of the quality.
- Deviation from the design standards.
- Change in the original conditions of the structure.

To facilitate the work of the manager, the structural members are grouped into a number of main members (about 20). The relation between members and main members is established in the system via the checklist for each structure. Because of this relation it is possible for the system to produce a short condition rating report for the main members and the whole structure. The system selects automatically for each main member the highest condition rating code given to the members which belong to the main member and subsequently for the structure the highest condition rating code given to the main members.

The condition rating for the structure can be adjusted manually, if needed, to avoid undue influence from the bad condition rating of a secondary member reflecting on the main members or on the structure.

Evaluation

The quality of the inspection results, the recommendations for maintenance and the condition rating depend on the experience and the knowledge of the inspectors. To ensure good inspection results and good quality of information stored in the system and subsequently to supply the users of the system and the managers of the structures with reliable information, it was decided to form a group of experts on design, inspection and maintenance of structures which have to judge the results of inspection. This group is called the 'evaluation committee' and is formed for each discipline (concrete bridges, steel structures including electrical and mechanical installations, water control structures). The task of the evaluation committee is:

- To judge the results of the inspection such as the deficiencies, the reason and the recommendations for maintenance.
- To judge the condition rating of the structures.
- To decide if special inspection and more research are needed if the results of the inspection are not satisfactory.
- To determine the inspection period.
- To determine the technical priority of maintenance.

The results of the inspection stored in the system must be adjusted according to the evaluation work. The interim deficiency report, condition rating report and the inspection report will be replaced by the definitive reports. It is also the task of the evaluation committee to expand, delete or adjust the standard information in the system such as:

- Structural members and main members.
- Types of deficiencies.
- Recommendations for maintenance.

MAINTENANCE

The inspection procedure is completed after the evaluation of the results and the production of the definitive reports. Then the maintenance procedure can start. The information system can support the maintenance procedure in a quick and efficient way by supplying the following information:

- Deficiencies report in priority order.
- Report with maintenance units.
- Planning of the maintenance units.
- Report with maintenance projects.
- Planning of maintenance projects and required budget.
- Report with postponed maintenance work.

Deficiencies Report in Priority Order

The condition rating of the structural members for safety and functioning after the observed deficiencies is given in the definitive deficiencies reports.

The information system can produce a report with all the observed deficiencies and rearrange them in priority order of the condition rating. It is also possible to produce a report with a certain condition rating or in between two condition ratings. This report is used to select the (similar) deficiencies which can be repaired in one maintenance action. This group of deficiencies is called 'maintenance unit'. In this report we obtain the administrative information, the identification and the description of the structure, the number of the deficiency, the impaired structure member, description and the reason for the deficiency, the required maintenance work and the final date by which the maintenance has to be carried out.

Maintenance Units

The deficiencies which form a maintenance unit are grouped manually on the deficiencies priority report. The maintenance cost of each unit has to be estimated. The information system can group the maintenance units per structure or per managing department.

These reports contain, besides the administrative information, the description of the maintenance units, the final date for maintenance, the estimated cost, and its state (rough or accurate estimation) and type of budget.

Planning of Maintenance Units

To support the managing of structures, the system can produce a report with expected maintenance and its estimated cost. The maintenance units per structure are given in the report. The necessary cost is planned for a period of 5 years. This report can be adjusted by the manager for practical reasons and because of the available budget for maintenance.

The following information is mentioned in the report: administrative information, maintenance units per structure per managing directorate, estimated cost per maintenance unit, type of budget required per year for a 5-year period and the total budget required per managing directorate.

Maintenance Project

For practical and efficiency reasons it is possible to combine several maintenance units within one structure to form a maintenance project. In certain circumstances, when the deficiencies are similar or can be repaired by the same contractor, the maintenance units of different structures in a certain area are also combined into one maintenance project. For economic reasons the manager can decide to accelerate the maintenance of certain deficiencies which were planned for later years and combine them with the maintenance project which must be carried out. The report of the maintenance projects contains, besides the administrative information, the project number, the number and the description of the maintenance unit, and the ultimate date for repair.

Maintenance and Budget Planning

The DISK system can support the various managing directorates of Rijkswaterstaat in the formation of their yearly work and budget planning.

The system can produce a maintenance planning report for each department. The report gives the total required budget for a 5-year period divided into the different budget types.

Postponed Maintenance

All maintenance projects to be carried out must be entered in the system. The real maintenance cost has also to be entered. These projects do not appear in subsequent maintenance projects reports. Other maintenance projects which are not carried out for one reason or another appear in the postponed maintenance report. This report can be produced per directorate.

HISTORICAL INFORMATION

It is essential for the designers, maintenance staff and the managers to know the history of the structures; therefore there are some activities which are entered and kept in the system, such as:

- Each inspection date.
- Each maintenance date.
- Condition rating of the main structural members after each maintenance.
- Inspection period.
- Real maintenance cost.

All this information is produced on the historical information report. This information accumulates and will never be deleted from the system.

REGISTRATION OF THE INFORMATION IN THE SYSTEM

The information which has to be entered in the system consists, as mentioned before, of two types:

- Constant data.
- Variable data.

The functioning of the DISK system depends on the registration of the above-mentioned data. To ensure the functioning and the reliability of the system, the following conditions should be met:

- The data must be complete and up to date.
- The data must be unambiguous.
- The results of inspection and maintenance must be registered in the system as quickly as possible.

To meet the first condition, an inventory of all structures under the management of Rijkswaterstaat has taken place. The inventory of other information has started and it will be gradually registered in the system. It is expected to be completed within 5 years.

The second and third conditions can be met by simplifying the registration method of the data in the system, especially the results of inspection and maintenance. Therefore it was decided to supply the users of the system (spread over the whole country) with personal computers, which can be connected to the central computer located at the bridge directorate, such that they can enter the data directly in the system. The advantages of this method are:

- Minimise the administrative work of the users.
- The user can directly control his input.
- The output is directly available.

HOW CAN THE SYSTEM SUPPORT THE MANAGEMENT?

The infrastructure of roads and waterways has great national economic value. The first task of the management is to ensure the safety of these structures for the public.

For an efficient and reliable management certain actions need to be taken, such as an inventory of all structures, inspection planning, periodic inspection of the structures, allocation of required maintenance funds and carrying out the maintenance on time.

With the development of the DISK information system many of these actions have been realised and others have been started and are in the development phase. The system provides the management with the following information:

- All structures under the management of Rijkswaterstaat are inventoried and stored in the system with many administrative and technical data. With this information, which is easily accessible, the manager can do his job efficiently, such as inspection, maintenance and budget planning.
- Inspection planning is facilitated by the list of structures which must be inspected in a certain period for a certain region. The manager of the discipline which is to be inspected makes his detailed inspection planning, such as the personal planning, the necessary inspection material and, if any, the required measurements.
- The system can supply the inspectors with technical data on the structures which must be inspected, such as identification, location,

results of preceding inspection, maintenance carried out and the technical history if needed.

- The standardisation of the structural members, main members, types of deficiencies, the reasons, etc., facilitates the work of the inspectors and minimises the administrative work. It also creates uniformity in the inspection and results in uniform reports of all inspection disciplines.
- The planning of maintenance depending on the technical priorities given by the inspectors and the expert evaluation group, as well as the planning of the necessary maintenance budget for a period of 5 years, are fundamental to the work plan and the allocation of the necessary funds.
- A complete and easily accessible databank is a good source of very useful information for designers and for the maintenance staff. They can observe the behaviour of certain members, certain types of structures or the effectiveness of maintenance undertaken in the past. It will also add basic information to studies on the life of structures and the determination of a suitable inspection period.

9

Optimization of Bridge Maintenance Appropriations with the Help of a Management System—Development of a Bridge Management System in Finland

ARI KÄHKÖNEN

Viatek Ltd (VIASYS), Ahventie 4A, SF-02170 Espoo, Finland

and

ALLEN R. MARSHALL

Cambridge Systematics Incorporated, American Twine Building, 222 Third Street, Cambridge, Massachusetts 02142, USA

ABSTRACT

Instead of building new infrastructure components such as bridges and highways, public works agencies nowadays are putting more effort into maintaining and repairing the existing structures under their charge. This also pertains to bridges. A basic management problem is how to direct the budget and in what order of priority repair actions to structures should be done. The priority can be determined based on a variety of different agency objectives. In Finland at present, it is particularly important to set a long-term minimum acceptable condition level for public bridges. As a matter of policy, Finland has sought to eliminate posting on public bridges but this objective is obviously more difficult to achieve as the bridge stock ages.

To solve these sorts of management problems many different countries have developed bridge management systems (BMS). Finland is following in these footsteps, drawing on their own experience with BMS and those of other countries such as Sweden, Denmark and the United States. The Road and Waterways Administration (RWA) is developing a two-level bridge infrastructure improvement management system which will serve the RWA and the 13 highway districts in Finland. The new management system necessitates the development of a new database of bridge information (the bridge directory in RWA terms). In order to support an effective deterioration modeling and optimization capability the BMS requires a more thorough bridge inspection system, including the collection of specific damage conditions for individual structural parts, as well as more comprehensive general condition data.

INTRODUCTION

In Finland the Road and Waterways Administration (RWA) builds and upkeeps bridges on public roads. Presently there are about 12 000 bridges on public roads. The bridge stock is quite varied and consists of bridges of stone, wood and more modern construction materials. A basic overview of the bridge stock is shown in Table 1.

The present value of the bridges is over 10 billion Finnish marks (Fmk). (There are approximately 4·2 Finnish marks to the US dollar, 6·9 to the £ sterling and 1·5 to the French franc.) Sixty million Fmk are used yearly for the maintenance and repair of the bridges (0·6% of the present value). Another 100 million Fmk/year are used for the rehabilitation (meaning widening, strengthening and replacement) of the bridges, which is 1% of their present value. The weight restrictions on the bridges in Finland will change at the beginning of 1990 in order to conform to EEC requirements. Presently there are 350 bridges with weight restrictions under current posting standards, but due to increasing vehicle weights it has been estimated that weight restrictions will need to be placed on another 560 bridges. If the new weight restrictions are to be removed by strengthening or replacing bridges within the next 5 years, an additional 80 million Fmk/year will be necessary.

TABLE 1
Statistics on Finland's bridges

Number	Length and deck area
9 496 bridges	177·1 km
3 216 culverts	2·2 million m^2

Material	Length (km)	Average span (m)	Deck (m^2)
Concrete bridges	1 501 408	106·7	15·0
Reinforced concrete bridges	273 055	22·3	46·0
Steel bridges	365 501	36·0	41·7
Timber bridges	88 004	10·5	12·6
Stone bridges	26 081	1·6	7·5

The cost of repairing, upkeeping and rehabilitating bridges in Finland will be in the neighborhood of 240 million Fmk/year. Determining the effectiveness of the current upkeep and repair system of the bridges requires that the following questions should be considered:

—What is the present condition of the bridges and what will it be in the future with:
 (a) no repair actions undertaken?
 (b) a mix of different action strategies?
—What is the optimum level at which bridges should be kept?
—What budget level is required so that:
 (a) bridges remain in their current states?
 (b) bridges are upgraded to the optimum level?
—What are the right targets, actions and timing at a given investment level?
—Are the chosen actions correct and economical, and have all alternatives been explored?

In order to provide answers to the questions mentioned above, the RWA began the development of a new Bridge Management System (BMS) in 1986–87. Previously the RWA kept detailed inventories of their bridges in a bridge directory, but it was felt that the directory was inadequate as a means to address the questions above. The original bridge directory was a fairly static inventory and did not effectively address changing bridge condition, loading impacts related to damages, repair and maintenance histories, and other crucial factors. The new bridge register is far more extensive and is designed to meet all types of information needs.

The function of the management system based on the enhanced directory information is to help decision makers and maintenance personnel to determine those goals and actions by which the safety, the level of service and the condition of bridges can be kept at the desired level. The system can also optimize and allocate correctly the available funding at the network level. On the network level, the system is also able to find the optimum condition level which the bridges should be kept at on a long-term basis.

In Finland it has been decided that the complete management system should perform its work at two levels: the network level, where bridges are examined as a system, and the project level, which deals with individual bridges or groups of bridges. The system at the network level will help the RWA and the upper level management in districts. The system can find that optimal condition level where bridges should be kept in the long term. The project level system is a tool for bridge engineers to use in development work programs (1-, 3- and 5-year programs).

SYSTEM DESCRIPTION

The Bridge Management System is a separate program that works in conjunction with two other modules: a bridge inspection system and a bridge directory. The general interrelationship of BMS components is shown in Fig. 1.

When work on the bridge management system in Finland began, we found that the directory did not satisfy the information needs of the management system. The RWA already had a database-type register on their DPS8 mainframe, but only the RWA could use it. The system was batch oriented and not very flexible in the types of reports that could be generated or the type of detailed analysis that could be performed (even if the data were available). Every year a routine report was sent to the districts telling them about the bridges in their own district. The register was composed of 37 different types of data (all of it was basically administrative information). The main problem with the register was the fact that all the information about bridges was centralized in the RWA, it was difficult to use and, in particular, it did not have any information about either the condition or the continuing repair/reconstruction costs of the bridges. These shortcomings have been corrected in the new directory.

In Finland there was a regular and thorough bridge inspection system in place long before the management system was even considered. The problem with this system was that during the inspection only subjective condition information was gathered. The system was changed and now objective damage reports are gathered, which the management system can use as is or after they have been converted into condition information. For more objectivity, damage class tables were produced to help the inspector. Bridge

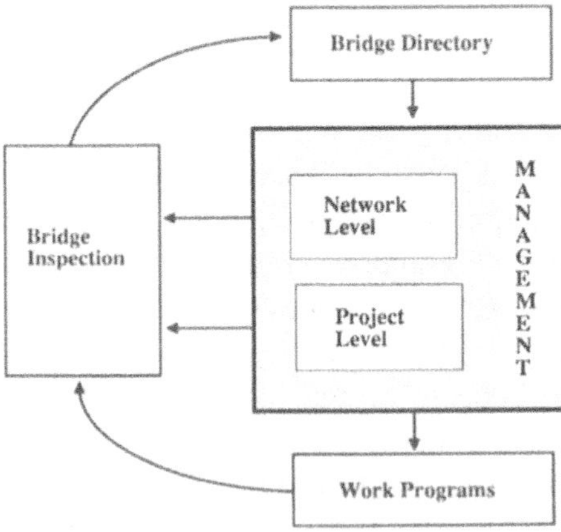

FIG. 1. General description of the Finnish RWA Bridge Management System.

damage, necessary repair actions and their costs are estimated with these tables.

The basic data for the management system are the damage and condition reports generated from bridge inspections. The inspection information is recorded in the directory, and the management system programs can access this information directly. The directory also contains other information the management system uses, such as traffic information, bearing capacity, repair history, utilities on bridges and other items.

BRIDGE INSPECTION PROGRAM (Fig. 2)

A systematic program of bridge inspections was initiated in Finland in 1970. The primary purpose of the inspections is to ensure acceptable levels of traffic safety. There are other purposes, too, such as making sure that bridges do not deteriorate unnecessarily and maintaining an acceptable structure appearance. The inspection system includes different kinds of inspections, such as:

—Final technical inspection.
—Yearly inspection.
—General inspection.
—Special inspection.
—Special control.

Final Technical Inspection

This is done as soon as the bridge is complete. The inspection rates the construction results, and notes any possible mistakes and damage due to construction. The inspection is done by a bridge engineer.

Yearly Inspection

This is done for the maintenance and upkeep of the bridge. The inspection is done by sight according to a set of maintenance instructions. This inspection is by the highway resident. The yearly inspections amend the general inspections.

General Inspection

A major inspection every 4–8 years which:

—Checks the condition of the bridge and its parts. The findings/recommendations of the bridge inspector are the first phase in project level programming for the bridge.
—Checks and completes the information in the bridge directory.

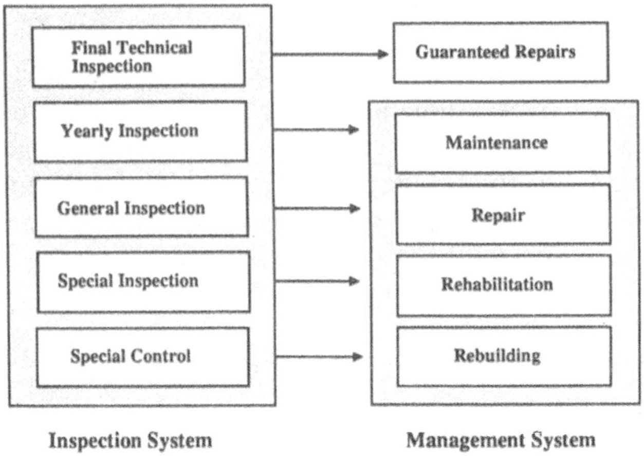

FIG. 2. RWA bridge inspection system.

The inspection is basically done by sight, but the investigation can be augmented with simple tests performed by specialized machines. This inspection is by a bridge engineer.

Special Inspection

This is done to fill in general inspections and when a repair plan is being made. Underwater bridge inspections are considered to be special inspections. This type of inspection is done by a bridge engineer and specialist from the RWA and the Technical Research Center, and a consultant also is present in most cases.

Special Control

Any changes, repairs or damage done to the bridge are checked on a regular basis 2–6 times a year. The purpose of these inspections is to make it possible to go over the calculated bearing capacity of the bridge or a part of it (e.g. if we want the weight restrictions on a bridge to be removed). This inspection is done by the highway resident or by a bridge engineer. This inspection basically deals with any form changes, cracking and erosion.

During inspections information is recorded onto a sheet. Each type of inspection has its own sheet, except for special inspections which are always different each time. The forms are quite detailed and will eventually be preprinted with the findings of the previous inspection so that the inspector can simply mark the items (condition items) that have changed between inspections. Information items obtained during a general inspection are shown in Table 2.

The information recorded by the district engineers on the field forms is put into the bridge directory by the engineers for later use. The central

TABLE 2
General inspection—Information collected

Inspection number	Inspection date
Inspection type	Inspector
Next scheduled inspection	Special equipment needed
Comments	Special control recommendation
Condition assessment of:	
Substructure	Railings
Edge beam	Expansion joints
Other substructure	Other fixtures
Surfacing	Other bridge site structures
Other surface structure	General condition
Calculated condition (rule-based) of:	
Substructure	Railings
Edge beam	Expansion joints
Other substructure	Other fixtures
Surfacing	Other bridge site structures
Other surface structure	General condition
Comments on general bridge deficiencies	
Specific assessment (for certain 'typical' bridges) of:	
Structural part type	Urgency class
Structural part material	(inspector's code for repair urgency)
Damage identification	Recommended treatment
Damage version number	(action proposal)
Cause of damage (coded)	Urgency of repair need
Type of damage (coded)	Unit cost
Extent (area, length, width) of damage	Effect on bearing capacity of damage (yes or no)
Damage location on bridge	Special inspection put into effect
Damage class (slight, moderate, severe)	Photograph of damage
	Inspector's comments about damages and treatments

government can obtain copies of the data via network links. The forms and photos are also stored in paper archives.

BRIDGE DIRECTORY

The bridge directory is where all information about bridges is kept. The directory is composed of technical and administrative information, any information (mainly condition and damage information) gathered during inspections, along with geometric and traffic data which affect the usability

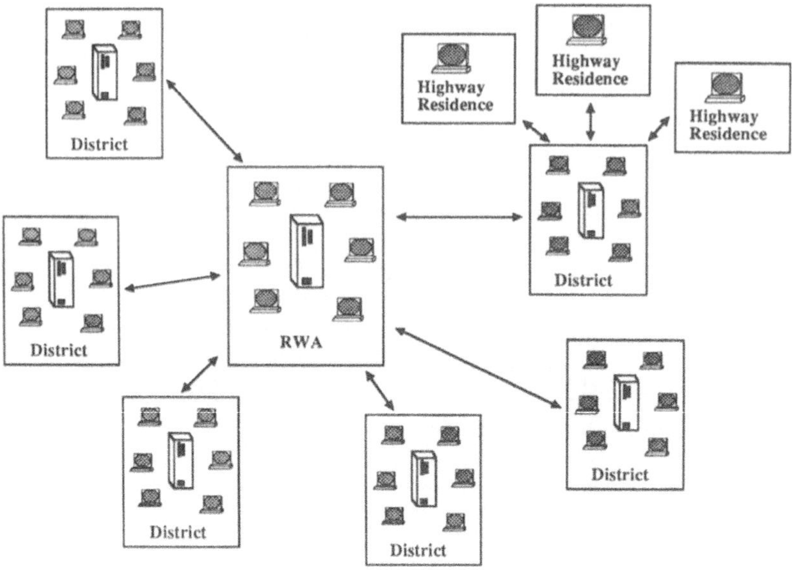

FIG. 3. The bridge directory information connections.

of the bridge. The directory is the database of the management system. A good directory, along with its programs for reports, is a useful tool for bridge engineers and any other people dealing with bridges.

The bridge directory in Finland is being completely redone, both the database structure and the information in it, as well as the user interface and reporting capabilities.* The new directory will be compiled so that it will suit the needs of different types of environments and users. The directory will be placed in all 13 districts and also in the RWA. Information about all the bridges in Finland will be given to the RWA, while the districts will only have information about the bridges in their own districts. The districts and the RWA are connected through a high-speed data network. Highway residences, which are smaller administrative units within districts, are connected to district offices with modems. The connections are pictured in Fig. 3.

It is not anticipated that the highway districts will have any access to the system other than to prepare reports and review data. The districts are responsible for updating and other maintenance of the bridge directory

* Cambridge Systematics Incorporated of Cambridge, Massachusetts, USA (617) 354-0167, is preparing the new register and transferring the information from the old database. The directory utilizes Oracle database software and operates on a 80286- or 80386-based microcomputer. More information can be obtained by writing to either of the authors of this paper.

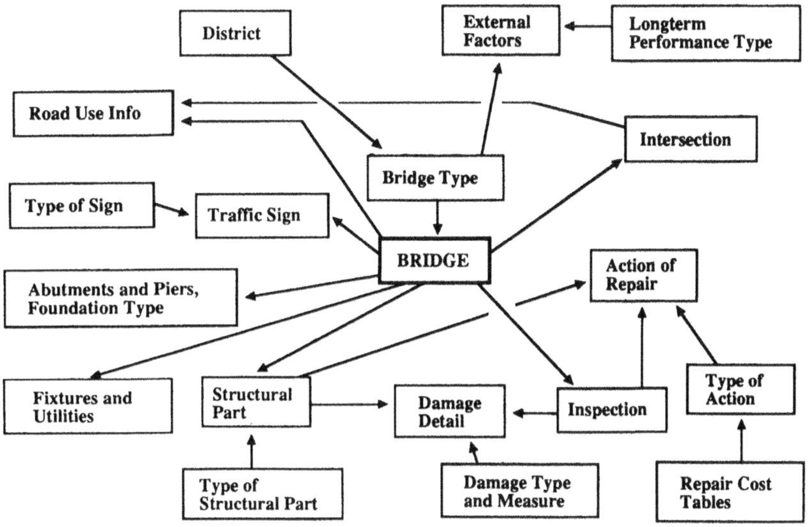

FIG. 4. The entity attribute relationship of the bridge directory.

(except for some bearing capacity and administrative information). This insures that the information is correct and current.

The general entity attribute model for the database is shown in Fig. 4. The prototype of the directory is being produced on a microcomputer, but the production version will be on a UNIX mainframe. The prototype is being tested now (June 1989) and the production version will be in place by early 1990.

MANAGEMENT SYSTEM

General

The main purpose of the maintenance and repair of bridges is safety. Bridges must be kept in good condition so that they can be used safely and so that they are not a danger to crossing traffic. After the safety requirements have been fulfilled, one must also think about damage and in what order they should be repaired. The appearance of a bridge is also important, and of course its ability to move traffic. Other problems are brought about by the budget, because there is never enough money for the necessary repair and maintenance actions. There is only a certain amount of money which must be used as effectively as possible. We must find a tool which can help decide what the budget should be, and what effect various spending scenarios would have on the bridge stock on both a long- and on a short-term basis.

Network Level System

One of the most important things the network level system does is to define the optimum condition level of the bridge system. It is evident that bridges in bad condition are safety risks and cause extra costs for the user (for example detouring the bridge because of weight restrictions) and also for society (higher accident costs, etc.). Operating expenses are higher due to more repairs. On the other hand, it is not cost-effective to maintain all bridges in exemplary condition, because more money would have to be poured into the maintenance of the bridge than would be economically beneficial to the users. The optimum level can be figured out through a benefit/cost analysis with some degree of confidence. In the simplest sense, the collective benefits that bridge users and society obtain from bridges should be equivalent to the bridge maintenance and repair expenses.

It is not simple to estimate the user and society costs and benefits, of course, and this poses one of the biggest challenges for the development of the management system. The system must seek to calculate costs and benefits objectively yet maintain a sensitivity to bridge system standards imposed by government policy makers that may make little or no use of formal analysis, e.g. 'all bridges on major public roads should be free of posting no matter what the cost implications may be'. The implications of the optimization scenarios can be determined and rationalized with respect to the vagaries of bridge 'politics' and policies.

The network level system analysis will also be geographically sensitive. This makes it possible to differentiate the optimization to recognize RWA district level costs and benefits. The basic principle is that money is put into areas which will generate the highest level of benefits. The network level system mainly serves the RWA and the management of the districts. With this system the RWA will be able to explain the higher budget needs to politicians who decide about such matters.

Project Level Systems

The project level system is a tool for bridge engineers when they are planning a work and action schedule. It is based on action recommendations received from the bridge system at the network level, condition and damage information gathered during bridge inspections, and on given budget limits. Based on a combination of job experience and help from the system a highway district bridge engineer can draw up a proposal for a repair program for the next year or even prepare a long-term action program.

The management system offers good tools for generating work and action programs. The engineer can immediately find out what effects adding, removing or changing an action will have on costs and condition. The system can, if necessary, show the condition and damage information on a particular bridge, information on previously performed actions, bridge aging factors, etc.

Other Management Capabilities

The programming of bridge inspections is also part of the management system. Historically, particularly in the United States, bridges have been inspected whether there is a need to inspect the bridge or not, for example, every 2 years. This fixed schedule has had some negative consequences for inspection programs, since visits are made to bridges in excellent condition while questionable structures are not seen until their turn in the fixed rotation. The Finnish RWA feels strongly that it is not an effective policy to inspect bridges on a set schedule. Instead the computer system can figure out what kind of inspection should be performed and schedule it based on the time since the last general inspection, the findings of that inspection, structure deterioration models and on aging factors.

The management system can also be used to help manage permits and routings for special transport. By determining absolutely critical bridges using high volume heavy transport routes and reviewing possible detour options stored in the register, it is possible to identify bridges that require special attention because of their site and situation.

An important part of the management system is the establishment of programs for strengthening and rebuilding bridges to satisfy new transportation system usages. The management system programs can help determine in what order weight restricted bridges, or bridges having some other problem with bearing capacity, should be rebuilt. This capability is very timely and important because of the new axle and total weight restrictions which will be put into effect in Finland at the beginning of the 1990s to meet EEC standards.

10

Highway Bridge Management

JOHN W. S. MAXWELL
*Grampian Regional Council, Woodhill House,
Westburn Road, Aberdeen AB9 2LU, UK*

ABSTRACT

Managerial, administrative, financial and technical information must be effectively translated into 'marketing' language to persuade client organisations to allocate adequate finance for bridges on the required time schedule.

Clients must produce policies and objectives for maintenance, improvement and replacement. Programmes of work and priorities will evolve from inspection and assessment information. Expenditure estimates will then enable programme revision as per financial constraints using relevant database and information systems.

Validity of traffic predictions must be checked and quality assurance systems applied to design, specifications and construction. Behaviour monitoring information has an important role to play, hence the urgent need to improve instrumentation and techniques.

If bridge management is a coin, 'heads' is the marketing dealt with in this paper and 'tails' is the production or inspection, assessment, maintenance and repair work (Fig. 1).

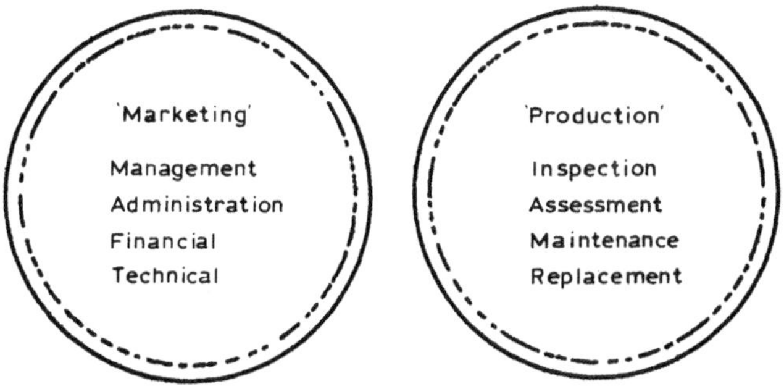

FIG. 1. Bridge management criteria.

INTRODUCTION

One hundred million pounds corrosion costs identified in 1960 increased to £600 million by 1970. No costs are available for 1980 but the new assessment code has placed severe weight restrictions on older steel, wrought iron and cast iron bridges.

Fifteen billion pounds was the estimated cost of replacing 120 000 UK bridges 10 years ago. In 1978 the Scottish Development Department suggested that estimated replacement costs of their trunk road bridges was £300 million with annual maintenance expenditure 0·28% of replacement cost. Non-trunk road bridges had a projected £1700 million replacement cost with an annual maintenance expenditure of 0·07%. Continuing deterioration suggests that not enough finance has been made available.

The design life for British bridges is 120 years whilst the French claim a 50-year life. Americans, because of salt attack, finance replacement on a 30-year cycle. Hence British bridges over 10 years old with no waterproofing may require an earlier and higher level of investment.

Access, inspection and assessment costs are now being addressed but expenditure on actual maintenance, repair or replacement work has not increased. The travelling public, commerce and industry find increasingly unacceptable any disruption to transportation and its associated costs. This paper identifies what needs to be done to ensure that the essential financial resources will be made available when required.

The main headings dealt with are as follows:

- Highway bridge management flow chart
- Policies and objectives
- Computerised database and information systems
- Technical issues for managers to consider
- Summary
- Conclusion
- References

HIGHWAY BRIDGE MANAGEMENT

A flow chart for highway bridge management is given in Fig. 2.

POLICIES AND OBJECTIVES

Central and local government strategy is to facilitate the transportation of goods and people within and through the highway networks which are their

Management of work, managers, workers and other resources, i.e.

1. Plan, identify, confirm objectives, priorities and programmes for long-term plans, 5-year programmes and emergencies.
2. Organise resources to achieve objectives, i.e. administrative procedures, information intelligence, staffing, finance, etc.
3. Direct, control, monitor, amend action from resources using control, coordination and communication techniques.

↓

Administration systems and procedures are essential to receive, record, arrange, analyse and give relevant information. Ensure that they are kept up to date and are always readily available, e.g.

1. Produce systems and procedures for inspection of bridges and recording of results with follow-up action necessary.
2. Acquire or replace basic information on contracts, drawings, schedules and instructions for maintenance and improvements.

↓

Finance

1. Annual estimated versus actual and projected expenditure budget sub-parts include estimates, carryovers, new starts, current actual, projected balance, engineering administration and final cost.
2. Correlate information from the sub-parts of the budgets and update on at least a 2-monthly cycle. Thereafter predict overspend/underspend and take corrective action required to achieve expenditure targets.

↓

Technical

1. Database and information systems should include inventory and location, inspection, testing and behavioural monitoring, structural assessment and loading capacities, programmes and costs for maintenance, repairs, strengthening, reconstruction and replacement.
2. Programme priorities usually motorways, trunk roads, principal roads, bus and emergency service routes and non-principal roads.
3. Interim measures vary from do nothing to temporary supports, single lane working, Bailey bridging or closure with diversion routes.
4. Planning, design, supervision of construction and maintenance works then implemented.

FIG. 2. Highway bridge management flow chart.

responsibility. Hence the policy that the first priority for road expenditure is to maintain and improve the structural fabric of the existing road network, including bridges, to a safe standard.

Objectives arising from this policy should include the following:

(a) achievement of requisite safety standards;
(b) reduction in rate of deterioration;
(c) cost-effective maintenance and improvement service;
(d) effective substantiation and marketing for increased finance;
(e) changing from unplanned emergencies to stable planned maintenance; and
(f) maximum in-service life from work done.

Different organisations will have to determine the policies and objectives for maintenance, improvement and replacement of bridges which are most relevant to their problems and circumstances. These should apply to existing bridges, those currently being designed or constructed, and those proposed for the future.

The problems of deterioration now with us suggest a failure in bridge management. A more realistic approach will be required to achieve and sustain optimum operational use of the highway network over the next few decades. Road Engineering Intelligence and Research (July/August 1989) highlighted £800m to repair Department of Transport concrete bridges over the next 15 years after a 2-year study. Similar sample surveys or studies to this and those previously carried out by the County Surveyors' Society on different types and size of bridges should be extended to statistically confirm the previous estimates of the scale and cost of the work to be done. Positive management should then implement the policies and objectives by translating the collated technical information from surveys, inspections and assessments into work programmes and estimated expenditure.

COMPUTERISED DATABASE AND INFORMATION SYSTEMS

Will computerised database and information systems fulfil their intended purpose for highway bridge management? The answer is yes and no.

For new bridges built since 1984 which have been subjected to the statutory system of cyclic inspections the answer should be yes. This assumes that reliable basic inventory and inspection data have been made available, checked as being correct, and used to substantiate the case for finance to initiate cyclic and structural maintenance where and when appropriate. In theory this should help to achieve maximum in-service life for the new bridges and safeguard the financial investment which enabled them to be built.

Prior to 1984 few systems had been developed and hence data were not necessarily kept in the format now preferred. The cyclic system of inspections introduced in 1978 had not been fully implemented, partly due to access problems and associated costs. Inventory information was either non-existent, incorrect or only partially available, requiring to be checked as valid and updated. Considerable expenditure and time would therefore be required for the collation of basic data before any system could effectively start to contribute to preventing collapse and minimising traffic disruptions, arresting deterioration and restoring bridges to a stable condition. This situation in particular applies to at least 75% of the regional and county bridges in Britain.

Most systems developed recently require large quantities of detailed information for the inventory and inspection data. Assessments of carrying capacity are also unlikely to be completed for some years yet. A more streamlined spartan approach is therefore required in the interim to produce draft programmes of work with cost estimates to substantiate the case for increased allocations of finance for bridges now before the deterioration identified for some time develops into an embarrassing crisis in the 1990s.

The author initiated a system with Dunsmore Data and Information Ltd some years ago[1] which was accepted by the County Surveyors' Society (Scotland) for their regions and subsequently adapted with the author by the Scottish Development Department for motorway and trunk road bridges.

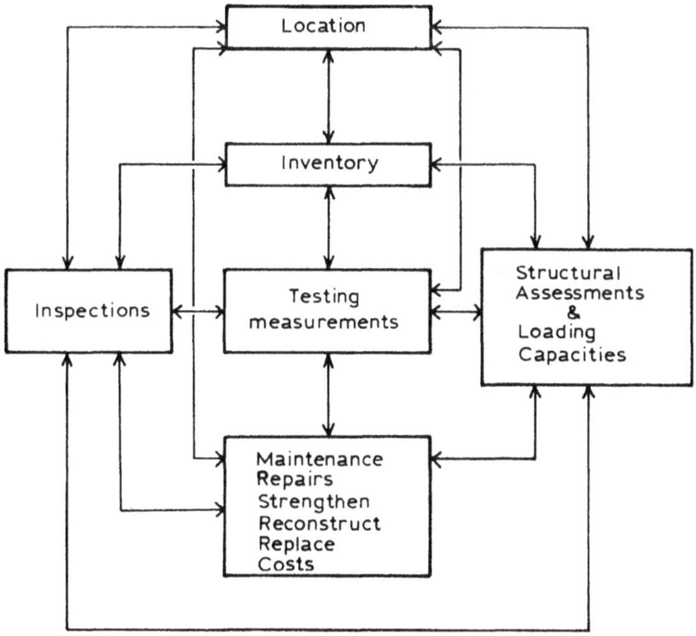

FIG. 3. Bridges database and information system.

In conjunction with Dunsmore Data and Information Ltd, Edinburgh, the author and a colleague, Mr Ron Lee, have now produced a streamlined spartan system more suited to the immediate purposes of the Scottish regions and English counties. It consists of the basic subparts shown in Fig. 3.

TECHNICAL ISSUES FOR MANAGERS TO CONSIDER

Too many repairs are required too soon or too often to relatively new road and bridge projects. This suggests that something may be wrong with our planning estimates, our design and specification criteria, or construction quality. Behaviour monitoring information is essential in such circumstances.

Traffic

Volume, mix and weight estimates require checking.

Volume

The increase in the number of licensed vehicles (UK) this century is as follows: 1900 (0·018m), 1920 (0·650m), 1940 (2·325m), 1960 (9·439m), 1980 (19·210m) and year 2000 (>30m?). New national road traffic forecasts in May 1989 superseding the 1984 figures suggest that growth in the 1988–2025 period will increase previous predictions by somewhere between 83% and 142%.

Mix

Having determined that the damaging power of commercial traffic was 30% heavier in 1982 than in 1974, predictions for 1990 suggested double for each category![2] British Road Federation (1987) determined an increase in lorry traffic 1977–85 of 10% in rural areas, 37% on motorways and then suggested a further 12% increase by the year 2000.[3]

Weight

Construction and use weight increased recently from 32 to 38 t with the suggestion of 40–50 t by 1996. More information is required on the number and frequency of vehicles exceeding these weights in addition to abnormal load notifications.

Design, Specification, Construction and Quality Assurance

Where current design and specification criteria may not accommodate the traffic volume, mix or weight, some ‘horses for courses’ intuitive allowance should be added to the computerised optimum solution. Materials and workmanship specifications must not be relaxed with records of compliance kept on database to facilitate any problem identification later. Too many new materials and methods are marketed on the basis of laboratory tests or applications abroad. Properly monitored on-site trials with recommended applicators must form the basis of any specifications written.

Construction quality should improve if client organisations are more careful with their start dates and contract periods. Selection of contractors requires a more professional approach also. Carefully control and limit the amount of subcontracting and do not necessarily accept the lowest tenderer.

Design or construction firms with accredited and properly implemented quality assurance systems should present less risk of mistakes being made than those without. Clients, however, should not be lulled into a false sense of security. They must continue to strive to satisfy themselves about the attitude of the people they are dealing with, and the pride they take in doing a good job.

Bridge Behaviour Monitoring

In addition to the foregoing concern expressed about planning, design, construction, etc., the improved braking efficiency of modern vehicles has significantly increased the longitudinal force on highway bridge articulation and substructures, i.e.

BS 153 HA maximum load 253 kN, i.e. 100 kN plus 17 kN/m span >3 m.
BS 5400 HA maximum load 700 kN, i.e. 200 kN plus 8 kN/m loaded length.

Thermal movements of curved or highly skewed decks can also introduce complexities. If unexpected forces, movements or deterioration in condition of materials arise, it seems sensible that instruments be installed to give a case history before, during and after structural distress and remedial works. This information should help confirm the extent to which the bridge is behaving in relation to current design and code parameters. Decisions can then be taken on the need to merely replace 'like with like', or strengthen. Such matters will, of course, greatly influence the estimated cost of remedial works.

The author has developed computerised systems[4] for monitoring (a) deck deflection and temperature with joint movement and column tilt, (b) wind speed and direction, and (c) tide rise and fall, current speed and water quality. Experience has also been acquired on structural vibration of 'lively' decks, monitoring retaining walls and load testing arch bridges to destruction. Such experiences confirm the need for the installation of instruments to be dealt with at the design stage to facilitate provision of information, giving advance warning of deterioration or imminent structural distress.

SUMMARY

The introduction highlights the serious deteriorating magnitude of the crisis now emerging. A method of alleviating this crisis is identified in the highway bridge management flow chart (Fig. 2). Organisational policies to be complied with and objectives to be met are then clarified and confirmed. An interim streamlined database and information system has now been developed to facilitate dealing with the aforesaid to resolve the crisis. In conclusion, the suggestion that something basic may be wrong is raised on technical issues for managers to consider.

CONCLUSION

Highway bridge management to be effective must include the 'marketing' and 'production' criteria identified in Fig. 1. Implementation of the Fig. 2 management, administration, finance and technical flow chart guidelines will stimulate the organisations and managers involved to comply with the policies and achieve the objectives required.

ACKNOWLEDGEMENTS

The author thanks George Kirkbride, Director of Roads, for his permission to contribute this paper, and Mrs Lesley Higazi and Mrs Kath Butler for their excellent assistance in the production of the paper.

REFERENCES

1. MAXWELL, J. W. S., A micro-computer database and information system for engineering management. *Civil-Comp 85, Proceedings Second International Conference Civil and Structural Engineering Computing*, Vol. 1. Civil-Comp Press, Edinburgh, pp. 115–22.
2. WRATHALL, D. and WALLIS, C. C., The practical impact of reduced investment in highway maintenance. *The Highway Engineer*, April 1982.
3. BRITISH ROAD FEDERATION LTD, *Fact*, London.
4. MAXWELL, J. W. S. and FYFFE, S., Essential monitoring for essential repairs—Tay Road Bridge, Dundee. *Civil-Comp 87, Proceedings Third International Conference Civil and Structural Engineering Computing*, Vol. 2. Civil-Comp Press, Edinburgh, pp. 253–9.

11

Bridge Management in Cyprus

P. H. MAY
John Burrow & Partners, Exeter, Devon, UK

and

S. VRAHIMIS
Department of Public Works, Nicosia, Cyprus

ABSTRACT

An attempt is being made in Cyprus to introduce an objective approach to bridge management through the development and setting up of a computerised bridge inventory system. A schedule of condition inspections has been introduced which picks up overdue routine maintenance and provides data for an objective assessment of repairs required. Work is costed and an order of priority established to assist budget preparation and planning of future work in various time scales. It is anticipated that by centralising responsibility for the condition inspection procedures and bringing the bridge inventory system under the control of a newly formed headquarters-based maintenance management unit the non-availability of skilled technical staff at district level can be overcome. Implementation of the system developed is now proceeding.

BACKGROUND

Cyprus is the third largest island in the Mediterranean with an area of 9250 km^2. It has an intense climate with very hot, dry summers (when the inland temperatures frequently exceed 40°C) and cooler, rainy, rather changeable winters, separated by short spring and autumn seasons. Snow falls frequently every winter in the Troodos mountains on ground over 1000 m above sea level. These falls usually commence in December and continue through until the end of April.

The island's population is currently estimated to be 673 000, of whom 77% are Greek Cypriots, 19% Turkish Cypriots and 4% other minorities. The political situation is complex and outside the scope of this paper, but it is sufficient to state that since the Turkish invasion of 1974 the Republic of Cyprus has effective control over only the southern portion of the island, representing some 63% of the land mass.

The Cyprus Department of Public Works (under the Ministry of Communication and Works) is responsible for the preparation and development of the government's road improvement programme as well as the design, construction and maintenance of roads, airports, fishing shelters, coastal protection works and government buildings throughout the island. In 1987 these responsibilities included the maintenance of 1898 km of bituminous-surfaced roads and their associated bridges and culverts in the southern portion of the island. Of this total, 209 km were within designated municipal areas.

The diverse history of Cyprus over the last 500 years has resulted in a fascinating variety of bridge and culvert forms and material types being used. Inevitably, on an island where workable limestones and calcareous sandstones predominate, traditional masonry arches are common. Several of these date back to the Lusignan and Venetian periods of Cypriot history (AD 1192–1571), and though not now taking heavy loads some are still in service on minor roads. There are, of course, many examples of masonry arches constructed more recently (particularly during the British colonial era), when skilled craftsmen were more readily available. Most of these are still giving excellent service even on roads carrying quite heavy traffic. More recent bridges have favoured beam and slab-type concrete construction (both reinforced and prestressed), and there are many composite structures of in-situ concrete decks on steel girders and, in some cases, on steel trough deck sections.

Since the mid-1970s considerable emphasis has been placed on the need to

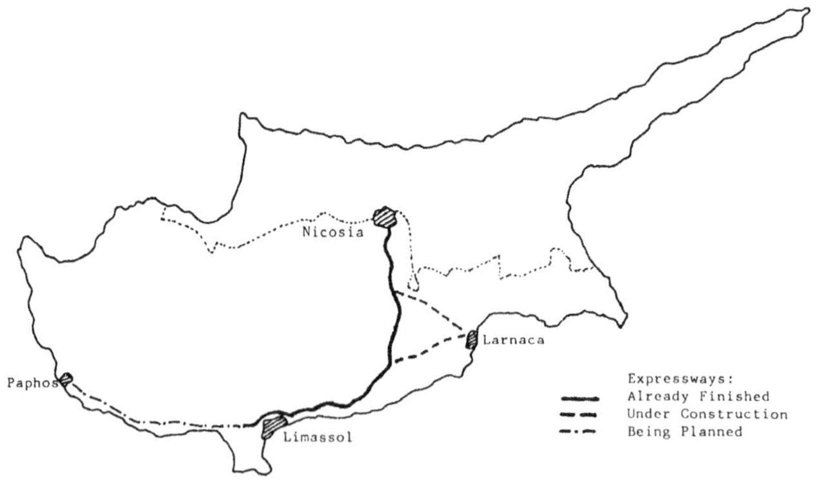

FIG. 1. Extent of current expressway development.

improve and extend the road network. Using financial assistance provided under four World Bank highway projects, as well as help from the Kuwaiti Fund and using internal resources, much of the primary trunk road network either has been or is being strengthened and/or upgraded to international standards. An indication of the present situation is shown in Fig. 1.

In 1984, under the Third Highway Project, attention was directed towards the need for improved road maintenance and a project was funded to develop and implement a computerised road maintenance management system. The system implemented adequately covered the department's requirement for inventory and condition information on road pavements and drainage but did not cover bridges or culverts in any detail. Accordingly, under the Fourth Highway Project, the same consultants were appointed to develop a complementary bridge inventory system which could cover these additional aspects.

ORGANISATION AND BUDGET ARRANGEMENTS

The Department of Public Works acts through a Nicosia-based headquarters and five operational district offices, each under the direction of an experienced qualified engineer. Though not all districts are of similar size (due to the present divided nature of the island), a typical district organisation is shown, together with the headquarters arrangements, in Fig. 2.

The establishment of a road maintenance unit in 1987 under the control of a senior executive engineer confirms the department's commitment to improve maintenance of roads and bridges. Though initially established to operate the road maintenance management system, it is now being developed to include responsibility for the bridge inventory system.

The construction of new roads and their associated bridges and culverts, together with the periodic maintenance of road pavements (resealing, overlaying, etc.), is carried out by private contractors from designs prepared by the department at headquarters level, sometimes with the assistance of external consultants. Only routine maintenance is, therefore, undertaken by the department's own direct labour force. The district engineers have the dual responsibility of supervising works being undertaken by private contractors and operating a small, but by no means insignificant, direct labour establishment. This is reflected in the organisation shown in Fig. 2.

No separate budget provisions are made for the maintenance of bridges and culverts (as distinct from the maintenance of roads). In 1987 the provisions for road maintenance were as follows:

Routine maintenance	\$1 700 000 (equivalent)
Periodic maintenance	\$2 800 000 (equivalent)

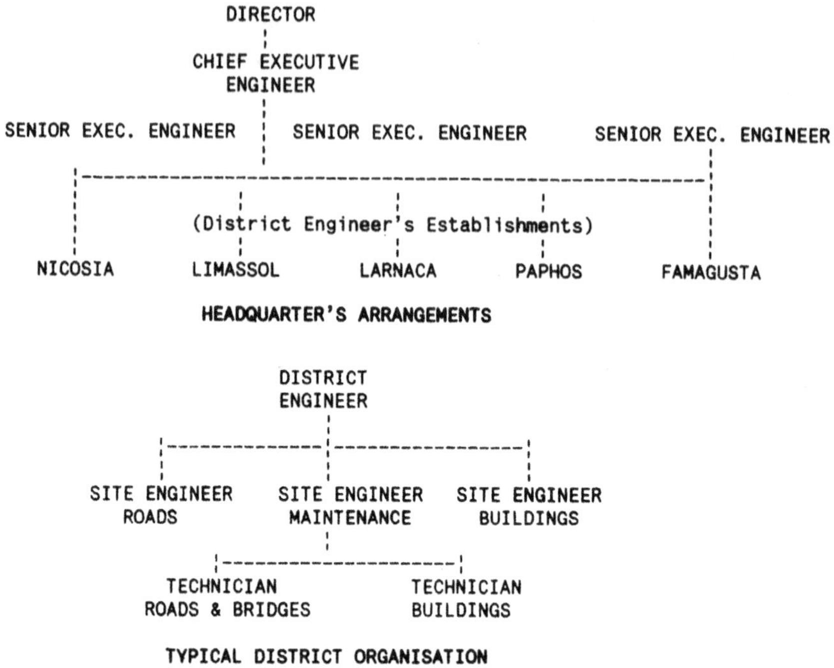

FIG. 2. Typical district organisation and headquarters arrangements.

Of the above figures, though no detailed breakdown of expenditure on bridges is available, less than 5% of the provision for maintenance is believed to have been spent on minor routine maintenance of bridges and culverts. This low level of expenditure reflects the priorities for attention perceived by engineers at the district level on the basis of only irregular reports on bridge condition given by district technicians (from inspections carried out during the course of their other duties). It is not an indication of the true condition of those structures. Until establishment of the bridge inventory system no objective surveys to establish bridge condition had been undertaken. Undoubtedly many of the district engineers had wished for some time to carry out such surveys but had found a shortage of skilled technical staff at the district level a major barrier. It is anticipated that the establishment of the bridge inventory system under a central road maintenance unit will circumvent this problem by centralising the staffing requirement and hence making better use of the trained personnel available.

INSPECTION PROCEDURES

It was known that in the UK, where skilled technical resources can perhaps be more readily assigned to the monitoring of bridge conditions, management is based on a cycle of 'general' inspections (from ground and deck level) at 2-year intervals with 'principal', more detailed, inspections at 6-year intervals. (More frequent inspections are required for bridges with cast iron members but these do not exist in Cyprus.) Similar inspection cycles appear to be adopted by other industrialised countries, though in several cases 'general' inspections are carried out on an annual rather than biannual basis.

The less developed countries of the world, on the other hand, have the combined problem of

(i) not having sufficient skilled technical resources available to carry out routine inspections, and
(ii) having generally adverse climatic conditions (high temperature variations and concentrated rainy seasons) which necessitate some form of inspection being carried out at least annually and in many countries at 6-monthly or quarterly intervals.

These problems may (arguably) be mitigated, to some extent, in the less developed countries by the reduced economic (and possibly social) consequences of bridge failure.

The Department of Public Works in Cyprus has a significant shortage of skilled technical staff brought about by factors outside their control. It was considered, therefore, essential that the time of those skilled technical staff that were available should be reserved for inspection of bridges which had been screened by less skilled staff for likely defects of importance. In addition, it was felt desirable, in view of the aggressive climatic conditions on the island, to carry out some form of quarterly checking procedure to pick up damage to bridges and culverts resulting from seasonal and, indeed, daily expansion/contraction cycles, from the effects of winter rains and from the spring thaw, which results every year in considerable erosion problems on rivers leading south from the Troodos mountains.

An inspection cycle has therefore been set up which provides for

—quarterly bridge and culvert checks (by district-based foremen from the direct labour force in the course of their other maintenance duties);
—follow-up bridge and culvert checks (by district technicians in response to a quarterly check showing the need for immediate further inspection);
—annual routine condition inspections (by headquarters-based trained technicians); and
—detailed condition inspections (by headquarters-based bridge engineers in response to routine condition inspections showing the presence of significant defects).

Initially resources will be allocated to enable routine and detailed condition inspections to be carried out on bridges but not culverts, the separation being taken at a combined clear span of 4 m. Eventually it is hoped that resources can be made available to extend this to include all crossroad highway structures irrespective of span, since the difference between bridges and culverts is often quite arbitrary.

Quarterly bridge checks are considered adequate to keep track of routine maintenance works which may be overdue or have been overlooked, and to pick up channel blockages or bank erosion before and after the winter rains. In addition, these frequent checks ensure that no bridge condition requiring urgent attention (to keep road users and general public safe or to preserve the structural integrity of the bridges and culverts concerned) goes unnoticed for any length of time. Responsibility for this element of the overall bridge management process is retained entirely at district level and is not computerised.

The format for routine and detailed condition inspections is basically the same, differing only in the extent of data collected. The format used is similar to that developed by the Northern Ireland Roads Service and includes identification of defects by codes according to the type of defect, the part, area and position affected, the extent of deterioration and its severity. Structures are inspected in a strict order to ensure no elements are forgotten and no defects overlooked. Concrete, steel, masonry and timber structures or a combination of these materials can all be accommodated.

The level of decision making required for the routine condition inspections is not considered beyond the scope of trained technicians and, indeed, such staff are preferred. The survey requirements for the detailed condition inspections are extended to include decisions on the type of treatment required to correct observed defects, the time scale within which the work should be undertaken, the degree of relief the repairs will provide and the estimated costs involved. This latter item can be either in the form of broad bands of cost (less than $1000, $1000–$3000, etc.) or in terms of a more exact estimate. It is intended that these decisions should be made by experienced engineers, wherever possible, on site.

BRIDGE MANAGEMENT APPROACH

The management approach adopted in Cyprus is based on the following sequence:

(i) bridge inventory collection,
(ii) quarterly checks (for urgent and overdue routine maintenance),

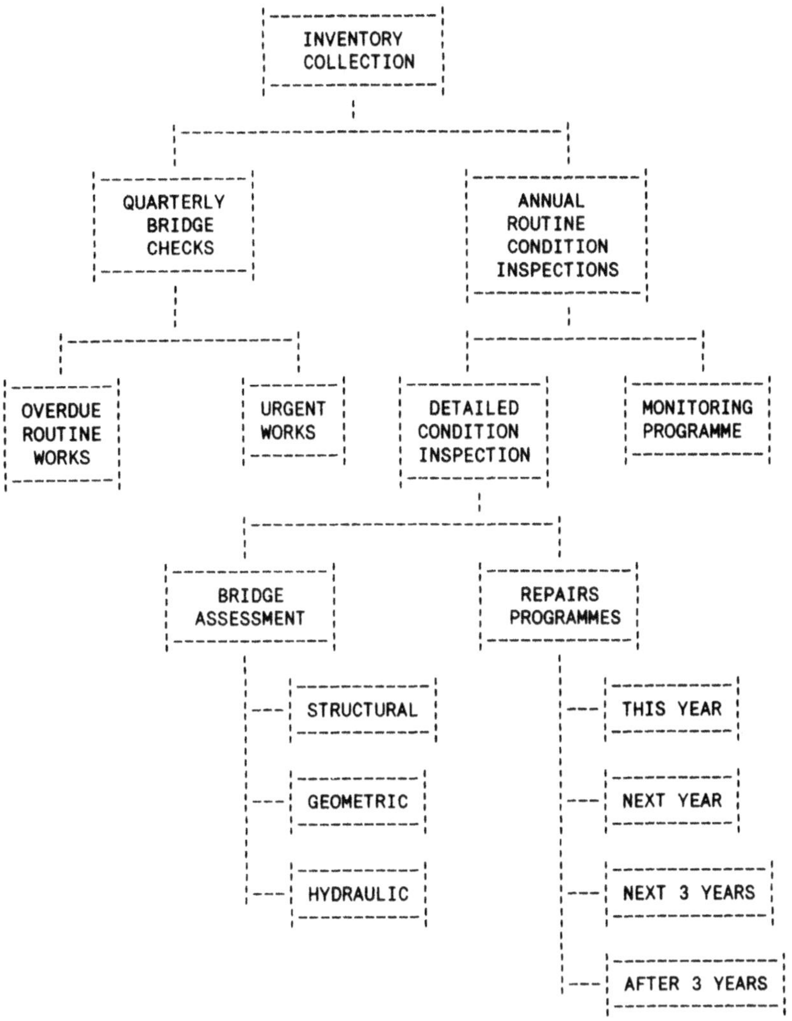

FIG. 3. Logic diagram of bridge management approach.

(iii) annual routine condition inspections,
(iv) detailed condition inspections (where required),
(v) prioritisation procedures,
(vi) repair works programming,
(vii) sufficiency assessments,
(viii) bridge rating procedures, and
(ix) bridge replacement programming.

A logic diagram showing the above approach is shown in Fig. 3.

From the outset there appeared a sound basis on which all the above could be computerised with the exception of the quarterly checks (the analysis of which, as outlined in Fig. 3, is best held at district not headquarters level and which cannot therefore be sensibly computerised in the Cyprus context at the present time) and the sufficiency assessments which form the basis of the bridge rating procedure and replacement programming. The sufficiency assessments, if done correctly, involve comprehensive investigations into the geometric, hydraulic and structural adequacy of the bridges concerned. At present these assessments appear better undertaken manually.

In looking round at available computerised systems which could handle the inventory and condition inspections, prioritisation procedures and repair works programming, it became apparent that the work undertaken by the Northern Ireland Roads Service was very much consistent with the approach it was intended should be adopted in Cyprus. The availability of this system on an IBM PC microcomputer proved to be an added attraction. In the event, the system made available from Northern Ireland was considerably amended and extended for use in Cyprus to the extent that complete rewriting of the software became necessary. None the less, the system being implemented in Cyprus is based on almost identical inventory collection and condition inspection procedures to those used in Northern Ireland and a senior engineer from the Roads Service was made available in Cyprus to assist with the training of local staff in the techniques of bridge condition inspection.

SYSTEM DETAILS

The bridge inventory system as developed for use in Cyprus is a computerised database of bridge information with associated interrogation and analysis programmes. The database contains

(i) a static library of bridge records which does not change with time,
(ii) a dynamic library of condition inspection records updated annually, and
(iii) an archive of repair records giving historical details of various works carried out.

These three elements of the database are linked by a common referencing system which allocates a unique reference number for each bridge and culvert in the network.

Interrogation programs included in the bridge inventory system allow the static and dynamic libraries of bridge and condition inspection records to be sifted and sorted according to a variety of procedures. These enable the user to obtain

—district bridge registers,
—individual bridge details,
—lists of bridges with specific physical characteristics,
—lists of bridges constructed within specified years,
—lists of bridges with load or height restrictions, and
—lists of bridges with defects that require monitoring.

The analysis programs allow prioritisation of defects noted in the routine and detailed condition inspections on the basis of the part of bridge affected, the type of defect observed, the importance of the defect (measured

PUBLIC WORKS DEPARTMENT — BRIDGE INVENTORY SYSTEM
NICOSIA, CYPRUS

BRIDGE RECORDS - DATA SHEET TYPE 2

DETAILS

BRIDGE NO. [_][_|_|_|_]-[_[_|_]/[_][_] RMMS REF. [_|_]

BRIDGE NAME [_|_|_|_|_|_|_|_|_|_|_|_|_|_|_|_|_|_|_|_|_|_|_|_|_]

DISTRICT [_] FUNCTION 2 - ROAD OVER RIVER

NUMBER OF ROAD OVER [_][_|_|_|_]

NAME OF RIVER UNDER [_|_|_|_|_|_|_|_|_|_|_|_|_|_|_|_|_|_|_|_]

ORIENTATION [_]/[_] YEAR BUILT [_|_|_|_]

DESIGNED BY: [_|_|_|_|_|_|_|_|_|_|_|_|_|_|_|_|_|_|_|_]

BUILT BY: [_|_|_|_|_|_|_|_|_|_|_|_|_|_|_|_|_|_|_|_]

G.A.DRG.NO. [_|_|_|_|_|_|_|_] LOAD RESTRICTION [_|_|_|_|_]

SURVEYED BY: [_|_|_|_|_|_|_|_] DATE OF RECORD [_|_]/[_|_]/[_|_]

DIMENSIONS

NUMBER OF SPANS (MAX 5) [_] SKEW ANGLE (DEG.) [_|_|_._|_]

SQUARE (1) [_|_|_._|_] [_|_|_._|_] [_|_|_._|_] [_|_|_._|_] [_|_|_._|_] [_|_|_._|_]

SKEW (2) [_|_|_._|_] [_|_|_._|_] [_|_|_._|_] [_|_|_._|_] [_|_|_._|_] [_|_|_._|_]

DETAILS OF ROAD OVER

PARAPET WIDTH [_|_|_._|_]
VERGE WIDTH [_|_|_._|_]
FOOTPATH WIDTH [_|_|_._|_]
HARDSHOULDER WIDTH [_|_|_._|_]
CARRIAGEWAY WIDTH [_|_|_._|_]
MEDIAN WIDTH [_|_|_._|_]
CARRIAGEWAY WIDTH [_|_|_._|_]
HARDSHOULDER WIDTH [_|_|_._|_]
FOOTPATH WIDTH [_|_|_._|_]
VERGE WIDTH [_|_|_._|_]
PARAPET WIDTH [_|_|_._|_]
OVERALL WIDTH [_|_|_._|_]
LENGTH OF PARAPETS N/W [_|_|_._|_]
LENGTH OF PARAPETS S/E [_|_|_._|_]

DETAILS OF RIVER UNDER

AVERAGE WIDTH OF RIVER [_|_|_._|_]
AVG. DEPTH (NORM. FLOW) [_|_|_._|_]
AVG. DEPTH (MAX. FLOOD) [_|_|_._|_]
BED TO SOFFIT (MAX.) [_|_|_._|_]
CONSTRUCTION DEPTH [_|_|_._|_]
BED TO SOFFIT (MIN.) [_|_|_._|_]
CONSTRUCTION DEPTH [_|_|_._|_]
LENGTH OF BED PROTECTION [_|_|_._|_]

FIG. 4. Typical bridge record data sheet.

```
PUBLIC WORKS DEPARTMENT                                    BRIDGE INVENTORY SYSTEM
NICOSIA, CYPRUS
                          INSPECTION RECORDS - DATA SHEET
==================================================================================

       BRIDGE NUMBER  [_][_|_|_|_]-[_|_|_]/[_]

     FUNCTION NUMBER  [_|_]

         ORIENTATION  [_]/[_]

     NUMBER OF SPANS  [_]

INSPECTED BY:  [_|_|_|_|_|_|_|_]          DATE OF INSPECTION  [_|_]/[_|_]/[_|_]

----------------------------------------------------------------------------------

PART     AREA   POSITION     DEFECT       SEVERITY        TIMESCALE         COST

                                      EXTENT       TREATMENT        RELIEF

1.  FOUNDATIONS
        |--------------------|                        |--------------------------
[_|_]   | [_]   [_|_|_|_]    |  [_|_]    [_]     [_]  |  [_]     [_]    [_]    [_]
        |                    |                        |
[_|_]   | [_]   [_|_|_|_]    |  [_|_]    [_]     [_]  |  [_]     [_]    [_]    [_]
        |                    |                        |
[_|_]   | [_]   [_|_|_|_]    |  [_|_]    [_]     [_]  |  [_]     [_]    [_]    [_]
        |--------------------|                        |--------------------------
           OPTIONAL INPUT                            ENGINEER'S INSPECTION ONLY
```

FIG. 5. Inspection record data sheet.

in terms of its extent and severity), the value for money the repair will represent (obtained from the degree of relief the work is likely to provide and the expected cost involved) and the importance of the bridge in the network (a function of road class).

The user can then obtain a costed list of urgent work, or of works to be carried out in the next financial year, or of work for longer term action in (say) the next 3 years each in order of priority. These listings can form the basis of objectively assessed budget requests and can be used for longer term financial planning.

By defining budget cut-off limits 'scalped' priority lists can be obtained of all works within the budget limit rearranged so as to be listed in bridge number order. Such listings are prepared to assist work planning.

Examples of typical system data sheets and some of the reports are given in Figs 4–7.

IMPLEMENTATION

The system developed is now complete and the programs have been installed on a computer in the Department of Public Works. Testing on a trial of

CYPRUS PUBLIC WORKS DEPARTMENT
BRIDGE INVENTORY SYSTEM — DISTRICT REPORT — Date : 22/ 6/89 Page No : 1

DISTRICT : 1

BRIDGE NO	FUNCT. :	YEAR	LOAD RSTR.	PLATED HEIGHT	SKEW ANG.	SKEW SPAN DIMS. SPAN 1	SPAN 2	SPAN 3	SPAN 4	SPAN 5	FOUND.	END SUPPORTS	PIERS	DECK	SURFACE	ANCILLARY
A0001-006/1	1		NONE		0.00	6.30	6.40	6.40	6.30		B A A	B E C	D E F	E E B	D A B	C C B E F
A0001-013/1	2		NONE		42.00	6.00	6.75				A A A	C E D	B E C	E G C	E A B	B A A F F
A0001-014/1A	2		NONE		30.00	9.70	9.80	9.80	9.80	9.80	B A A	B E D	D E F	E F C	D A B	D D B C E
A0001-014/1B	2		NONE		30.00	9.80	9.60				B A A	B E D	D E F	E E C	D A B	D D B C E
A0001-017/1	2				21.44	19.30	19.70	19.70	19.30		A C A	B E D	B E G	E E B	C A B	C G B A A
A0003-004/1	2				0.00	6.50	6.50				A A A	B E D	B E C	D E B	D A B	A A A A A
B0001-006/1	2		NONE		40.00	5.00					A A A	B F D	A A A	D E D	E A B	A A E F F
B0001-008/1	2		NONE		0.00	6.00					A A A	B F G	A A A	E G B	E A B	B B C F F
B0001-009/1	2		NONE		15.00	6.50	6.90	6.60			A A A	B E D	D E D	E G B	E A B	B A B F F
B0001-013/1	2		NONE		0.00	9.40	9.40	9.45	9.45	9.45	A A A	B E D	D E E	E E B	D A B	B D E F F
B0001-013/1B	2		NONE		0.00	9.40	9.40	9.45			A A A	B E D	D E E	E E B	D A B	B D E F F
B0001-017/1	2		NONE		30.00	3.60	3.60				A A A	B F E	B F C	D E B	E A B	B A A F F
B0001-018/1	2		NONE		45.00	6.30					A A A	B E D	A A A	D E B	E A B	A B A F F
B0001-019/1	2		NONE		42.00	8.50					A A A	B E D	A A A	E G B	E A B	A A A F F
B0009-021/1	2		NONE		18.00	5.70	6.00	6.00	6.00	5.90	A A A	B E D	B E C	E G B	E A B	B A B E F
B0009-025/1	2		NONE		0.00	7.70	8.60	8.60	8.60	8.60	A A A	B E D	B E C	E E B	D A B	C A E F F
B0009-025/1B	2		NONE		0.00	7.70					A A A	B E D	B E C	E E B	D A B	C A E F F
B0009-026/1A	2		NONE		0.00	6.05					A A A	A A E	A A A	C C A	D A B	A A A F F
B0009-026/1B	2		NONE		0.10	8.10					A A A	B G D	A A A	E G B	D A B	A A A F F
B0009-026/1C	2		NONE		0.00	8.40	6.30				A A A	B E D	B E C	E E B	D A B	A A B E F
B0009-027/1	2		NONE		0.00	5.50					A A A	B E D	A A A	E G B	E A B	A A A F F
B0009-029/1	2		NONE		0.00	9.25	9.45	9.50	9.45	9.35	A A A	B E D	B E C	E E B	D A B	B A B E F
B0009-035/1A	2		NONE		0.00	4.60					A A A	B E D	A A A	E E B	D A B	A A B F F
B0009-035/1B	2		NONE		0.00	4.60					A A A	A A D	A A A	C C A	D A B	A A A F F
B0009-035/2A	2		NONE		26.00	3.40					A A A	A A D	A A A	C C A	D A B	A A A F F
B0009-035/2B	2		NONE		28.00	3.50					B C A	B E D	A A A	D E D	D A B	A D E F F

FIG. 6. District bridge register.

CYPRUS PUBLIC WORKS DEPARTMENT Date : 8/ 6/89
BRIDGE INVENTORY SYSTEM Page : 1

PRIORITY LISTING BY BRIDGE NUMBER

From Timescale : A
To Timescale : D
Budget Limit : 45000
District :

BRIDGE NO	PART	AREA	DEFECT	EXTENT	SEVERITY	TREAT.	TIMESC.	COST	CUMULATIVE COST	BALANCING FACTOR
1A0001-006/1	PO	B	CB	A	B	A	B	800	800	315.000
1A0001-006/1	JV		HB	B	A	E	B	300	1100	243.000
1A0001-006/1	DN	U	CA	B	C	A	A	1200	2300	220.500
1A0001-013/1	JU	Q	DA	B	A	D	C	400	2700	729.000
1A0001-013/1	JV		HB	B	A	F	B	450	3150	243.000
1A0001-014/1	DR	T	GB	C	B	A	B	500	3650	364.500
1A0001-017/1	JV	R	HA	C	C	A	B	1360	5010	1012.500
1A0001-017/1	HF	Q	CA	B	B	A	B	700	5710	850.500
1A0001-017/1	CR	E	GA	B	B	A	B	200	5910	656.100
1B0001-008/1	JR	T	GB	C	B	F	B	350	6260	220.500
1B0001-008/1	JR	U	GB	C	B	F	B	250	6510	220.500

FIG. 7. Defect listing in bridge number order.

about 40 bridges and culverts near Nicosia proceeded smoothly and preparations are now being made for wider implementation.

ACKNOWLEDGEMENTS

The bridge inventory system prepared for use in Cyprus is based on a similar system developed by the Northern Ireland Roads Service under its Director, Mr I. Joiner. The assistance of the Roads Service in allowing use of the system and in providing the services of a senior engineer to assist with the training of local staff is gratefully acknowledged. The authors also wish to thank the Director of the Department of Public Works in Cyprus, Mr M. Christodoulides, for his support during the course of the project and permission to publish this paper.

MAINTENANCE STRATEGIES

12

Bridge Rehabilitation: Department of Transport's Fifteen-Year Strategy

D. A. Holland and P. H. Dawe
Department of Transport, London, UK

ABSTRACT

The Department of Transport has embarked on a 15-year programme for the rehabilitation of motorway and other trunk road bridges in England. This paper gives the background to the programme and describes the different types of remedial work to be carried out. It outlines the various studies which have been undertaken to determine the extent of the problems and the research and development work which is being undertaken to support the remedial work. The organisation and management of the programme is described. Finally, the paper looks at the problems which face a major public bridge owner seeking to keep an important stock of bridges in a serviceable condition in the face of ever-changing demands and with limited resources.

INTRODUCTION

In November 1987 the Minister for Roads and Traffic announced a 15-year programme for the rehabilitation of bridges on motorways and all-purpose trunk roads. This is an indication of the importance which is now being given to maintaining our existing and ageing facilities and systems in the face of ever-increasing demands. It is a problem which affects all parts of our infrastructure and which is being faced by the developed countries throughout the world.

This paper describes the background to the 15-year rehabilitation programme for the highway bridges belonging to central government in England and gives details of the various items in the programme. The paper does not aim to provide a model of bridge management to be followed by other bridge owners but shows how one country with its own particular problems and circumstances is setting about the task of managing its stock of bridges. Moreover, the programme is still in its early stages so there is much to be learnt about the best ways of tackling the various tasks. In particular, the problems associated with the assessment and strengthening of existing structures, some often quite old, are entirely different from those connected with new design and construction. The paper also gives details of the studies which have been undertaken to determine the size and scope of the programme, and the research which is being carried out to assist in the efficient and effective execution of the programme.

CENTRAL GOVERNMENT BRIDGES

The transport departments of England, Scotland and Wales are responsible for 9500 miles of motorways and other trunk roads. Although these represent only about 4% of the total road network in Great Britain, they carry 30% of all traffic and 60% of all heavy goods traffic. They are therefore heavily used routes and vital components of the country's road network.

The Department of Transport in England owns about 8500 highway bridges and about 3000 other structures, including retaining walls, sign/signal gantries and tunnels. About 75% of the bridges are concrete, 15% steel (mainly steel/concrete composite) and 10% masonry arches. They range from the Severn Bridge with a main span of 988 m down to small culverts with spans of 3 m. A large number of the bridges are over 100 years old, the oldest being built in 1185, though the majority are modern bridges built within the last 20–30 years. Details of all the department's structures are held on a computerised data base, which also holds information from inspection reports and details of maintenance expenditure.

There are about 100 000 highway bridges altogether in the UK. The majority belong to local authorities although British Rail owns over 10 000 highway bridges. About 50% of these are concrete, 15% steel (or steel/concrete composite) and 35% masonry arches. Since most masonry arches were built over 100 years ago, there is thus a high proportion of older structures within the national bridge stock, most of these being on local roads.

FIFTEEN-YEAR REHABILITATION PROGRAMME

The various items of work included in the programme have been grouped under the following main headings:

(a) Steady-state maintenance.
(b) Assessment and strengthening.
(c) Upgrading substandard features.

Steady-State Maintenance

This forms the core of the ongoing programme of repair and replacement of the various elements of bridges which have deteriorated or been damaged as a result of time or use. It includes such things as the repair of reinforced concrete, painting of steel structures, replacement of bearings, expansion joints, etc. Most of this work is carried out under agreement by agent authorities and their annual bids for this work have risen considerably over recent years, reflecting increasing rates of deterioration, particularly of concrete structures.

Assessment and Strengthening

This part of the programme is concerned with structures which were designed for loading criteria which are no longer adequate for current traffic. A programme for assessing the older short span structures has started and is mainly concerned with bridges which were built before the introduction of national loading standards in 1922. Many of these bridges are masonry arches built in the late 1800s. A subsequent programme will deal with long span bridges, with spans greater than 50 m, where traffic loading has increased due to the increased numbers of heavy goods vehicles on the roads. It may also be necessary to look at some of the more modern short/medium span bridges which could be deficient in shear resistance.

Upgrading Sub-standard Features

This is concerned with rectifying deficiencies in certain structures where current design standards and specifications are not being met, mainly those involving safety and durability. It is also intended to deal with particular problems which have been identified for prestressed concrete bridges. The following are some of the topics to be dealt with under this heading:

(i) Waterproofing unprotected bridge decks.
(ii) Rehabilitation of post-tensioned psc bridges.
(iii) Repairs to prestressed precast concrete beams with deflected tendons.
(iv) Replacement of sub-standard vehicle parapets.
(v) Countermeasures (structural) to the 'bashing' of low headroom bridges.
(vi) Strengthening of piers and columns to resist higher impact forces.
(vii) Health and safety aspects of access to structures.

It should be noted that some of these items, although of a 'one-off' nature, can be carried out as part of the on-going steady-state maintenance programme.

Programme

A comprehensive 15-year programme has been developed to cover all the items of work identified above. The length and formulation of the programme has been determined by the need to even out the demands on resources and to avoid too much disruption to the road network at any one time. But this has had to be balanced by the need to complete certain items of work where safety is at risk in a reasonable time. It is estimated that the total cost of the programme will be between £1000m and £1500m (1987 estimate). Other bridge owners have similar problems to those of the department and will need to take similar steps to restore the state of their bridges. However, the total costs of their work cannot be estimated at present.

BACKGROUND STUDIES

By the late 1960s it had become clear that the great expansion of road transport which had taken place as a result of post-war economic development was putting the UK's bridge stock under considerable strain. 'Operation Bridgeguard' was an interim programme to identify and deal with a limited number of the weakest bridges, pending implementation of a more thorough programme of assessment and strengthening. 'Bridgeguard' commenced around 1970 and by the early 1980s the bulk of the work had been completed. Preparations were then put in hand for the basis of the present 15-year programme. The first problem was the need to assess and strengthen the older short span bridges as a result of the introduction of a new Bridge Assessment Code in 1984. The code was produced by a working party representing all the major public bridge owners and was to replace an earlier code which had allowed the use of reduced margins of safety. There was also a need to organise, in a rational way, the work necessary to tackle a number of diverse bridge problems which had by then come to light.

Bridge Assessment Code

This is a comprehensive document which covers all aspects of the assessment of the load-carrying capacity of a highway bridge. It comprises a mandatory standard which is supported by a complementary advice note. The code adopts the limit state format and deals with inspection, loading and strength assessment. The code is intended to be used with a rationalised system of weight limit signing for those structures which are found to be incapable of carrying the full traffic loading. The code at present makes reference to existing design codes for the strength assessment of the various structural elements.

Because it is intended for older structures the code deals with such materials as cast iron, wrought iron and early steel. It also contains an empirical method for assessing the capacity of masonry arch bridges. In addition, it has simplified, but conservative, methods for the assessment of beam and slab-type bridges. The loading was re-derived from scratch and takes account of overloading and lateral bunching of vehicles and is intended to be fully representative of the effects of current traffic.

Bridge Census and Sample Survey

In order to assess the implications of applying the code to the UK stock of bridges a study was undertaken with the help of the other major bridge owners. This consisted of a census to determine the number of bridges likely to be affected by the code together with the assessment of a sample of about 550 structures randomly selected to be representative of the bridge stock as a whole. The results of the study showed that about 50 000 bridges in the UK were likely to be affected by the code, of which about 11 000 would need strengthening to meet current standards. No DTp bridges were included in the study but from the results and knowledge of the DTp stock it was estimated that about 2000 trunk road bridges would need to be assessed, of which about 1000 might need strengthening or replacement.

Long Span Bridges

The loading developed for the assessment code has been adopted as the basis for a revision to the short span end of the loading for design. At the same time a new loading for long span (greater than 50 m) bridges has been developed since it was realised that the existing loading did not fully represent the effects of current traffic which now contains a much higher proportion of heavy goods vehicles than originally assumed. The loading was derived using probabilistic methods operating on up-to-date vehicle and traffic data and showed that the design loading needed to be doubled for the longer spans. There is therefore a need to carry out fairly urgently a programme of assessment of existing long span bridges to the new loading, though it must be remembered that for these structures the dead load provides a fairly high proportion of the total load. There are about 150 long span bridges belonging to the department which need investigation. The first of these is the Severn suspension bridge, which is nearing the end of an extensive repair programme to bring it up to current standards.

Condition of Concrete Bridges

Over the last 5–10 years most bridge owners will have become aware of the increasing signs of deterioration being exhibited by fairly modern concrete bridges. Much of the deterioration can be attributed to the widespread use of deicing salt causing chloride damage. Other causes include carbonation, alkali–silica reaction, sulphate attack and frost damage. Deterioration can also be due to poor design and detailing, poor materials, poor workmanship or inadequate maintenance.

The department owns about 6500 concrete bridges and these are subject to a general inspection every 2 years and a more rigorous principal inspection every 6–10 years. However, such inspections involve close visual inspection and thus only record damage or deterioration which has manifested itself. Such inspections can only give a snapshot view of the state of the bridge stock at any moment and do not give a clear picture of potential deterioration and future maintenance needs.

In order to obtain better information on the overall condition of its concrete bridges the department commissioned a study by consultants of 200 randomly selected but representative concrete bridges. Besides a visual examination of each structure, half-cell potentials, depths of cover and depths of carbonation were measured. Samples were taken for analysis of the cement, chloride and sulphate contents, and cores were examined petrographically to check for the presence of or susceptibility to alkali–silica reaction.

The study did not reveal any new problems but indicated that the general state of the concrete bridge stock was worse than had been anticipated. The report estimates the level of maintenance funding likely to be needed to tackle these problems over the next 10–15 years. The report also recommends a programme of maintenance works, including a crash programme to replace or repair damaged expansion joints and the impregnation with a hydrophobilising material of some existing structures. Much of the work identified in the report had been allowed for in the 15-year rehabilitation programme but some adjustments have had to be made to ensure that the recommendations are covered on a priority basis.

RESEARCH AND DEVELOPMENT

The studies described above have been mainly concerned with identifying the extent of various bridge problems and developing assessment criteria. Other work is taking place which will help in finding solutions to the problems or help to avoid them in the future. For instance, there is a comprehensive programme of research being conducted by the Transport and Road Research Laboratory concerned with concrete repair methods and materials. This is complemented by work which is aimed at improving the durability of reinforced concrete in new construction.

The assessment of the carrying capacity of existing structures is a much more difficult and demanding task than the design of new structures, since the financial consequences of any decision can be serious. It is a matter which demands the application of considerable judgement by the engineer, especially when dealing with older structures whose forms of construction are very different from modern practice. TRRL have nearly completed a comprehensive investigation into the real strength of masonry arch bridges. This has involved full-scale collapse tests as well as model tests and the development of analytical methods. The outcome will be a more realistic method for assessing the safe carrying capacity of this form of structure which represents about a third of the UK bridge stock. Further full-scale testing is being done on structures incorporating steel trough decking and on beam bridges with brick jack arches between the beams. Work is also being done on testing elements such as precast beams salvaged from existing structures. All this work will enable the real strength of existing structures to be determined with greater confidence and prevent them from being unnecessarily replaced or strengthened.

In assessing the strength of the elements in existing structures it is realised that design codes may not provide the most appropriate criteria, even though modern codes are based on the latest research. Very often the criteria in design codes have been subject to simplifying and conservative adjustments which are of little consequence in the design but may be critical for assessment. Design code models may also be associated with a partial safety factor which is intended to give the required margin of safety over a range of element sizes and dimensions, whereas for the structure in question the margin may be greater than intended. The department has therefore sponsored the production of an assessment version of the national concrete bridge design code, BS 5400: Part 4. This has involved a clause by clause examination of the code and the production of alternative assessment clauses where appropriate and worthwhile. It is intended to carry out a similar procedure for the steel design code, BS 5400: Part 3. Here the main problem is that detailing practices, provision of stiffeners, etc., in the older structures may not be covered by current codes.

In 1987 a post-tensioned prestressed concrete segmental bridge in Wales collapsed due to corrosion of the prestressing tendons. The corrosion originated from the use of deicing salts which had percolated into the joints between the segments. As a result TRRL have instigated a study of nine typical segmental bridges to see whether there are similar signs of corrosion. The study will also develop procedures for the inspection of this type of bridge so that if it is felt necessary to examine all post-tensioned bridges the programme can be instituted without delay.

MANAGEMENT OF PROGRAMME

The maintenance of the department's bridges and roads is coordinated, programmed and funded by the Highway's Maintenance Division. The work is supervised by staff in the nine regional offices and is normally carried out by agents acting for the department. These maintenance agents are usually the local authorities (county or metropolitan) although British Rail and private consultants act as agents in some cases. The maintenance agents are also responsible for handling the annual round of bridge inspections. Based on the results of these inspections each agent prepares, for the region's approval, bids for the work it proposes to undertake in the following year. The bids are assigned priorities according to a fairly crude and subjective five-point rating system and the available funds allocated accordingly. Through the use of a system of standard forms all this information is stored in the structures database and can be interrogated as desired.

Many of the items identified in the 15-year programme will be handled under the procedures described above for steady-state maintenance. For instance, a certain amount of replacement of sub-standard parapets will be carried out each year as part of the general maintenance cycle. However, other items require a separately identified programme and dedicated funding, as for the assessment and strengthening of the older short span bridges. Here the bridges concerned have been identified and programmes prepared to complete their assessment in 3–5 years. Any strengthening work found necessary will be carried out on a priority basis and the aim is to complete all the assessments and strengthening work on trunk road bridges within about 10 years. Completion of this work is expected to have been achieved prior to the admission of the heavier 'European' lorries, both domestic and from other community countries, in 1999.

With the possibility of a structure suffering from more than one shortfall it will be important to ensure that all the required work is carried out at one go. This will be coordinated at a local level by the regional offices, who will also determine priorities based on advice produced by HQ divisions. The recent announcement by the government of an expanded road programme includes the widening of many existing heavily used routes. It is expected that a good number of the structures earmarked for remedial work under the 15-year programme will also be affected by the widening programme. The regional offices are also likely to be required to liaise with the local authorities over the assessment and strengthening of the older bridges on local roads. The object would be to provide a comprehensive network of the more important roads for use by the heavier vehicles from 1999.

MANAGING A STOCK OF BRIDGES

Although this paper has concentrated on the department's 15-year bridge rehabilitation programme it has in a sense just been describing some of the typical problems faced by the owner of a large stock of highway bridges. Because bridges are long-life structures it is difficult to forecast accurately the loading they will be required to carry throughout their entire life. Design standards may change as the results of new research become available. Public expectations in terms of safety and reliability may change. Procedures for ensuring the free movement of traffic in all weathers may have unforeseen effects on rates of deterioration. Thus, in addition to the normal on-going maintenance required to keep structures exposed to the elements in good condition, the owner also has to be prepared for discrete programmes of work to bring his stock up to current standards.

The problems of the owner are exacerbated by the need to minimise traffic disruption while he carries out this work. In the case of the department's bridges the problem is even more severe due to the heavy volumes of traffic carried on the national roads. The owner also has limited resources at his disposal and the nature of the work is often time consuming and labour intensive. All this demands great professionalism and expertise from the bridge owner in looking after his bridge stock. He needs to be sure that the money spent on remedial work is spent in the most cost effective manner, taking account of traffic delay costs, and that repairs are carried out at the opportune time without compromising public safety.

In order to improve the way that it tackles its bridge management the department is undertaking a study of the economics of bridge maintenance. Part 1 of the study has been completed and has reviewed the present structures management procedures. The report has recommended the development of a computerised bridge management system which would optimise maintenance strategies based on best value for money and rank maintenance activities according to benefit/cost ratios. The BMS would be based on the existing structures data base and would be operated by the regional offices and agent authorities on a network basis. The report contains a specification for a proposed system which will form the basis of the development work to be carried out in Part 2 of the study. At the time of writing, it is expected that the work will be started in 1989 for completion in 1990. Meanwhile, management decisions are continuing to be made on the basis of the existing evaluation system. Besides optimising the effectiveness of bridge maintenance expenditure, the system will provide the information necessary to justify the bids made for bridge maintenance funding in competition with the other demands upon the government's purse.

13

Comparative Maintenance Costs of Different Bridge Types

DAVID LEE

Chairman, G. Maunsell & Partners,
Yeoman House, 63 Croydon Road, London SE20 7TP, UK

ABSTRACT

Absolute maintenance costs, excluding those for routine cleaning, painting and resurfacing activities, are not generally available. In practice maintenance of bridges is virtually non-existent and bridge owners are only driven to take action when serious deterioration becomes apparent.

This paper discusses aspects of bridge design and construction which from experience have an impact on deterioration. It is important to take stock of these aspects so that both maintenance strategies and design in future are improved. (See The Assessment of Highway Bridges and Structures. Bridge Census and Sample Survey, Department of Transport, January 1987; Report on the State of Roads and Bridges in the United Kingdom—The Case for Action. The Institution of Civil Engineers, January 1988.)

INTRODUCTION

This paper attempts to review various problems of maintenance in different bridge types and the related costs involved. Inevitably the obtaining of maintenance cost information is difficult.[1] Some of the comments made must therefore be subjective in nature with the expectation that discussion of the subject may be enriched by the experience of maintenance engineers willing to support or, in some cases, refute some of the assertions in the paper.

In general, there are three main maintenance aspects arising from different types of bridge structure.

(a) Those designed and built in the hope that maintenance costs will be non-existent or very low.

(b) Those low first cost structures built without regard for ongoing maintenance issues.
(c) Structures of either (a) and (b) above but which are let down by design deficiencies and construction defects.

It follows that for the three types (a), (b) and (c) noted above there is a different maintenance cost involved in each class. In some cases the long-term application of experience is necessary to improve structures for the future. At the present time many maintenance programmes require an element of firefighting action to refurbish as well as maintain structures in a reasonable condition.

Whilst there may be exceptions it is a fair generalisation, in the author's opinion, to say that bridges were not designed in the past to be maintained and even if a maintenance schedule was prepared it was never followed. The painting of steelwork generates a sense of realism, normally through the appearance of ugly rust on bridge steelwork. Repainting is skimped by not properly removing corrosion material and in some cases painting over dirt and bird droppings.

The lack of maintenance on expansion and rotation joints in bridges is particularly noticeable. In these and many other details it is better to spend more up front to achieve quality to reduce maintenance costs. To expect inferior products to perform well over a long life is usually expensive.

THE DE-ICING PROBLEM

For concrete bridge decks, columns and abutments the primary form of attack in the UK is by de-icing salts which lead to an irreversible chloride degradation of the concrete.[2] The maintenance cost of a structure has to balance the cost of de-icing damage by chlorides. The alternative of using urea is ten times the cost of salt. The additional cost can be justified when in highly significant areas of the road system. Urea is being used on the Midland Links Viaducts and the Severn and Avonmouth Bridges, for example. It is difficult to imagine that the increase in the cost of de-icing material is acceptable where there are only limited numbers of bridges per kilometre of road. Another alternative is the use of calcium magnesium acetate, which is in fact 40 times the cost of salt.

THE WATERPROOFING PROBLEM

Much damage follows from inadequacy of the waterproofing membrane on the bridge deck. In some cases relatively efficient waterproof membranes are subject to faulty application or damage. Currently spray-applied plastic waterproof membranes are coming into favour and certainly offer the advantage of full bonding over the whole deck without the seam joints. The drainage of bridge decks usually suffers from the difficulty of sealing membranes around gullies.

This country has been very slow to adopt the use of concrete surface treatments and with hindsight it can be seen that if these had been applied at the time of construction much of the maintenance costs now arising might have been substantially reduced.

PERFORMANCE OF CONCRETE BRIDGES

The Department of Transport is responsible for about 8900 bridges on the motorway and trunk road network in England. Some 5900 of these are nominally of concrete with a span of at least 3 m and excluding masonry arches. Over 75% of them have been built since 1960. Two hundred bridges were chosen at random and the maintenance costs of concrete in them has been studied.[3] By establishing the performance of the concrete in service the potential repair and preventative maintenance costs have been estimated.

Of the 200 sampled, 30% were classified as in good condition, 50% as fair and 20% as in poor condition. In the next 10 years at present-day prices some £21 million needs to be spent on the 200 bridges in the survey.

The good bridges indicate that investigation, surface treatment and repair of joints requires on average £16 150 per bridge. For bridges in fair condition it is necessary to monitor condition as well as the above items, leading to a mean cost of approximately £69 150 per bridge. For the bridges in poor condition maintenance strategies would also embrace cutting-out and repair, cathodic protection where appropriate and replacement of whole or part of the bridge, leading to a mean cost of £314 950 per bridge. The average cost of a typical unit bridge would be approximately £600 000 at present-day prices.

Concrete bridge repairs expenditure has been running in England at the rate of about £20 million per annum but the study recommends that an average rate of £60 million per annum be implemented over the next 10–15 years, with a peak expenditure in any one year of nearly £140 million. The rate of expenditure is thus very approximately £10 000 per bridge per annum.

MAINTENANCE STRATEGIES

An enormous influence on dealing with actual or potential maintenance costs is how one determines the past, present and future strategies. It has been previously stated that much maintenance was ignored which might be termed the 'do nothing' strategy. Such a strategy is just one of the options which may be set out in a form of array table where the maintenance option can be selected on one axis and the time and date runs along the other. This technique has been used in an unpublished joint working paper for the Midlands Links Motorway Viaducts. In view of the assistance this array method gives to optimising maintenance costs and the date they are carried out, a brief description will be given below for a particular reinforced concrete portal frame. There are five potential classifications.

1. Uncorroded—no sign of deterioration.
2. Corroding but not delaminated.
3. Corroding with delamination of cover.
4. Severe corrosion with delamination.
5. Delamination under the main steel causing loss of bond and structural integrity.

It is possible to split each classification into sub-categories. For example, item 2 might have a division into the percentage of area of the element contaminated by chlorides such as 25%, 50% or 75%. For structural integrity conditions 1, 2 and 3 might lead to no action. Condition 4 is not satisfactory, however, and a condition 3 element requires maintenance or refurbishment to ensure it does not deteriorate to 4 or 5. In classification 1 there will be some elements which have been contaminated by chloride and are therefore suspect but which can be determined to be unlikely or likely to further deteriorate seriously in the future.

The maintenance cost must be applied to each option such as in the following example for a concrete crossbeam.

(a) Do nothing, apart from inspection. The maintenance costs are therefore related to inspection and access costs.
(b) The application of surface coating or impregnation. This option may be adopted if the crossbeam is not already deteriorating and can be of surface coating such as urethane or acrylic paint types or silane or siloxane impregnations.
(c) Application of cathodic protection.[4] There is possibly still a reluctance to take this method of protection seriously but the trials and tests that have been carried out over a number of years suggest that it has a use in arresting further deterioration. The cost of cathodic protection is substantially less than options (d) and (e).
(d) Repair of concrete (including patching).
(e) Replacement of the structural member.

The costs of such maintenance options have been estimated for various

TABLE 1

Maintenance option	Number of cost units	Remarks
Inspection only and no action	1	
Application of coating	3	Maintenance of coating—10 years 2 units
Apply impregnation	1	Maintenance not known
Install cathodic protection	30–40	Propping not required
Install cathodic protection	210–360	Props required (one use only)
Install cathodic protection	110–260	(ten re-uses)
Maintenance of CP	5	Over 10 years
Replacement of concrete 25%	200	
100%	340	

elements of the Midland Links Viaducts and for work on the Tees Viaduct. The quotation of costs is fraught with misinterpretation. To consider the order of magnitude, assume that one unit represents approximately £1000 and Table 1 suggests the units that various options will generate. Similar figures can be evaluated from strategies of strengthening and repair, and cathodic protection in combination.

For larger viaducts (such as the Tees Viaduct) the crossbeam replacement figures have been estimated to increase by 100–150 units per item. Such estimates indicate that refurbishment may require the expenditure of sums commensurate with or exceeding the first cost of construction.

DESIGN STRATEGIES

The following suggestions are offered to improve the design of bridge schemes.

Durability, Simplicity and Cost

There is room for improvement in many bridge designs by adoption of simple solutions. Examples include the greater use of simple, straight right spans rather than unnecessarily curved and skewed. The saving in cost of building simple structures which act and flex in a predictable fashion have to be measured against any theoretical saving of a difficult design.

Spans and Articulation

Where appearance allows it is preferable to use larger spans to reduce the number of intermediate columns. Given a free choice every additional column detail is a recipe for additional maintenance costs.

Continuity is preferable provided the potential shrinkage and creep cracking in concrete decks is adequately combatted. Greater consideration could be given to simply supported spans if they are linked over the column supports by a continuity detail of the deck which would resist leaking.

The use of continuous bridge decks has always been a useful criterion to assist waterproofing and by minimising the number of expansion joints in a bridge deck. In the case of the Tees Viaduct the simply supported composite steel spans are being changed in the refurbishment contract to allow continuity over each pair of spans with an improved joint between the pairs.

Maintenance as a Design Criterion

We do not put on one shirt and wonder what to do when it needs laundering. Yet this is precisely what we do for bridges. Much greater provision for inspection and access for maintenance is required.

All bridges require adequate safe access. Expensive scaffolding is frequently necessary and the cost of providing this often delays and deters inspection programmes. The cliché that prevention is better than cure is particularly forcefully demonstrated in the maintenance of bridge construction.

Prestressed Bridges

It seems that external tendons for prestressing will be preferred in the future to ducts within the concrete section to facilitate simple inspection and maintenance. Internal ducting will presumably follow the design of concrete nuclear pressure vessels where tendons are protected with special grease and also provide the non-bonded action.

Prestressed concrete bridge beams produced on the long line system have a relatively good record with high durability. Owing to the low capital investment in our prestressed concrete factories, however, there have been instances of poor quality through lack of adequate quality control.

In reviewing composite steel construction this paper mentions the use of cathodic protection but with prestressed concrete this is not recommended in case the generation of hydrogen leads to hydrogen embrittlement of the prestressing tendons.

The quality of long line precast beams should be matched by a higher quality of erection and on-site completion.

Cable-supported Bridges

The primary support of cable-stayed bridges is naturally the cables themselves and it is logical to locate these so that inspection, maintenance or replacement can be performed clear of any traffic on the deck. A similar consideration would be applied to the hangers of suspension bridges.

The current state of the art for cable-supported bridges has not established a particular standard method of tower design and the cables connected to it. It would seem logical to splice cables at the tower in appropriate cases so making them easier to replace. Of course, in other smaller bridges continuity of the cable stays may not present an access or weight problem.

Composite Steel Bridges

Steel beams are supplied from the fabricator and he is not generally involved in the quality of erection and completion of the composite deck and other details. The danger of such division of responsibility is usually compounded if alternative elements or designs are accepted whether in steel or concrete.

Many steel details in existing bridges have not properly identified the requirement of water-shedding and weathering. It is fair to claim in principle that with a proper painting schedule and an allowance for loss of parent metal owing to corrosion a steel bridge can be brought back to the as-new condition. This is a lesson that should also be applied to concrete elements of bridges. One of the dramatic features of the current maintenance and refurbishment programme is the realisation that concrete elements very often perform extremely well and are virtually maintenance free but that many suffer from chloride attack or cracking and in these cases it is extremely difficult to bring the structure back to an as-new condition. At the very least protective coatings should be considered far more seriously.

For composite construction the development of a glass-reinforced plastic system to enclose the steelwork, and at the same time provide access, is generating considerable interest currently on the refurbishment contract for the Tees Viaduct.

Bridge Furniture

It would seem logical to provide more space to allow robust detailing and maintenance of parapets and bridge edges, lighting masts and sign gantries. The detailing of safety barriers to allow tidal flow, repair lanes and access of that type has not had any attention. With major routes having at least four lanes in each direction, investigation should proceed on intermediate safety barriers between groups of lanes.

Regional Influences

The international nature of much bridge design is self evident. Each region of the UK should have design criteria to enhance robustness and durability. The British weather requires designs which reflect a response to the service conditions. It is important to clarify attitudes before the free for all onslaught of post-1992.

BRIDGE PROCUREMENT

On bridge construction in the UK it is somewhat unfortunate that contractors have fallen into the habit by the competitive tendering and claims procedures of only aiming at the lowest possible standard of construction that could possibly be approved by the client or his agents. This is abetted by the nature of the standards being used which encourage the notion of only just requiring them to be exceeded to achieve compliance. It is not economical to demand high quality when it is not required but equally when engineers demand a certain standard of quality there is usually an outburst by the suppliers and contractors in that unnecessary restrictions are being placed and unnecessarily high quality is being requested.

Whilst good relations between contractors and site supervision is the rule rather than the exception this happy state of affairs seems to be only due to the essential good nature of personal working relationships. In reality these relationships are strained because contractors do not expect to supervise their own work; they leave it to the resident engineer and his staff to do all the enforcing of the specification. This accounts for the negative image consultant site staff present to many workmen on site. Another adverse feature of bridge construction contracts is that frequently the time programme is paramount and with the lowest possible cost puts quality well into third place.

These remarks may be regarded as over-critical but the upshot is not conducive to lowering maintenance costs during service life of bridges.

CONCLUSION

It will be seen that in the author's view maintenance costs for bridges are similar to estimating the length of a piece of string. Thorough inspection will elucidate the quality of the materials of the string and a rough idea of its length but costs would clearly not be reportable in an absolute manner. For this reason this paper has attempted to indicate areas where maintenance costs may be both increased or reduced according to type and detailed design of the bridge, and with the ultimate aim of seeking bridge designs which may have a forecastable maintenance cost in the future.

REFERENCES

1. Evaluation of maintenance costs in comparing alternative designs for highway structures. Departmental Standard BD36/88 and Departmental Advice Note BA28/88, Department of Transport, 1988.

2. Boam, K. J., *Concrete Highway Structures—Current Experience of Investigation and Repair. Concrete Repair—Problems, Questions and Answers.* Palladium Publications, London, March 1988.
3. Wallbank, E. J., The performance of concrete in bridges. *A Survey of 200 Highway Bridges.* HMSO, London, April 1989.
4. Boam, K. J., Impact of cathodic protection on civil engineering. Second International Conference on Cathodic Protection, Stratford upon Avon, June 1989.

14

Programmed Maintenance of Motorway Bridges: Italian Experience in the use of 'Expert Systems'

G. CAMOMILLA,[a] A. DRAGOTTI,[a] G. NEBBIA[b] and M. ROMAGNOLO[a]

[a]*Autostrade SpA, Via A. Bergamini, 50-00159 Rome, Italy*

[b]*SPEA—Ingegneria Europea, Via Cornaggia 10, Milan 20123, Italy*

ABSTRACT

The maintenance management program is designed to provide the motorway network manager with the tools needed to program maintenance of an entire aggregate of structures of widely varying structural, environmental and 'generational' characteristics. Using specially designed management software one can then develop overall assessments of structure reliability and thus obtain a classification on the basis of their state of conservation. The reliability assessment automatically expressed by the computer (the 'expert' system) is of a global *nature, i.e. aimed at creating an objective list of intervention priorities for the entire population of the structures under management.*

INTRODUCTION

The Autostrade Company of the IRI–ITALSTAT Group, responsible for managing some 3000 km of motorway containing around 3000 bridges and viaducts exceeding 10 m in length, utilises a series of control and maintenance systems coordinated within the framework of the partially automated SAMOA program (surveillance, auscultation, maintenance of structures).

Under this program, which requires classification and recording of motorway structures according to their different structural components, one can obtain either a detailed analysis of their construction characteristics or an accurate survey of the defects present, selected according to ease of identification or their importance in assessing the state of health of the structure. To avoid subjective interpretations by different inspectors, data forms are compiled according to codified methods spelled out in special operator manuals, and then input into a system of compatible computers located throughout the territory covered by the Autostrade network.

The maintenance works are always of a preventive nature, ranging from ordinary 'conservative' maintenance to more elaborate repair interventions. In some cases this may be preferable to the effect of complete restoration or a functional improvement, to upgrade the existing structure to cope with increased load conditions or to implement antiseismic protection utilising dissipative systems of more recent development. At this point one can also assess the advisability of modifying the original static configuration, with the advantage of extending the presumable working life of the structure.

AUTOMATED MANAGEMENT PROGRAM

The Autostrade Company has organised its maintenance of motorway bridges and viaducts within the framework of the so-called SAMOA project, which is aimed at developing the tools to achieve completely programmed operations.

The high average age of the structures and their growing numbers have reached such a point that surveillance based solely on visual inspection of the state of the structures and diagnosis based on individual inspector experience has become increasingly unreliable. It thus became necessary to devise specific criteria to ensure more uniform and sound assessment of the state of 'health' of certain structures, but also to predict the probable evolution of this state over time.

Starting from this global assessment or diagnosis of the entire 'population' of structures under surveillance, it is possible to extract three distinct subgroups of structures (or structure components):

—certainly reliable;
—certainly requiring maintenance; and
—not perfectly defined in terms of reliability.

The two subgroups in precarious or suspect condition, presumably comprising a very limited number of items with respect to the total 'population', are then subjected to more detailed inspection (and thence diagnosis) using instruments and tests more sophisticated than simple visual inspection.

The specific aims of the SAMOA program are as follows:

(a) creation of a data base and related software for the management of the morphological data on the structures and the maintenance interventions performed on them;

(b) research and development of rapid, non-destructive control systems for automatic acquisition of such data; and
(c) development of structural verification programs to assess the level of safety and need for intervention.

Implementation of this project permits, starting from the comparative analysis of the various data contained in the data base, the rational definition of intervention needs and priorities. The flow chart of this procedure (which constitutes the entire SAMOA project) is shown in Fig. 1.

The first part of management activity consisted in recording the registration data on all the structures in the network. Drawing on design and cost data from existing files, combined with information obtained on possible site visits, the next step was to compile special 'morphological records' for each of the structural elements comprising the structure. Various type groupings were defined such as, for example, piers, foundations and all other elements making up the structures, indicating the dimensions and compositions occurring in the individual structure in question.

Each structure was thus sub-divided into its constitutive elements and these latter reported, case by case, on the number of records required. General summary records refer to these latter and combine them with other information either pre-existing or acquired subsequently such as, for example, general data on the environment where the structure is situated, the construction methods employed, static tests conducted, subsequent maintenance and repair interventions.

These data, subsequently memorised on computer, constitute the historical data base of the structures. Employing software able to select/sort on the basis of the historical data, the structures can then be classified on the basis of element types in such a way as to visualise the appearance in the manner in which the human operator is more accustomed to viewing it. As far as inspections are concerned, the surveillance is performed essentially by means of close visual examination or binoculars of the individual structural parts, to spot and take note of any defects which may be present, with particular care to observe the development of those already noted previously. Visual checks, despite their limitations with respect to the difficult but not always impossible task of identifying hidden defects or accessing parts concealed from view (e.g. parts below ground level, inaccessible heads of prestressed concrete beams, etc.), remain the fundamental method of surveillance able to provide at least general indications of the overall state of conservation of the structures.

In the case of those structures for which further detailed examination is considered advisable, however, recourse is made to more sophisticated methods based essentially on local and/or global non-destructive type tests.

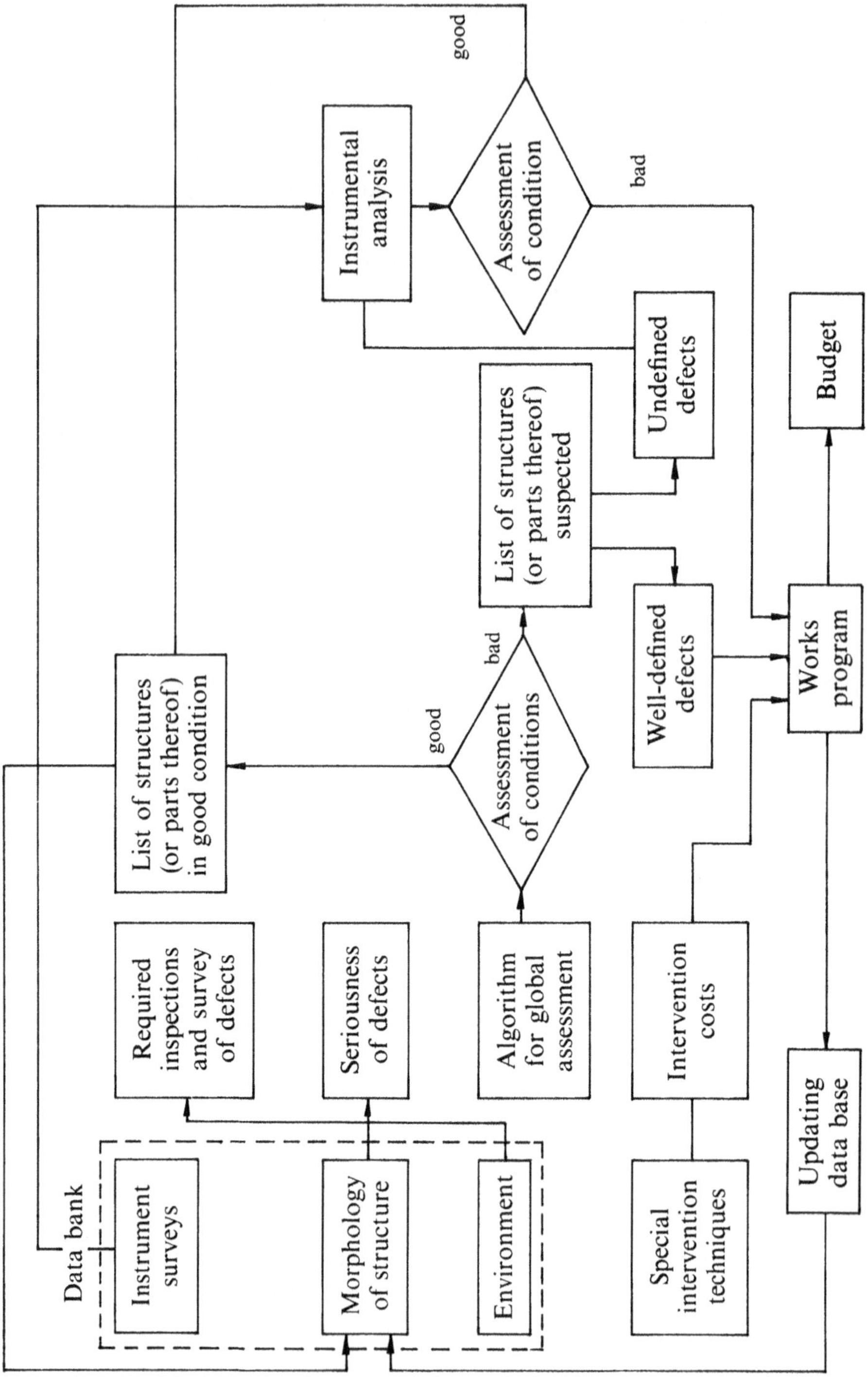

FIG. 1. Flow chart.

Considerable advances are occurring and are under study in this sector, but there already exist a number of different well known and widely utilised techniques, which will be described in a following section. In order to guarantee the greatest possible uniformity in classifying the different types of defects and the most important parameters for their description and control, use is made of 'defect charts' on which such characteristics are defined in univocal fashion.

In fact the inspections are conducted in accordance with a special manual on damage survey methods containing not just the record forms for recording the defects but also detailed explanations of these defects (including descriptions and photographs), as well as instructions on how to fill out the forms reporting the locations and extent of the deterioration observed. By breaking down the structure into its individual structural components it is possible to conduct an accurate survey of all the defects present, using precise and codified methods such as to limit subjective interpretations by the inspectors.

The record forms, as can be seen from the example shown in Fig. 2 of a deck consisting of sliding cable prestressed reinforced concrete beams, are organised in such a way as to divide the structural parts into further specific elements to which the defects observed are referred. In the case in question, the deck is first divided into its constitutive elements (beams, crossbeams and slab) and each of these is then divided into fields: for each beam, for example, a field consists of a section located between two consecutive crossbeams.

After the forms have been compiled in the field, the data are then entered into the computer, where they are processed according to special interpretative algorithms. The resulting output can take the form of an overall but at the same time objective assessment of condition of the single element, of the single structural component or of the structure as a whole, depending on the information desired.

The processing software contains defect assessments of varying degrees of seriousness in relation to the type of structure involved (supported beam, continuous beam, framework, arch, etc.), the component materials (reinforced concrete or prestressed reinforced concrete, steel, masonry, etc.), as well as the extent and location of the deterioration. This is combined with instrument measurements and/or considerations regarding the evolution of certain physical parameters measured geometrically (e.g. inertia moments) or instrumentally (e.g. measuring vibration modes or using other non-destructive methods) on the structures themselves.

The processing criteria are such as to permit a 'global assessment' of the state of the structure or of its single component parts, which serves as the basis for establishing criteria for intervention priority. In fact the global assessment, which is quick to use and yet reasonably accurate, serves to

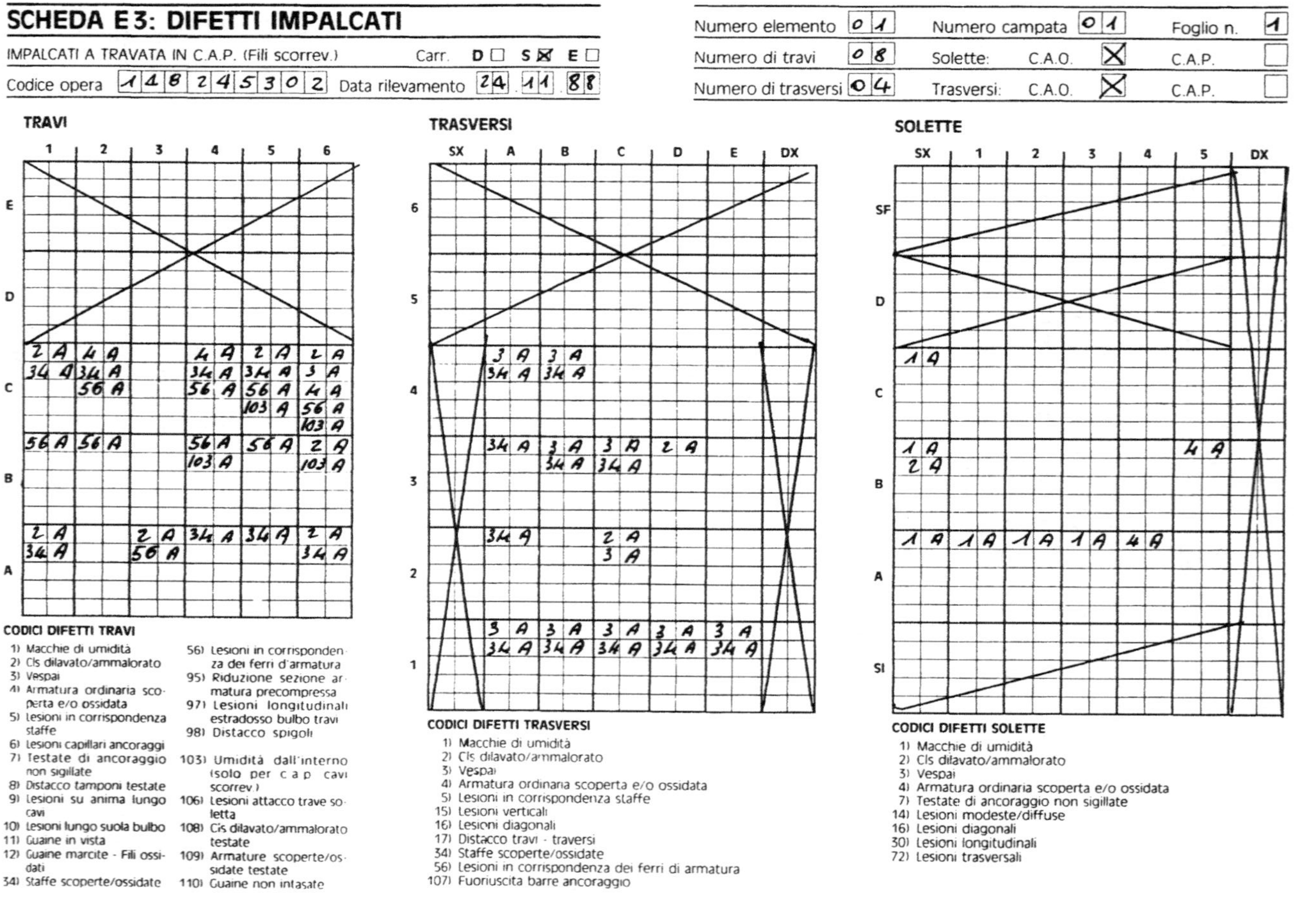
SCHEDA E3: DIFETTI IMPALCATI
IMPALCATI A TRAVATA IN C.A.P. (Fili scorrev.)
Carr. D S E
Codice opera
Data rilevamento
Numero elemento
Numero campata
Foglio n.
Numero di travi
Solette: C.A.O. C.A.P.
Numero di trasversi
Trasversi: C.A.O. C.A.P.
TRAVI
TRASVERSI
SOLETTE
CODICI DIFETTI TRAVI
1) Macchie di umidità
2) Cls dilavato/ammalorato
3) Vespai
4) Armatura ordinaria scoperta e/o ossidata
5) Lesioni in corrispondenza staffe
6) Lesioni capillari ancoraggi
7) Testate di ancoraggio non sigillate
8) Distacco tamponi testate
9) Lesioni su anima lungo cavi
10) Lesioni lungo suola bulbo
11) Guaine in vista
12) Guaine marcite - Fili ossidati
34) Staffe scoperte/ossidate
56) Lesioni in corrispondenza dei ferri d'armatura
95) Riduzione sezione armatura precompressa
97) Lesioni longitudinali estradosso bulbo travi
98) Distacco spigoli
103) Umidità dall'interno (solo per c.a.p. cavi scorrev.)
106) Lesioni attacco trave soletta
108) Cls dilavato/ammalorato testate
109) Armature scoperte/ossidate testate
110) Guaine non intasate
CODICI DIFETTI TRASVERSI
1) Macchie di umidità
2) Cls dilavato/ammalorato
3) Vespai
4) Armatura ordinaria scoperta e/o ossidata
5) Lesioni in corrispondenza staffe
15) Lesioni verticali
16) Lesioni diagonali
17) Distacco travi - traversi
34) Staffe scoperte/ossidate
56) Lesioni in corrispondenza dei ferri di armatura
107) Fuoriuscita barre ancoraggio
CODICI DIFETTI SOLETTE
1) Macchie di umidità
2) Cls dilavato/ammalorato
3) Vespai
4) Armatura ordinaria scoperta e/o ossidata
7) Testate di ancoraggio non sigillate
14) Lesioni modeste/diffuse
16) Lesioni diagonali
30) Lesioni longitudinali
72) Lesioni trasversali

FIG. 2

separate the *certainly reliable* structures from those which require more detailed examination.

These latter must be examined with special attention (measurements, determination of restoration measures to be undertaken) and hence will be subject to more frequent surveillance or, if necessary, constant monitoring. The reliable structures, on the other hand, will only be subjected to ordinary routine surveillance, save for the repetition every 2 years of the reliability inspection related to the 'global assessment'.

Before examining the data processing methods in detail, it would be advisable to look first at the special techniques employed in structure surveillance and the criteria employed in determining restoration interventions, so as to have a better understanding of the procedures followed in coping with the various problems of structure management.

SPECIAL TESTS

In presenting the surveillance methods it was mentioned earlier that, contrary to visual inspection procedures, the specialised control techniques are employed on an *ad hoc* basis when the need arises to have more detailed information on a single structure or on those located on a particular section.

Such needs may be dictated by different circumstances:

(a) visual inspection and subsequent data processing may leave room for various interpretations as to the actual state of the structure and the seriousness of the situation;
(b) planning maintenance or restoration operations may require more specific measurements; and
(c) special research projects.

Examples of situations (a) and (b) are quite similar insofar as the problem is to define either the intervention priority or the manner of intervention; in the case of point (c), however, the situation is much more varied, and we shall limit ourselves here simply to a few references. In all cases, however, it is advisable to deal with the subject starting from the single tests or, better, from the description of the criteria for the utilisation of these in an integrated manner, without lengthy treatment of the single methods. Proper integration of the various methods used is fundamental for valid interpretation of the data measured; indeed only in this way is it possible to effect a type of iteration which, taking into account the various approximation factors proper to each of the systems, permits one to obtain a realistic 'quantification' of the state of conservation.

In particular, for type (a) and type (b) situations the following tests are normally conducted either on specific parts or sample areas.

—Ultrasonic: for the following purposes:
- to supply indications of the homogeneity of the concrete by means of transparency and/or indirect tests;
- to provide indications of the depth of cracks with indirect tests across the cracks or alongside them; and
- to determine, via correlation of the data obtained, the ultrasonic transmission velocities along with the recoil index obtained at the same points using sclerometric tests, concrete compression strength values.

—Sclerometer: this traditional non-destructive test provides information on the compression resistance correlated with the recoil index, but if used by itself it is limited to providing information tied very closely only to the 'cortical' layer. For this reason it must always be used in conjunction with ultrasonic tests. Such integration has proven to be optimal in 'filtering' the results from the influence of a number of variables (granulometry, moisture, binder dosage, type of aggregate) which in many instances register contrary values with the two methods.

—Pull-out: this is utilised mainly in cases where it is impossible to effect ultrasonic tests by transparency (on 'chunky' elements), or where it is deemed useful to obtain compression strength data for comparison with that obtained from other non-destructive tests; these data are always combined with pacometer data regarding the position of the superficial reinforcing, so as to ensure that the piece is inserted in a suitable position such that extraction is not blocked by possible presence of reinforcing rods or mesh.

—Windsor probe test: in using this test the same applies as was said in the case of the pull-out test; the combination of this test with the preceding one is always useful, especially if there is a lack of information regarding the aggregate grain size and characteristics.

—Measurement of carbonation depth: this test is always employed, as it provides an important index of the aggressivity of a particular environment on a given composition of concrete, and thus provides extremely useful information for detailed studies of mix design for durable concretes; it is superfluous to stress the value of a test which permits the designer to know whether or not one can still hypothesise passivation around the reinforcing.

—Measurement of chloride penetration: this is used generally only on exposed parts: pulvinos, heads of beams, heads of crossbeams. It may be direct (samples taken from the structure) or indirect, based on electrochemical potential measurements conducted on accessible structure surfaces and linked together by isopotential curves (mapping).

Obviously this list of tests is far from complete, but mention is made here only of those which constitute part of the routine system which, as set out above, automatically comes into play once certain 'alarm thresholds', defined in assembling the visual inspection data, are exceeded.

Besides these tests, which can be defined as 'local' insofar as they provide punctual results on the state of the structure, we also utilise global-type systems which yield information on the characteristics of the structure as a whole, obtained from static loading tests or measurements under dynamic excitation.

Dynamic analysis is the most commonly used test, aimed at determining the vibration modes of the various parts of the bridge under a known source of vibration. The method is not very sensitive to deterioration of the structure itself and cannot, at the present time, be used to assess the actual state of structures. It can, however, be used to memorise the state of the structure at a certain moment, so as then to draw a comparison with the results of other dynamic tests conducted subsequently in time. Certain recent developments in survey systems and finite element modelling would seem to be promising for the practical use of this method. Besides their use in maintenance design, these measurements can provide new criteria for assessing 'sample defects' which can subsequently be incorporated in the global assessment and applied in expert systems to determine SAMOA intervention priorities.

In this way the bridge management system is continually improved over time: constantly increasing its data base and consequently improving the work of the human operators who continue to manage well the more traditional tasks; thus we have surveillance, auscultation and intervention as integrated moments of a single process, with the substantial but hardly 'tyrannical' assistance of the computer.

In concluding this section, mention should be made of the various research objectives we will be aiming at in the use of the above tests and others. On the one hand, we will be trying to understand better the limits of the various systems utilised and obtain more detailed information on those not routinely used and, on the other hand, we will also attempt to develop hypotheses regarding the process of deterioration with respect to predetermined types of environment and structure.

In the case of the first of these two aspects we are already conducting careful analyses of the data as they are gathered, so as to improve design and programming of interventions, also by increasing the number and distribution of these tests and checking the results by means of destructive tests and controls employed during partial demolitions conducted in the course of repair operations.

So as to gain a better understanding of other systems (e.g. potential mapping, release of tensions, etc.), trials are conducted in conjunction with both destructive and non-destructive tests on a case by case basis in specific situations.

Work continues toward the development of 'deterioration curves over time' (a sort of measurement of the speed of deterioration under standard conditions—of crucial importance for a company responsible for the maintenance of such a large number of structures) as a function of the environment and the structural and technological characteristics. We have begun to conduct on-site tests to determine the depth of carbonation on sample elements of selected structures of particular significance. These tests will naturally be repeated at regular intervals (around every 2 years, as already stated), so as to construct curves of the type indicated, also with a view toward future projections.

INTERVENTIONS ON STRUCTURES

Starting with the list of priorities provided automatically by the computer program, and following whatever specialised testing may be required, we proceed to the design of the intervention in accordance with defined general reference criteria.

Distinctions are made among different possible interventions, which can be grouped as follows:

—Ordinary or extraordinary maintenance: these are distinguished according to the type and importance of the intervention to be performed, and are intended to maintain the structure at full efficiency in accordance with its original design characteristics.

—Static restoration: when one wishes to restore the original bearing capacity of the structure.

—Retrofitting (functional maintenance): when one decides to maintain unchanged the original geometry and static scheme while at the same time permitting the structure to be subjected to actions either greater or differing from those for which it was initially designed.

—Restructuration: this entails the alteration of the original static scheme of the structure and/or its geometric characteristics so as to restore or increase its bearing capacity.

Naturally the last two items can be adopted also where the state of conservation of the structure is judged to be good.

At this point we shall illustrate several special intervention techniques employed on various parts of a bridge structure.

With regard to protective techniques (to restore the protective function of certain parts of the structure, which may have deteriorated or not have been supplied in the first place), the following may be employed:

For piers:
—plating with metal mesh and rheoplastic mortars to restore protective covering of exposed reinforcing;
—light hooping where it is necessary to provide a certain additional strength to columns (the reinforcing is rendered continuous by suitable overlapping beyond the edges); and
—localised repairs, followed by application of protective paint to prevent carbonation of the concrete cover layer over the reinforcing.

For slabs:
—surface repairs with stuccos or small castings and/or waterproofing of the deck with synthetic membranes or other systems;
—more extensive repairs and waterproofing of specific points (drains, heads of prestressed beams, etc.); and
—diffuse anode cathodic protection of outer reinforcing to eliminate chlorine ions present in the existing concrete.

For beams:
—application of protective paints with or without repairs of unfilled sheaths.

In the case of repairs to restore bearing or bonding function which has been lost or not suitably provided initially, on the other hand, operations might include the following:

For piers:
—substitute lining (new castings which by themselves support the total load on the pier) of columns with reinforcing and form casting using non-shrink rheoplastic concretes;
—transverse strengthening of piers with additional septa; and
—upgrading according to more strict criteria to better resist transversal seismic action.

For slabs:
—reconstruction of the upper layers of the slab, including elimination of degraded material by high-power hydrodemolition, application of new reinforcing and substitution with non-shrink rheoplastic concrete.

For beams:
—strengthening with steel plate elements glued and/or bolted to the existing structure (thin plate reinforcing); and

—addition of external cables to existing prestressed or non-prestressed structures to restore lost prestressing or provide it anew (positioning of the cables with respect to the beam cross-section, and that of anchor and drawing points, will vary depending on the space available).

The preceding provides an overview of possible interventions normally adopted. It must be stressed, however, that in the majority of cases the aim is essentially to obtain a functional improvement of the structure as a whole, as well as of its component parts.

Types of intervention which could be cited as particularly successful examples are integral decks (jointing of decks), the introduction of special unidirectional and multidirectional bearing devices as well as seismic protection devices, and lastly the installation of breakthrough-proof guardrails capable of dissipating impact energy.

The problems faced by the designer of maintenance interventions on existing structures are considerably more complex than is the case with the design of new ones. Some of the additional constraints encountered include the need to operate in the presence of traffic; the need to select materials which yield reliable results within a very short period of time and in the presence of traffic-induced vibrations; the very strict limits on intervention time; the need to assess material strengths in situations where degradation is present; and the need to achieve higher performance standards than those adopted in the original design of the structure.

It is easy to see from the above considerations that the concept of maintenance thus assumes the more 'noble' connotations of the term 'project' (i.e. that of 'projecting the mind into the future so as to foresee in advance' what the work will require), thus superseding the rather simplistic notion of those who see no need for a specific design stage in the concept of maintenance.

DATA PROCESSING METHOD

At this point we would like to provide an example of how one goes about setting up an expert system of classification for purposes of controlling the state of the art of the structures under management.

Given the fact that the assessment is conducted separately for each component part of the structure, we shall concentrate our attention on a specific application, in this case that of a deck.

As shown in the attached sample flow chart for decks (see Figs 3 and 4), starting from the analysis of the morphological characteristics, the

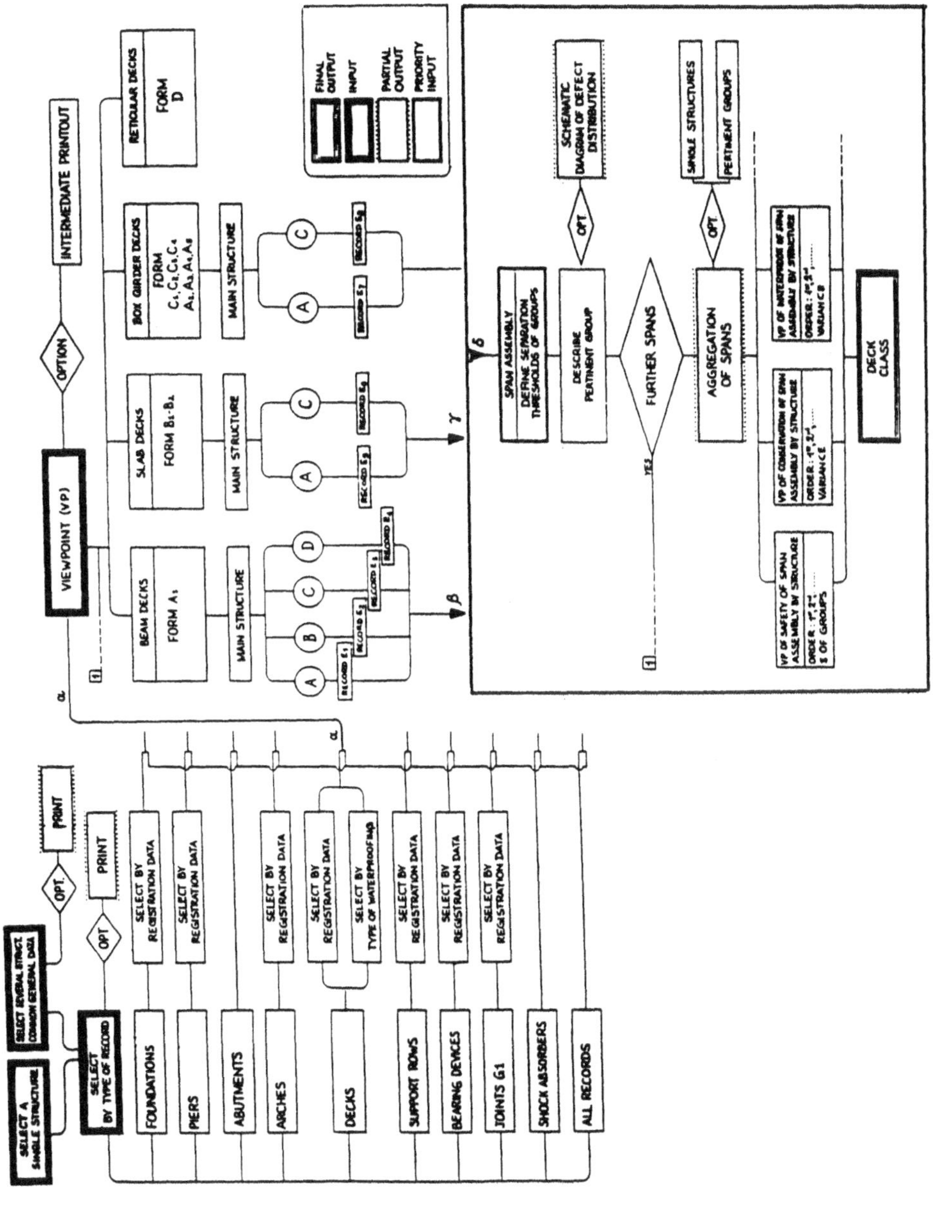

SELECT A SINGLE STRUCTURE
SELECT SEVERAL STRUCT. COMMON GENERAL DATA
OPT.
PRINT
SELECT BY TYPE OF RECORD
OPT
PRINT
FOUNDATIONS
PIERS
ABUTMENTS
ARCHES
DECKS
SUPPORT ROWS
BEARING DEVICES
JOINTS G1
SHOCK ABSORBERS
ALL RECORDS
SELECT BY REGISTRATION DATA
SELECT BY TYPE OF WATERPROOFING
VIEWPOINT (VP)
OPTION
INTERMEDIATE PRINTOUT
BEAM DECKS
FORM A1
SLAB DECKS
FORM B1-B2
BOX GIRDER DECKS
FORM C1,C2,C3,C4 A2,A3,A4,A5
RETICULAR DECKS
FORM D
MAIN STRUCTURE
FINAL OUTPUT
INPUT
PARTIAL OUTPUT
PRIORITY INPUT
SPAN ASSEMBLY
DEFINE SEPARATION THRESHOLDS OF GROUPS
DESCRIBE PERTINENT GROUP
OPT
SCHEMATIC DIAGRAM OF DEFECT DISTRIBUTION
FURTHER SPANS
YES
AGGREGATION OF SPANS
SINGLE STRUCTURES
PERTINENT GROUPS
VP OF SAFETY OF SPAN ASSEMBLY BY STRUCTURE
VP OF CONSERVATION OF SPAN ASSEMBLY BY STRUCTURE
VP OF WATERPROOF OF SPAN ASSEMBLY BY STRUCTURE
DECK CLASS

FIG. 3

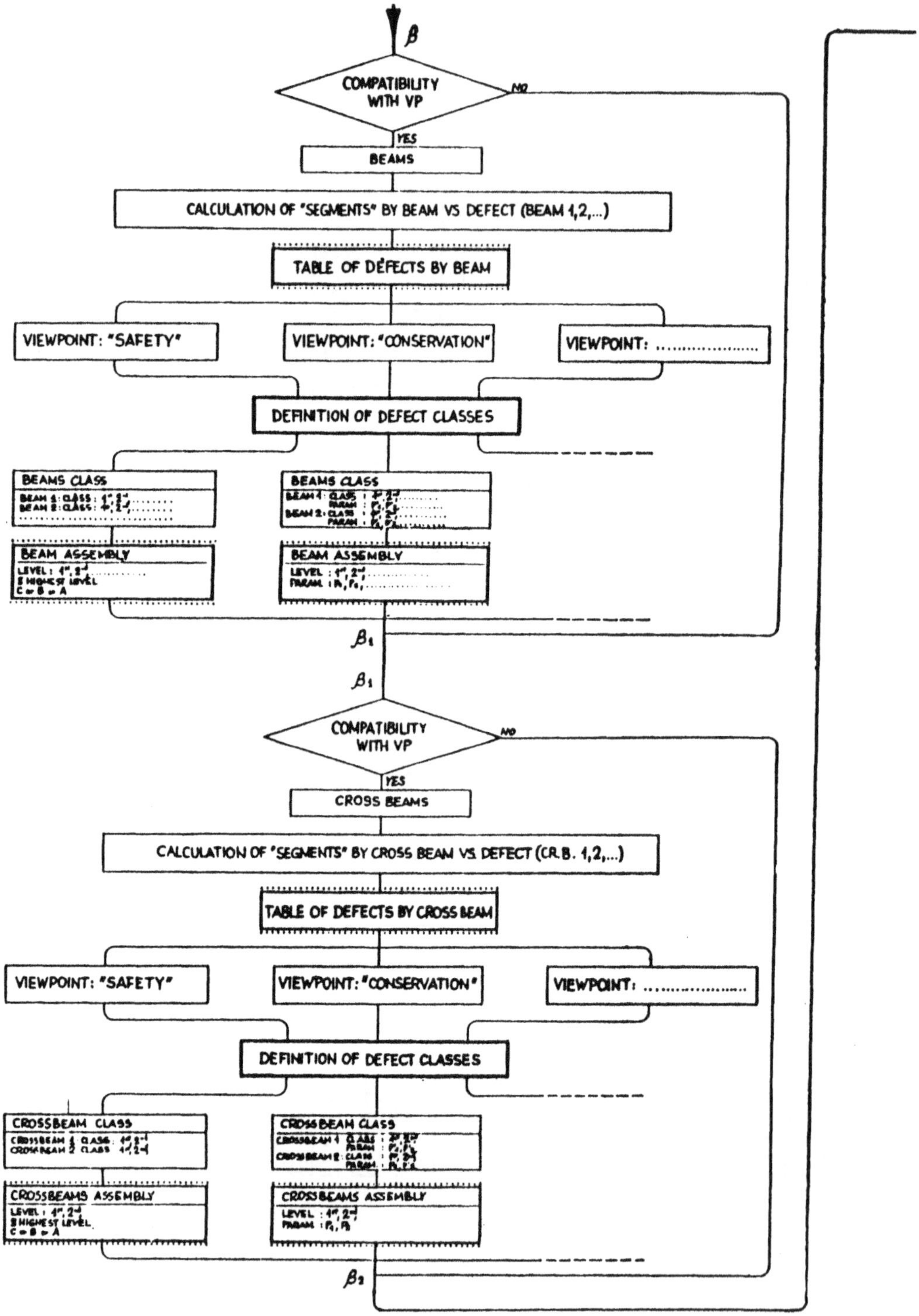
β
COMPATIBILITY WITH VP
NO
YES
BEAMS
CALCULATION OF "SEGMENTS" BY BEAM VS DEFECT (BEAM 1,2,...)
TABLE OF DEFECTS BY BEAM
VIEWPOINT: "SAFETY"
VIEWPOINT: "CONSERVATION"
VIEWPOINT: ....
DEFINITION OF DEFECT CLASSES
BEAMS CLASS
BEAMS CLASS
BEAM ASSEMBLY
BEAM ASSEMBLY
β1
β1
COMPATIBILITY WITH VP
NO
YES
CROSS BEAMS
CALCULATION OF "SEGMENTS" BY CROSS BEAM VS DEFECT (CR.B. 1,2,...)
TABLE OF DEFECTS BY CROSS BEAM
VIEWPOINT: "SAFETY"
VIEWPOINT: "CONSERVATION"
VIEWPOINT: ....
DEFINITION OF DEFECT CLASSES
CROSSBEAM CLASS
CROSSBEAM CLASS
CROSSBEAMS ASSEMBLY
CROSSBEAMS ASSEMBLY
β2

FIG. 4

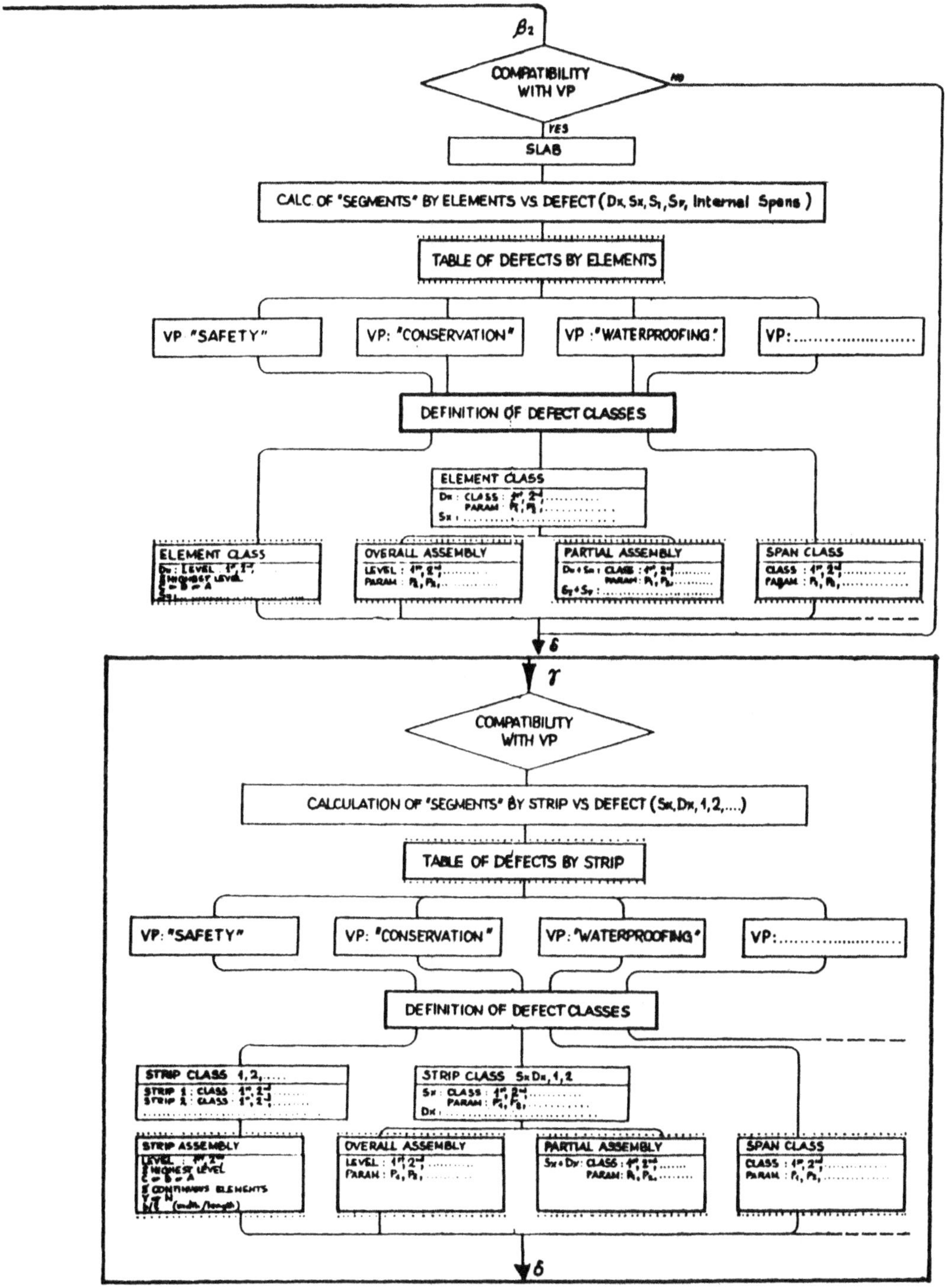
β2
COMPATIBILITY WITH VP
NO
YES
SLAB
CALC. OF "SEGMENTS" BY ELEMENTS VS DEFECT (Dx, Sx, S1, Sr, Internal Spans)
TABLE OF DEFECTS BY ELEMENTS
VP "SAFETY"
VP: "CONSERVATION"
VP: "WATERPROOFING"
VP: ....
DEFINITION OF DEFECT CLASSES
ELEMENT CLASS
ELEMENT CLASS
OVERALL ASSEMBLY
PARTIAL ASSEMBLY
SPAN CLASS
δ
γ
COMPATIBILITY WITH VP
CALCULATION OF "SEGMENTS" BY STRIP VS DEFECT (Sx, Dx, 1, 2, ...)
TABLE OF DEFECTS BY STRIP
VP: "SAFETY"
VP: "CONSERVATION"
VP: "WATERPROOFING"
VP: ....
DEFINITION OF DEFECT CLASSES
STRIP CLASS 1, 2, ...
STRIP CLASS Sx Dx, 1, 2
STRIP ASSEMBLY
OVERALL ASSEMBLY
PARTIAL ASSEMBLY
SPAN CLASS
δ

FIG. 4—*contd.*

processing algorithm performs different specific tasks depending on the type of deck itself. Selecting as our example the case of a deck consisting of simply supported beams of prestressed reinforced concrete (form E3) with sliding cables (a structure quite common on the Autostrade network), we now proceed to a detailed description of the procedure.

The steps followed are not always the same but will depend on the objective one wishes to achieve. The choice among the different analysis viewpoints takes the form of a selection from among the distinct 'viewpoints' presented: structural safety, state of conservation, waterproofing, etc. It is necessary to select the viewpoint (VP) at the very beginning, as in this way the processing operation can be carried out automatically without interruption.

At this selection level it is also possible to choose whether one wishes to have only the final output, or also the intermediate outputs corresponding to the various partial processing stages. The intermediate outputs constitute a partial processing of the codified defects reported on the forms compiled during the site inspections. These partial outputs, e.g. the 'table of defects by beam' (or by crossbeam or by slab element) provide, in the case of each element (beam, crossbeam or slab field), a 'photograph' of the associated defects and related intensity/extent parameters.

The assemblies of the various elements are more complex, as they will differ for each of the above-mentioned viewpoints, and will provide data strings indicating a level of 'seriousness' and function parameters of the intensity/extent of the defects for all of the beams, crossbeams or slab elements taken together.

Another important partial output useful for further assessment of a non-automatic nature is the schematic diagram, which summarises the defect distribution of an entire span (beams plus crossbeams plus slab).

The final partial output consists of the aggregation of all the spans of a structure. These are sub-divided into groups, each of which is distinguished by appropriate separation thresholds. These latter serve to filter the parameters defined in processing each of the spans, even where these are not morphologically similar (simple reinforced concrete beams, prestressed reinforced concrete beams, box girders, etc.).

The groups associated with each span are organised according to two different output possibilities:

—aggregation of spans by single structure, with related indication of group; and
—aggregation of spans by group, with related indication of the structure to which they belong.

These outputs provide a sufficiently complete indication of the state of the structure, always in accordance with the viewpoint selected. In this case, to obtain a more synthetic unit view, we have opted here for assembly of all the spans of each structure. In fact the final output is represented by an overall assessment of the conditions of all the decks of a structure, and the ranking of this latter in a condition classification in relation to the condition ratings of various structures of a selected population. The assembly operation takes into account spans having different structural characteristics, and hence different processing procedures. 'Homogenisation' of the spans is achieved by attribution of thresholds which are different for each type of record form; the thresholds thus constitute the point of equivalence in comparing spans of different types.

The operation proceeds as follows for the different viewpoints:

'Safety': The group of spans of a structure is classified in a determined order equivalent to the preceding groups. If at least one span is classed as group 1, the whole group of spans is of first order, and so on. Furthermore, each order is attributed percentages relating to the spans contained in each group, with respect to the total number of spans in the structure as a whole.

'Conservation': The group of spans of a structure is classified in a determined order resulting from processing the data on the groups to which each span belongs. First, the frequencies of the groups are calculated. The order of a given structure corresponds to the mean of the groups present weighted according to their relative frequencies as in the following expression:

$$\bar{g} = \frac{\sum fg}{n} \tag{1}$$

The order value is linked to a second parameter consisting of the variance of the frequency distribution, calculated as follows:

$$S^2 = \frac{\sum f(g-\bar{g})^2}{n} \tag{2}$$

Depending on the viewpoint, each structure thus remains associated with an order and certain percentages or a variance.

In classifying all the structures of a selected population, these are first sorted in descending rank, from those of highest order to those of lowest order. In the case of the safety VP the ranking is determined by relative percentages in reference to the order, whereas for the conservation VP the position is determined by the variance, in inverse proportion to the value.

15

Engineering Management of the Tamar Bridge

W. I. HALSE
*Bridges Division, Mott MacDonald Civil Ltd,
20/26 Wellesley Road, Croydon, Surrey CR9 2UL, UK*

and

R. L. C. STEPHENS
*County Surveyor's Department, Cornwall County Council,
County Hall, Truro, Cornwall TR1 3BE, UK*

ABSTRACT

The Tamar Bridge (Fig. 1) was constructed within the period 1959–61 as a private toll crossing for the Tamar Bridge and Torpoint Ferry Joint Committee, comprising representatives of the Cornwall County Council and the Plymouth City Council. Since the end of the maintenance period, the engineers responsible for the design and supervision of construction (Mott, Hay and Anderson) have been retained by the joint committee to undertake an annual inspection of the bridge and its immediate approaches, and to report and make recommendations on any necessary maintenance. Apart from small or ad hoc *items arranged through the joint engineers, the county surveyor of Cornwall and the City Engineer of Plymouth, subsequent works have invariably been arranged and supervised by those same consultants who, through close liaison with the bridge owners and their staff, have effectively provided a professional service for the long-term management of maintenance activities by forecasting financial commitments and arranging the operations around the increasingly restrictive limitations imposed by traffic densities.*

INTRODUCTION

When opened to traffic on 24 October 1961 the Tamar Bridge (Fig. 1), with its 335 m main span, became the longest span road bridge in the UK and 26th equal in the league table for 'suspension bridges of the world'. Since then it

FIG. 1. Tamar Bridge.

has been successively relegated in the UK by the Forth, Severn and Humber road bridges, and has disappeared without trace from world rankings.

However, the joint enterprise with the Torpoint Ferry still maintains a unique record amongst UK tolled estuarial crossings,[1] with no capital debts (discharged by 1981), a regular operating surplus without subsidies and the lowest tolls.[2] Since 1980 tolls on the bridge have been payable in one direction only, and whereas the charge per car was 3 shillings (15p) each way in 1961 the current cost is 40p cash to east-bound travellers, effectively reducing to 20p for regular users taking advantage of the concessionary vouchers available for pre-purchase.

PRE-CONSTRUCTION

Vehicular ferry crossings of the River Tamar existed at Torpoint and Saltash from the early 19th century, but by the 1920s local agitation had already begun for a fixed crossing to supplement or replace those increasingly inadequate facilities and as early as 1931 a scheme for a high-level bridge between Torpoint and Devonport was prepared but abandoned on objection from the Admiralty. The post-war boom in traffic produced further impetus and in 1950 the Cornwall County Council and the Plymouth City Council formed a committee to pursue the matter with the Ministry of Transport.

By 1955 it became obvious that, despite a favourable report the previous year from the technical panel appointed by the minister, the matter was unlikely to be pursued in the immediate future and the two councils determined to investigate a private joint enterprise.

The consultants, appointed in September 1955, reported on bridge and tunnel alternatives in April 1956, recommending a fixed crossing on the present alignment, and a firm decision to proceed was taken by the councils in July. Such was the considered degree of urgency and of the cooperation afforded by all parties that the consultants and parliamentary agents prepared and deposited a bill in Parliament by November 1956 and Royal Assent was obtained at the end of July 1957.

Unfortunately the then current financial climate delayed Treasury consent to even private borrowing, and it was not until the autumn of 1958 that the engineers were appointed to prepare the design, following which, continuing the afore-mentioned cooperation, tenders were invited in March, a contract was let in June and work commenced on site on 7 July 1959.

The construction details have been described elsewhere.[3] Suffice it to state that the 152 mm thick deck in 43 N/mm^2 concrete is supported on steel stringers carried on cross-girders at 9·17 m centres slung by suspenders from the main cables spaced at 15·24 m centres, with traditional steel Warren truss stiffening girders. Extensive use was made of box sections in high tensile (BS 968) steel, shop-fabricated by welding, with site connections made with friction-grip Torshear bolts. The 10·06 m wide three-lane carriageway and twin 1·83 m footways were surfaced in hand-laid mastic asphalt at 38 and 20 mm thickness respectively.

TRAFFIC CONDITIONS

Hindsight is a wonderful gift and a golden opportunity to the local media, who from time to time, prompted by some traffic delay through accident or other, pontificate on the folly of the planners in building a three-lane bridge.

Figure 2 provides a histogram of recorded annual traffic flows across the Tamar Bridge, rising sharply and inexorably from 1·8 million vehicles in 1962 (the first full year of trafficking) to the anticipated total of 12 million vehicles in 1989. The authors venture to suggest that had the engineers included such a prediction in their original report then at very least their professional judgement would have been called into question, if not their report rejected outright.

The effect of the surprisingly regular increment of some 375 000 vehicles per annum exhibited every year for 27 years, although marginally alleviated overall by the introduction of tidal flow in 1974 and one-way toll collection

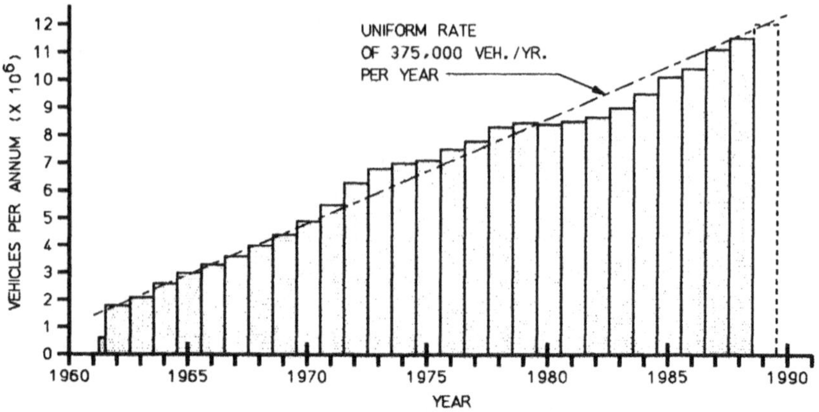

FIG. 2. Histogram of annual traffic flows.

in 1980, has increasingly contributed to problems in arranging maintenance works requiring lane possessions. Whereas up to the late 1970s it was possible to consider and implement closure of one lane for up to a week, for example to construct and cure replacement epoxy nosings at damaged deck joints, a decade later lane closures for the major refurbishment works subsequently described had to be stringently restricted to avoid traffic chaos and maintain reasonable public relations.

These latter operations had to be executed outside the commuter peak hours, within working sessions from 09.30 to 15.30 hours and again after 18.30 hours, on weekdays. The main opportunity for progress ostensibly lay on Saturdays and Sundays when a dawn-to-dusk lane closure was acceptable, unless Plymouth Argyle were playing at home (most of their supporters appear to reside in Cornwall) and the returning fans merged with their compatriots returning from Saturday shopping in Plymouth at around 16.30 hours, or unless Sunday morning dawned dry and sunny, at which a substantial proportion of south Devon residents travelled to Cornwall, spreading their departures through the morning but always reappearing at the west end of the bridge around teatime.

Again, come the first weekend in July, the summer visitors arrive in droves and it becomes impossible to even consider lane closures until school holidays end in mid-September but latterly extended into early October by a high proportion of senior citizens. Recuperating from the combination of grandchildren and school holidays? Any work during January, always a hostile month weather-wise, can be discounted. Leave out 10 days at Easter and a week each at the May Day and spring bank holidays to cater for local holiday-takers and a proportion of early tourists, mix in a few wet weekdays in east Cornwall, when shopping in Plymouth becomes an attraction to those early visitors, and garnish, as in 1987, with the wettest June for 70 years and the overall mixture becomes a recipe for ulcers to the programmers charged with limiting traffic disruption to a bare minimum.

ROUTINE INSPECTION

A longitudinal walkway is provided each side of the bridge immediately below deck level extending to some 600 m from back-to-back of the anchorages and two more at bottom chord level across the suspended structure. These, supplemented by hand ropes along the main cables, ladders from deck level up to the saddles and down to the caissons inside both main towers and ready access at both river banks for low tide inspection and binocular viewing, provide adequate permanent facilities for the routine annual inspections carried out by the engineers.

A reduction in the main span camber became apparent early on, and in 1964 an extensive set of annual measurements was instituted to record relative levels of 'fixed' points, e.g. anchorages, side and main towers; cambers of the side and main spans; out-of-plumb of the main towers (by nearly 70 m long plumb lines, permanently installed within the north legs); and relative dimensions between the ends of the suspended structure steelwork and the concrete faces of the towers.

After a decade and with the only significant recorded variations being a continuing loss of main span camber and inward leaning of the main towers, it was concluded that the effect was due, despite repeated prestressing by the

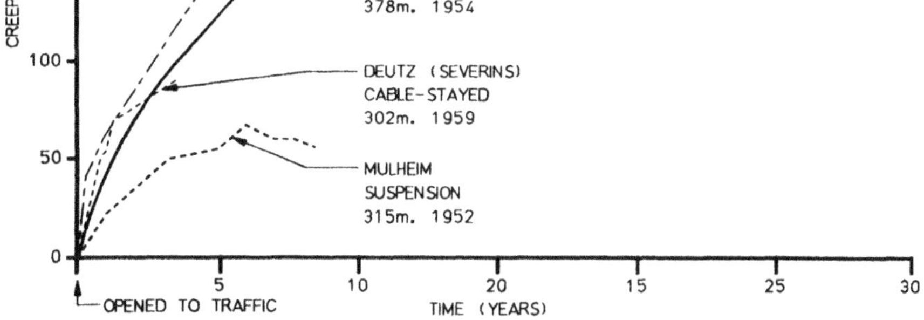

FIG. 3. Creep–time relationship (locked-coil ropes).

rope makers prior to delivery, to permanent stretch (creep) under load in the locked coil wire ropes forming the main cables and suspenders. Little published data were available, but such as could be obtained appeared to agree in reasonable measure with these early observations. The effect has continued but at a reducing rate, as illustrated by the derived creep–time relationship plotted in Fig. 3, together with comparable data.[4]

ROUTINE MAINTENANCE

Apart from weekly sweeping, quarterly gulley cleaning and 6-monthly charging of the grease nipples to the pins of the suspended structure, arranged by the joint engineers, repainting of the structure has been undertaken every 7 years (1968, 1975, 1982 and 1989) on the recommendation of the engineers, virtually as a routine maintenance operation.

The choice of the initial protection system, consisting of grit-blasting, zinc metal spray, PVB etch primer followed by two undercoats and one gloss coat (each of phenolic white lead), had to be made at an early stage of the trials, which eventually resulted in the development of the 'Forth Bridge' system and which subsequently reigned unchallenged in bridgeworks protection for over a decade. Unfortunately, despite a good early performance in these trials, the white lead coats proved less impermeable than the eventually chosen micaceous iron oxide top coats, and by 1968 some considerable areas of the steelwork had to be blast cleaned to remove the underlying zinc corrosion from the metal spray coating. These and defective areas in subsequent repaints were two-coat primed with zinc epoxy initially and latterly zinc phosphate, followed by two coats of MIO, and completed in 1968 and 1975 with white lead undercoat and finish, varied in 1982 to an alkyd medium and in 1989 to a silicone alkyd.

As a result of this regular sequence, repainting has largely become cosmetic, and the cost of remedial treatment of isolated defective areas, principally on the north side below deck through lack of exposure to drying by the sun, does not normally exceed 10% of the total contract sum. The close proximity of houses and parked cars at each end of the bridge and moorings in the river beneath preclude spray application of paint, but the extensive use of box sections in the construction allows under- and finish-coating to some 90% of the area of the suspended structure to be applied by roller, reducing time and hence labour and scaffold hire cost.

Conversely, the successive repaints have highlighted the increasing pressure for enhanced safety measures in the construction industry. The cost of providing and maintaining the necessary scaffolding and safety nets below deck, expressed as an on-cost percentage to other (labour, plant and materials) charges, has risen from 15% in 1975, through 47% in 1982 to 64% in 1989, and actually reached 91% in one unsuccessful tender for the latter.

1960s

The early years proved free of major problems other than that of zinc corrosion, affecting the 1968 repaint.

By this date the annual traffic density had already more than doubled and preliminary consideration was given to some upgrading of the facility. A widening of the carriageway would be restricted to 11·6 m by the main tower legs, providing 2·9 m wide substandard lanes, and there seemed little advantage in any alteration on the bridge without attention to both approaches, in the responsibility of Cornwall and Devon County Councils.

1970s

Various schemes were pursued, none of which proved acceptable to all three parties. Eventually a tidal flow system was introduced on the bridge in 1974, controlled by signals on new overhead gantries, more acceptable than sub-standard lanes, equally effective in commuter peak hours and less expensive.

By 1978 densities had again more than doubled in a decade and thoughts were directed to further alleviation. The toll booths formed a restriction but tolls had to be maintained to pay off the (rapidly diminishing) construction loan, provide for maintenance and subsidise the uneconomic unified toll charges on the Torpoint Ferry. Accordingly an enabling act was prepared, receiving Royal Assent in 1979, by which in 1980 toll collection was limited to one way (eastbound), thus relieving afternoon congestion on the Plymouth side which, on occasion of accident or other mishap, had been known to back up some 5 miles to the city centre.

Over the first half of this decade the 60 no. air-gap articulation joints in the concrete deck of the suspended span, closed by rubber bitumen filler in the depth of the surfacing, began to cause increasing problems with local asphalt break-up. In 1975 the worst affected areas were cut out and provided, in lane widths, with nosings in a recently developed epoxy mortar (Febplate SLS), which proved more durable than traditional rigid mortars. The remainder were similarly treated within a 3-year programme, completing in 1979.

The severe winter of 1978/79 created some problems with kerb and asphalt damage, and raised attention to the not-too-distant need for resurfacing. Soon after, coinciding with the onset of ASR problems in the

South-West, a series of some 60 cores were cut from the deck. When tested these demonstrated that, as a result of the use of sea-dredged coarse and fine aggregates in the construction, the reactive content was too high to sustain an alkali reaction under normal conditions but that chloride contamination from de-icing salt had already penetrated some millimetres into the concrete, although not yet to top reinforcement level. Both conclusions indicated the need to provide an effective waterproof membrane below any future resurfacing.

THE EARLY 1980s

These were marked by the continuing incidence of local damage to the surfacing and a resurgence of earlier problems at the joints, although generally to localised areas within wheel-tracks, not directly through failure of the epoxy mortar but as a result of damage in the underlying concrete under the ever-increasing speeds and shear gross weight of traffic.

The former was checked in 1983 by a surface dressing with chippings and the latter contained by virtually biannual spring and autumn *ad hoc* repairs, effectively 'buying time' before the now obvious major refurbishment could be programmed.

In May 1981 one such repair was undertaken using a magnesium phosphate concrete (Febset 45) on a trial basis and this appeared encouraging. Without coarse aggregate its flow characteristics obviated traditional compaction operations and its fast curing under exothermic reaction subsequently proved it capable of direct trafficking within 2·5 h. By extending its use in subsequent patch repairs, often under conditions and by procedures not encouraged by the manufacturers, it was demonstrated to be virtually idiot-proof, subject to rigid control of water content and mixing time, which was simplified by the purchase of domestic stainless steel measuring jugs and a cooking timer from a local branch of Boots.

In late 1983 the engineers were formally instructed by the joint committee to prepare a report which eventually provided detailed proposals and estimates for a staged refurbishment of the bridge. These comprised, at Phase 1, the permanent reconstruction by break-out and reconcreting over a 600 mm width of each of the 64 no. deck articulation joints including, as a result of the reducing rate of cable creep, the 'closing up' of two in every three over the suspended deck sections. Phase 2 would then comprise carriageway resurfacing and provision of a deck waterproof membrane and include attention to the footways and kerbing, and drainage improvements.

Following discussion with the joint treasurers on finance availability, these phases were programmed around February to June working in 1987 and 1988, within the lane possession limits previously outlined, and the work extended to include, at Phase 3, the repaint due in 1989. The package dovetailed into the projected mid/late 1988 completion data for adjacent DTp works on the Saltash and St Budeaux bypasses immediately either side of the river.

Fortuitously, the preparation of this report was aided firstly by the cooperation and practical experiences of the Forth Road Bridge Joint Board and their consultants, then engaged on a resurfacing of the side spans of that bridge, including the provision of a sprayed MMA acrylic membrane (Eliminator) originally developed by British Rail and applied to some 200 of their bridges before its existence became more widely known. Secondly, by an unexpected and sudden deterioration of the main articulation joint at the western side tower at Tamar, requiring the rapid arrangement of a permanent full-width reconstruction of that joint in late 1984. This interruption was unwelcome at the time but in the event provided a valuable insight into procedures, materials, plant and labour content of such reconstruction, leading to proven recommendations and more realistic estimates for the Phase 1 operations.

A MAJOR REFURBISHMENT

The report was presented in May 1985 and formally accepted by the joint committee in July, with a commission to proceed with arrangement of the necessary contracts on an unusual basis, as subsequently described.

As recommended in the report, three 9·17 m bays of one-lane width and the adjacent footway were stripped of surfacing for a trial in the autumn of 1985, overlain with the acrylic membrane, regulated with a nominal 20 mm of dense bitumen macadam and surfaced with 20 mm of pervious bitumen macadam. The membrane was continued up the back-of-kerb and across the footway area, which was then treated with four alternative proprietary 'thin' surfacings.

Although not previously used on a full-scale road surfacing in the UK, pervious macadam wearing course had a long-standing track record on airport runways and was then being laid extensively on highways in Hong Kong. Its reduced spray characteristic and virtual freedom from risk of aqua-planing were proven, but perhaps of more immediate advantage to the maintenance operation was the lack of need for a chipping spreader, the charging of which always disrupts traffic in the adjacent lane and the machine itself inevitably appears prone to breakdown or other malfunction, disrupting the whole surfacing operation.

The joint committee then accepted a recommendation by the consultants to undertake, under a pilot works in April 1986, reconstruction of eight joints in the Saltash side span. This was principally to relocate the starting end of the 1987 (Phase 1) works some 80 m further eastwards, isolating them from works adjacent to the west end of the bridge, due to commence in March of that year under the Saltash bypass scheme. After consultation, it was agreed that this work should be executed under the direct control of the engineer, by the same 'bridge gang' from Cornwall DLO as had undertaken most of the previous *ad hoc* deck repairs. This large-scale 'trial' added further realism to the assessment of optimum sequences, methods, plant and labour requirements for the main works in the following year, and incidentally provided sites for trials of three proprietary types of continuous joint surfacing then being considered for use in Phase 2.

Movement joints are a perpetual headache to maintenance engineers. There is no universal solution, as witnessed by the multiplicity of proprietary types on offer at any one time which inevitably go off the market within a decade or so. The authors believe that for small movements the asphaltic plug joints as finally used in Phase 2 are currently the best value for money, despite some defects which occurred during the maintenance period and which may recur. Laid with reasonable care the riding quality is superb, they are relatively inexpensive compared with any alternatives on offer and, more importantly, defective areas can be cut out and replaced within a matter of a few hours.

The joint reconstruction and the resurfacing works subsequently followed as planned. The Phase 3 repaint is in progress at the time of drafting and it is a matter of some satisfaction that the overall refurbishment cost is anticipated to be just less than the 1985 estimate, after adjustment for subsequent price fluctuations by published indices.[5]

THE APPROACH TO MAINTENANCE

Maintenance works can be tedious and to some engineers may not be as attractive as heading-up some multi-million pound scheme, but can provide job satisfaction if approached in the right manner.

A differing degree of involvement is required from that for new works in a green field. The engineer must become more deeply concerned with minor details, operational planning and in site supervision. He should be prepared to consult with the contractors, but must specify the programme in some detail and maintain a rigid control. Where the works are on a time-and-materials basis, he must go further by providing and insisting on compliance with a detailed operational programme to make the most efficient and economic use of plant and operatives. By avoiding sophistication and limiting site works to simple, clearly defined operations in a strict sequence, small contractors relatively inexperienced in bridgeworks can be used to advantage and with economy. Common sense and a near instinct to foresee potential problems are a prerequisite, along with an ability to adapt and improvise as difficulties are encountered. With a lesser degree of supervisory assistance than is normal in new works, he must be prepared to spend time when problems are encountered, observing and suggesting alternative sequences, procedures or plant. In maintenance works it is neither acceptable nor economic for the engineer to sit back and wait for a contractor to put forward solutions.

With new works there is some scope for the engineer to drive the contractor to provide a superior job at no extra over the contract sum, but in maintenance this will rarely happen and the client will only get what he is prepared to pay for. What the engineer cannot fully see, he cannot describe and quantify, and what is not fully specified in the tender document cannot be fairly priced by a tenderer, leaving a potential for later, expensive, claims. Hence much of the work must be on a time-and-materials or provisional sum basis, and the experience and judgement of the engineer will be critical in providing realistic assessments of the appropriate amounts to be included in the document. Competitive tendering should and can be financially advantageous but for maintenance it can also be disastrous. Wherever possible the contractors should be locally based, and preference should always be given to those who have previously worked on the site to the satisfaction of the engineer.

CONTRACT PROCEDURES

Over the years and on the basis of results, the joint committee has gradually accepted the somewhat unusual procedures recommended by the engineer. Competitive tendering on a national basis has been restricted to repainting and even then the same local firm has won the contract on the last three occasions. Selective tendering by a few local contractors has been limited to types or extent of works not previously or recently undertaken on the bridge, and on a time-and-materials basis where appropriate. More often negotiated tenders have been arranged with single contractors, either with specialists after adequate trials or with those of previous proven ability and experience of working conditions on the bridge.

For the recent refurbishment, the joint committee further extended this enlightened approach, in acknowledgement of their public responsibility to road users. With the certain knowledge of the level of traffic due on the first weekend in July, extensions of time were not permissible and contracts had a completion date rather than a commencement and a contract period. Whilst under a normal arrangement the main contractor would take contractual responsibility for any overrun, the issue would as usual be clouded by alleged defaults of his subcontractors, weather, deliveries and myriads of other excuses, and meanwhile traffic chaos could still ensue. Accordingly, the joint committee determined that the works should be arranged around individual contracts, with the engineer effectively undertaking the contract management and the resident engineer being additionally charged with responsibility for ensuring the works were finished on time each year.

The contractual implications *vis-à-vis* one contractor's work and another's, or due to default or poor workmanship, were daunting. The logistics were no less so. The Phase 2 works finally involved 25 separate contracts ranging in value from £600 to nearly £300 000. Of these, eight (after selective or negotiated tender) were sealed and bonded under modified ICE Conditions of Contract whilst the remainder were against written order following prior quotations. Six were for the supply of specialist materials, leaving the engineer directly responsible for ensuring that 80-odd operatives from the other 19 contractors were present on site at the right time and executing the work in order and to programme, between mid-February and end of June 1988.

MAINTENANCE COSTS

Expenditure on all maintenance has been researched, separately accounting the costs of repainting, other contract works and a combined allowance for minor works, fees and supervision costs, which latter has been averaged for simplicity.

These have been expressed in histogram form in Fig. 4 as a percentage of the then current bridge valuation, being the original construction cost of £1·8m at 1959 prices annually updated by output price indices,[5] which has also been plotted for information on a separate scale. These do not include 'running costs', e.g. cleansing, lighting, insurance, etc., nor special items such as maintenance of toll equipment or buildings.

From the histogram it can be deduced that the mean annual cost of minor maintenance initially increased but over the past decade has remained constant at some 0·35% of the annual bridge valuation; that the corresponding cost of repainting on a 7-year cycle has been 0·3% per annum; and that other contract maintenance, including a major refurbishment after more than a quarter of a century, has amounted to some 0·5% per annum. It should perhaps be noted that these percentages have been related to the updated construction cost rather than the annual insurance valuation,

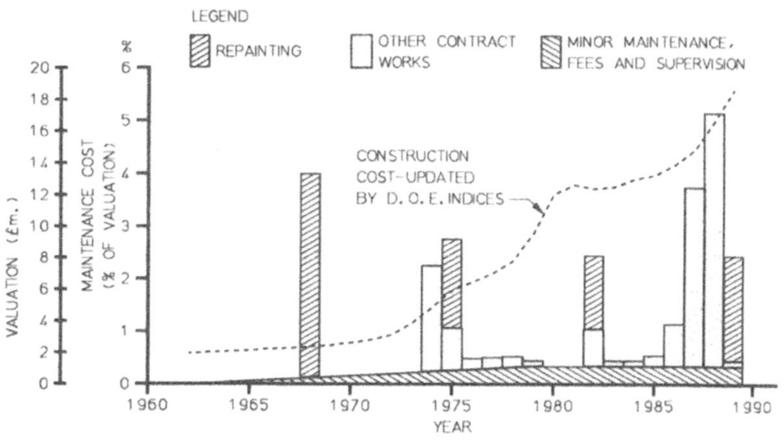

FIG. 4. Maintenance costs and bridge valuation.

which includes allowance for demolition, fees and supervision, and which in the instance of Tamar is assessed some 40% higher.

AN OVERVIEW

Above all else, experience over the 28 years of life of the bridge has demonstrated the value of long-standing relationships between the owners, operators, contractors and the engineer in executing, managing and financing maintenance works. This sharply contrasts to difficulties encountered elsewhere, arising through the alternative of regularly seeking competitive tenders for all works, where the sole criterion for acceptance is on the basis of minimum cost in the short term, and where successful tenderers have little interest and no incentive to provide a good working relationship.

Previous mention has been made of such cooperation in respect of the short gestation period between the first serious initiatives in 1955 and commencement on site in mid-1959. Subsequently the joint committee has almost invariably accepted the recommendations of the consultant, which have always been formulated after close, often informal, consultation with the joint engineers and the bridge and ferry manager. Again the joint committee has always been prepared to authorise expenditure on trials of new or unfamiliar materials and has actively encouraged use of more expensive products or procedures where these would lead to a lessening of traffic disturbance.

Over the years this rapport has extended down through the engineers to the various contractors and their operatives, and the permanent bridge staff. The toll collectors are at the 'sharp end' when any traffic problems occur and can then be subject to verbal, and sometimes physical, abuse. Invariably their reaction has been to laugh it off rather than shift the blame to the engineers and contractors.

Problems can and do arise from time to time during any works but always, within the harmonious environment that has evolved, these have been overcome. The number of persons who, through past involvement, now feel they have a personal stake in the success of the Tamar Bridge is legion.

ACKNOWLEDGEMENTS

The authors wish to acknowledge the cooperation afforded them in preparation of this paper by the joint engineers, Mr B. W. Mansell, County Surveyor of Cornwall, and Mr R. Fairclough, Acting City Engineer of Plymouth, and the bridge and ferry manager, Mr A. R. Warren, together with members of their respective staffs.

REFERENCES

1. Tolled Crossings, Second Report from the Transport Committee, Session 1985–86. House of Commons 250-1. HMSO, London.
2. Truscott, R. P. and Warren, A. R., The Tamar crossings. *Highways and Transportation*, **35**(5) (May 1988) 83–8.
3. Anderson, J. K., Tamar Bridge. *Proc. Instn Civ. Engrs*, **31** (August 1965) 337–60.
4. Moser, K., Time-dependent response of suspension and cable stayed bridges. *Proc. 8th Congress IABSE*, New York, 1968, Final Report, pp. 119–29.
5. Department of the Environment, Housing and Construction Statistics Part 2, Quarterly, HMSO, London.

16

Modelling and Predicting Bridge Repair and Maintenance Costs

MOHAMED BOUABAZ[a] and R. MALCOLM W. HORNER[b]
[a]*Department of Civil Engineering,* [b]*Department of Engineering Management, University of Dundee, Dundee DD1 4HN, UK*

ABSTRACT

Rational decisions about cost-effective bridge designs, optimum replacement ages and recurrent cost budgets are hampered by the absence of reliable data on which to base forecasts of repair and maintenance costs. Research into simple models for predicting the new-build cost of bridges has led to the development of equally simple models for predicting the costs of repair contracts exceeding £10k in value. These models, based on the principle of cost-significance, contain only 17 or 18 elements yet are accurate to within 10%. Analysis of historical data has allowed us to propose a tentative relationship between the area of a bridge deck and the cost of repairs. The results lend weight to the view that repair and maintenance costs for masonry bridges are less than those for reinforced concrete bridges.

INTRODUCTION

Repair and maintenance now account for more than 50% of the construction industry's turnover, yet very little data are available on which to base predictions of future costs. Local authorities with large building stocks have therefore considerable difficulty in justifying maintenance budgets. Nowhere is this more true than in the roads and bridges departments. In many cases historical records are incomplete or non-existent, especially before local government reorganisation. Even when records do exist they are not consistently structured. The inability to quantify maintenance and repair costs, however, has further ramifications. First, it makes the prediction of total life cycle costs impossible. Thus investment decisions are based solely on the criterion of initial capital cost, a particularly dangerous practice when the authorities responsible for capital and recurrent costs may not be one and the same party. Second, there is no rational basis on which to choose between replacing and repairing a road or bridge.

This paper describes some first faltering steps towards the identification of those elements of a bridge which give rise to the majority of repair work and towards the prediction of its cost. Eventually it is hoped that the work will lead to a better understanding of the relationship between capital and recurrent costs. Tentative relationships between deck size, repair costs and age are also reported.

PREDICTING NEW-BUILD COSTS

Recent work at Dundee University has demonstrated the feasibility of developing a simpler model of the new-build costs of bridges than those which currently exist.[1] The models are based on the principle of cost-significance. It has been known for many years that 80% of the value of a bill of quantities is contained within only 20% of the items; those which are 'cost-significant'. The cost-significant items are easily identified as those whose value is greater than the mean. The relationship between value and

TABLE 1
Cost model for feasibility design stage (bridges)
(cost model factor = 0·73)

CSWP number	CSWP description	Unit
1	Supply and driving and load testing of piles for main piling, including establishment and moving of piling equipment	Item
2	Vertical (85–90°) formwork >300 mm wide for end supports and intermediate support	m^2
3	Horizontal (0–5°) formwork >300 mm wide for deck	m^2
4	Curved formwork at any inclination >300 mm wide for intermediate supports	m^2
5	In-situ concrete	m^3
6	Precast concrete members for deck	m^3
7	Bar reinforcement	t
8	Paving in paved areas to surfaces >10° to the horizontal	m^2
9	Waterproofing on surfaces >300 mm sloping up to 45° to the horizontal	m^2
10	Supply of parapets	Lin. m

number of the cost-significant items is closely approximated by a Pareto curve.

Analysis of several hundred bills has shown that projects can be categorised in such a way that the cost-significant items within any one category are more or less the same. Moreover, the cost-significant items can frequently be grouped into packages exhibiting two important features:

(1) they closely resemble a contractor's site operation; and
(2) a single unit rate can be applied to them.

Thus, within each category of project, cost-significant work packages (CSWPs) can be identified whose value is a constant proportion (typically 80% or thereabouts) of the total project cost. This has spawned the development of a new-build cost model containing only ten elements which

TABLE 2
Cost model for detailed design stage (bridges)
(cost model factor = 0·82)

CSWP number	CSWP description	Unit
1	Establishment of piling equipment for piles in the main piling	Item
2	Moving piling plant for piles in the main piling	No.
3	Supply and driving of piles for main piling	m^3
4	Load testing of piles in main piling	No.
5	Vertical (85–90°) formwork >300 mm wide for end supports	m^2
6	Vertical (85–90°) formwork >300 mm wide for intermediate supports	m^2
7	Horizontal (0–5°) formwork >300 mm wide for deck	m^2
8	Curved formwork at any inclination >300 mm wide for intermediate supports	m^2
9	In-situ concrete for end supports	m^3
10	In-situ concrete for intermediate supports	m^3
11	In-situ concrete for deck	m^3
12	Precast concrete members for deck	m^3
13	Bar reinforcement for end supports	t
14	Bar reinforcement for intermediate supports	t
15	Bar reinforcement for deck	t
16	Paving in paved areas to surfaces >10° to the horizontal	m^2
17	Waterproofing on surfaces >300 mm sloping up to 45° to the horizontal	m^2
18	Supply of parapets	Lin. m
19	Drainage of end supports	Item
20	Imported fill deposited adjacent to structures, including around structural foundations	m^3
21	Void formers	m^2

can be used at the feasibility design stage to predict the cost of a new bridge to 10%. Refinement of the model to include an additional 11 elements allows the cost of a bridge at the detailed design stage to be predicted with an accuracy of 5%. The feasibility and design stage models for reinforced concrete bridges are shown in Tables 1 and 2 respectively.

The ratio of the value of packages in the model to the total bill value is known as the cost model factor (CMF).

On behalf of the Scottish Development Department, and in collaboration with Babtie Shaw and Morton, the model has been built into a computer package called BRIDGET which allows the price of a new bridge to be calculated in less than 15 min.[2] BRIDGET is now available to all Scottish Regional Councils.

MODELLING REPAIR COSTS

Success in modelling new-build costs encouraged us to investigate the possibility of modelling maintenance and repair costs using similar techniques. Earlier work[3] had shown that the system was most likely to work on bills of quantities containing at least 50 and preferably over 100 items. Accordingly our study was confined to repair contracts worth more than about £10k. From our experience of new-build bills we decided to divide repair work into three categories: masonry arches, masonry concrete arches (i.e. masonry arches with an infilled concrete deck) and reinforced concrete bridges. Bills of quantities were obtained from Tayside and Lothian Regional Councils. The total number of bills available were for masonry arches 14, masonry concrete arches 13 and for reinforced concrete bridges 24.

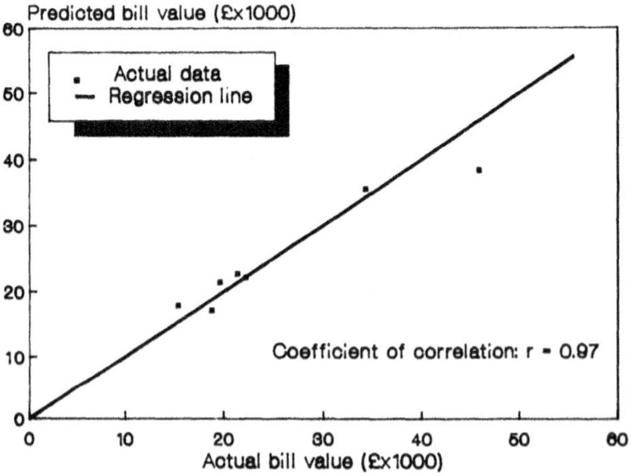

FIG. 1. Predicted versus actual bill value for masonry bridges.

TABLE 3
Unit rates for rolled asphalt wearing courses of different thicknesses

Thickness (mm)	Unit rate (£/m^2)	Unit rate (£/m^3)
40	4·10	102·50
50	5·00	100·00
60	6·24	104·00

Methodology

The methodology was similar to that used in the new-build analyses. Cost-significant items were identified for each bill, and the results for bills within each category were inspected for consistency. A variety of techniques were then used to determine the minimum number of cost-significant work packages which represented a constant proportion of the total bill value.

Changing units of measurement

In some cases it was found that the unit rate was linearly proportional to a unit of measurement different from that recommended in the Method of Measurement for Road and Bridge Works (MMRB). For instance, typical items for rolled asphalt wearing courses are shown in Table 3. The unit of measurement for rolled asphalt wearing courses was therefore changed to cubic metres.

Minor differences in unit rates

Differences in rates for items of marginally different specification were known from our earlier work to be statistically insignificant. Table 4 shows typical rates for formwork providing different qualities of surface finish.

Trial and error

The cost-significant work packages contributing the smallest average

TABLE 4
Unit rates for horizontal formwork of different classes

Class	Unit rate (£/m^3)
F1	15·28
F2	16·19
F3	16·53

proportion of the total bill value were progressively deleted from the models until a balance was struck between size and accuracy of the model. In general, an accuracy of 10% was our goal.

Rates applied to the model

The accuracy of an estimating system depends on two factors: the accuracy of the cost model itself and the accuracy of the rates applied to the model. Because in the first instance our objective was to test only the accuracy of the model, the rates used to develop and test the models were abstracted from the bills of quantities. In those cases where items with slightly differing rates were combined into a single work package the weighted mean unit rate was used.

Testing

The models were developed from an analysis for 37 bills of quantities and tested on a further 14.

RESULTS

Models

The cost-significant work packages for masonry and masonry concrete arches proved to be identical, so the two were combined into one single category, 'masonry'. The resulting models for masonry and reinforced concrete bridges are shown in Tables 5 and 6 respectively.

Accuracy

For each bill the ratio of the value of the cost-significant work packages to the total bill value was calculated. The results for masonry bridges are shown in Table 7.

The mean value of the ratio (value of CSWPs)/(total bill value) is the cost model factor (CMF). The CMF for masonry bridges is thus 0·76. Similar analysis of reinforced concrete bridge projects yields a CMF of 0·82 with a standard deviation of 0·08.

Testing

Table 8 shows the results of testing the cost model for masonry bridges on the seven bills retained for that purpose.

Figure 1 shows the results of linearly regressing predicted on the actual bill value.

For reinforced concrete bridges the mean error of prediction was 2·27%, with a standard deviation of 8·66%. The coefficient of correlation for the regression of predicted on actual bill value was 0·98.

TABLE 5
Cost model for repair of masonry bridges

CSWP number	CSWP description	Unit
1	Preliminaries	Item
2	Excavate any material except rock or reinforced concrete on bridge superstructure	m^3
3	Rolled asphalt wearing course in carriageway	m^3
4	Dense bitumen macadam base course in carriageway	m^3
5	Precast concrete	Lin. m
6	Horizontal formwork more than 300 mm wide	m^2
7	Vertical formwork more than 300 mm wide	m^2
8	In-situ concrete	m^3
9	Gunite 40 N/mm^2 to soffit and vertical surfaces	m^3
10	Repair of concrete surfaces	m^3
11	Bar reinforcement of any diameter	t
12	Tie bars of any diameter	No.
13	New masonry with battered or vertical face	m^3
14	New general random rubble masonry previously set aside	m^3
15	Existing general random rubble masonry	m^3
16	Hand pointing on arch and soffit	m^2
17	Waterproofing more than 300 mm wide	m^2
18	Dayworks	Sum

TABLE 6
Cost model for repair of reinforced concrete bridges

CSWP number	CSWP description	Unit
1	Preliminaries	Item
2	Excavate unsuitable material in flexible surfacing on bridge deck	m^3
3	Disposal of unsuitable material in tips off site	m^3
4	Rolled asphalt wearing course in carriageway	m^3
5	Dense bitumen macadam in carriageway	m^3
6	Horizontal formwork more than 300 mm wide	m^2
7	Cut out and scabble off unsound concrete in deck	m^3
8	Application of epoxy bonding aid to reinforced concrete	m^2
9	In-situ concrete in screeds	m^3
10	In-situ concrete	m^3
11	Precast prestressed beams	m^3
12	Movement joints to bridge (expansion joints)	No.
13	Waterproofing more than 300 mm wide	m^2
14	Dayworks	Sum

TABLE 7
Percentage value of cost-significant work packages for masonry bridges

Project number	Actual value (£)	Value of CSWPs (£)	Value of CSWPs / Actual value (%)
1	21 709·70	14 159·50	65·22
2	7 631·55	6 242·75	81·80
3	20 240·00	15 351·00	75·84
4	20 087·12	14 161·21	70·50
5	50 725·90	32 208·00	63·50
6	20 437·92	14 827·66	72·55
7	54 884·50	44 550·00	81·17
8	20 600·00	14 480·31	70·29
9	51 070·60	38 302·80	75·00
10	15 900·00	11 426·00	71·86
11	18 152·35	15 388·05	84·77
12	53 285·00	35 661·95	66·93
13	49 592·00	43 770·67	88·26
14	16 632·87	13 196·81	79·34
15	23 621·50	21 016·00	88·97
16	30 401·15	20 386·86	67·06
17	13 621·75	10 362·00	76·07
18	22 595·82	18 509·44	81·91
19	27 269·00	24 205·00	88·76
20	14 670·95	12 210·25	83·23
Mean			76·00
Standard deviation			8·10

TABLE 8
Model tests on masonry bridges
(CMF 0·76)

Project number	Value of CSWPs (£)	Predicted value (£)	Actual value (£)	Difference (£)	Percentage difference (%)
1	29 196·90	38 415·78	45 796·90	−7 381·12	−16·11
2	16 841·80	22 160·26	22 231·79	−71·54	−0·32
3	16 267·09	21 404·06	19 625·96	1 778·10	9·05
4	13 032·15	17 147·56	18 791·25	−1 643·69	−8·74
5	13 600·00	17 894·73	15 391·50	2 503·23	16·26
6	17 266·63	22 719·25	21 421·49	1 297·76	6·05
7	27 011·30	35 541·18	34 283·30	1 257·88	3·66
Mean					1·40
Standard deviation					10·94

DISCUSSION

The average number of items in the bills of quantities analysed was 40. The simple models developed for repair work therefore represent a reduction in model complexity of some 60%. Their accuracy, however, is still of the order of 10%. This is of the same order as the accuracy of quantity surveyors' estimates of the cost of new construction, which is reported to be about 13% (Ref. 4).

If the models are to be used to predict the cost of future repairs, the estimator will be obliged to insert a single unit rate against each item. The extent to which this is possible, and the resulting change in accuracy, are still to be tested. Nevertheless, there is reason to believe that no great loss of accuracy will ensue. Clearly the estimator will not be able to calculate the weighted mean rate for, say, formwork of classes F1 and F3, since the quantities of classes F1 and F3 formwork will not be differentiated. However, preliminary tests using the arithmetic rather than the weighted mean unit rates indicate no significant loss of accuracy. We anticipate that the central limit theorem will work to our advantage, and that positive and negative errors will cancel out.

ALTERNATIVE MODELS

It would clearly be an advantage if the cost of bridge repairs could be predicted before any detailed design was executed. Figure 2 shows the relationship between costs per square metre of deck area and deck area for the two categories of bridges studied. Best fit curves were derived from linear regression analyses of the logarithms of the variables. The coefficient of correlation for masonry bridges was −0·96 and for reinforced concrete −0·84.

The geometry of only 22 bridges was available to us, but although much more data would be required before predictions could be made with certainty, the curves do provide at least a starting point for the prediction of the cost of repairs expected to exceed £10k in value.

It is of interest to note that whilst the average age of the masonry bridges at the time of repair was 63 years compared with 18 years for the reinforced concrete bridges, the cost per square metre of repairing masonry bridges, since we know that each bridge in our sample underwent major repairs only once in its life, is no different from the cost for reinforced concrete bridges. This is further confirmation that maintenance costs of masonry bridges are less than for reinforced concrete bridges, a belief which led Tayside Regional Council to specify in 1987 the first masonry arch to be built in Scotland for more than 50 years.

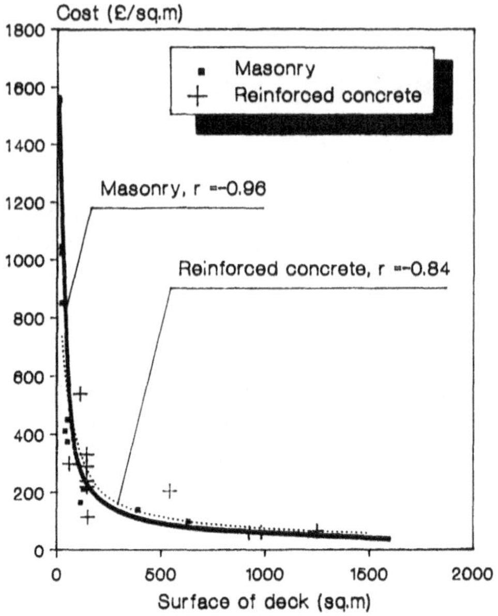

FIG. 2. Relationship between repair cost and area of deck for masonry and reinforced concrete bridges.

CONCLUSIONS

1. It is imperative that the movement by local authorities to record repair and maintenance costs on a consistent basis is sustained. Without good quality historical data it is impossible to justify repair and maintenance budgets or to make rational decisions about the cost effectiveness of new bridge designs or about the optimum time at which old bridges should be replaced.
2. Bridges can be categorised in a way which reduces by 60% the number of elements in the repair cost model and yet which still allows the cost of repairs to be predicted to an accuracy of about 10%.
3. The tentative relationship between repair cost and surface area of bridge deck proposed in the paper may provide the basis of an even simpler way of predicting the costs of repairs expected to exceed £10k in value.
4. Our data lend further support of the view that the repair and maintenance costs of masonry bridges are significantly less than those for reinforced concrete bridges. There is therefore good reason to thoroughly reappraise the relative merits of masonry and reinforced concrete for new bridges of small spans.
5. Much work remains to be done before the relationship between the age of a bridge and the cost of maintenance and repair can be reliably established.

REFERENCES

1. HORNER, R. M. W. and ASIF, M., Economical designs using simple cost models. *Proc. 4th Int. Conf. on Structural Faults and Repair*, London, 1989 (in press).
2. MURRAY, M., HORNER, R. M. W. and MCLAUGHLIN, A., BRIDGET—a cost estimating suite for highway structures. *J. Inst. Highways and Transportation* (in press).
3. DULAIMI, M. F., Towards simple contracting and estimating procedures. MSc thesis, University of Dundee, 1986.
4. ASHWORTH, A. and SKITMORE, R. M., Accuracy in estimating. CIOB Occasional Paper No. 27, 1983.

17

Bridge Operation and Maintenance Costs

Hans Ingvarsson*

Swedish National Road Administration, S-781 87 Borlänge, Sweden

ABSTRACT

Within the framework of the Swedish Commission on Maintenance and Costs (DKU), an autonomous alliance between six city authorities, a comparison study with regard to maintenance costs has been carried out by the bridge committee of the commission.

The comparison study was concerned with operation and maintenance costs, and their relation to the following parameters:

—amount of de-icing salts spread out.
—climatic and geographical conditions,
—age and size of bridge stock, and
—traffic intensity.

An explicit analytical function was found which estimates the annual operation and maintenance costs probably needed with respect to the parameters mentioned above.

INTRODUCTION

Within the framework of the Swedish Commission on Maintenance and Costs (DKU), an autonomous alliance between six city authorities, a comparison study[1] of bridge maintenance costs has been carried out by the bridge committee of the commission. Members of this committee were

*Adjunct Professor at the Royal Institute of Technology, Department of Structural Engineering, S-100 44 Stockholm, Sweden.

Messrs J. Gustavsson (Gothenburg), I. Karlsved (Västerås), A. Malmberg (Stockholm), C.-H. Silfwerbrand (Stockholm), W. Skottke (Solna) and G. Wegrell (Västerås). In order to make the cost comparison study by the committee more comprehensive, not only the street authorities of Stockholm, Gothenburg, Västerås and Solna participated in the commission but also the Swedish National Road Administration was included. The author of this paper was therefore requested to join the committee, which also acted as an advisory group to Ingvarsson and Westerberg[2] when this state-of-the-art report on operation and maintenance of bridges was prepared.

This paper comprises a brief summary of the committee report[1] mentioned above and its conclusions. This report, comprising cost statistics from 1977 to 1984, was published in 1986. In this paper corresponding statistics from 1985 to 1988 are also included on behalf of the committee.

RECORDED OPERATION AND MAINTENANCE COSTS

In the study of bridge operation and maintenance costs,[1] it was stated that the following parameters were of interest:

—Total bridge deck area, D (m^2).
—Age of bridge stock on average, A (years).
—Average number of freeze–thaw cycles per year (F). In this case F denotes the number of days during which the maximum and minimum temperatures recorded are above and below the freezing point (0°C) respectively.
—Amount of de-icing salt (sodium chloride) spread out per year, S (g/m^2).
—Traffic intensity (T), described as total amount of vehicle-kilometres per m^2 street area.
—Percentage (P) of bridge deck area constructed before 1965. This year is of special interest from the Swedish point of view as previously no air-entraining agents were used and the typical bridge concrete had a water–cement ratio of about 0·6. After 1965 this ratio was normally about 0·5 and the concrete was air-entrained. Accordingly, it can be noted that the durability of the concrete with regard to freezing and thawing was significantly increased in 1965.

For each city, as well as the Swedish National Road Administration (SNRA), the parameters listed above are shown in Tables 1(A) and 1(B). With these basic facts as a background the annual bridge operation and maintenance costs recorded, from the year 1977 until 1988, are shown in

TABLE 1(A)
Size of bridge stock and other basic facts

City	Number of bridges	Total bridge deck area, D (m^2)	Age on average, A (years)	Number of freeze–thaw cycles per year (F)
Solna	83	81 000	18	37
Västerås	177	57 000	19	30
Gothenburg	580	260 000	20	34
Stockholm	745	677 000	27	37
SNRA[a]	11 600	2 780 000	27	37[b]

[a] SNRA = Swedish National Road Administration.
[b] Average value for the whole of Sweden.

TABLE 1(B)
Size of bridge stock and other basic facts

City	Amount of sodium chloride spread out per year, S (g/m^2)	Traffic intensity, T (vehicle-kilometres per m^2 street area)	Percentage of bridge deck area constructed before 1965 (P%)
Solna	260	240	9
Västerås	200	90	37
Gothenburg	765	170	22
Stockholm	600	230	40
SNRA[a]	220	60	67

[a] SNRA = Swedish National Road Administration.

Fig. 1. In this figure all costs are adjusted in order to correspond to the 1988 price level.

DISCUSSION

From Fig. 1 it is rather difficult to draw any firm conclusions concerning the maintenance costs as these vary considerably because of different circumstances. In order to solve this problem it may be convenient to define a bridge maintenance cost index as described below.

The recorded operation and maintenance costs are primarily due to maintenance and repair of concrete bridge deck slabs suffering from waterproofing systems no longer being watertight. Consequently, if the

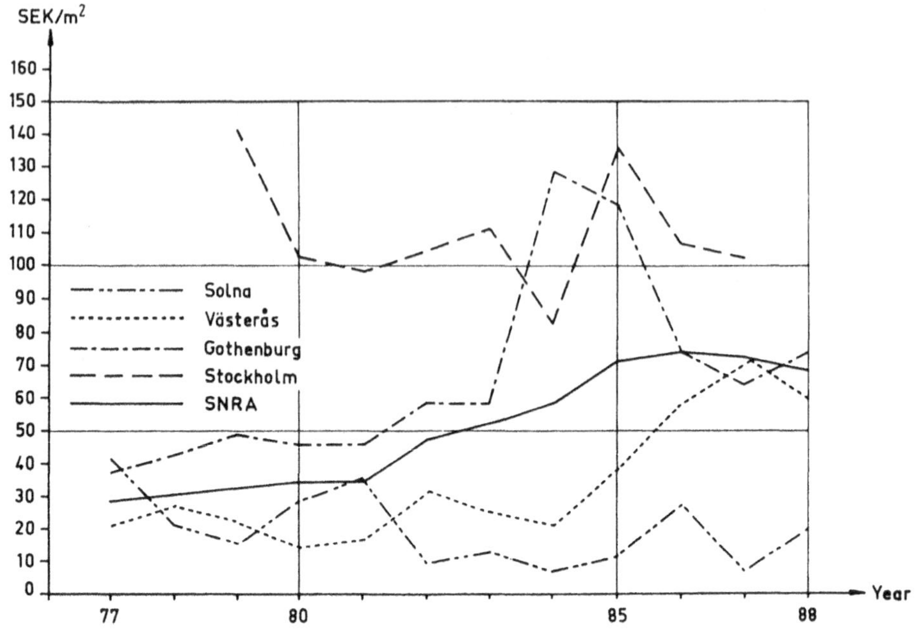

FIG. 1. Annual maintenance and operation cost per m^2 bridge deck area (SEK/m^2) for each city between 1977 and 1988 (1988 price level, 1 SEK ≈ 0·15 US$).

concrete is not air-entrained, it will be damaged by freezing and thawing. Bearing this typical deterioration mechanism in mind, the following bridge deterioration index (B_{DI}) is proposed in (1):

$$B_{DI} = AFSTP/100 \tag{1}$$

where A, F, S, T and P are defined as described above. In Table 2 the resulting bridge deterioration index for each city as well as that of the Swedish National Road Administration (SNRA) are shown. In order to make it possible to compare this index with the recorded operation and maintenance costs the average costs (C_A) valid for 1983–87 are also shown (see Table 2 and Fig. 1).

As found by DKU,[1] the expected annual maintenance cost (SEK/m^2 bridge deck area) can be calculated as follows:

$$B_{MCI} = 72{\cdot}5\sqrt{\log B_{DI} - 6{\cdot}45} \tag{2}$$

This cost refers to the 1984 price level. In order to correlate it to 1988 it must be increased by 15%, obtaining the bridge maintenance cost index (B^*_{MCI}):

$$B^*_{MCI} = 83{\cdot}4\sqrt{\log B_{DI} - 6{\cdot}45} \tag{3}$$

As can be seen from Table 2, where this index is shown, the bridge

TABLE 2
Bridge deterioration and maintenance cost indices

City	B_{DI} ($\times 10^6$)	B^*_{MCI} (SEK/m^2)	C_A (SEK/m^2)	C_A/B^*_{MCI}
Solna	3·7	28·7	12·5	0·44
Västerås	3·8	30·0	42·7	1·42
Gothenburg	19·4	76·3	88·8	1·17
Stockholm	55·1	94·8	110·6	1·17
SNRA	8·8	58·6	65·7	1·12

B_{DI} = Bridge deterioration index (eqn (1)).
B^*_{MCI} = Bridge maintenance cost index (eqn (3)).
C_A = Average recorded operation and maintenance cost 1983–87.

maintenance cost index (SEK/m^2 bridge deck area) roughly represents the need for annual funding for operation and maintenance of bridges, if the recorded costs are regarded as the proper level of funding.

The results obtained by DKU[1] can probably be used by other local or national authorities in the financial planning of operation and maintenance of bridges in other municipalities. This can be done either by strict use of the bridge maintenance cost index (B^*_{MCI}) or by modification thereof if the typical bridge deterioration is different from that in Sweden. If, for example, the major problem being dealt with is reinforcement corrosion, the parameter P can be defined based on a critical age or service life with respect to this phenomenon. Furthermore, the parameter F is perhaps irrelevant in this case.

CONCLUSION

Through the cost comparison study[1] carried out by the bridge committee of the Swedish Commission on Maintenance and Costs (DKU) an explicit analytical function was found which estimates the annual operation and maintenance costs probably needed with respect to the following parameters:

—total bridge deck area,
—age of bridge stock on average,
—average number of freeze–thaw cycles per year,
—amount of de-icing salt spread out per year, and
—traffic intensity.

The results of this study can thus probably be used by other local or national authorities in the financial planning of operation and maintenance of bridges in other municipalities.

REFERENCES

1. DKU/7, Driftkostnadsutredningens konstbyggnadsgrupp, Specialrapport Maj 1986. Kommunförbundet, Stockholm (in Swedish).
2. INGVARSSON, H. and WESTERBERG, B., Operation and maintenance of bridges and other bearing structures. Publ. No. 42 from the Swedish Transport Research Board, Stockholm, Sweden, 1985 (in English).

18

Clifton Suspension Bridge: An Historic Monument that Earns its Keep

DAVID MITCHELL-BAKER

Howard Humphreys and Partners, Thorncroft Manor, Dorking Road, Leatherhead, Surrey KT22 8JB, UK

and

STUART CULLIMORE

Clifton Suspension Bridge Trust, 66 Queens Square, Bristol BS1 4JB, UK

ABSTRACT

Completed in 1864, the Clifton Suspension Bridge now carries annually over 3·6 million vehicles at speeds not then envisaged. It is as much an important communications link as part of Bristol's heritage. As the integrity of the design, materials and construction have become more fully understood and appreciated, the Clifton Suspension Bridge Trust have evolved a policy of, in effect, indefinite preservation. The main historic events of damage and repair are described, together with various tests and analyses which have encouraged the trust to their present policy. The origins and organisation of the trust itself are also of interest, offering a solution particularly suited to such a responsibility as this bridge.

INTRODUCTION

The foundation stone to a bridge designed by Isambard Kingdom Brunel was laid in 1837, almost a century after a bequest of £1000 by William Vick in 1753. It remained unfinished at Brunel's death in 1859 and shortly after the Institution of Civil Engineers led a commercial initiative to complete it to remove 'what was considered a slur upon the engineering talent of the country'. They appointed Hawkshaw and Barlow to design and construct a bridge on Brunel's abutments using ironwork, bought at a favourable price, from Brunel's footbridge across the Thames at Hungerford. The Clifton Bridge was finally opened in 1864 and was operated by a company financed by tolls. After 80 years, accrued dividends were used to redeem the shares and the bridge was vested under the 1952 Clifton Suspension Bridge Act[1] in the present Clifton Suspension Bridge Trust, comprising up to 12 trustees. One trustee is appointed by each of the riparian local authorities—currently Avon County, Bristol City and Woodspring District; the remainder have to reside within 20 miles of the bridge. They bring to the direction of its affairs a wide range of responsible experience in management, commerce, engineering and public service.

Because of the bridge's historic importance, its basic integrity and a growing interest in its conservation, the trust has developed a policy of indefinite preservation. The objectives of this policy may be summarised as follows.

Technical

Maintain the bridge to a high standard by

—regular and frequent inspection,
—careful protection against corrosion and rot,
—protection against overload, impact and fire,
—timely and careful repair and maintenance,
—analysis and testing using latest techniques, and
—application of experience from elsewhere.

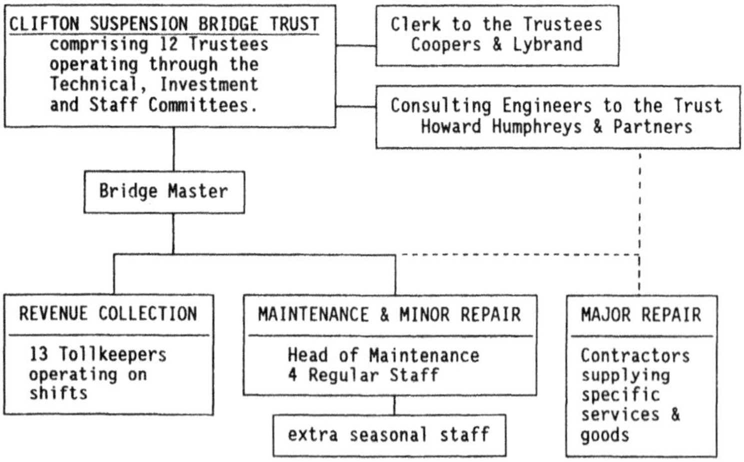

FIG. 1. Organisation of Trust.

Financial

Provide adequate resources to maintain and replace by

—efficient toll collection,
—effective financial and investment management, and
—insurance against external risks.

The organisation set up by the trust to achieve these objectives is shown in Fig. 1.

EARLY HISTORY

The construction of the bridge, which skilfully incorporated the chains and saddles from the original Hungerford bridge, is well described by Barlow.[2] It has been possible, without serious problems, to keep what is virtually the original structure in a serviceable condition because of the inherent soundness of the original design and construction procedures, the low stresses in most members, and the high quality and nature of the wrought iron.

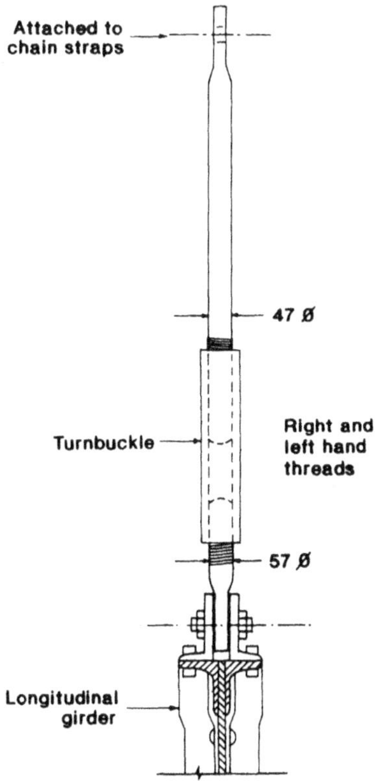

FIG. 2. Suspender rod connection to longitudinal girder.

The only serious damage ever reported was caused by extreme wind conditions in 1877 and again 10 years later, when a total of five suspender rods failed between the turnbuckle and eye connection to the longitudinal girder (Fig. 2).

The transverse timber planks on the deck were replaced in 1884, in 1897 when the deck was first surfaced with mastic asphalt and again in 1948. Some longitudinal baulks were replaced at various times and the whole deck was replaced in 1958. Concern about the strength of the anchorages led to reinforcement with additional chains in the Leighwoods anchorages in 1925 and in the Clifton anchorages in 1939. Corrosion of the chain links at ground level in the days of horse-drawn transport made necessary strengthening at the land saddles in 1932.

These and subsequent major works are more fully described by Mitchell-Baker and Cullimore.[3]

THE POST-WAR PERIOD

Use during the Second World War by heavy vehicles, over the 6-t limit set under the 1861 Act, exacerbated damage to the already deteriorating deck surfacing and timbers. Continuing restrictions on labour to maintain the bridge and the shortage of replacement timber led Howard Humphreys & Sons, who had been responsible since 1910 for engineering inspections of the bridge, to place limits of $2\frac{1}{2}$ t per axle and 4 t maximum vehicle load on traffic but keeping the overall distributed load of 28 t unchanged from the 1864 value. The former was based on the poor conditions of the deck timbers and was vigorously resisted by the company. It was at this stage that the present trust assumed responsibility for the bridge and ordered the major inspection which was carried out in 1953 and supported by an extensive testing programme, which has been described by Flint and Pugsley.[4] Concern over the extent of corrosion of the lattice cross-girders resulted in the end lattice girders, which were the worst affected because of inaccessibility of the back face, being replaced by rolled steel joists. These lattice girders were then tested to failure in the laboratory at Bristol University, with the result that the earlier fears regarding loss of strength from corrosion were allayed. The extent of corrosion on the remaining cross-girders, which with the timber had been coated with pitch, was easily determined when they were grit blasted in 1955 by Bristol Metal Spraying Company preparatory to zinc metal spraying. The extent of corrosion was found to be less than expected and not to be significant.

The trustees then considered alternatives to timber for the deck. These included the concept of an aluminium deck which would have eliminated the problem of providing a waterproof running surface on the timber deck and so remove the problem of rot and generally reduce maintenance. The resulting reduction in self-weight of the deck might have allowed an increase in the permitted vehicle weight. This was, however, rejected in favour of an improved timber deck, without increasing the permitted traffic loading. A number of important benefits have resulted from this decision—repairability by locally available building industry tradesmen has been retained and the character and original design of the bridge has not been compromised. It is believed that the damping imparted by interface friction in the heavy timber deck, by limiting vibration under dynamic loading caused by wind and traffic, reduces fatigue damage to the ironwork.

A small number of longitudinal baulks and cross-planks on the Leighwoods end of the deck, thought to have been damaged by water penetrating through the mastic asphalt, were replaced in 1988. In relaying this section of the surfacing a polymer-modified mastic asphalt was used. It was considered that this material would be more resistant to the continual flexing of the deck under traffic and assist in maintaining a waterproof seal.

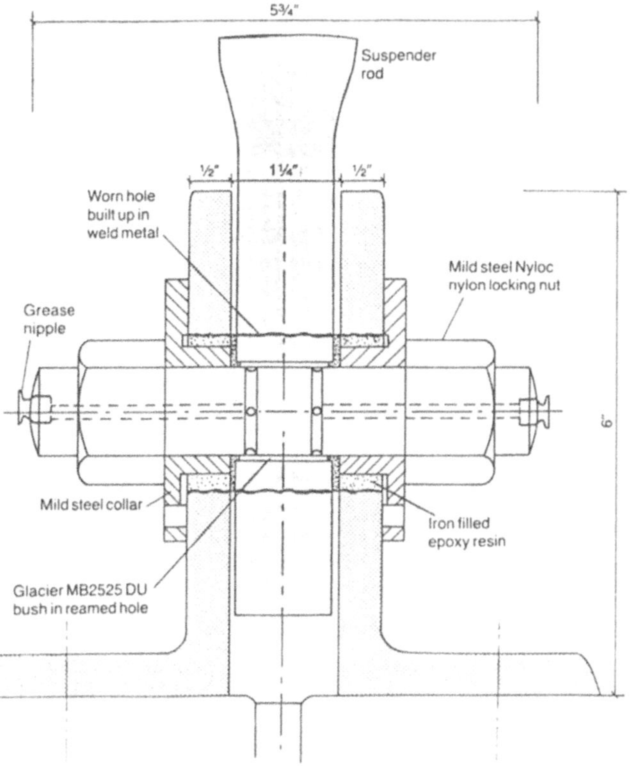

FIG. 3. Revised bolt assembly on suspender rod.

Longitudinal differential movement between the chains and the deck is taken by bending of the longer suspender rods and by rotation at the upper and lower bolts of the shorter ones. The resulting wear in the bolts, rod eyes and suspender straps which led to replacement of bolts at various times had by 1970 become serious in the shorter rods. Following extensive experiments the revised arrangement, shown in Fig. 3, was adopted for fixing on the 33 shorter rods on each catenary spanning the centre of the bridge.

An important, and unexpected, environmental benefit has resulted from the retention of the weight restriction. This limits traffic to private cars and light commercial vehicles, compelling heavier through traffic to use the alternative Cumberland Basin and other routes. The 'Clifton Village' has therefore been largely spared the effect of such traffic, a significant factor in maintaining its character.

OPERATION

Toll collection provides an effective means of traffic control, limiting the total number of vehicles on the bridge and, with a weighbeam system identifying overweight vehicles, is a crucial factor in preserving the bridge. The automated toll collection system installed in 1975 was based on widely used car park control barrier equipment, using electromechanical coin rejector 'slug' mechanisms. These were very satisfactory as they could operate in exposed conditions, so avoiding unacceptable canopies near the listed structure. As electronic replaced electromechanical devices, severe operating problems developed. New coinage with recessed faces, as opposed to embossed detail, adhered to plastic chutes in the presence of even light rain. Minor dimensional variations in coins, such as between earlier silver and later cupronickel 10p pieces, led to high rejection rates and ingress of small amounts of dirt and foreign matter caused unacceptably frequent cleaning of chutes.

Festoon lighting has illuminated the bridge for many years. Originally used only for festive occasions, it is now provided every evening as a contribution to local civic amenities. The life of the tungsten robust filament lamps used in a horizontal position and subjected to vibration in these exposed conditions is well below the rated value. The labour costs for replacing lamps, particularly in positions in the upper parts of the structure, is high and forms a significant part of the overall maintenance labour costs.

In recent years there has been a considerable investment in new equipment to improve safety and increase efficiency of maintenance operations. This has included two demountable cradles to traverse the chains and these and the cradle for underbridge inspection are now electrically driven. Steel cables on each chain for use with a running harness now greatly improve safety for those inspecting and painting the chains. Gantries and cradles for work on faces of towers and abutments have been improved for safety of operatives and the structure. Improvements in access arrangements to the anchorage chambers are in progress.

TESTING AND ANALYSIS

The consulting engineers, in advising the trustees on major repairs and maintenance, have been able to base their recommendations on extensive programmes of testing and analysis commissioned by the Trust. The first of these, reported by Flint and Pugsley,[4] comprising the load testing of the bridge deck together with tests to destruction of two cross-girders removed from the bridge, followed the major survey of the bridge in 1953. It yielded information about the structural action of the superstructure which confirmed the interim weight limit and was useful in designing fixings for the new timber deck.

A thorough investigation of the foundations of the masonry abutment on the Leighwoods side was ordered in 1969 as a precautionary measure because of reported 'slips' in other parts of the Avon Gorge and of the geological conditions which were known to exist in the area. Such possible risks as were found to exist were extremely small and considered to be negligible.

An appraisal of the risk of fatigue failure, initiated in 1972, was later extended to examine susceptibility to brittle fracture and, subsequently, to the propagation of existing cracks by fatigue. The work, extending over some 12 years, is described by Cullimore and Mason.[5] It included strain and deflection measurements on the bridge under controlled loading, laboratory testing of specimens and of full-size replicas of elements of the structure and a fracture mechanics-based research programme supported by the Science and Engineering Research Council.

The most fatigue-sensitive element of the structure was found to be the eye of the chain link joined to the tower saddle and regular inspection of this area by fibre-optic endoscope was instituted. It was concluded that there was a satisfactory margin of safety against fatigue crack initiation in the chain eye and a negligible risk of the propagation of a pre-existing crack. In reaching this conclusion account was taken of the compressive residual stresses resulting from the initial proof loadings of each chain link to over thrice its design load and, after construction, by 500 t of stone distributed over the deck. Encouraged by these results, and bearing in mind the increasing pressure of conservation interests, the Trust has adopted the policy based, in effect, on achieving indefinite preservation described above.

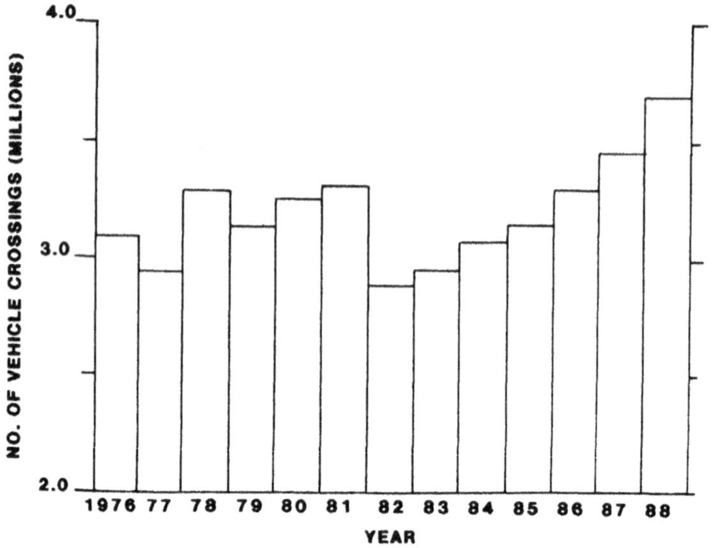

FIG. 4. Vehicle crossings 1976–88.

The traffic over the last 13 years, during which the toll equipment has permitted accurate records to be kept, is shown in Fig. 4 and reflects the growth in size and population of the local catchment areas and of vehicle use generally. The drop in 1981 coincides with the opening of the Avonmouth Bridge on the M5 to the west of Clifton. Prior to this the trend was upward, the peaks corresponding to periods when the A4 portway was closed for remedial works necessitated by rock falls in the Avon Gorge. There were 3·7 million crossings in 1988, up 7% on 1987 and up 12% on 1986.

These increases, and particularly the peak hour throughputs, now over 1200 vehicles in one lane, have causes the trustees to examine the adequacy of the traffic control in these conditions. Consequently Howard Humphreys have undertaken a large deflection analysis of the bridge using the LUSAS finite element program. This method of modelling the behaviour of the structure gives a better assessment of the actions in its component elements than was previously possible. This is confirmed by the comparisons (Table 1) with Flint and Pugsley's 1953/54 measurements. It is noted that the linearised deflection theory calculations, which although not as close, still give reasonable values. Having obtained a satisfactory calibration of the model with isolated loads, the most severe lengths of distributed live load have been identified and a quasi 3D analysis has been adapted to model the more heavily loaded of the two chain/girder systems under eccentric loading. All analyses are being carried out with total working loads and their effects will be assessed against the strength, rotation and deflection capacities of the relevant components of the structure.

TABLE 1
Comparison of measured and calculated deflections

Location of symmetrical load	Deflection measured at	Deflection in main girders caused by $10\frac{1}{2}$-t vehicle (mm)		
		Large deflection analysis	Flint and Pugsley 1953/54	
			Measured	Calculated
$\frac{1}{4}$ span	$\frac{1}{4}$ span	81	84	78
$\frac{1}{4}$ span	$\frac{3}{4}$ span	55	55	62
mid-span	mid-span	58	66	54

CONCLUSIONS

Because of its splendid setting and its historical associations with Brunel the Clifton Suspension Bridge has become an object of local pride and national interest. It is a tourist attraction and a symbol for the promotion of Bristol commerce. Its importance as a working element of Bristol's communication system is confirmed by the large and steadily increasing usage. For these reasons it is incumbent on the trustees to keep the structure fully functional and, as far as possible, in its as-built condition. Therefore in repairing or replacing a component its form is retained and the original material used whenever possible. Structural quality wrought iron is no longer obtainable and mild steel is used instead. Similarly, in repairing a joint, bolts might have to replace rivets but arc welding would not be used.

In the event of major structural damage, making the bridge unserviceable, it is considered that conservation interests would strongly favour the replacement of the bridge in its present form. Consequently the trustees have insured the bridge against accidental damage in a sum which would provide for building a replica using modern materials.

The trustees are able to practise such policies because they have financial autonomy and, unlike a public authority, they lack competing demands on their resources. Income is buoyant and capital reserves, although insufficient to build a new bridge—a duty laid on the original trustees—are adequate to provide a very high standard of maintenance and to carry out major repairs.

The prestige of this part of the local and national heritage attracts as trustees persons who are able to bring to its affairs a high level of relevant expertise and, being locally resident, are inescapably accountable for the engineering and environmental effects of their decisions. Over the years therefore an active and responsive organisation has evolved to manage and maintain this unique civil engineering structure, providing an interesting example of how an important, useful and historic structure, taken out of direct public control, may be effectively maintained for public benefit and in the public's interest.

ACKNOWLEDGEMENTS

The authors wish to thank the trustees of the Clifton Suspension Bridge for allowing them access to the Trust's records and Howard Humphreys and Partners for assistance in preparing the paper. The views expressed in the paper are wholly those of the authors.

REFERENCES

1. Clifton Suspension Bridge Act 1952.
2. BARLOW, W. H., Description of Clifton Suspension Bridge. *Mins Proc. Instn Civ. Engrs*, **26** (1867) 243–57.
3. MITCHELL-BAKER, D. and CULLIMORE, M. S. G., Operation and maintenance of the Clifton Suspension Bridge. *Proc. Instn Civ. Engrs, Part 1*, **84** (April 1988) 291–308.
4. FLINT, A. R. and PUGSLEY, A. G., Some experiments on Clifton Suspension Bridge. Correlation between calculated and observed stresses and displacements in structures. Institution of Civil Engineers, London, 1955, Preliminary Vol., pp. 124–34.
5. CULLIMORE, M. S. G. and MASON, P. J., Fatigue and fracture investigation carried out on Clifton Suspension Bridge. *Proc. Instn Civ. Engrs, Part 1*, **84** (April 1988) 309–29.

19

A Systematic Approach to Future Maintenance

A. van der Toorn and A. W. F. Reij

Department for Structural Research, Ministerie van Verkeen en Waterstaat, Rijkswaterstaat, PO Box 20.000, 3502LA Utrecht, The Netherlands

ABSTRACT

If bridge management is the art of ensuring a good connection between two opposite sides, the technical aspects are just a small (but essential) part.

One of these technical aspects is the prediction of the future behaviour of the structure in the light of ageing mechanisms, and a second related aspect is how to react to this in the form of preventive or corrective maintenance measures.

This paper deals with models which take into account the ageing of the structure, the possible consequences of failure and the maintenance required to return the bridge back to an acceptable condition.

INTRODUCTION

Up until 10 years ago the maintenance of civil engineering structures was a low-profile requirement based on experience without any theoretical background and executed by technician engineers working with small financial budgets.

Forced by the fast-growing number of structures which have fundamental and costly maintenance problems, nowadays there is a strong push to upgrade our knowledge about maintenance or, even broader, to have a good overall bridge management system in which all relevant factors are presented and can be weighted to assist in taking the right decisions.

In Holland the Rijkswaterstaat (State Public Works) is responsible for the maintenance of about 4000 structures that vary from simple viaducts to huge infrastructural works like tunnels, bridges across the River Rhine and storm surge barriers along the North Sea.

All these structures together represent a replacement value of about 10 billion US$. Assuming a maintenance budget of about 1% of the original cost, the future maintenance budget will be of the order of 100 million US$.

With the aid of a good bridge management system this amount of money will be spent rationally, so that short-term economies do not give long-term problems.

BRIDGE MANAGEMENT

In general, bridge management should contain the following basic topics:

1. Registration and description of the individual structures.
2. Initial determination of their actual condition.
3. Judgement of (some) structures in terms of safety, etc.
4. Prediction of their future behaviour resulting from ageing.
5. Definition of specific maintenance strategies.
6. Allocation of the (limited) budget to specific structures.
7. Execution of these strategies with adequate men and means.
8. Registration and evaluation of findings.
9. Back coupling with maintenance strategies and design.
10. Scenarios for the replacement of structures for economical, technical or other reasons (increased traffic, new demands).

Although for the management of 4000 structures each of these ten topics is a problem in itself, in this paper particular attention will be given to points 4 and 5.

MAINTENANCE MODELS

Maintenance models are intended to predict the time and scope of future maintenance actions such as inspection, repair or replacement. In the first instance these actions follow from an historically based maintenance strategy, but later these models give the opportunity to come up with a more or less optimised maintenance strategy.

Maintenance models need the following structure-related inputs:

1. A characterisation of the actual condition.
2. A prediction of future behaviour.
3. An estimation of maintenance-related costs such as inspection, repair, replacement, loss of production and damage.

With the help of maintenance models the life cycle cost for a (historically) given maintenance strategy can be determined.

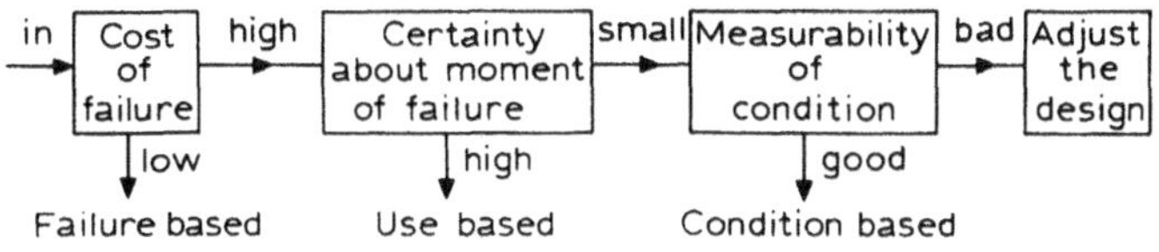

FIG. 1. Qualitative decision tree for maintenance strategies.

The right choice of maintenance strategy (see Fig. 1) and, within that strategy, the right adjustment of steering variables (inspection intervals, action boundaries, cluster size and time of replacement) make it possible to optimise the strategy by minimising the cost.

Depending on the period of use, the cost can be expressed as cost in a fixed period of time, per life cycle, per unit of time or as 'market value'.

Sometimes decisions about the maintenance strategy are not based on a pure economical weighting but also on external demands such as a minimum serviceability level, regulations, aesthetics, etc.

So far on this global level all maintenance models are the same, but as soon as one is predicted for specific categories of structure the models diverge. A basic difference occurs between the civil technical and electromechanical parts of the bridge structure.

Civil technical parts are long-term (50–100 years), usually complete elements, and their lifetime can only be estimated if the underlying ageing mechanism is known and described parametrically.

Electromechanical parts have a medium life span (10–50 years) and are more or less standard components with a known rate of failure but without a precise or measurable underlying ageing mechanism.

So the future behaviour of these two categories of construction are organised differently in the maintenance models (see Fig. 2).

For both types of component (civil technical and electromechanical) there

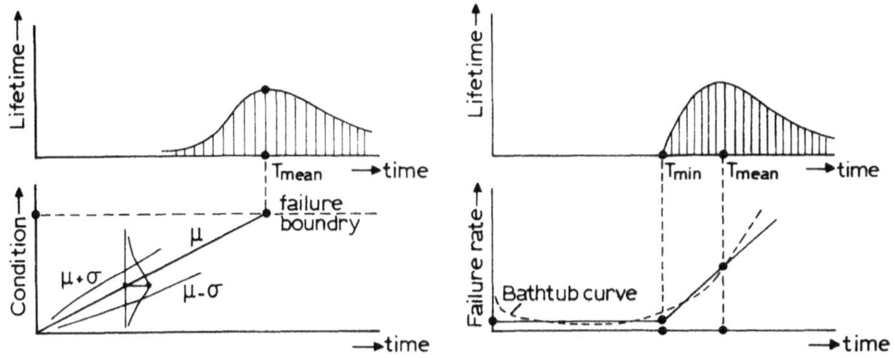

FIG. 2. Ageing of civil and electromechanical parts.

are now computer models which give the life cycle cost for a single ageing element which has only one renewal maintenance action (see Appendix).

Although sometimes the degeneration of a structure can be simplified to that level (by considering only dominant mechanisms), most of the practical situations are more complicated:

1. There are more different elements in one structure.
2. There is more than one ageing mechanism working.
3. The ageing mechanism is not directly affecting the function of the element but has two stages.
4. Malfunctioning of an element does not always lead to malfunctioning of the system as a whole because of redundancy.
5. There is more than one repair action possible, varying from a limited local repair to total replacement of the element or structure.

When there are more ageing elements in a structure, the maintenance strategy for the structure as a whole will normally differ from the sum of the individual maintenance strategies. On the one hand, inspection and repair actions clustered together will give lower cost (less mobilisation cost, less production loss, etc.), but alternatively individual elements are not always optimally maintained.

In many civil technical structures the load-bearing construction is protected against the aggressive environment by some form of protection layer. For example, coatings on a steel structure, the concrete cover on the reinforcement and the asphalt wearing layer on a bridge deck provide protection. The necessity of maintenance in the first stage of ageing is not because of the threat of production loss or the cost of damage but for prevention of extra maintenance cost in the next stage.

Although for a single element an optimal repair cycle can be found, the way in which an individual element which is part of a system is maintained depends on the redundancy, the condition of the accompanying elements and the savings that can be achieved by joint action. To find the real optimum is very difficult.

APPLICATION

To check the validity of maintenance models and assess their shortcomings, an attempt was undertaken to apply the models on a 25-year-old steel bridge crossing the River Rhine (see Fig. 3).

First of all, an analysis was undertaken considering the present maintenance costs. After a considerable amount of work the cumulative

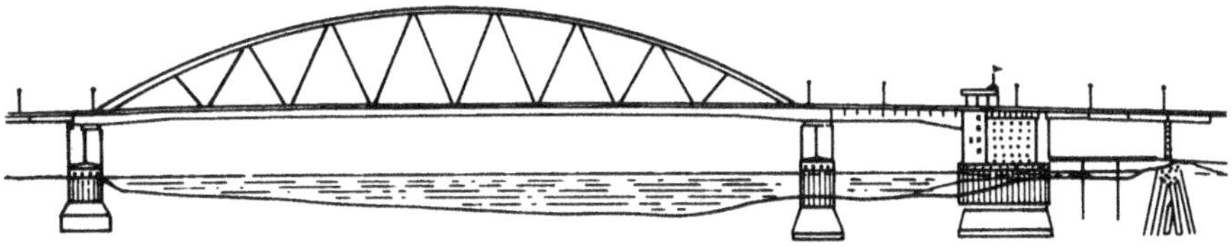

FIG. 3. Bridge used for maintenance analyses.

maintenance level seemed to be between 1% and 2% of the initial cost of the structure (see Fig. 4).

A second analysis was done of the different ageing mechanisms relative to the amount of money spent to maintain the structure. In the top 12 maintenance categories, the repainting of the bridge was first, followed by repair of the wooden deck adjacent to the movable part of the bridge. The third highest category was replacement of the expansion joints.

If the use of maintenance models can reduce cost, it is clear that the most benefit can be expected for the highest absolute contribution to the total maintenance cost. It was therefore decided to first model the painting of the bridge.

Ageing Mechanism

Considering a small area of the bridge structure, the ageing mechanism consists of two stages. First, the degeneration of the paint layer occurs without further consequences for the functioning of the bridge. Secondly, the corrosion of the underlying steel structure results, which leads to a decrease in structural safety.

Considering the total bridge structure, elements with different degradation velocity can be distinguished. There are three main parts with their own paint system:

1. The carriageway
2. The arches
3. The basculing bridge

} Total bridge area 80 000 m^2.

Within the main parts degradation differences are caused by: sharp edges and bolt nuts have minimal layer thickness; vertical web plates can easily

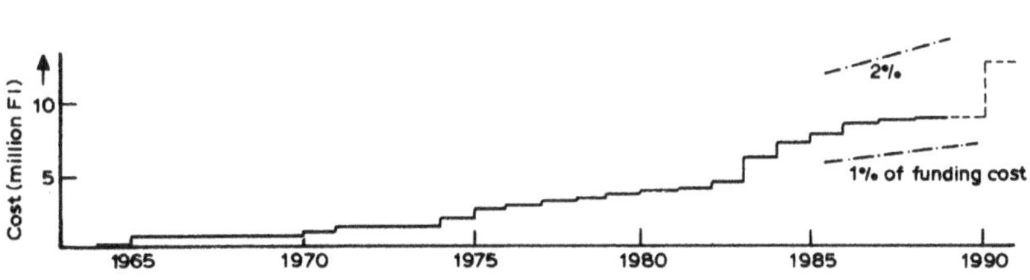

FIG. 4. Development of the maintenance cost.

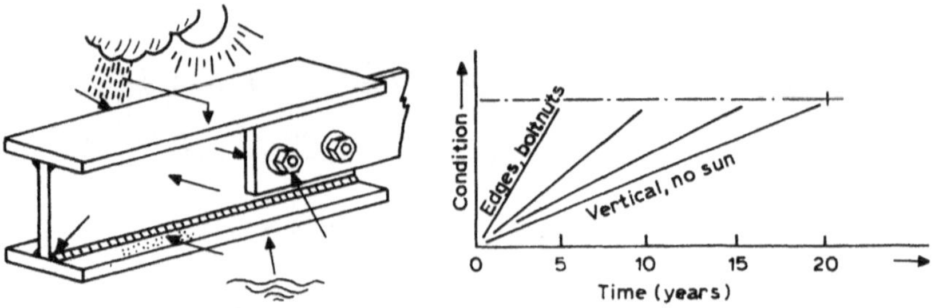

FIG. 5. Elements with different degradation velocity.

dry; horizontal plates can accumulate (salt)water; welded joints have less suture, etc. (see Fig. 5).

Actual Condition

Although there are some techniques for measuring the initial condition (layer thickness, etc.), in practice the condition or damage parameters in the second stage are used, namely the percentage of corroded area and if needed the loss of material in a cross-section.

Maintenance-related Cost

The cost related to the painting of the bridge consists of the cost of scaffolding, inspection, (blast) cleaning if a certain degradation is detected, cost of environmental measures, the paint itself, application, the processing of the waste and the replacement of steel parts if considerable degradation is detected.

Maintenance Model

Because of the number of different elements and the different but correlated ageing mechanisms considered, the maintenance model used here was based on Monte Carlo simulation.

Inspection of the condition (corrosion percentage and loss of steel) could take place after a 'certain' period of time. Local painting was prescribed if inspection of an element gives a 'certain' percentage of corrosion, and total painting was prescribed if the group of elements which needed painting exceeded a 'certain' level.

Replacement of heavily corroded construction parts took place if a 'certain' safety level was exceeded.

By varying the 'certain' values of the steering variables in the Monte Carlo simulation a few optimal strategies were selected (see Fig. 6).

For the first strategy the aim is to prevent the high cost of (blast) cleaning and cost of other measures by means of frequent inspection and considerable local intermittent painting.

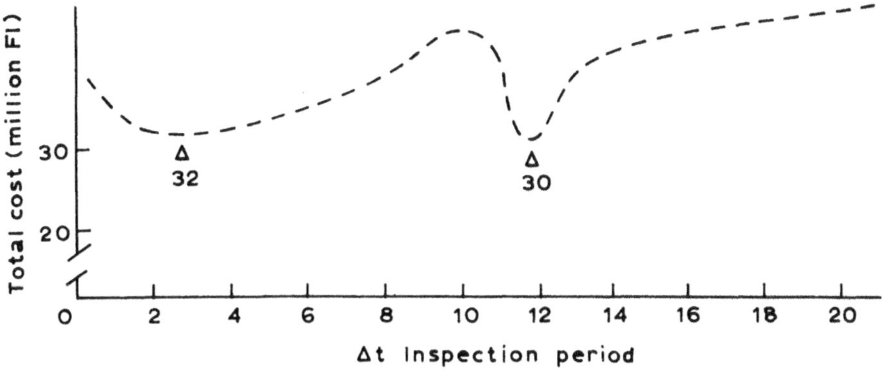

FIG. 6. Total maintenance cost versus inspection period.

The second strategy is directed at prevention of the replacement of steel parts by means of low cycle inspections coupled with the total painting of the bridge, including blast cleaning, etc.

Although the second strategy is somewhat better, there is only a small difference between the schemes. The cumulative effect of steel loss during the lifetime of 100 years is not critical.

CONCLUSIONS

1. Modelling of maintenance procedures to optimise a maintenance strategy for a given structure is possible but still needs to be extended when there are more elements and ageing mechanisms.
2. Maintenance models force decisions to be made which were not previously considered by inspectors.
3. Maintenance models require condition parameters as input data which demand good measuring instruments and protocols.
4. Maintenance models demand the consideration of maintenance alternatives and require consideration of their price and period of effectiveness.
5. Maintenance models allow consideration of the consequences of bad maintenance, or alternatively 'doing nothing'.

APPENDIX: MAINTENANCE MODELS

On the basis of a knowledge of the undisturbed future behaviour of an element in a structure (input by means of ageing mechanisms or failure

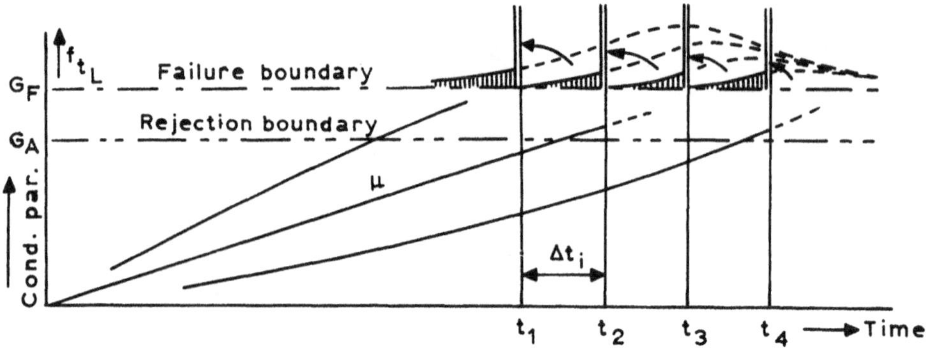

FIG. A1

rates), these models account for the probability of failure and rejection within certain time intervals, depending on the maintenance strategy.

Failure-based maintenance

$$E(c) = (C_v + C_s)/t_L$$

where

$$t_L = t_{mean}$$

Use-based maintenance

$$E(c) = (C_v + C_s \times P_f)/t_L$$

where

$$P_f = \sum p_f(i) \quad \text{for } i = 0 \rightarrow t_0$$

and

$$t_L = \sum \{p_f(i) \times i\} + \{1 - \sum p_f(i)\} \times t_0 \approx t_0$$

Condition-based maintenance

$$E(c) = (C_v + n \times C_i + C_s \times P_f)/t_L$$

where

$$t_L = \sum \{p_a(i) + p_f(i)\} \times i \quad \text{for } i = 0 \rightarrow \infty$$

in which

C_v = the cost of repair or replacement of the construction (part),
C_i = the cost of an inspection,
C_s = the cost of damage (direct or indirect),
$E(c)$ = the cost assessment parameter,
$p_f(i)$ = the probability of failure in the ith interval,
$p_a(i)$ = the probability of rejection in the ith interval,
t_L = the expected value of the lifetime,
t_{mean} = the lifetime in case of mean ageing,
t_0 = the previously determined moment of maintenance, and
n = the number of inspections during the lifetime.

20

Management of Bridgeworks Maintenance in the UK

N. J. SMITH

Project Management Group, Department of Civil and Structural Engineering, University of Manchester Institute of Science and Technology, PO Box 88, Manchester M6O 1QD, UK

ABSTRACT

This paper reviews current methods of managing the process of bridgeworks maintenance in the UK at a time when there is a national movement in the allocation of construction funds away from new works towards maintenance. Bridgeworks maintenance has been the subject of particular attention in recent years with the combined effects of increased deterioration in recent concrete structures and the recent increases in the permissible loadings for highway vehicles. These demands outstrip the available budget and hence many owners of bridgeworks stock have adopted formal bridge management systems and cooperated in research programmes.

INTRODUCTION

Bridges form a key part of the infrastructure of the United Kingdom (UK), which facilitates the movement of people, goods and services vital to the national economy. The total UK bridgeworks stock, probably one of the most diverse bridge stocks in the world, is estimated to be about 150 000, of which about 8900 are the responsibility of the Department of Transport (DTp)[1] and 129 000 the responsibility of the local authorities. In general terms, the stock consists of about 70 000 masonry and brick arch and culvert structures, mostly constructed prior to 1922, and about 60 000 concrete and 25 000 metal bridges, largely constructed post-1922. Over 90% of the bridgeworks stock consists of small bridges with a span of less than 10 m, although due to the motorway and trunk road networks the majority of the major structures are within the DTp stock. The major bridges tend to have more serious maintenance problems and in terms of percentage of initial capital cost require higher maintenance expenditure.[2]

Bridges have traditionally been designed to operate for long periods of time without requiring major maintenance, indeed the current design life for concrete and steel road bridges is 120 years.[3] However, bridges which were constructed before 1922, and several which were constructed shortly afterwards, were not designed to standards which would be acceptable today. These older structures are now subjected to much greater loadings than could have been envisaged at the time of their design. In addition, they are subjected to the same problems as all other bridges, for example poor workmanship, poor materials and accidental damage, which accelerate the need for maintenance work.

The design of bridges is a process of development and new structural forms, new materials, new codes of practice and new methods of maintenance all influence the final structure. Consequently the bridge stock contains many structures which would not be constructed in a similar way today, and which are disproportionately expensive to maintain using current techniques. Many of these structures are unique and some are listed buildings, hence the materials, working procedures and workforce skills required to undertake maintenance are specialised and costly. Indeed the cost of reconstruction has been shown to exceed the cost of initial construction by a factor of between 1·5 and 8·0.[4]

Further to these existing maintenance problems, the decision to permit by 1999 increased maximum vehicle loading in line with other European countries[5] has caused new problems. In 1987 the DTp announced a programme of strengthening works for the 38-t vehicle lasting 15 years and costing £2000m. It is estimated that the increase from 38 t to 40 t would cost an additional £100m for DTp trunk roads and £600m for local authority roads. Although the total vehicle weight has been the subject of political attention, it is the axle weight that is important for bridge loading. Therefore the existing 30·5-t four-axle vehicle provided a maximum axle loading of 10·5 t and more recently a partially loaded 38-t vehicle would provide an axle load of 11·5 t, which exceeds the axle loading for the 40-t vehicle and causes the maximum load case for short span structures.

Research work into the causes of deterioration and alternative methods of physical repair has been undertaken by academia and the industry for many years. In 1981 the OECD Report[6] divided bridgeworks maintenance into the objectives of safety, serviceability and economy. For public safety it is estimated that in the UK the chances of a fatality due to a bridge failure are 1 in 10 000 000. Serviceability is defined as the optimal amenity for traffic with minimal interference and the economy consideration is a measure of financial decision of the trade-off between an existing structure with high maintenance costs and the capital cost but predicted low maintenance costs of a new structure.

The Maunsell Report published in 1989[7] was the most recent attempt to try to determine the causes and the extent of the deterioration of the concrete bridgeworks stock. This report investigated some 200 concrete bridges, many of which were less than 25 years old, demonstrating considerable deterioration due to chloride attack, alkali–silica reaction and carbonation. However, the increasing pressures on maintenance budgets and the new requirements have concentrated current research on the investigation of the management of the maintenance of the bridge stock.

EXISTING SITUATION

A bridgeworks maintenance programme has to meet certain objectives, one of the most important being the statutory requirements for public safety. This has to be achieved by making the most effective use of limited resources, by efficient planning and selection of priorities, and by using the most appropriate contract strategy, and all within strict budgetary constraints.

Bridgeworks maintenance can be defined as ensuring the safe, unrestricted passage of people, animals and vehicles as specified in the construction and use regulations without limitations.[8] The work is a combination of planned maintenance, refurbishment, replacement and the repair of accidental damage. The first three items are regarded as traditional bridgeworks maintenance and would usually be funded from the annual budget. However, the last item is often regarded as a special case, sometimes requiring immediate action, and therefore special funding arrangements usually exist for these cases.

The expenditure on bridgeworks maintenance can be divided into costs for structural elements, physical access elements and elements associated with continued usage of the bridge during maintenance. The structural elements are not always easy to quantify precisely as many abutments form part of large retaining walls and substructure foundation works may not be immediately adjacent to the structure. Access to the bridge is often a major item of expenditure and sometimes inspections have been combined with routine maintenance work. OECD indicate that the maintenance costs should be of the order of 3% of the asset value whereas recent UK local authority bridgeworks maintenance budgets have been about 0·3%,[1] however these figures are too global to be of much value for assessing the maintenance cost of a particular bridgeworks stock.

Maintenance work could be regarded as the action taken to prolong the useful life of a bridge at a minimum cost with least interference to its operational function. This approach contains contradictory aims as the minimum cost of maintenance is often only achievable if the bridge is closed. Frequently the cost of diversions of pedestrians and vehicles exceeds the cost of the physical bridge repair. The expertise in the management of the maintenance process is in deciding upon cost-effective minimum works which permit the bridge to be used without restrictions.

It appears that the rate of deterioration of bridges is not standard or uniform. Although there are few published data to justify this statement, there are many examples of arch bridges over 200 years old in good condition whilst some bridges less than 25 years old require extensive refurbishment or replacement.[9] It is suspected that the relatively old and relatively new bridges account for a higher proportion of the maintenance budget than would be expected.

In the case of the older bridges, particularly pre-1922, this could be due to poor materials, but the introduction of the DTp's new loading code means that all 70 000 of the pre-1922 masonry and brick arch bridges and culverts, and about 30% of the 15 000 pre-1922 metal bridges, require attention or loading restrictions.[10] The more recent bridges, post-1960, were constructed at a time when the specification had been revised and alternative bids were allowed. This system, used in conjunction with tendering procedures which concentrated on one parameter, the minimum capital cost, may have affected durability adversely. Additionally, more sophisticated designs were being produced with novel uses of prestressed concrete, leading to minimum weight and hence low priced bridge decks but with a tendency for high maintenance costs if problems should arise.[11] The implication of these factors is that more attention should be given to the consideration of the life cycle cost of the bridge. A bridge which is easily and cost-effectively maintained is not likely to be the bridge with the lowest capital cost at tender stage. This raises the question as to whether or not the cost of maintenance should be treated as capital or revenue expenditure.

MANAGEMENT STRATEGY

Traditionally maintenance has been regarded as an 'after-the-fact' activity and was something done in response to an unacceptable condition arising. However, the life cycle costing approach includes the total management of the maintenance process, consisting of maintenance planning, design maintenance considerations, maintenance in construction and maintenance in operation.[12] For new structures it is possible to consider the complete cycle, but for the existing bridgeworks stock only maintenance in operation can be implemented.

The fundamental requirement for the management of the maintenance in the operation stage is relevant base data. These data are largely obtained from visual general and principal inspections, and it is the frequency of these inspections and hence the cost of collecting the data, and the reliability and accessibility of the data, which influence the effectiveness of the management process. There are a number of existing computer-based maintenance databases operated by the DTp and several local authorities and consultants,[13] however it should be noted that a database records the data available at the time of inspection or works and cannot be completely up to date. There are also several computer-based maintenance management systems for bridgeworks;[14] these mainly originated in the United States but are being adopted in the EC countries in increasing numbers. These systems provide instant access to bridge data to monitor rates of deterioration, or assess damage, or consider the maintenance spend profile on a structure over time to assist in the decision-making process and also to facilitate feedback to develop the database as work progresses.

On the basis of the bridge record data one of the key management decisions is to determine the level of intervention, which can be described as the appropriate time for particular types of maintenance work: repair, refurbishment or replacement. This aspect is the central problem for management and illustrates the difficulties of decision making under conflicting constraints. The decisions are influenced by the quality of data available, by the experience and expertise of the staff of the maintenance section, and by the importance of the route using the bridge. The budget will not allow all the works identified to be treated at the same time and hence the expenditure has to be made in such a way as to produce the 'outcome of least regret'.

The importance of knowledgeable staff should be emphasised because in order to save money on site and use the budget efficiently the expertise must be available. Apparently similar problems may occur on two bridges. The decision has to be made as to which can be patched and which requires major work. This decision should reflect not only which structure is a priority due to structural safety considerations but also if a temporary repair is carried out whether the structure is likely to deteriorate rapidly, incurring excessive maintenance costs in future years. The cost of employing qualified engineers is relatively small compared with the magnitude of the cost savings in maintenance work which can be achieved through good pre-planning and assessment of the works.

The interrelationship between the available budget and priority spending, the age and type of the bridgeworks stock, statutory requirements and the vehicular flows using the bridges is not well defined. Detailed planning is difficult as, particularly on the older bridges, the full extent of the maintenance works required may not become known until the work has started on site. Problems can also occur on some of the minor elements of the structure, which whilst not serious for the structural safety of the bridge can block or obstruct footpaths or highways, thus necessitating immediate action. These types of uncertainty mean that management has to be flexible in the scheduling of routine maintenance work and will need additional funding from time to time.

IMPLICATIONS FOR NEW BRIDGES

Maintenance procedures have significant implications for the design of new works, design being considered in its widest sense, incorporating feasibility, technical assessment and detailing. There are two main areas of interest: first, design which prolongs the life of the bridge and prevents deterioration, and secondly, design features which make routine inspections and routine maintenance, such as replacing bridge bearings, simple and low cost operations. The main conceptual design is usually well prepared and considers the 120-year design life, however it is in poor detailing, usually carried out by junior engineers, that the source of future maintenance problems can be found. A particular problem in recent years has been adequate drainage and waterproofing of decks, a problem which becomes more serious in box beam bridges.[15] The cost of including features in a bridge design to facilitate maintenance are small if these requirements are identified early in the design process. Many of the larger bridges have included access ways, lighting, water and other services in the original designs.[16]

Recent sophisticated structural designs, frequently using new materials, have been successful in reducing the initial capital cost of the bridge but are now seen to be absorbing a disproportionate amount of the repair and maintenance budget, and causing problems in the operation of the facilities. Two factors may be of significance: first, there has been a trend to separate the bridge design from bridge maintenance, with different firms being responsible for each phase. This has the disadvantages of a lack of feedback to the designers of the maintenance performance and of a lack of maintenance considerations during the construction on site phase.

CURRENT RESEARCH

The increasing problems of the management of bridgeworks maintenance and strengthening programmes required for the 40-t vehicle are reflected in the growing investment into research in this area. A major study is being carried out by UMIST funded by the Repair, Operations and Maintenance Programme of the Science and Engineering Research Council.

The research work at UMIST is concerned with the management of the maintenance process for bridgeworks and gives particular attention to information from a specific subdivision of this sector, the bridge maintenance sections of Sheffield City Council and Manchester City Council. The research will commence in 1989 and is due for completion in 1991.

The principal objectives of the work involve four main areas of interest, which are described below:

(a) To investigate the existing management procedures, particularly the organisation, planning and priority, contract strategy and legal requirements associated with bridgeworks.
(b) To examine the criteria used in the decision-making process to determine the level of intervention, which can be described as the appropriate time for particular types of maintenance work: repair, refurbishment or replacement.
(c) To study maintenance budgets, the cost breakdown of maintenance work into structural elements, physical access elements and elements associated with continued usage of the bridge during maintenance. The specific problem of mechanisms for the funding of emergency maintenance work would also be included.
(d) To consider methods of assessing potential maintenance requirements in both the technical appraisal of a design of new works and in the evaluation of contract tender bids for new works.

ACKNOWLEDGEMENTS

The author gratefully acknowledges the assistance of the bridges sections of the City of Sheffield and the City of Manchester in the current research work.

REFERENCES

1. Jones, C. J. F. P., Bridgeworks maintenance. Lecture to the Yorkshire Association of the Concrete Society, 24 November 1988.
2. Andrews, J. C. and Grindall, A. M., Maintenance and reliability congress discussion: maintenance and renewal. *Proc. Instn Civ. Engrs*, **74** (1983) 562.
3. British Standards Institution, Code of Practice for Design of Steel Bridges, BS 5400, 1982.
4. Mallett, G. P., Bridge Maintenance—Value for Money. National Workshop

on Bridge Maintenance Initiatives, Leamington Spa, Institution of Highways and Transportation, 8 April 1986.
5. More bridge cash for 40-t lorries. *New Civil Engineer* (8 June 1989) 14.
6. OECD Report, Bridge Maintenance, 1981.
7. Department of Transport, *The Performance of Concrete in Bridges.* Prepared by G. Maunsell & Partners, HMSO, London, 1989.
8. Leadbeater, A. D., The Practical Use of Bridge Assessments. National Workshop on Bridge Maintenance Initiatives, Leamington Spa, Institution of Highways and Transportation, 8 April 1986.
9. Jones, H., New lease of life for Severn Bridge. *New Civil Engineer* (17 March 1988) 16.
10. New loading code to put squeeze on old bridges. *New Civil Engineer* (31 March 1983) 5.
11. Jones, C. J. F. P., Preventative Maintenance. National Workshop on Bridge Maintenance Initiatives, Leamington Spa, Institution of Highways and Transportation, 8 April 1986.
12. Blanchard, B. S., Total maintenance management. *Tertoecchnia,* **2**(2) (1981) 139.
13. BRIDGET bridges data to cut out labour. *New Civil Engineer* (25 February 1988) 48.
14. Ho, C. K., Management of maintenance. MSc dissertation, University of Sheffield, 1988.
15. Welsh, N., Bridges affected by concrete problems. *Construction News* (3 October 1985) 12.
16. Rufford, N., Team assembles for Severn start. *New Civil Engineer* (16 April 1987) 30.

PROTECTION

21

Crack Bridging by Surface Treatments to Concrete

J. G. KEER and B. H. LE PAGE
Department of Civil Engineering, University of Surrey, Guildford, Surrey GU2 5XH, UK

ABSTRACT

Cracks offer an easy path for the ingress of water and chlorides into concrete. In certain situations on bridge structures, coatings which are capable of bridging over cracks and maintaining a continuous film are desirable. Although the relationship between crack width and corrosion has been the subject of debate, it is argued that a crack-bridging coating is preferable in critical situations. Current tests in the UK which can be used to assess crack-bridging ability are reviewed. West German tests and specifications are also summarised and form a good basis for the development of better UK test methods and specifications for crack bridging. The need for further work on the effect of cracks on hydrophobic surface treatments is highlighted.

INTRODUCTION

Surface treatments are applied to a concrete structure to improve its appearance or to protect the structure from potentially aggressive agents. In the latter case, past emphasis has perhaps been on the protection of concrete against chemical attack when concrete is used in non-typical, rather special situations.[1,2] In recent years, however, attention has focused on the application of surface treatments to improve the performance of concrete structures in typical environments which form common uses of concrete. The principal objective has been to reduce the occurrence of reinforcement corrosion and associated damage. In bridges, the depth of cover and concrete quality is normally sufficient to eliminate problems due to carbonation, although corrosion initiated by carbonation has been observed.[3] The main corrosion problem in concrete bridge structures results from the ingress of deicing salt solutions. The distinction in the source of the problem is important, because some treatments may be effective only as anti-carbonation coatings, others only as anti-water/chloride treatments.

There are two principal types of treatment. Coatings/sealers rely upon the formation of a pinhole-free film of finite thickness over the concrete surface, which can act as a barrier to the diffusion into concrete of CO_2 and/or solutions. Those materials described as sealers claim some penetration into the pores of the concrete. Coatings would normally have a dry film thickness of 150–500 μm, although thicker 'high-build' coatings are also used. Pore-lining treatments are hydrophobic materials which line surface pores of the concrete and repel moisture. The most widely used materials are commonly called silanes, a term which embraces monomeric silane and oligomeric siloxane formulations. Whilst among coating materials both anti-carbonation and anti-water/chloride treatments can be found, the pore-lining materials are not effective as anti-carbonation coatings, only for keeping water/chloride out.

Preventing carbonation and water/chloride ingress is clearly beneficial for the durability of concrete bridge structures. Keeping concrete in a dryish state is also beneficial in restricting ASR and freeze–thaw damage. Equally clearly, cracks which cause breaks in coatings or which locally reduce water repellency of the surface become paths for the ingress of moisture and

TABLE 1
Parameters affecting coating selection[4]

Protection	Durability
Diffusion of CO_2, chloride, O_2, sulphate Acid resistance Water vapour permeability Water permeability	Proven use on concrete Case histories Ability to span over passive cracks Ability to seal live cracks Durability under strong UV, rain/wind condensation, temperature and immersion cycles, freeze/thaw, salt crystallisation
Application	**Cost**
Tolerance to surface preparation, moisture on application and during cure Resistant to alkaline conditions Ease of application, toxicity, flammability	Special application needs Relative cost/benefit at applied thickness Maintenance period Ease of recoating

potentially destructive agents. A crack-bridging ability is therefore among the many parameters influencing coating selection in Table 1 suggested by Browne.[4] It is, however, worth reviewing the role of cracks in corrosion of reinforcement and concrete deterioration to establish the necessity of crack-bridging ability. Crack bridging is a term generally applied to coatings when the film spans across a crack, but the term is used here to include the ability of hydrophobic treatments to repel water from penetrating down cracks.

THE NEED FOR CRACK BRIDGING

The role of cracks in the corrosion of reinforcement has been debated for many years. Beeby has argued that the checks carried out on crack width intended to control corrosion are based on 'no sound foundation of data relating crack width to corrosion'.[5] While crack width may have an effect on the time it takes for the depassivating agents (CO_2 or chlorides) to reach the steel, once corrosion has started, the rate of corrosion is not very sensitive to crack width. If the difference in time to initiation of corrosion between cracked and uncracked concrete is small compared to the design life of the concrete, crack width can have an insignificant role in corrosion damage. The reason why the role of corrosion may not be sensitive to crack width is that the corrosion process may be controlled by the cathodic reaction requiring the diffusion of oxygen and water to steel in sound concrete and by the electrical resistance of the path between cathode and anode. In this respect treatments which are non-crack bridging may nonetheless be beneficial in restricting oxygen and moisture penetration between cracks. Another point in the argument about the influence of crack width is that the crack widths referred to are surface crack widths. Crack widths decrease toward the bar surface and the width at the bar surface can be considerably less than at the external concrete surface.

The argument above applies to cracks running perpendicular to the reinforcing bar and is about the influence of crack width, not of the presence of cracks. Thus crack-bridging coatings which negate the presence of cracks, irrespective of width, are likely to be beneficial, particularly if the protection is such that the initiation period is significantly extended beyond the design life of the structure. Also, as Table 1 suggests, a continuous film will enhance the durability of the film itself, preventing peeling at the edges, such that the vital role of the coating between cracks is maintained. For cracks parallel to reinforcing bars, both anodic and cathodic regions may be more easily accessible and such cracks are potentially very dangerous.[6] It may be true again, however, that the danger is not very sensitive to crack width (e.g. whether the crack is 0·1 or 0·3 mm wide), merely to whether a crack is present or not. It is worth remembering that a loading crack perpendicular to one reinforcing bar is likely to run along another. Also cracks resulting from plastic shrinkage and settlement may follow the lines of reinforcement. If it is then beneficial to use surface treatments which are capable of crack bridging, the engineer must be able to relate his requirements for crack bridging in practice to the results of laboratory tests which effectively assess crack-bridging ability.

TEST METHODS FOR CRACK BRIDGING OF COATINGS

The crack-bridging ability of a coating is principally related to the extensional characteristics of the coating material, the film thickness and the adhesion to the substrate, all of which may change with ageing. While these parameters can be measured independently (for example the extensibility of a free film of coating can be assessed using ASTM D2370[7]), their complex interaction means that tests modelling films bridging cracks in mortar/concrete substrates are essential. Tests can be broadly divided into three types: static tests which measure the crack width at which coatings 'fail'; dynamic tests which subject the coating to cycles of measured crack movement; compliance pass/fail tests where the crack widths at failure are not explicitly measured.

UK Tests

In a test developed at BRE,[8] a coating is laid on a mortar block with a weak section at its centre (Fig. 1). The block is slowly extended until a crack forms, and is controlled so that the initial crack width does not exceed

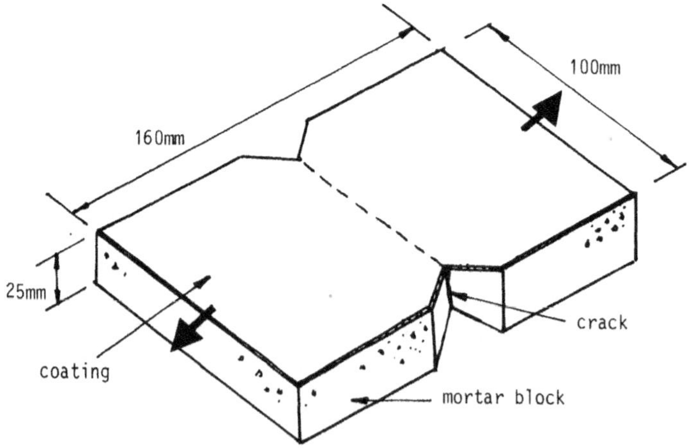

FIG. 1. BRE crack-bridging test.[8]

0·02 mm. The extension is then continued until the film cracks or small splits develop. The crack width at failure is recorded. This approach has been used by the British Board of Agrement to assess masonry paints for extensibility, although MOAT No. 33[9] refers only to specimens of the coating on aluminium strips which are extended in a tensile testing machine such as an Instron. The coating is examined microscopically for signs of failure. By either method extensibility is determined on unaged and artificially weathered specimens. A styrene acrylate copolymer textured coating breaks at a crack width of 1·0 mm when tested in the unweathered condition in the BRE test. The crack width at coating failure is reduced to 0·5 mm after a period of artificial weathering.

In MOAT No. 24,[10] for plastic renderings drawn up by a body including the British Board of Agrement, the test piece substrate comprises three fibre cement segments carefully aligned and clamped so that the substrate has two 'cracks' less than 0·1 mm wide. After application and curing of the coating, the clamp is removed and the cracks are widened by the introduction of a wedge between the segments. The crack width at failure of the coating is recorded.

A laboratory test was developed by TRRL to assess the resistance of waterproofing membranes for concrete bridge decks to substrate cracking.[11] The membrane is laid on a small reinforced slab, which is then loaded to produce cracking (Fig. 2). Loading is increased to widen the crack width, with pauses of 30 min at crack widths of 0·25 and 0·60 mm to allow for delayed fracture and at other stages where there were indications that fracture of the material was commencing or developing. Three test temperatures, 20, 3 and −10°C, were used. A satisfactory performance for a membrane was withstanding a crack width of 0·60 mm. The basis of this requirement was that the membrane should withstand fracture at the permitted crack width of 0·25 mm factored by 1·2 to allow for widening under repetitive load and 2·0 to allow for variability in membrane performance and for some crack widths in excess of 0·25 mm. (The maximum observed width of crack has been reported to be approximately double the average crack width.[12]) There are proposals to update this test to include a dynamic effect.

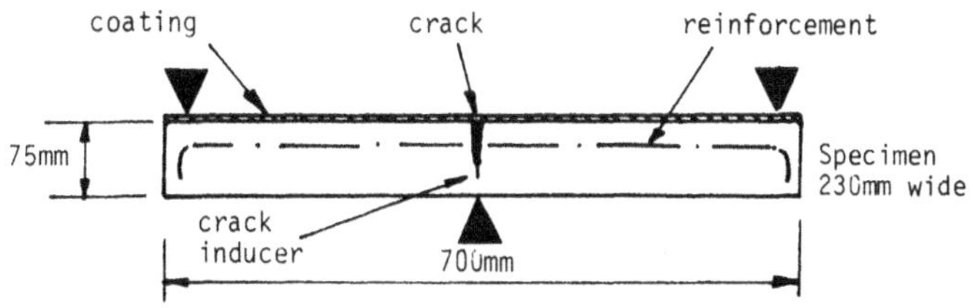

FIG. 2. TRRL crack-bridging test.[11]

West German Tests

The protection of concrete by surface treatments is popular in West Germany. In the very harsh climate of Berlin, bridges are given a protective covering including a polymer deck membrane, silane or siloxane on soffits and an elastic crack-bridging coating on abutments and columns which can be subject to salt spray. Crack movements of 0·4 mm over a 12-h period have been recorded.[13] The Federal German Institute for Materials Testing (BAM) have developed tests to assess the crack-bridging ability of coatings

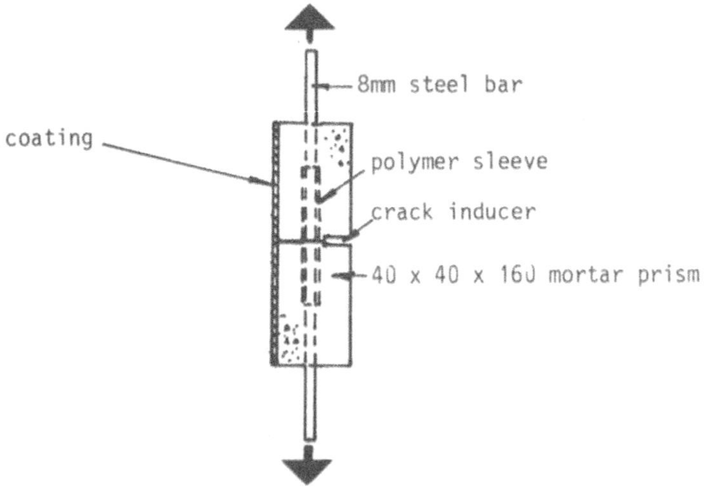

(a)

(b)

FIG. 3. Crack-bridging test to West German specifications.[14] (a) Specimen type and (b) specimen under static test crack width ≈ 0·5 mm.

over static and dynamic cracks in concrete. The test procedures have largely been incorporated into regulations introduced by the Federal German Transport Ministry for bridge surfacings on concrete and for the protection and maintenance of concrete structures generally.[14]

The specimen used is a small concrete prism reinforced centrally in which a crack is induced (Fig. 3(a)). One face of the prism is coated. The prism is cracked by a tensile load applied through the ends of the steel bar. Figure 3(b) shows a specimen during a static test at the University of Surrey. Cyclic loading/unloading in tension results in crack opening/closing. The test is carried out at $-20°C$ and coatings may be subjected to accelerated weathering prior to test. Another specimen design used in West Germany allows greater widths of coating to be examined.[15] The coating is applied to a layer of mortar, itself laid on top of two abutting steel plates. The plates are moved apart gradually, inducing a crack in the mortar over which the coating bridges. Static and dynamic tests can be carried out.

Draft German proposals classify coatings according to their performance under a number of test regimes which relate to crack movement caused by temperature cycles or temperature and load cycles. The lowest class of coating, to cover a crack subject to thermal cycling only, would have to withstand 1000 cycles at 0·03 Hz between crack widths of 0·1 and 0·15 mm, i.e. 0·05 mm movement. The highest class is required to resist 10^5 cycles at high frequency (5 Hz) to simulate traffic effects superimposed on the low frequency thermal movement, with a maximum crack movement of 0·3 mm. Also, under static conditions, a crack opening of 1·0 mm must be accommodated without film failure.

It is suggested that these approaches should be the basis of improved test procedures and specifications in the UK. The specimens in Refs 14 or 15 have advantages over larger reinforced slab specimens in ease of manufacture, handling and testing, particularly when tests are conducted at low temperatures and when specimens are to be artificially aged prior to test.

CRACKS AND HYDROPHOBIC TREATMENTS

There is little published literature on the effects of cracks on the performance of hydrophobic surface treatments in keeping moisture from penetrating below surface layers. The Department of Transport Advice Note BA 23/86[16] suggests that cracks up to 0·3 mm wide may be treated by impregnation of the concrete surface with a silane treatment to prevent chloride ingress. The basis of the 0·3 mm width is not clear, but it may be some US research[17] in which a silane and a methyl methacrylate treatment provided added protection for embedded bars in cracked reinforced concrete slabs. It was suggested that these two materials penetrate existing cracks and could provide added protection for bridge surfaces containing cracks of up to 0·25 mm. There is likely to be a significant difference in performance between horizontal and vertical treated surfaces, since the possibility of ponding in the former is likely to be influential.

A simple technique has been used at the University of Surrey based on specimens of fibre-reinforced mortar in which fine cracks (<0·1 mm) can be induced.[18] Untreated specimens laid on a moist surface draw water up through fine cracks very quickly so that the cracks are highlighted on the top 'dry' surface. A silane treatment applied pre-cracking is not effective when crack width exceeds about 0·05 mm. When applied post-cracking, the treatment is much more effective and can prevent moisture penetration through the crack up to about 0·1 mm and possibly beyond this, although an upper bound has yet to be established. Further work is proceeding at Surrey to develop tests to relate crack width and moisture penetration for hydrophobic treatment.

CONCLUDING REMARKS

1. Surface treatments for concrete are likely to play an increasing role in maintaining concrete bridge structures.
2. In certain situations it will be desirable to use coating films which are capable of bridging over cracks or, in the case of hydrophobic treatments, preventing by surface effects water or chloride ingress via cracks.
3. Tests for coatings should include cycles of crack opening/closing at low and high temperatures representative of site conditions, in addition to the assessment of the maximum crack width at coating failure, following cyclic movement. Performance after accelerated ageing should be assessed.
4. For hydrophobic surface treatments, further work is needed to establish the effect of crack width on the performance of an otherwise water-repellent surface.
5. Results of laboratory tests to assess crack-bridging ability should be correlated with field performance of treatments.

ACKNOWLEDGEMENT

The authors would like to acknowledge the financial support of the Science and Engineering Research Council.

REFERENCES

1. ACI, Committee 515. *J. Am. Concr. Inst.*, **63**(12) (1966) 1305–91.
2. American Concrete Institute, *Manual of Concrete Practice*, Part 5, ACI 515 IR-79, 515 IR-1, 515 IR-41, 1985.
3. Vassie, P., Reinforcement corrosion and the durability of concrete bridges. *Proc. Instn Civ. Engrs, Part 1*, **76** (1984) 713–23.
4. Browne, R., Building deteriology—the study and prediction of building life and performance. *Chem. Ind.* (December 1986) 837–44.
5. Beeby, A. W., *Cracking and Corrosion.* Concrete in the Oceans Tech. Report No. 1, Cement and Concrete Association, 1978, 77 pp.
6. Darwin, D., Debate: crack width, cover and corrosion. *Concrete International* (May 1985) 20–35.
7. American Society for Testing and Materials, ASTM D2370-82, Standard Test Method for Tensile Properties of Organic Coatings, 1982.
8. Whiteley, P. and Rothwell, G. W., Appearance and performance factors in coatings for buildings. *J. Oil Colour Chem. Assoc.*, **54** (1971) 855–78.
9. MOAT No. 33: 1986. The assessment of masonry coatings. British Board of Agrement, 1986.
10. MOAT No. 24: 1983. Directives for the assessment of plastic renderings. European Union of Agrement, 1983.
11. McDonald, M. D., Waterproofing concrete bridge decks: maintenance and methods. TRRL Laboratory Report 636, Transport and Road Research Laboratory, Crowthorne, Berks, 1974.
12. Illston, J. M. and Stevens, R. F., Long-term cracking in reinforced concrete beams. BRE Current Paper CP14/73, Building Research Establishment, May 1973.
13. Middelboe, S., Berlin bridges take cover. *New Civil Engineer*, No. 639 (9 May 1985) 14–15.
14. Herold, C., Bridging of cracks in concrete using liquid plastic coatings. *Kunststoffe German Plastics*, **78**(7) (1988) 30–2 (English translation of paper in *Kunststoffe*, **78**(7) (1987) 631–4).
15. Engelfried, R., Requirements profile and suitability examinations for materials for concrete repair. *Proc. Int. Colloq. Materials Science and Restoration*, Esslingen, September 1986, pp. 175–82.
16. Departmental Advice Note BA 23/86. The investigation and repair of concrete highway structures. Department of Transport, 1986.
17. National Cooperative Highway Research Program Report 244. Concrete Sealers for Protection of Bridge Structures. Transportation Research Board, Washington, 1981.
18. Keer, J. G., Xu Guodong and Filip, R., Cracking and moisture penetration in fibre cement sheeting. *Proc. Int. Conf. on Recent Developments in Fibre Reinforced Cements and Concretes*, Cardiff, September 1989. Elsevier Applied Science Publishers, London (in press).

22

Keeping Water Out of Concrete—The Key to Durability

M. B. LEEMING

Arup Research and Development, 13 Fitzroy Street, London W1P6BQ, UK

ABSTRACT

All deterioration mechanisms to reinforced and prestressed concrete bridges and other structures are influenced or promoted by water. Chloride ingress, corrosion of the reinforcement, freeze/thaw, alkali–silica reaction and sulphate attack can be controlled by restricting moisture movement. Carbonation being dependent on the diffusion of a gas into the concrete is the oddity yet even this only causes corrosion of the reinforcement at certain moisture states of the concrete. The rate of carbonation is also moisture dependent. Moisture has a similar effect on other deterioration mechanisms and a proper understanding of these sometimes conflicting influences and the way concrete takes up water can lead to a proper strategy for dealing with deterioration. The whole philosophy of dealing with water ingress will be covered from design details, material specification, construction sensitivity, non-destructive testing for diagnosis and refurbishment. Durability has a greater influence than strength on the design of concrete mixes and on construction for most severe exposures. There are many methods and materials which can be used to waterproof concrete but few of these are failsafe and most fail at joints, junctions and edges. The economics of various approaches, both preventative and remedial, will be discussed.

INTRODUCTION

Water is at the very heart of civil engineering; houses and factories provide a dry environment, bridges and tunnels are needed to cross rivers and canals, dams provide drinking water, pipes convey water and sewage, the sea requires harbours, docks and coastal defences, while roads provide all-weather surfaces. When considering the deterioration of concrete, water is found to be a major, if not the main, cause of the problem. This statement will be studied in detail in the remainder of the paper. Perhaps with some understanding of the mechanisms involved there will be a better chance of controlling deterioration.

THE DETERIORATION OF MATURE CONCRETE

The deterioration of concrete can be divided into two categories, the chemical or physical deterioration of the concrete itself and corrosion of the reinforcement. These two forms of deterioration have essentially different mechanisms yet moisture still remains a common factor.

Chemical/Physical Deterioration of the Concrete Itself

These forms of attack can be due either to internal or to external sources.

Sulphate attack

An external form of deterioration when sulphates[1] leached out of the ground or waters high in sulphates react with the cement matrix to weaken it and cause disintegration. Often dealt with by the use of sulphate-resisting cement low in C_3A or alternatively by dense high-strength concretes with high cement contents and/or the use of pfa or ggbs as a cement replacement which additionally helps to lower the C_3A content. The action of water is to provide the supply of sulphate.

Freeze/thaw damage

An external environmental form of attack by freezing of water within the pores of the concrete.[2,3] Normally air entrainment is used to combat the problem. However, hydraulically pressed concrete such as paving slabs are generally immune to freeze/thaw damage due to their inherent density and a coarse granular structure. Dense well compacted concretes which have a degree of saturation below a critical level usually survive freeze/thaw conditions well.

Alkali–silica reaction

An internal reaction as a result of mixing certain reactive aggregates with cements of high alkali metal content.[4] This phenomenon has been known for a number of years but research is still going on as to the exact mechanisms involved and the structural consequences of the resultant deterioration. The concrete swells and cracks in a characteristic pattern when unreinforced. The presence of moisture is essential to the disruptive swelling.

High alumina cement conversion

An internal reaction within high alumina cement concretes.[5] This cement has not been used in structural elements for many years and this form of deterioration is now rarely experienced. However, the conversion of the cement causing considerable loss of strength was most common in areas of high humidity.

Salt crystallisation

The surface of the concrete just above ground or water level can deteriorate due to the capillary rise of salt-bearing water which evaporates at a higher level leaving the salts behind which crystallise and disrupt the concrete.

Salt scaling

A similar phenomenon to that above but found in horizontal slabs and caused by the application of deicing salts. The dissolved salts increase the severity of freeze/thaw action. The salt finds its way into the concrete in solution.

Popovics[6] lists six classifications of deteriorating mechanisms in concrete and in all but one mechanism, mechanical deterioration, water is cited as an influence.

The Corrosion of Reinforcement

Moisture can have a complex influence on corrosion, as shown in Fig. 1.[7] Concrete provides the ideal alkaline environment for reinforcement where a

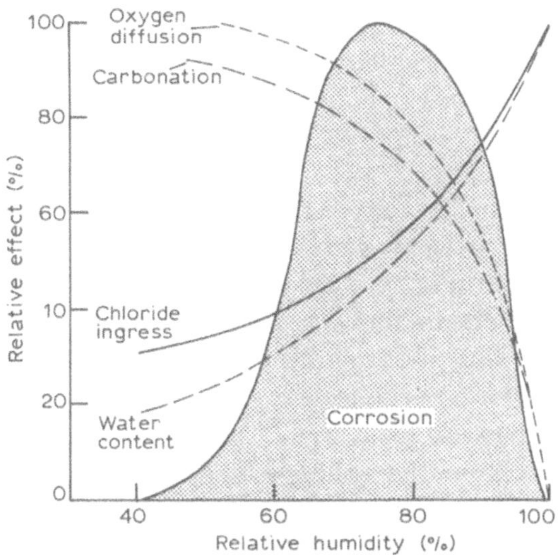

FIG. 1. The influence of moisture on corrosion of reinforcement.

stable film of corrosion products forms on the surface of the steel, strongly adhering, which stifles further corrosion (passivity). This situation can remain indefinitely and requires a certain amount of moisture and oxygen to sustain the situation. However, two forms of attack can upset the balance.

Chloride ingress

Chloride ions at the surface of the steel which breaks down the passive film causing corrosion in pits. The chlorides in concrete can result from one or both of two causes. First, chlorides can be cast into the concrete due to salt contamination of the aggregates or be added as an accelerator, a practice no longer permitted. Secondly, chlorides can diffuse into the concrete in solution either from seawater in marine situations or from traffic spray containing deicing salts.

Carbonation

Carbon dioxide gas from the air diffuses into the concrete and reacts with the cement in the presence of water to make the concrete less alkaline. When the pH of the concrete falls below about 9–10, the passivity breaks down and general corrosion can occur.

However, there are two further ways in which moisture influences corrosion.

Resistivity

Corrosion is an electrochemical process and requires an electrolyte between the anodes and cathodes on different parts of the reinforcement. This electrolyte is the pore water in the concrete which contains various dissolved salts. A large amount of pore water provides a low resistance to corrosion currents, hence corrosion can occur apace. Dry concrete, on the other hand, is highly resistive and corrosion is limited.

Gas diffusion

Saturated concrete severely restricts the diffusion of gases, in particular carbon dioxide and oxygen. Wet concrete restricts the rate of carbonation (loss of alkalinity) of the concrete, delaying the onset of corrosion. Oxygen is required at the cathode to allow the electrochemical reaction to proceed. In saturated concrete lack of oxygen controls the rate of corrosion in spite of a low resistance in the electrolyte.

Hence, as can be seen from Fig. 1, little corrosion occurs below about 50% humidity because the concrete is too dry to provide an adequate electrolyte to the corrosion cell in spite of high carbonation rates and an ample supply of oxygen to the cathode. When the concrete is saturated as stated above, the lack of oxygen limits corrosion to very low rates. At relative humidities between these limits rapid corrosion can occur. Alternative wetting and drying can aggravate the situation with rapid carbonation and oxygen ingress during dry periods followed by corrosion at high relative humidities.

THE INFLUENCE OF WATER ON FRESH CONCRETE

Water is a vital constituent of concrete. Furthermore, concreting is a 'wet trade'; control of water content is required at early ages to influence both strength development and durability. Most designers seem to understand that water chemically reacts with the cement (hydration) yet many practitioners seem to think that it is a drying process and as a result the concrete does not get properly cured.

Strength is usually the prime consideration when specifying concrete. The grade number usually relates to the required 28-day strength and the main compliance requirement is that cubes made of the concrete reach the specified compressive strength. Modern codes,[8] however, now recognise that durability is an equally important criterion and give guidance on the choice of the correct cement content, water/cement ratio and cover to the reinforcement for the various exposures. There are no generally recognised tests for fresh concrete that indicate whether the required durability criteria have been met other than the mix has been correctly proportioned. As a result the cube test remains the sole pass or fail criterion yet a strong concrete is not necessarily a durable concrete and vice versa. Hence it often happens that when the cubes exceed the required strength the contractor seeks to reduce the cement content or to increase the water/cement ratio, paying scant attention to the durability requirements. This point will not be got across to the contractor until the drawings state not only the strength requirements but also the durability requirements.

The main factors at casting which influence the durability of concrete are as follows.

The Water/Cement Ratio

Literature is full of references to the fact that the lower the water/cement ratio the more durable the concrete. More water is required in the mix to allow proper mixing and placing than is required for hydration of the cement. The excess water eventually dries out of the concrete and leaves a coarse structure behind.

Workability/Compaction

The amount of water in the mix determines the ability to place the concrete (workability) around the reinforcement in a dense homogeneous mass. If the concrete is too dry voids and honeycombing of the concrete results. If it is too wet it tends to segregate and is more prone to plastic cracking.

Curing/Maturity

The concrete needs to be kept moist for sufficient time to allow the cement to hydrate properly. Curing[9–12] has greatest effect on the durability of concrete as the surface layers are most affected. The inner parts of concrete are virtually self-curing. Maturity is part of the same mechanism although it is normally affected by age and strength development, the striking time of formwork being the main criterion for fast construction. Monitoring the heat of hydration is used in the maturity meter[13] as a diagnostic tool. The 'Capo' test and other pull-out tests monitor the strength of the concrete at early ages. It is also possible to monitor the resistivity of the concrete where significant changes can be seen as the hydration of the cement occurs.

MECHANISMS OF INGRESS OF WATER INTO OR THROUGH CONCRETE

The movement of water into a porous medium such as concrete is highly complex and is well described in detail in Ref. 14 and will only be briefly summarised here before concentrating on what the author believes to be the most important factor as far as durability of concrete is concerned.

Permeability is a term which is often used loosely to cover a number of mechanisms by which water passes through concrete, but has been defined[14] as that property of a porous medium which characterises the ease with which a fluid will pass through it under the action of a pressure differential. Permeability is quantified in Darcy's law.

This mechanism assumes the concrete to be saturated for the section under consideration with a head of water at least on one face under a steady state. The tests which measure this property are usually done with the specimen submerged. The property is of importance when estimating the amount of water passing through concrete in structures such as dams, water-retaining structures and offshore oil platforms.

Diffusion is the process by which a liquid, gas or ion can pass through concrete due to a concentration gradient. Fick's first law describes the process. Because a concentration gradient is required the process is important with respect to concrete for solutions of different strengths, vapour diffusion under differing relative humidities, differing ion concentrations in saturated concrete and the diffusion of carbon dioxide and oxygen.

Adsorption occurs when molecules of water adhere to the surface of the concrete, held there by Van der Waals forces. It is a transitory process and not of much interest in the mass transfer of significant amounts of water in the deteriorating process.

Capillary flow or absorption occurs in narrow channels when a meniscus forms and capillary forces are set up drawing the liquid through the channel. The rate of flow is given by Washburn's equation.

Osmotic forces can be set up across thin semi-permeable membranes subject to concentration gradients. These forces can be quite high but are of importance only as far as water transport into concrete is concerned for waterproof membranes and surface coatings.

Of the above mechanisms, the most important with regard to the deteriorating mechanisms of concrete is believed to be capillary flow. The pore structure of concrete is highly complex and consists of many voids interconnected by micropores and microcracks up to cracks visible to the naked eye. Many of these pores/cracks can be of capillary size and will therefore suck free water on the surface of the concrete into the body of the concrete. However, the variable size of these pores/cracks does not lend itself to simple estimation of forces and flow rates involved. Capillary pressures of up to 2 bar (20 m depth of water) have been observed due to capillary suction.[15]

Diffusion and permeability are of importance in fully submerged concrete and for concrete which retains or resists the ingress of water. However, research has shown that under fully submerged conditions corrosion is severely limited due to lack of oxygen diffusion even with high chloride levels. Diffusion and permeability are of more importance for sulphate attack and for alkali aggregate reactions where the reaction can continue in saturated conditions.

However, the deteriorating mechanisms of concrete under fully submerged conditions are relatively simple as they are steady state and are subject to analysis to obtain a reasonable estimate of the problem. Most deterioration problems occur in the aerial zone of the structure where the damage is more readily seen. It is well known that offshore and marine structures are most vulnerable in the splash or tidal zones. Onshore structures have a similar zone which is much less well defined. The problem is no longer steady state and the worst affected zone is subject to wetting and drying on an entirely random basis. It is believed that the important mechanisms are absorption of surface water due to capillary suction and subsequent drying by vapour diffusion. Water take-up is also possible by vapour diffusion at times of high humidity. However, this effect is not significant for most forms of deterioration except in swimming pools.

Capillary absorption and vapour diffusion have been studied by the author with respect to the use of surface coatings on concrete to limit carbonation[16] and ingress of deicing salts into concrete. These same mechanisms are also influenced by changes in mix design and other measures to make concrete more impermeable, hence more durable.

CONTROLLING MOVEMENT OF WATER IN CONCRETE

Mix design. Correct specification and good workmanship are the most important methods of achieving a durable concrete of low permeability. It requires a low water/cement ratio which is consistent with adequate workability, allowing the concrete to be well compacted, providing a dense structure to the concrete. It needs to be properly cured and have the right cover. These measures, combined with the choice of the right cement content, are recognised by CP 8110[8] as the means of achieving the required durability. When the exposure requires a stronger concrete than is needed for structural purposes, and when workmanship is not of the best quality, further methods to improve durability are required.

Chemical admixtures included in the mix can help to improve the durability of concrete. Water-reducing admixtures help in maintaining a low water/cement ratio with high workability. Air-entraining admixtures are mainly used in providing freeze/thaw durability. The precise action of air entrainment in achieving this durability is still open to debate, however it also affects workability to some extent and also the absorption of water.

Cement replacements such as pulverised fuel ash, ground blast furnace slag cement and silica fume can produce more impermeable concretes. However, the gain of strength of these cement replacement materials is slower than normal OPCs and as a result they generally require longer curing to achieve the improvement in durability.

Surface treatments have been extensively studied with a view to inhibiting ingress of chlorides and carbon dioxide. They come in many forms from penetrants through sealers to coatings. Treatments can have various properties and it is necessary to be clear as to what they are required to do and to choose the right material accordingly. Most treatments can significantly reduce the water absorption of an average concrete by a factor of at least 20 and for the most impermeable coatings by a factor of 100. The penetrant materials generally have little effect on carbon dioxide ingress while sealers and coatings can give increasing resistance so that further carbonation is limited to very low levels. Treatments generally have a lower resistance to water vapour than to carbon dioxide and this property can allow the concrete to breathe. Resistance to oxygen lies between that of carbon dioxide and water vapour but the effect is insufficient to have any significant influence on deterioration processes.

Surface treatments are, however, applied in very thin layers up to 500 μm and are subject to deterioration themselves, such as embrittlement, weathering due to ultraviolet light and lack of adhesion. These systems have much shorter lives than the structure they are applied to and therefore need recoating. These materials have not been in use long enough to generate sufficient reliable data on their useful lives and on the change of protective properties with time. There is evidence that ingress of chlorides, sulphates and carbon dioxide is greater when the concrete is young so that this is the period when treatments are most needed. However, to rely on a thin coating for durability in lieu of normal covers of sound concrete seems poor engineering judgement.

Mortar renderings have been used from very early times to provide a durable skin to brickwork and other building materials. However, bond with the substrate can be a problem and in the UK their use has strong regional variations. Generally, in this country, durability requires concretes of a similar strength to that required for structural purposes and therefore to use fair-faced concrete with the same mix throughout the member makes economic sense. But where structural requirements are modest in severe environments it seems sensible to use a durable skin to the minimum requirement for the structured core. This philosophy, used in many Mediterranean countries, has a certain logic. The inherent durability of thin ferrocement members constructed by plastering techniques using high cement contents and low water/cement ratios surely points to possible methods of achieving durability through renders. The corrosion protection applied to the steel external columns of the Hong Kong Shanghai Bank[17] is an extreme example of the use of renders.

Cladding. There are many other finishes, such as stone cladding, tiles, mosaics, dry cladding, etc., which can be effective in keeping water out of concrete. Where these systems provide an air gap at the face of the concrete, consideration must be given to the higher rate of carbonation in the dryer concrete. Corrosion could possibly occur unnoticed behind the covering in periods of high humidity.

Waterproof membranes are widely used in roofs, for lining tanks and other water-retaining structures. In bridges their use in the UK, in marked contrast to the USA, has limited the amount of corrosion occurring in bridge decks. However, problems have occurred when they begin to leak at vulnerable points. These membranes have been found to have a finite life and need renewing after 20 years or so.

Structural detailing can be a source of water ingress into concrete. A Department of Transport report[18] found that most bridge joints leak, causing damage to piers and abutments due to chloride ingress. This damage can be eliminated by making bridge decks continuous. If structural economy dictates simply supported spans, with a little ingenuity the deck slab can be made continuous. Service troughs complicate waterproof membrane details and are also a source of leakage of chloride-bearing water on to piers and abutments. Again some ingenuity could be applied to eliminate these by building in sufficient electricity and telephone ducts for all possible future needs and building in gas and water pipes below the deck with extra wall thickness so that they require no maintenance. Drips at cantilever edges are correctly detailed but in many cases they channel the water along the edge where it runs to the joint over a pier and down the beam face to be concentrated on the pier top. It is essential that the top of waterproofing membranes have an escape path for the water that collects above them. This is one of the causes of failure and corrosion damage at bridge expansion joints. The same problem can also be seen at drain outlets to buildings. Drainage systems need to be simple, easily maintained and preferably requiring no maintenance at all. Drainage systems on bridges usually give trouble due to the difficulty of accommodating standard gulley outlets and lack of adequate falls. Consideration should be given to combining the expansion joint and the drainage in one robust cross carriageway drainage system.

Having prevented the ingress of water into one concrete face, consideration must be given to all other possible water paths in the structure. Many causes of the failure of surface coatings on concrete are due to moisture on the rear side of the coating, causing loss of adhesion. Having stopped the main source of water ingress, other minor sources become apparent. Any method of controlling water ingress on the surface of concrete is more likely to fail if vapour pressure at the concrete/system interface cannot be relieved. This is why 'breathability', i.e. high vapour permeance, of a surface coating is an advantage.

One of the main objectives of controlling the ingress of water into concrete must be to try and achieve a steady and uniform state of moisture in the concrete, particularly as far as corrosion is concerned. For instance, corrosion in a tunnel lining could be controlled with a vapour barrier on the inside so that the concrete will remain in a permanently saturated state where lack of oxygen restricts reinforcement corrosion. Maintaining the concrete in a dry condition equivalent to an internal office environment again limits corrosion.

RELEVANT TESTS

Absorption

As capillary flow or absorption has been identified as the primary parameter with regard to the durability of concrete it is necessary to consider how this parameter can be measured. The simplest method of doing this is to take a sample of concrete, seal the sides, place it face down on capillary matting or in a shallow depth of water and to weigh it at predetermined intervals. This is the method adopted by DIN 52617.[19] The results are expressed in a graph as water gain plotted against the square root of time which usually gives a straight line, the slope of which is the water absorption coefficient. This method was used by the author to compare the relative performance of surface coatings on concrete in inhibiting the ingress of water-containing deicing salts for the Transport and Road Research Laboratory. In this instance a 15% salt solution was used and all faces of the specimen except the test surface were sealed. The test ran for 21 days and measurable gains of weight were recorded provided that an accurate balance was used. The uncoated plain concrete control specimens, however, rapidly gained weight to near saturation within a few hours. This method is only suitable for use in a laboratory but has the advantage that the specimens can be preconditioned relative to a known humidity before the test. Other similar tests[20] measure the depth of penetration of water against time.

There are other methods of test which monitor the volume of water entering the test face. The simplest of these is the Karsten tube which is fixed to the test surface and measures the absorption of water to the nearest 0·1 ml over an area of 450 mm^2 at a head of 115 mm. The coarse measurement over a small area makes it only useful for very absorptive materials. A more advanced test is the initial surface absorption test (ISAT)[21] which uses a narrow capillary tube on a test area of 5000 mm^2 at a head of 200 mm. This test, however, gives results in units of $ml/m^2/s$ at fixed intervals of 10 min, 30 min, 1 h and 2 h. These four figures are not very meaningful except in a relative sense for comparison with other results on concrete. However, these results can be expressed as a water absorption coefficient related to weight of water absorbed per square metre of surface area related to the square root of time by the following relationship:

$$X \times 60/[(\sqrt{}(t+1)/60) - (\sqrt{}t/60)] \quad (g/m^2/\sqrt{}h) \qquad (1)$$

where X is the ISAT reading in $ml/m^2/s$ and t is the time of reading in minutes from start of test. The units of $g/m^2/\sqrt{}h$ have been used as they give relatively sensible figures on concretes which have been coated of up to 20 $g/m^2/\sqrt{}h$. An average concrete[14] would have a value between 750 and 1500 $g/m^2/\sqrt{}h$.

The tests on surface coatings have been carried out by setting up the test as specified in BS 1881: Part 5 until the point when the valve is closed after 10 min. From that point on the test is modified so that the valve to the reservoir remains closed and the total movement of the meniscus in the capillary tube is recorded from the start point to the various recording times.

The water absorption coefficient can be calculated in $g/m^2/\sqrt{h}$ as follows:

$$X_1 \times 0{\cdot}6/(\sqrt{t_2} - \sqrt{t_1}) \quad (g/m^2/\sqrt{h}) \tag{2}$$

where X_1 is the movement in the capillary tube in numbers of units of 0·01 $ml/m^2/s$ in the period from t_1 h (normally 10 min) to t_2 h.

Tests have been carried out using the ISAT apparatus on the same specimens that were used for the capillary test and reasonable correlation of the results were obtained. However, the capillary test is basically a uniaxial flow test while the ISAT method includes radial components. As a result the simple root time relationship may be less valid. The ISAT method can be used both in the laboratory and on site, although it is not widely used as the equipment is cumbersome; it can take up to 2 h to carry out a test and achieving a good seal with the concrete surface can be difficult. However, the principle of the test is sound and some of the above problems can be overcome. More work is required in improving the test to make it more convenient for site use.

There are other versions of the initial absorption test, some of which apply various pressure heads to the water which are reviewed in Ref. 14. As capillary forces of up to 20 m head of water have been measured in concrete,[15] the application of small heads to the water on the surface are not significant. A simple test that has simple equipment and is easy to do has an advantage.

The Figg water permeability test[22] works on much the same principle except that the absorption is measured in a 10 mm diameter hole in the concrete from 20 to 40 mm deep. This gives a relative value for absorption within the cover zone of the concrete, but as the flow is radial in all directions the results are not directly comparable with other methods of absorption testing. This test has another difference—it tests the absorption into a freshly cut surface of concrete, which has been found to give different values to the absorption into cast surfaces.[23]

Vapour Permeability (Diffusion)

Having measured the rate at which water is absorbed into the surface of a concrete, the other parameter of interest is the rate at which it dries out. In the tests described earlier on surface coatings, the specimens were taken off the capillary matting and placed face up in a chamber at 50% RH. Weight loss over the next 21 days was recorded. This loss was found to be up to five times slower than weight gain for some specimens. The test is not rigorous as the specimens did not all start from the same initial state of saturation.

This parameter is of importance in understanding the mechanisms of water transport into and out of concrete in the environment but there are few tests to measure it. It is clear from the above that it can take longer for concrete to lose the water it gains. Vapour diffusion is a mechanism obeying Fick's first law. A water vapour diffusion test is described in Ref. 14, Section 4.6—the dry cup test. Another variant of the same test is the wet cup test, which may be more relevant to actual conditions in that the relative humidity difference is from 100% RH to, say, 50% RH as opposed to the dry cup test where the difference is from, say, 50% RH to 0% RH. The test can only be carried out in the laboratory on specially prepared specimens which may have cut or cast surfaces or a combination of both. The difference between cut and cast faces mentioned above may well apply to these tests, as has been found for oxygen diffusion[24] which was found to be about ten times lower through a specimen with two cast faces than through a specimen with two cut faces.

The Figg air permeability test[22] gives a relative measure of the air diffusion within the cover zone of concrete, but for the same reasons given above for the Figg water permeability test it cannot be related to an absolute measure of vapour diffusion for comparison with the rate of water absorption. Schonbin and Hilsdorf[11] describe a similar test which is applied to the surface of the concrete.

Moisture Content of Concrete

All the above tests are dependent on the moisture content of concrete. In a laboratory situation the specimens can be conditioned to a known relative humidity, allowing some comparison between various results. Determining moisture content in the laboratory can only be done accurately by weighing and oven drying, which is destructive with respect to further tests. On site the moisture content of concrete is not easily determined. Moisture meters are available which work on the principle of resistance or capacitance of the concrete but these are far from accurate. Part of the problem of these meters is that they do not measure water directly and can hence be affected by other constituents. Parrot[25] uses a method based on the relative humidity of the air in a sealed hole in the concrete but this method cannot be quickly used on site. A method based on the attenuation of microwaves is under development and offers some hope of more accurate measurements.

SUMMARY

In all forms of deterioration of concrete water is a major influence, either encouraging the chemical action or conveying salts causing the damage. It is a major influence in the corrosion of reinforcement in concrete. As a major constituent of the concrete mix it influences the durability in several ways. The movement of water into concrete is a complex mechanism but, where concrete is not permanently submerged, capillary absorption is the dominant factor in water take-up followed by vapour diffusion on drying.

The movement of water into or out of concrete can be controlled in many ways by basic mix design, addition of cement replacements and other chemical admixtures, surface treatments, renderings and other claddings. Inadequate detailing can allow water to enter the fabric of the structure, causing local damage. There are tests to measure the capillary absorption of concrete, both in the laboratory and on site, which are not used extensively. Cube strength is the main acceptance criteria for newly cast concrete but durability rather than structural strength often determines the mix criteria. Yet strength and durability do not necessarily go hand in hand. The drawings should state the durability requirements as well as those for strength.

The initial surface absorption test (BS 1881: Part 5) is based on a sound principle and has the possibility of being a useful measure of the durability of concrete but as presently configured is cumbersome, lengthy and prone to leakage. Improvements in the method and equipment are possible, leading to a greater versatility with meaningful comparisons of results from both the laboratory and the field.

The vapour diffusion of concrete is much less easy to measure in the laboratory and is not possible to measure on site. Research has shown that water can be taken up by concrete much more quickly than it can lose it by vapour diffusion. Alternate wetting and drying can be more damaging to concrete than a steady state of moisture content, particularly with regard to corrosion. However, the environmental regime at the surface of concrete is little known.

The moisture content of concrete is an important factor in the mechanisms of deteriorating influences yet its measurement in the laboratory is destructive. Moisture meters are available for site use but are very inaccurate. Microwave attenuation may provide a more accurate means of measurement.

REFERENCES

1. Concretes in sulphate-bearing soils and groundwaters. Building Research Establishment Digest 250.
2. SAWAN, J., Cracking due to frost action in Portland cement concrete pavements—a literature survey. The Katharine and Bryant Mather International Conference on Concrete Durability. American Concrete Institute, SP100, Vol. 1, 1987, SP100-44, p. 781.
3. BJEGOVIC, D., MIKULIC, D. and KRAINCIK, V., Theoretical aspect and methods of testing concrete resistance to freezing and deicing chemicals. The Katharine and Bryant Mather International Conference on Concrete Durability. American Concrete Institute, SP100, Vol. 1, 1987, SP100-53, p. 947.

4. *Proceedings 8th International Conference on Alkali–Aggregate Reaction*, Kyoto, Japan, 17–20 July 1989. The Society of Material Science, Japan, and proceedings of several earlier conferences in the series.
5. BATE, S. C. C., High alumina cement concrete—An assessment from laboratory and field studies. *The Structural Engineer*, **58A**(12) (December 1980).
6. POPOVICS, S., A classification of the deterioration of concrete based on mechanism. The Katharine and Bryant Mather International Conference on Concrete Durability. American Concrete Institute, SP100, Vol. 1, 1987, SP100-10, p. 131.
7. NURNBERGER, V., Chloride corrosion of steel in concrete. *Betonwerke und Fetigteil-Technik*, Part 1, No. 9 (1984); Part 2, No. 10 (1984).
8. Structural use of concrete, Part I. Code of practice for design and construction. British Standards Institution, BS 8110: Part 1, 1985.
9. MEYER, A., The importance of the surface layer for the durability of concrete structures. The Katharine and Bryant Mather International Conference on Concrete Durability. American Concrete Institute, SP100, Vol. 1, 1987, SP100-5, p. 49.
10. SENBETTA, E. and MALCHOW, G., Studies on control of durability of concrete through proper curing. The Katharine and Bryant Mather International Conference on Concrete Durability. American Concrete Institute, SP100, Vol. 1, 1987, SP100-7, p. 73.
11. SCHONBIN, K. and HILSDORF, H., Evaluation of the effectiveness of curing of concrete structures. The Katharine and Bryant Mather International Conference on Concrete Durability. American Concrete Institute, SP100, Vol. 1, 1987, SP100-14, p. 207.
12. POTTER, R. and HO, D., Quality of cover concrete and its influence on durability. The Katharine and Bryant Mather International Conference on Concrete Durability. American Concrete Institute, SP100, Vol. 1, 1987, SP100-25, p. 423.
13. NAIK, T. R., Concrete strength prediction by the maturity method. American Society of Civil Engineers, ASCE Convention, Boston, April 1979, preprint 3576.
14. Permeability testing of site concrete—A review of methods and experience. Concrete Society, Technical Report No. 31, 1987.
15. GUNTER, M. and HILSDORF, H. K., Stresses due to physical and chemical actions in polymer coatings on a concrete substrate. *ISAP 86. Adhesion between polymers and concrete*, RILEM, Aix-en-Provence, September 1986, ed. H. R. Sasse. Chapman and Hall.
16. Protection of reinforced concrete by surface treatments. Construction Industry Research and Information Association (CIRIA), Technical Note 130, 1987.
17. ZUNZ, G. J., GLOVER, M. J. and FITZPATRICK, A. J., The structure of the new headquarters for the Hong Kong and Shanghai Banking Corporation, Hong-Kong. *The Structural Engineer*, **63A**(9) (September 1985) 255–84.
18. The performance of concrete in bridges. A survey of 200 highway bridges. Department of Transport, HMSO, April 1989.
19. Determination of the water absorption coefficient of building materials. DIN 52617, December 1984, Deutsches Institute für Normung e.v.
20. BAMFORTH, P. B., POCOCK, D. C. and ROBERY, P. C., The sorptivity of concrete. Our World in Concrete and Structures Conference, Singapore, 27–28 August 1985.

21. Methods of testing hardened concrete for other than strength. British Standards Institution, BS 1881: Part 5, 1970.
22. FIGG, J. W., Methods of measuring the air and water permeability of concrete. *Magazine of Concrete Research*, **25**(85) (December 1973) 213–19; also **36**(129) (December 1984).
23. KREIJGER, P. C., The skin of concrete, composition and properties. *Materiaux et Constructions*, **17**(100) (1984) 275–83.
24. GJORV, O. E., VENNESLAND, O. and EL-BUSAIDY, A. H. S., Diffusion of dissolved oxygen through concrete. NACE Corrosion 76, Paper 17, Houston, March 1976.
25. PARROT, L. J., Moisture profiles in drying concretes. *Advances in Cement Research*, **3** (July 1988).

23

Reinforced Concrete Bridge Protection in Northern Ireland

F. R. MONTGOMERY
Department of Civil Engineering, the Queen's University of Belfast, Belfast BT7 1NN, UK

and

A. McC. MURRAY
Department of the Environment Roads Service, Commonwealth House, 35 Castle Street, Belfast BT1 1GU, UK

ABSTRACT

Of the 7700 road bridges in Northern Ireland, approximately 1300 are of reinforced concrete construction. Some of these date back to the early years of this century but many large reinforced concrete bridges were constructed in the 1960s and early 1970s during the development of our motorway system. Since 1973 all of these bridges have been maintained by the Roads Service of the Department of the Environment for Northern Ireland. In 1987 the DoE (NI) commissioned the Civil Engineering Department of the Queen's University of Belfast to investigate the current condition of the bridges and to make recommendations on a programme for protective measures. This has concentrated on surface treatments and on cathodic protection. It is planned that a coordinated maintenance management strategy will result from this work.

INTRODUCTION

The Northern Ireland road network consists of some 23 700 km of adopted road carriageway on which there are more than 7700 bridges with spans greater than 2 m. Prior to 1973 the maintenance of the bridge stock was organised along similar lines to the system in Great Britain, with the various councils undertaking or arranging the maintenance and the Ministry of Development fulfilling a role similar to that of the Department of Transport.

When local government was reorganised in 1973, responsibility for the maintenance of the province's road network and the associated bridges was transferred to the Roads Service of the Department of the Environment for Northern Ireland. Roads Service has a headquarters office in Belfast and six divisional headquarters throughout the province. It employs almost 1400 professional, technical and administrative staff, and around 2000 industrial staff.

This paper includes details of the existing bridge stock in Northern Ireland and reports on the initial results of a research project which is currently being undertaken by the Department of Civil Engineering of the Queen's University of Belfast in conjunction with DoE (NI). The research has included an appraisal of the conditions of the reinforced concrete bridges in the province and an investigation of the various means of protecting them against further damage, so that an effective maintenance strategy can be developed.

THE NORTHERN IRELAND BRIDGE STOCK

The relatively high number of bridges in Northern Ireland is accounted for by the dense network of roads. However, of the 23 700 km of carriageway less than 10% is either motorway or class A while almost 60% is unclassified. One consequence of this largely rural and static network is that a high proportion of the bridges, for which the department is responsible, are both old and minor. For example, approximately 75% of the bridges are of masonry arch construction and the vast majority of these have short single spans. Of the remaining 1970 bridges, 1300 are of reinforced concrete construction and 165 are a combination of reinforced and prestressed concrete.

The reinforced and prestressed concrete bridges, which are the subject of this paper, are fairly evenly distributed around the province but the larger and newer structures are heavily concentrated in the greater Belfast area or on the motorway network. While there are some relatively old reinforced concrete bridges in the province, more than half of the total number were built during the 1960s and 1970s.

The condition of the reinforced concrete bridges is exceptionally good and so far large-scale repairs have only been required to a very small number of structures. In recent years, however, Roads Service bridge inspectors have increasingly noted the early warning signs of problems, such as rust staining, limited cracking or spalling of concrete. These observations, along with some more detailed reports on the worst affected structures, led to the view that more widespread problems could be on the horizon and that a formal maintenance policy would be required to limit their impact.

INSPECTION PROGRAMME AND DATA HELD

When the department took responsibility for the bridge stock in 1973, records and inspection data were held in inconsistent forms, were of poor quality in many areas and often the local foreman was the best source of information. As the importance of holding bridges data was recognised, divisions endeavoured to improve their systems, but it was not until 1980 that a standard procedure was adopted for the whole province. This procedure, which involved general inspections every 2 years and principal inspections every 6 years, was updated in 1985 to take account of the Department of Transport's record and inspection systems. The data, which are now largely complete for the province, is held on DEC Rainbow computers using the sensible solution database manager. By 1991, however, the data will have been transferred to the integrated computer system which Roads Service is currently installing at a cost of approximately £8 million. This system will offer greatly enhanced interrogation facilities.

DEVELOPMENT OF A MAINTENANCE STRATEGY

As the routine inspection programme built up a databank of results, it became clear that some of the older reinforced concrete bridges in vulnerable positions were beginning to show signs of distress. Detailed tests carried out on both damaged and sound parts of these bridges revealed the need for some form of protective systems either to limit further chloride ingress or to prevent rebar corrosion from taking place by some other means. An initial assessment of the available literature revealed that there was little consensus of opinion on the preferred protective system. It was, however, clear that the most cost effective form of action would depend heavily on the amount of chloride ingress or damage which the bridge had already suffered.

While some policy decisions were made to commence modest programmes of silane treatment and deck joint repairs, it was decided that a maintenance strategy could not be formalised until the condition of the Northern Ireland bridge stock had been objectively assessed and until the various methods of protection had been appraised. It was this background which led in 1987 to the Department of Civil Engineering of the Queen's

University of Belfast being commissioned to undertake a 3-year research project to examine the Northern Ireland bridge stock and make recommendations on its future maintenance, especially in relation to protection against further damage.

Although the funding available for this project was relatively modest, the aims were broadly similar to those described to G. Maunsell and Partners for their survey, the results of which have recently been published[1] and are being examined with interest. The Queen's University project will, however, concentrate on the Northern Ireland context.

PROTECTIVE SYSTEMS AVAILABLE

Much has been learned from hard experience in the last few years on how to construct concrete bridges to make them durable, but only some of the techniques available for new construction are of use to improve existing structures. The idea of applying liquid sealants or penetrants to the surface of reinforced concrete to provide protection against the ingress of marine or deicing salts has been common for a number of years in West Germany and North America. The results of research which examined the performance of a range of concrete sealants for the protection of bridge structures were published by the National Co-operative Highway Research Program of the USA as Report 244[2] in 1981. Subsequently other work has been reported, much of it in the USA. A useful recent summary of this work is contained in Ref. 3.

It seems that three systems, a monomeric alkylalkoxy silane, a silane-siloxane overcoated with a methacrylic polymer and an epoxy casting, are being particularly recommended. The silane system is getting more use in USA than the others, although the two-coat silane and acrylic is being used for special high quality work only, due to its greater expense. Both of these systems are being seriously considered for Northern Ireland's bridges, indeed the silane has already been applied ahead of the results of this present study. However, it is envisaged that they will be used only for recent construction or where there has not been much chloride penetration. In the case of silane it is quite likely that the criteria detailed in recent DTp draft publications[4,5] will be adopted.

For those structures showing mild to modest amounts of chloride or carbonation-induced rust staining or spalling one of a number of proprietary concrete cover replacement systems is envisaged, together with removal of the cause of the problem, if possible, often found to be badly maintained deck expansion joints.

As a protective measure for those structures showing severe rebar corrosion and spalling from chloride penetration a last resort is the installation of an induced current cathodic protection system. Most of our problems where this may be warranted are on bridge substructures. Even in the USA the experiences of this technique on substructures is limited[3] and very much in the developmental stage. One such system was installed during 1988 on a motorway bridge in Northern Ireland. All three piers have been protected after first having had all loose concrete and rust removed and then been gunnite repaired. The cathodic protection system used comprises a conductive polymer strip bonded vertically, at 24-in centres, to the pier faces. They are powered by a conventional controller rectifier. However, problems with this initial installation have been discovered during routine monitoring. It has emerged that high current concentrations in persistently wet areas have led to acid formations at the anode/concrete interface. This has caused partial debonding of some anodes. At present an assessment of alternative anode systems is being finalised and it is hoped that the installation of a new system will begin shortly. Experience gained from this trial project will enable more valid decisions to be made on any further use of the technique.

LABORATORY INVESTIGATIONS

The sealants and penetrants mentioned previously appear on the market in many forms and under many names. It is possible to apply them in a number of ways, such as single coat, two coat, full strength or dilute. They may be applied to dry surfaces or wet surfaces and anything in between. To some extent their performance is influenced by the quality of the concrete to begin with. All of these factors are being investigated in an extensive test programme designed to assess change of surface permeability due to treatment, improvement in chloride penetration resistance and, very importantly, the likely duration of any derived benefit.

The laboratory investigations are currently being extended to field work where similar, though less all-embracing, measurements are being made on trial structures already treated to detect any effects due to variability in field applications.

CONDITION SURVEY

It is obviously impossibly expensive to perform a very detailed survey and analysis of all our concrete bridges but they are all subject to the periodic inspection mentioned previously.

A second level of inspection is presently being performed where a sample selection of bridges is being submitted to a range of test techniques. The sample has been chosen to represent the range of bridge types by size and construction form, by type of environment and by age. Within the sample there are those with no known problems and others with obvious problems. It is hoped to include about 20 bridges in this survey, which is about 50% complete at the time of writing and is tending to confirm the belief that the bridge stock is in reasonably good condition.

The test techniques being used include chloride measurements on exposed faces, surface permeability and strength measurements, rebar cover assessment and half-cell potential mapping. Visual observation of sources of trouble, if any, and photographing and measuring local factors such as distance from carriageway, etc., help to place all measurements in perspective.

A third level of inspection has been performed on a small number of bridges, all of which are showing varying degrees of distress. Basically the same techniques as above have been used but here many more readings have been taken. The information gained in this series will enable threshold values to be placed on measurable parameters above which damage is likely to occur.

CONCLUSIONS

Much information has yet to be properly analysed and correlated but already some factors are beginning to emerge. It is felt that, in broad terms, likely performance can be quantified by combinations of measurable parameters.

Knowing these parameters and having available on the data base many factors which will enable the grouping together of structures of similar age, similar construction, similar environment and similar maintenance history will, it is hoped, enable a programme of planned protection to be carried out.

The results available from the early laboratory tests of the various protective systems indicate differences of performance on relatively new concrete. Accelerated weathering is proving some data on the likely long-term performance, but just how some of the applied protective systems will actually fare under real conditions is not yet clear. There is a lack of information in the literature on the useful life of most applied protective systems. This is probably the aspect which we find most difficult to incorporate in our programme of planned maintenance.

However, with the best information we can find or can generate a strategy is being devised and the hope is that time will prove it to have been profitable.

REFERENCES

1. WALLBANK, E. J., The performance of concrete in bridges. A survey of 200 highway bridges. A report prepared for the Department of Transport by G. Maunsell and Partners. HMSO, London, April 1989.
2. Concrete sealers for protection of bridge structures. National Co-operative Highway Research Program, Transportation Research Board, National Research Council, Washington, DC, December 1981, Report 244.
3. TASKER, J., HUMPHREY, M., MCANOY, R. and MONTGOMERY, F. R., *Coatings for Concrete and Cathodic Protection.* Thomas Telford, London, 1989.
4. Criteria and material for the impregnation of concrete highway structures. Departmental Standard, Department of Transport, London, 1989 (in draft).
5. Impregnation of concrete highway structures. Departmental Advice Note, Department of Transport, London, 1989 (in draft).

24

Rebar Corrosion—FBECR: The Fight to Cure the Problem

JAMES A. READ
W. S. Atkins Consultants Limited, Woodcote Grove, Ashley Road, Epsom, Surrey KT18 5BW, UK

ABSTRACT

The recent history of rebar corrosion on bridges in the UK and North America due to ingress of chlorides from road deicing is outlined. The UK investigation into the solution using fusion-bonded epoxy-coated reinforcement (FBECR) by both TRRL and the author's company are referred to. The main problems and faults to the reinforcement and concrete on bridges is outlined together with general remedial measures.

An example of a proposed replacement cross-beam for a bridge substructure is put forward using FBECR as a replacement for uncoated bars. Finally, the main methods of FBE application are described with surface preparation and homogeneity of the cured film being discussed. A résumé of UK application using FBECR is given at the end of the paper.

THE RECENT HISTORY OF REBAR CORROSION ON BRIDGES

The UK situation on concrete deterioration of bridges constructed with uncoated reinforcement and subjected to road de-icing using chlorides and in the marine environment is increasing. All parts of the UK are experiencing these problems.

The recent report on the Midlands links motorway viaducts reported in the *New Civil Engineer* (February 1989)[1] is a typical example, with the DTp reported as saying 'it is a latter-day Forth Bridge requiring monitoring, maintenance and repair for the rest of its useful life'.

Another example is the Tay Bridge, where salt from the marine waters has penetrated the concrete piers of the bridge in the splash zone, leading to severe attack on the uncoated reinforcement.

The North American situation is as bad if not worse. The recent study mission organised by the Institution of Civil Engineers and supported by the Department of Trade and Industry[2] to study surface coating, cathodic protection and epoxy coating of reinforcement to concrete reported that

> 'the overall picture of highway bridges, however, is one of deteriorating condition ... Figures quoted for the USA of between £9000 and £14 000 million for necessary repair and replacement of bridges....
>
> One state alone—Pennsylvania—has 22 500 bridges with spans in excess of 8 metres and 35 percent classified as structurally deficient or functionally obsolete.'

The report goes on to state after field surveys:

> 'Epoxy coating was found to be effective in protecting the reinforcing steel in chloride-contaminated concrete during the nine years of exposure, and even where the coating was damaged there was only superficial corrosion.'

In New York State:

> 'Confidence about the performance of epoxy coatings is so high that no evaluation of the technique is being included in the current Strategic Highways Research Programme.'

Babaei and Hawkins, in their 'Evaluation of bridge deck protective strategies',[3] after examining ten different strategies used throughout the USA and noting that a number of states had discontinued using certain types of protective strategy because of

> 'problems such as cracking in and debonding of overlays, wear and stripping of asphalt overlays, or the ineffectiveness of some types of sealers....'

goes on to state:

> 'Among the strategies used as standard practice, epoxy coating of bars is the most popular. Forty-one states use this method, either alone or in combination with other strategies.'

The TRRL Report 667 on 'FBECR in bridge decks' by J. Willis,[4] published in 1982, concluded that

> 'epoxy coating of the top steel in addition to current waterproofing practice would provide—at relatively little extra cost—additional assurance that the reinforcement would be adequately protected throughout the life of the bridge.'

An investigation by W. S. Atkins into the specification performance of

FBECR and the merits of ASTM A775 was reported in the author's paper presented at the CIRIA/BSE Conference in Bahrain in October 1987.[5]

The work carried out by R. R. Bishop of TRRL on behalf of the Department of Transport is published in their Application Guide 6: The specification of epoxy-coated reinforcement bars, 1987.[6] This concludes that:

> 'It is not practical to use the American Standard ASTM 775 (in the UK) because the division of tests between those intended for quality assurance of coated bars and those for certification of powder is not satisfactory.'

The work carried out by both Atkins and TRRL reach similar conclusions on the shortcomings of the American specification in regard to its suitability for use in a UK or European context. Including as reasons the different types of steel used for rebar between North America and the UK/Europe, bond/adhesion tests, measurement of coating thickness, coating damage and repair, etc., Bishop makes the point that:

> 'This is not to be construed as a criticism of epoxy-coated bars. On the contrary, there is much evidence of the value of epoxy-coated reinforcement bars in concrete exposed to a corrosive environment.'

MAIN PROBLEMS WITH CONCRETE DETERIORATION ON BRIDGES

The main cause of premature deterioration in the reinforced concrete road and bridge stock in both the UK and North America is penetration of chlorides, used for deicing, into the concrete and reaching the uncoated reinforcement, depassivating the normally alkaline layer around the surface of the bar after which, provided moisture and oxygen are present, corrosion of the rebar takes place, leading to loss of bond between steel and concrete, concrete cracking and eventually spalling of the concrete and loss of structural integrity. Whilst different waterproofing strategies have been employed between and within the UK and North America, to a large extent, the above effects have occurred, are still occurring and will certainly continue for the foreseeable future on those existing structures having uncoated reinforcement and where chlorides are used for road deicing. Secondary causes may be attributed to a number of other factors, such as reduced cover to the reinforcement, new design techniques, changes in raw materials, the constant search for more slender structures coupled with commercial pressures associated with new types of contract, performance-type specifications, and an increasing tendency towards a lack of on-site impartial and independent inspection and checking during construction.

The methods and techniques of inspection, surveying and monitoring of concrete deterioration in bridge structures have been developed over the past 20-odd years but there is still a lot to learn. They include visual surveys of surface condition, permeability and porosity measurements, testing for chloride levels and carbonation of the concrete at various depths, half-cell potential surveys to locate sites of active reinforcement corrosion, petrographic examination of aggregates to check for alkali–silica reaction, delamination surveys and measurement of ultrasonic pulse velocities. The interpretation of electric potential methods and non-destructive permeability and porosity measurements are particularly difficult. They are sensitive to the methods used when carrying out the tests and results can vary due to changes in environmental conditions.

MAIN FAULTS TO CONCRETE ON BRIDGES

Notwithstanding the differing designs and techniques being utilised, the concrete deterioration on bridge structures, which nearly all use concrete with uncoated reinforcement, is remarkably similar and the main faults exposed over the last 10–15 years can be listed as follows:

- Cracking and spalling of concrete columns together with corrosion of reinforcement due to chloride ingress either from spray from passing vehicles, leakage down columns from above or within the splash zone when columns are standing in marine waters. This is often associated with lack of specified cover and a loss of cementitious material in the concrete matrix (see Fig. 1).
- Cracking and spalling of concrete in cross-beams and shear walls, often with efflorescence and corrosion of reinforcement, again due to chloride ingress from spray and condensation from below, leakage from above via construction/expansion joints, etc., and often associated with less than specified cover (see Figs 2–4).
- Cracking and spalling to deck slabs tending to be limited to ends of slabs at construction and expansion joints; also along kerb lines, particularly where there are drainage outlets. Deterioration damage has also occurred where waterproof membranes have been damaged either during initial construction or during resurfacing operations.
- Contamination and deterioration of concrete to plinths and bearings supporting deck beams due to leakage from above and bad drainage around bearings.
- Disintegration of central reserve upstands which often carry a safety fence and lighting columns. Due to initially frost damage and then

chloride ingress followed by cracking and spalling. This also occurs to precast concrete edge parapets (see Fig. 5).

Often associated with the above, most of which can be detected because of their surface visual effects, is widespread localised pitting corrosion. This can be much more dangerous as it can proceed undetected because the products are not expansive in the early stages and therefore the usual tell-tale signs of cracking and spalling do not occur.

GENERAL REMEDIAL MEASURES

If detected early enough many remedial measures may be effective in halting and preventing further deterioration progressing. These include providing better drainage, sometimes by installing extra guttering and downpipes;

FIG. 1. Spalling at the base of a column.

FIG. 2. Typical cross-beam with cracking and rust stain.

FIG. 3. Leakage of water onto a cross-beam.

FIG. 4. Shear wall with cracking and efflorescence.

FIG. 5. Disintegration of the central reserve upstand.

cutting out damaged concrete and replacing with or without additives; and the use of various coatings, some surface, some of the impregnation type and sometimes both. Remaking joints using better sealants and mastics and cathodic systems may also be effective for certain situations.

REPLACEMENT OF CONCRETE MEMBERS

However, if the deterioration has progressed beyond the stage of any repair remedial measure being suitable, the replacement of the concrete members becomes necessary. If this is the case then the author would recommend that coated reinforcement could be used in the replacement members even if other measures are incorporated. The extremely high disruptive costs associated with repairs to road-running surfaces and replacement of substructure members whilst traffic is allowed to continue overhead make the small extra cost to coat the rebar insignificant when weighed against the possibility of recurrence in the future.

The replacement of road-running members such as parapets, central reserve upstands and parts of the deck are relatively straightforward although highly disruptive. The replacement of substructure elements, however, is even more costly and therefore a solution is given here for a typical composite concrete deck supported on steel longitudinal beams carried by reinforced concrete cross-beams and columns, and where the cross-beams have to be replaced (see Fig. 6). Figure 6 shows a typical cross-beam layout.

The major operation necessary before the existing cross-beam and/or

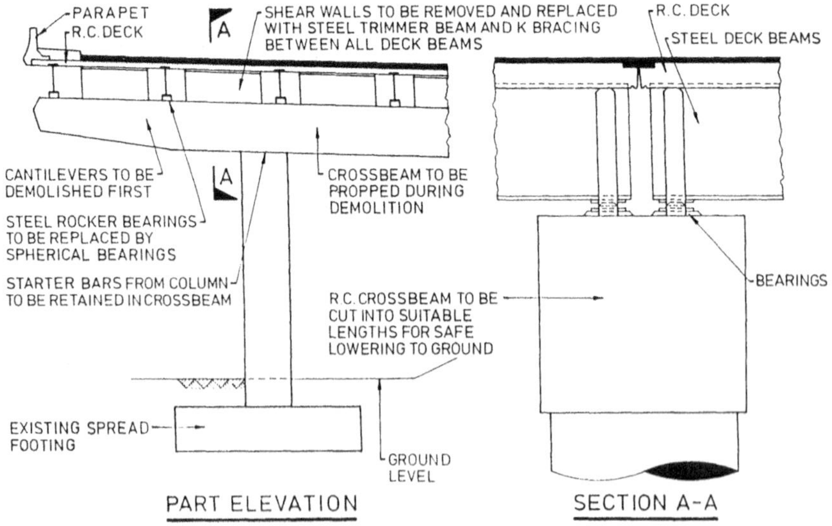

FIG. 6. Typical cross-beam layout.

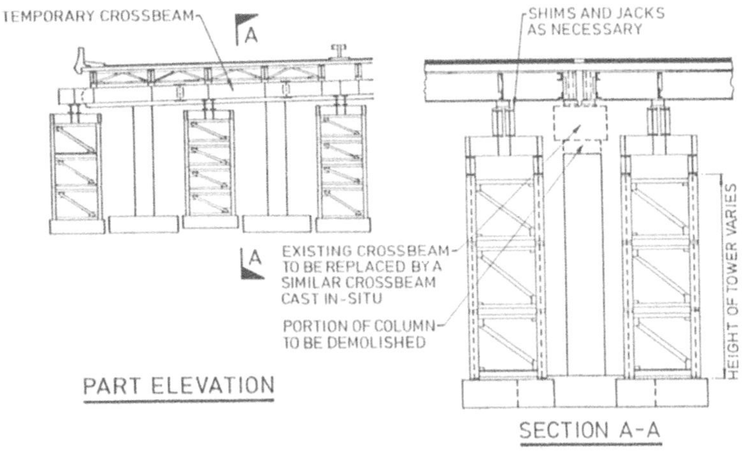

FIG. 7. Proposed propping solution.

columns can be removed/repaired is to relieve the loads from the running deck and this has to be done by temporary supports. Figure 7 shows a proposed solution for propping the carriageway. This is obviously very expensive, but if a number of beams have to be replaced then the propping can be designed to be largely reusable.

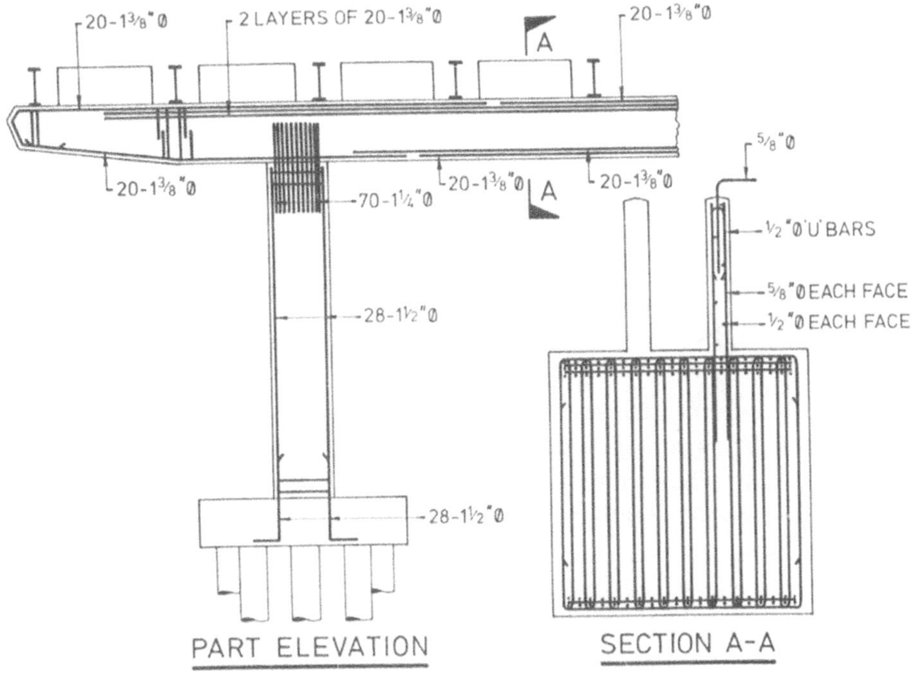

FIG. 8. Reinforcement of a cross-beam.

A typical uncoated rebar in a cross-beam is shown diagrammatically in Fig. 8.

CHANGES NECESSARY IF FBECR IS USED

Assuming a cross-beam of 33 m overall length and 1·5 m cross-section with reinforcement as in Fig. 8 using mainly 32, 20 and 12 mm rebar, then:

- All rebars to be changed to high yield deformed type II to BS 4449, coated to meet the draft British Standard but amended to meet DTp and W. S. Atkins Consultants Limited specification, cut and bent to the new BS 4466 complete with patch kits, tying wire and plastic end caps.
- Lap lengths to be increased by 20%, in order to allow for the present doubts on bond criteria.

Bond

The following results (Table 1) have been produced from current unisteel deformed bars coated by Allied Bar Coaters of Cardiff and tested in accordance with BS 4449, Appendix B1, requiring six bars to be tested with none exhibiting free end slip greater than 0·2 mm when loaded.

TABLE 1

Bar diameter (mm)	Free end slip (mm)	
	Uncoated	E-bar
32	0·017	0·082
20	0·005	0·062

DISCUSSION ON COSTS OF USING FBECR

It is estimated that the cost of replacing a cross-beam of the type described above can be as much as £500 000 if the full cost of temporary works is included. The rebar tonnage is about 25 t. Replacement of uncoated rebar would cost about £9000 delivered to site. The cost of coating would add about £7000, also cut, bent and delivered to site. There should be no difference in fixing costs as the handling underneath a live roadway will have to be done with care and by skilled steel fixers anyway.

It is obvious from these figures that the small increase in the overall cost of replacement is insignificant when weighed against the increased assurance that the use of coated bar will give for durability. Whilst this is a small percentage of the total cost of replacement for new works with no temporary works the percentage additional cost will be significantly higher but still justifiable in terms of the confidence in the future durability and life.

SOME POINTS ON APPLICATION/QUALITY OF FBECR

There are a number of points on the manufacture of FBECR which if not properly controlled and monitored can lead to the possibility of disastrous breakdown in the protective properties of the coating. A few of the main ones (in the author's experience/viewpoint) are given as follows.

Method of FBE Application

The three main methods of application of FBE on rebar are:

(i) Electrostatic spraying—for straight bars.
(ii) Tribostatic spraying—for straight bars.
(iii) Fluidised bed dipping—for complex shapes, fittings and welded mesh reinforcement cages.

The first two involve spraying FBE granules electrically charged from specially designed nozzles on to the preheated bars. The bars have to be temperature controlled between 220 and 260°C at the time the granules touch to allow proper melting, flowing and then curing. The electric charge induced in the granules is considerably different by the two methods. In electrostatic spraying it is induced by injecting epoxy powder granules into an airstream around an electrically charged discrete electrode which imparts a charge of about 45 000–50 000 V to the granules. For tribostatic spraying the charge is induced by circulating the powder in a spiral mode between PTFE surfaces, the granules becoming charged by friction; in this case the charge imparted is about 25 000–30 000 V.

The plant, equipment and nozzles for the two methods are considerably different, but what is even more important is that the granules themselves have to be specially formulated for the two different methods. For fluidised bed dipping the granules are held in suspension in a bath and the preheated cage dipped into the bath and held for a time sufficient to allow the fusion to take place. For this method the granules do not need to be electrically charged and as there is usually no requirement for bending of the bars after application the formulation of the granules is again different to the other two methods. (See Fig. 9 showing an example of a cage coated by this method.)

If the granules formulated for one of these processes is used for another then it is likely that the fusion bonding may not be effected properly.

FIG. 9. A prefabricated welded reinforcement cage just having been fusion-bonded epoxy-coated by the fluidised bed dipping technique at Webb Coatings Systems, Birmingham, on 29 November 1988 (courtesy of 3M UK).

Surface Preparation

The surface preparation of the bar prior to coating is perhaps one of the most potentially dangerous. If this is not carefully monitored then defects result which will often only become apparent in use. The three main points to watch are as follows.

Surface defects such as shelling, rolling laps, blowholes, fissures, slag inclusions, slivers, scabs, etc., can be detrimental to the adhesion of the coating and its subsequent protective properties. These defects are liable to be more prevalent on rebar because much reinforcement is manufactured from steel of a 'lower quality' than would be permitted for other structural uses and the rolling processes used to produce the deformations will often promote these defects. The abrasive blasting used for cleaning can expose these defects. The profile and amplitude of the blasted surface will be dependent on the type of grit and shot used, pressures, number of times the material is recycled and the equipment used. It is suggested that an angular profile produced by a mixture of grit and shot abrasive blast cleaning to give an amplitude in the range 50–75 μm should give a suitable surface for subsequent coating.

Of course the surface must be clean of rust, oils, greases, chemical contaminants and dust residues, and the cleanliness should be checked immediately prior to the coating process.

Figures 10 and 11 show two examples of the above-mentioned defects.

FIG. 10. Blast profile showing corrosion deposits (× 200).

FIG. 11. Blast profile showing fine cracking of substrate (× 200).

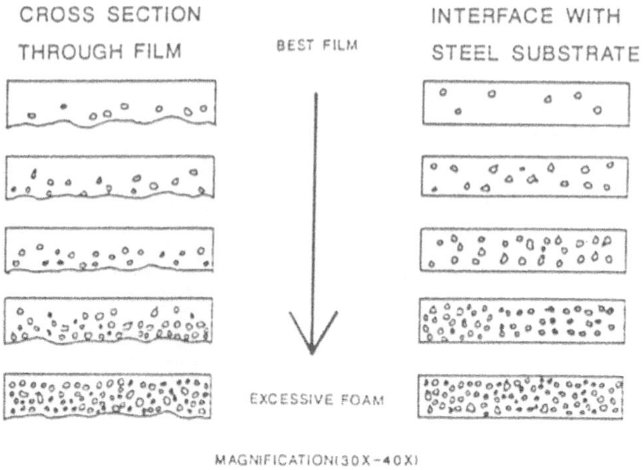

FIG. 12. Foam Evaluation Guide (voids/porosity).

Obviously cure of the coating must be effected and the only method of satisfactorily checking this is by differential scanning calorimetry to ensure that the powder has been taken through its glass transmission temperature. This has been amply dealt with elsewhere.[7]

Homogeneity and density of the coating are often missed and are of equal importance to the satisfactory performance of the finished coating.

Voids within the film can affect its adhesion, elasticity and protective qualities. In electrostatic and tribostatic spraying the speed at which the

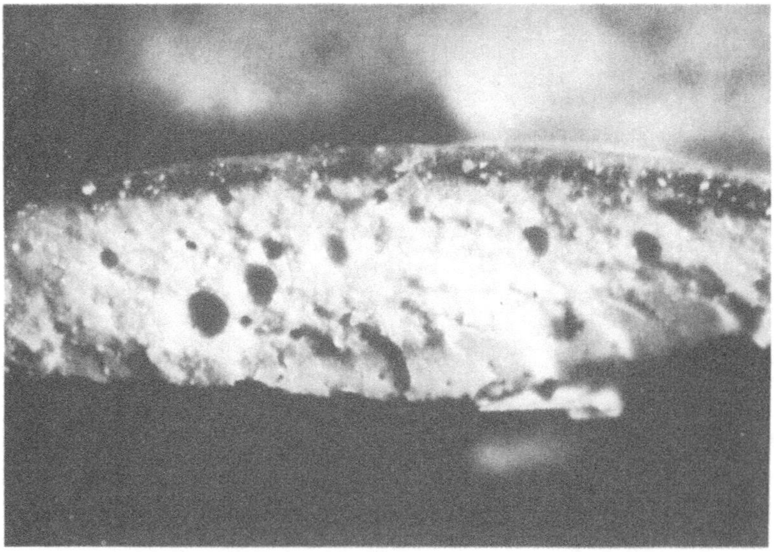

FIG. 13. Cross-section through coating showing voids.

rebar is passed through the spraying area and time allowed for the powder to gel are critical.

A rating system was developed in America by Bell & Stephens Laboratories of Houston, Texas, called a Foam Evaluation Guide, to form a basis for evaluation on a five-point system. Both the through film and coating substrate interface foam level are given (foam = porosity/voids). Figure 12 shows the Bell & Stephens Foam Evaluation Guide, produced by courtesy of 3M UK, and Fig. 13 shows an example of the voids in a coating. It is suggested that this is a further significant point affecting the quality of coating and is worthy of consideration as an addition to the testing of coatings.

UK APPLICATION USING FBECR

The first UK applications were in the early 1980s using FBECR imported from North America in order to evaluate the material. Some were used on a bridge on the Colwich loop road as a trial organised by Nottinghamshire County Council and 42 t was used on a test section of concrete reinforced pavement on the M18 organised by the DTp. This material was American rebar coated in the USA.

The first major application was in 1987 for the 'Cardiff peripheral distribution link road' built by South Glamorgan County Council. Two hundred tonnes of FBECR were used in the parapets of a 1·5 km long dual carriageway viaduct. FBE starter bars were cast into the precast deck segments and then further FBECR was fixed to the startèr bars once *in situ* to provide a cast in-situ bridge parapet. This material was mainly 16 mm diameter produced by Allied Steel and Wire shipped to the USA and coated on Mid-West Pipelines coating plant with 3M powder.

In 1988 the first two railway/road bridges extensively using FBECR were built. One carrying rail traffic over the new Worcester southern link road spanning 15·0 m used 26 t of FBECR. The other allowing a dual carriageway to be built under the London to Cardiff main line at Reading spanning 27·2 m used 130 t of FBECR. Both bridges used the FBECR in the soffit of the deck and parapets, i.e. the areas of the bridge most liable to chloride penetration from road traffic spray. The lower mat of reinforcement in the decks, all links between upper and lower mats of reinforcement, all chairs and spacers to support prestressing ducts, and all parapet steel used coated bar. Diameters used went up to 40 mm for some of the straight bar with links out of 10 mm. British Rail designed and specified both bridges, with the coated bar being supplied by Allied Bar Coaters (ABC) of Cardiff and HD34 epoxy powder coating by International Paints plc. These are the first UK bridges using all UK-produced coated bar (see Ref. 8 and Fig. 14).

FIG. 14. First UK bridge using all UK-produced coated bar.

In 1989 seven box culverts were constructed to carry a road across the River Itchen at Winchester; one of the seven has been built using FBECR, allowing a direct comparison with other box culverts using uncoated rebar in the same project. These were by Hampshire County Council with material supplied by ABC. Nottinghamshire County Council specified epoxy-coated dowel bars for concrete carriageway repairs, and during the summer of 1989 over 16 000 dowel bars were used in the new construction of expansion and contraction joints on sections of the A1 north road.

Two projects specified to use FBECR are:

(i) Dornie Bridge for the Scottish Development Department, a replacement bridge near the Kyle of Lochalsh in western Scotland crossing Loch Long with ten spans each of 26 m. Work started in the summer of 1989 and will use over 150 t of coated bar in the piers and edge beams. FBECR was chosen because of the severe marine environment and the poor durability of the existing structure as a result of chloride contamination.

(ii) The Ness Viaduct for Scottish Railways, being built in 1989/90 to replace a bridge washed away during flooding in the spring of 1989, is using 30 t of coated bar.

In addition to road and bridge works, a number of other applications where coated bar has been used in the UK are worthy of note, namely:

- The Craibstone–Dyce link road phase 3B has used 28 t of coated bar in a structure to carry the new road on an embankment over the 38-in diameter oil pipeline from the Forties field to the Grangemouth refinery. Grampian Regional Council chose this solution to protect the steel in the structure from stray currents in the ground, which might result because the pipeline is cathodically protected. If the structure collapsed and the pipeline was damaged, the resulting environmental and oil production losses could be catastrophic.

- Coated bar is also being used on electrical sub-station foundations where there is a danger of inducing stray current corrosion in the reinforcement.
- Two 22-m high reinforced concrete salt silos were built in 1988 at the Weston Point Salt Works, Runcorn, Cheshire, for ICI Chemicals and Polymers Ltd. Used 70 t.
- Treated effluent outfall for the NWWA built in 1989 at Sandon Docks in Liverpool. Used 50 t.
- Water storage tanks built in 1989 for the Central Scottish Water Board at Blairlinnans, Scotland. Used 50 t.
- Sea defences for the Anglian Water Authority.

DISCUSSION

It is the author's personal view that unless and until we as engineers are able to produce impermeable concrete that does not crack, and whilst chlorides are used for deicing, then protection of the reinforcement at the surface of the reinforcement will be necessary. This is in addition to any other measures employed, including surface coatings and cement additives.

The science/art of tetrology, which has progressed since the days of Brunel and the great engineers of the past, is demonstrably still unable to allow the engineering and accounting professions to predict with any reasonable expectation of accuracy the true maintenance costs of major structures. Yet qualitative and financial decisions affecting the basic durability and engineering designs continue to be used based on false premise.

Civil engineering construction still relies on labour-intensive operations carried out by all levels of skills, in all types of weather, and often against heavy time and cost pressures.

It has always been necessary to employ safety margins and extra protective measures, often in parallel, to give added assurance to the safety and integrity of our structures.

Engineers learned many, many years ago that the best method of preventing corrosion was to put the protection where the corrosion occurs. This is why fusion-bonded epoxy-coated reinforcement should be our answer to the current problem.

ACKNOWLEDGEMENTS

The author would like to express thanks to his company, W. S. Atkins Consultants Limited, plus many outside organisations that have been involved in the research into the production of FBECR and its testing which have led to the comments and recommendations within this paper. He would also like to express his special thanks to Mr N. Woodman, Allied Bar Coaters; Mr B. Cruse and Dr P. Langford, 3M UK; Dr K. McLeod, International Paint Powder Coatings; and Mr J. Wallace, NEI Research and Development.

REFERENCES

1. Record repair planned for Midlands Links, *New Civil Engineer* (28 February 1989). Repair and Maintenance of the Midlands Links Viaduct. Working party report 1988 by Department of Transport, W. S. Atkins & Partners and G. Maunsell and Partners.
2. Coatings for concrete and cathodic protection of reinforcement, report of a mission sponsored by the Institution of Civil Engineers and supported by the Department of Trade and Industry held at ICE, London, 25 January 1989.
3. Babaei, K. and Hawkins, Neil M., Evaluation of bridge deck protective strategies. Concrete International Design and Construction. *J. American Concrete Institute* (December 1988).
4. Willis, J., Epoxy-coated reinforcement in bridge decks. Transport and Road Research Laboratory Supplementary Report 667, 1982. Department of the Environment, Department of Transport.
5. Read, J. A., Examination of FBECR and ASTM A775 for use in the UK and Middle East. CIRIA/BSE Conference, Bahrain, October 1987. FBECR, The Need for Correct Specification and Quality Control, University of Sheffield, May 1989.
6. Bishop, R. R., Application Guide 6: The specification of epoxy-coated reinforcement bars. Department of Transport. Transport and Road Research Laboratory, 1987.
7. Manley, T. R. and Scurr, G., Thermal analysis of epoxy anti-corrosive coatings. In *Coatings and Surface Treatment of Corrosive and Wear Resistance*. Ellis Horwood, London, 1984.
8. Horseler, J., British Rail Western Region, epoxy-coated reinforcement—A designer's viewpoint. University of Sheffield, May 1989.

INSPECTION AND MONITORING

25

Experiences with the First Generation of Prestressed Concrete Bridges in Germany

BERNHARD GÖHLER

Leonhardt, Andrä und Partner, Lenzhalde 16, 7000 Stuttgart 1, FRG

ABSTRACT

Four major bridges built in the years 1950–1953 had to be inspected and refurbished by the author. The bridges and their main problems are described. Thin webs and slabs and small construction depths were typical of the designs.

The main defects encountered were: uncompacted carbonated concrete, ungrouted tendons and defective waterproof membranes. These are presented and compared for similar bridges. Conclusions are given regarding requirements for maintenance, guides for inspection and lessons for the design of new bridges. The refurbishments described give a general indication of the 'state of the art' of bridges in Germany.

DESCRIPTION OF THE BRIDGES AND THEIR MAIN PROBLEMS

Bridge over the Danube at Ulm—Gänstorbrücke—built in 1950 (Fig. 1)

The bridge system is a frame with a deck span of 82·4 m. Ties anchor the cantilever sections to the footings.

With a depth of 1·2 m at mid-span the deck is very slender. Remarkable oscillations were registered by the pedestrians as the first trucks passed over the new bridge, and discussions started on how long such a bridge would last.

Similar discussions have been held about temperature effects in such a wide spanned continuous structure. The information gathered within the last 40 years has proved encouraging, both in relation to oscillations and temperature.

The concrete proved to be excellent and the ultimate cube strength is now

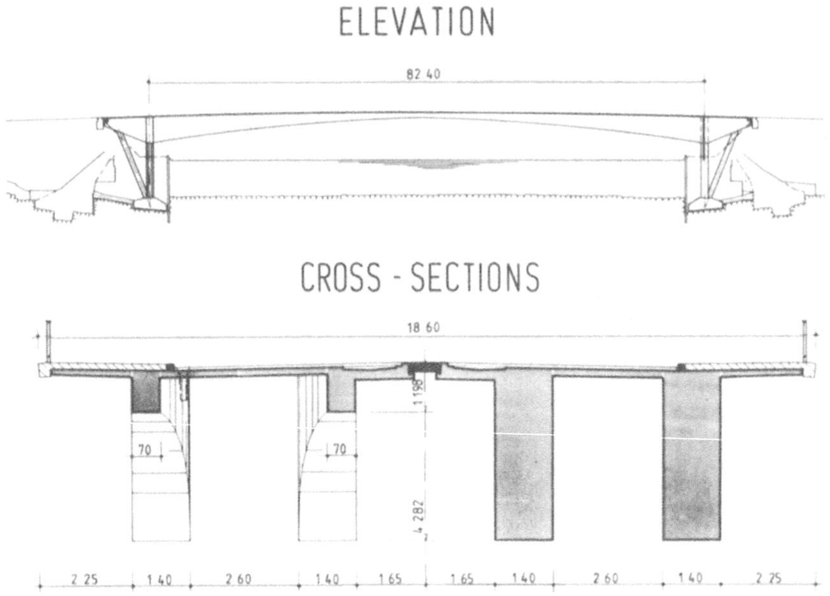

FIG. 1. Bridge over the Danube at Ulm, Gänstorbrücke, elevation and sections.

about 70 MPa. The carbonation was only a few millimetres. In the deck more than 30% of the ducts are ungrouted or partially grouted. The ducts in the ties running down to the footings were filled with water up to the level of the Danube. One of the 26-mm diameter bars was taken for testing and we were pleased that the original designers had called for an extremely good quality steel with an ultimate strength of 900 MPa. The test showed no significant losses in strength but it was not possible to determine the percentage of deterioration which had taken place as only the outer bars could be inspected within limited areas.

The amount of reinforcement in the bridge is extremely small, as in all bridges of that time, so there was no risk of a serious ultimate limit state occurring. After some discussion it was decided to strengthen the ties but not the deck as it could be monitored more easily and possible cracks would indicate losses of load capacity.

Bridge over the Main-channel at Bamberg, built in 1953 (Fig. 2)

The prestressing system used was the same as in the first bridge. The span is 68·0 m but the cross-section of the deck is a twin box girder.

At first it was thought that similar defects would be found, but the ducts in the ties were carefully grouted and there were only a few instances of insufficient grouting in the deck. It was obvious that the bridge was heavily deteriorated. Rusted rebars and ducts could be seen in different places from

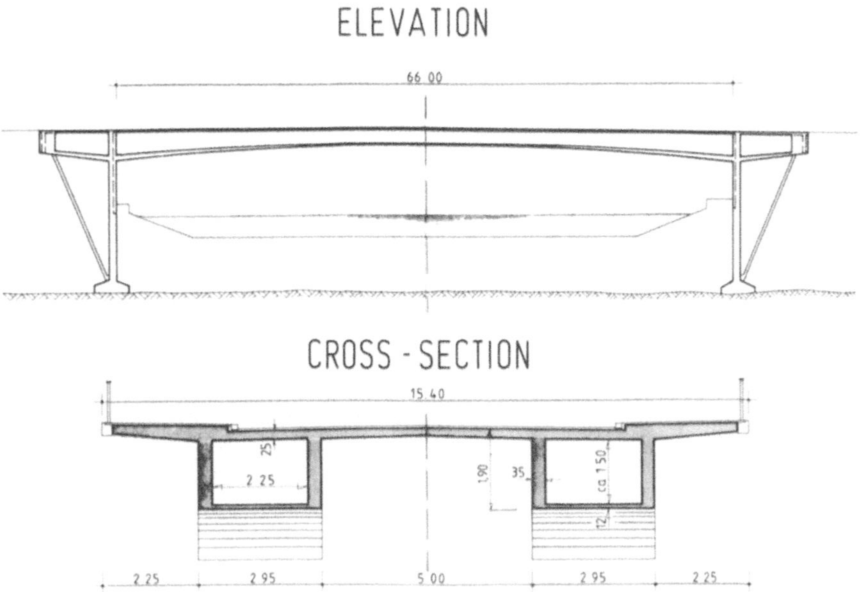

FIG. 2. Bridge over the Main-channel at Bamberg, elevation and section.

underneath, and water was dripping through top and bottom slabs. The concrete was fairly sound but large parts of the bottom slabs showed evidence of incomplete compaction.

Carbonation was up to 10 mm deep but in the uncompacted areas carbonated zones were found up to about 150 mm deep. Several honeycomb areas were found at the couplings of the bars on the bottom of the webs, as shown in Fig. 3.

The waterproofing was defective and water mixed with deicing salt dissolved the cement at the top of the deck slab. Pieces of wood left in the slab were rotten and along these most of the water dripped into the box and then through the bottom slab. In some parts of the surface the concrete had changed into sand to a depth of 30 mm. Fortunately there was not much deterioration at the bars as they were mostly arranged in the webs outside the wet zones.

If the defective waterproofing and uncompacted concrete zones had allowed water to run along ungrouted ducts the result of the inspection would almost certainly have been to recommend demolition.

Bridge over the Danube at Untermarchtal (Fig. 4)

This bridge, designed by Professor Leonhardt and built in 1952–53, was one of the first and for some time the longest continuous concrete girder with a total length of 375 m. The end spans are 62 m long and the mid-spans are

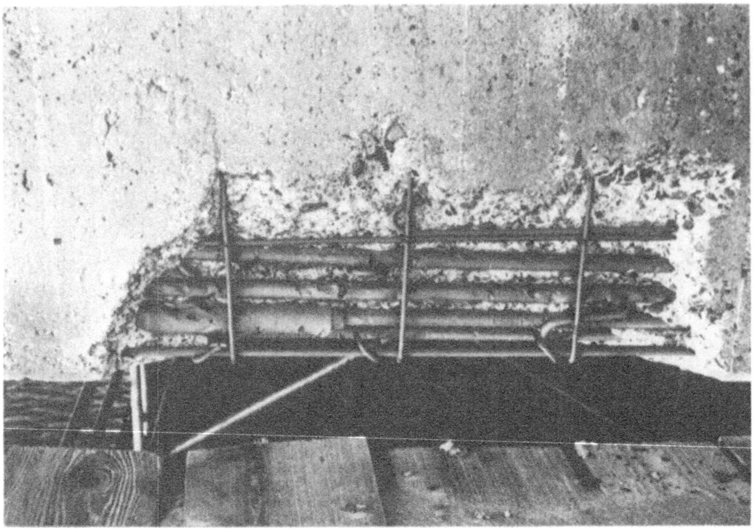

FIG. 3. Bridge over the Main-channel at Bamberg, honeycomb zones.

70 m long. The cross-section is a double T-beam with a depth of 4·1 m and therefore quite slender.

There were long discussions about how a continuous concrete girder could withstand settlements. After nearly 40 years the bridge showed no problems due to settlements or temperature but there were other problems.

First, it was obvious that something was wrong at the top of the deck. There were wide cracks in the asphalt. The waterproofing was destroyed in places and in other areas the upper 20–30 mm of the concrete had disintegrated. It had changed into sand and parts of the reinforcement were

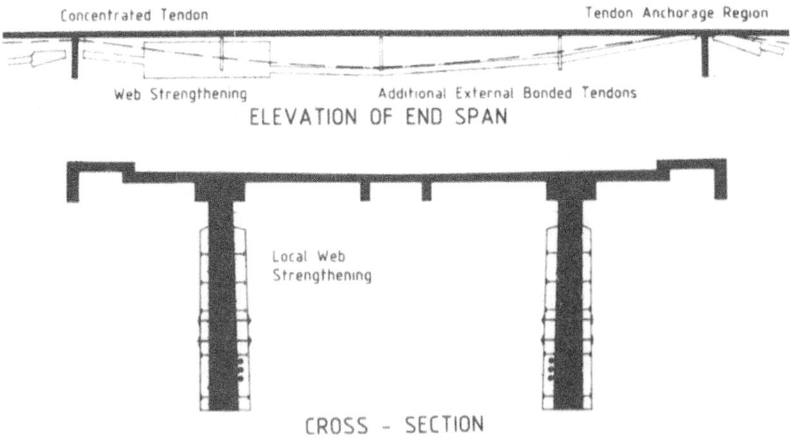

FIG. 4. Bridge over the Danube at Untermarchtal; strengthened end span.

shining as if sandblasted, as the sand moved by the traffic was rubbing against the steel.

After repairing the concrete surface and installing a new waterproof membrane and asphalt layer the bridge was inspected carefully from underneath. In the first span a long crack was found on the inner side of the web along the line of the cable. The crack had been injected some years ago and had opened again about 1 mm.

By partially removing the concrete cover to the cable we found that on a length of about 4–5 m the cable was only half grouted. The ungrouted strands were heavily corroded—some of them broken—and corrosive chemicals such as chlorides could be found on the surface of the steel.

Drainage of water from the deck was not possible and the excess water in the grout had dissolved the chemicals from the grout and started corrosion on the surface of the cables. About 50% of the ungrouted strands had deteriorated and the damage was so severe that the cable had to be inspected on its total length. At least the same defect in the same position and with similar deterioration was found at the other web.

Bridge over the Neckar at Stuttgart, Rosensteinbrücke (Fig. 5)

This bridge, designed by Leonhardt in 1952, is quite similar to that of Finsterwalder but the cross-section with longitudinal boxes, transverse beams and again longitudinal beams looked more vulnerable. The concrete quality again was excellent and only minor defects were in evidence, such as water draining through the deck and some corrosion of bars and tendons.

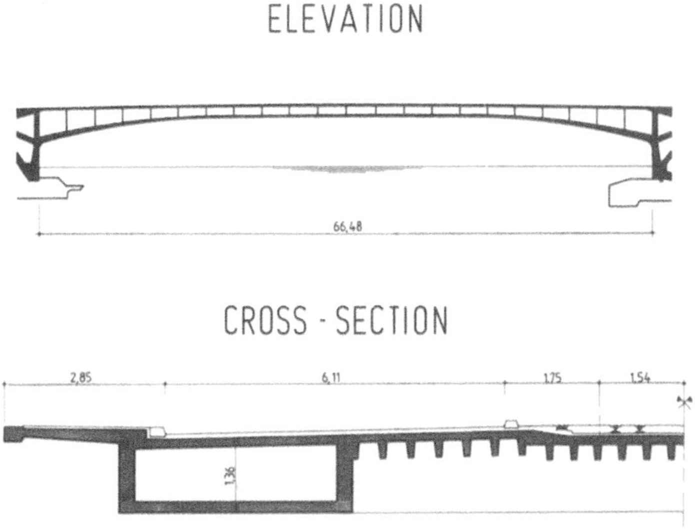

FIG. 5. Bridge over the Neckar at Stuttgart, system and section.

The main reason for this was the difficulty in fixing the tramrails waterproofing system on to the thin deck slab.

COMPARISON OF THE DIFFERENT BRIDGES AND THEIR MAIN DEFECTS

In Table 1 the problems found at the four bridges are compared, supplemented by experiences with similar old bridges. Table 1 shows that the most common problem was that of defective waterproof membranes. Honeycomb concrete areas which carbonate after some decades are in combination with ungrouted tendons. In some cases the resulting corrosion was so severe that some smaller bridges had to be demolished.

TABLE 1

Bridge over river	Main problems		
	Uncompacted carbonated concrete	Ungrouted corroded tendons	Defective waterproofing
Danube at Ulm	Rare	Often	None
Main-channel at Bamberg	Often	Some	Often
Danube at Untermarchtal	Rare	Some but one heavy	Often
Neckar at Stuttgart	Rare	Some	Often
Other similar bridges	Some	Some	Often

This deadly combination of leaking membranes, porous concrete and ungrouted tendons put bridges at severe risk. The smaller the tendons the more often there will be ungrouted areas but corrosion will be slow.

Bigger tendons will have less ungrouted areas, but if there is one near the concrete surface deterioration will be relatively fast. Through small cracks wet air will be sucked inside the tube and continue the corrosion process. With modern rehabilitation techniques it is not difficult to repair such local defects, as shown in the next section.

It is difficult to find and rehabilitate small ungrouted ducts, especially if there are several layers. Mostly they are used for transverse post-tensioning of deck slabs and therefore additional safety in such parts of bridges is recommended.

Figure 6 shows ungrouted transverse tendons in a cantilever slab, which was cut off with a diamond saw.

Concerning the fear of carbonation, the experience with these bridges, and

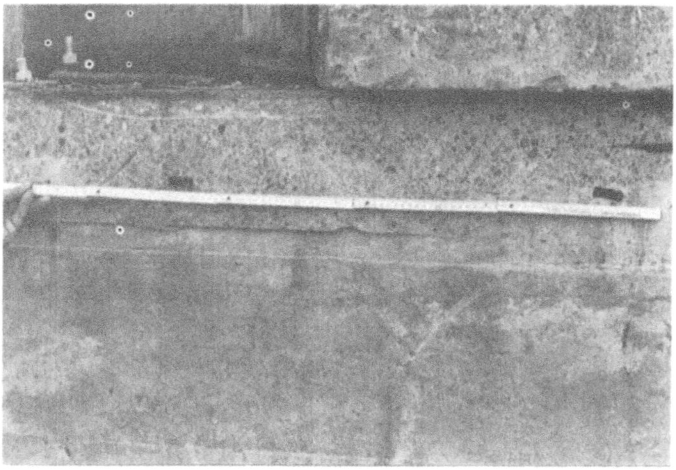

FIG. 6. Ungrouted tendons in a broken-off bridge.

in general, is that carbonation in bridges is not a problem if the W/C ratio is below 0·45 and compaction is done carefully.

REHABILITATION OF THE BRIDGES

The state of the art nowadays in Germany is to use an elastic waterproof membrane glued to the concrete, either a bitumen or a polyurethane membrane. Such membranes were installed on all four bridges. In three cases it was a bitumen membrane and for the bridge at Untermarchtal it was polyurethane.

Replacement of deteriorated concrete zones is well known, but materials based on epoxy resin are avoided as it is difficult to get a mixture which is simultaneously dense and with similar properties to the parent concrete. We prefer materials which are cement based, probably modified by polycarbonates.

Ungrouted tendons are a special problem of prestressed concrete demonstrated on all four bridges. The first aim is to keep moisture away from the tendons. Waterproofing on the top is vital, and possibly coating of vertical surfaces. Grouting of all hollow spaces is also vital to avoid the sucking of moisture inside the ducts by temperature effects.

For the two bridges designed by Finsterwalder tensioned with the 26-mm diameter bars it was not possible to grout all the small empty ducts. The bridge over the Main therefore had to be coated outside.

For the ties of the Danube bridge at Ulm grouting of the empty ducts was attempted and additional load capacity was installed by assembling ground anchors beneath and parallel to the ties drilled from the deck and through the spread footing into the rock.

Typical rehabilitation of corroded tendons is illustrated in the Danube bridge at Untermarchtal. The remaining load capacity of the concentrated cable had to be calculated at 75% of the original. Inside each web three additional cables were installed, each with 12 no. 12-mm diameter wires. They were assembled inside an additional thickening of the web (see Fig. 5) and anchored at additional corbels at both ends of the span.

Rebars grouted with epoxy resin into drilled holes transfer shear forces from the new concrete to the old. At the corbels additional post-tensioned 26-mm diameter bars transfer the anchor forces to the webs.

To avoid bursting due to the corroded and broken strands additional heavy reinforced concrete thickenings were assembled at the defective zone.

Summing up, it can be said that all rehabilitations were successful and it can be expected that all bridges will survive the next four decades.

26

Movable Bridge Machinery Inspection and Rehabilitation

CHARLES BIRNSTIEL

Consulting Engineer, Forest Hills, New York 11375, USA

ABSTRACT

Age and marine environment, exacerbated by inadequate maintenance, have deteriorated movable bridge machinery and controls. The scope of six levels of field inspections performed to quantify and document the deterioration is described. For one of these, an intermediate level inspection recently adopted for the biennial bridge inspection program of New York State, the work items are outlined in detail. A discussion of the current approach to machinery and control rehabilitation in New York State concludes the paper.

INTRODUCTION

Approximately 1700 movable bridges span navigable channels in the continental United States. Of these 1100 are highway bridges and 600 carry railroad traffic. Eighty percent are at least 40 years old. Within New York State there are 60 movable highway bridges, of which over 70% are 40 years or older.

It is seldom that the agency that caused a bridge to be built is still fully responsible for that bridge. Many of the older movable highway bridges were built as privately-owned toll facilities and after some years under private operation were acquired by a state or other governmental agency. Railroad bridges also changed ownership as a result of railroad company mergers and through acquisition by states. In some cases the states took over the railroad passenger service and the property. In other cases the states acquired the bridges but other agencies operate the trains running over them.

The responsibilities of local governments for the older movable road and street bridges has also changed. A bridge built, rehabilitated and maintained by a city for many years may have been transferred to a state highway system, in which case the responsibility for operation and maintenance now lies with the city but rehabilitation would be by the state, with federal assistance.

The machinery of many of the older movable street bridges in the northeast United States has deteriorated at an accelerated rate since 1950 for two reasons. First, the replacement of solid timber decking by open steel grating in the 1950s results in more debris accumulating on the machinery. In winter the debris often contains deicing salts. Moisture falling through the open grating is retained in the accumulated roadway debris and promotes corrosion of the machinery. The second reason for the accelerated deterioration of movable bridge machinery in the northeast was the reduction in maintenance forces two decades ago. The social upheavals in the 1960s created conditions in which budgets for maintenance of bridges and other transportation facilities had to be drastically reduced in order to make funds available to mitigate social unrest. It is only recently that maintenance forces are being rebuilt.

Before describing the scopes of machinery and electrical inspections, and discussing the rehabilitation of bridge machinery, the different types of movable bridges and the kinds of mechanical and electrical equipment found on these bridges will be reviewed.

MOVABLE BRIDGE TYPES, MACHINERY AND CONTROLS

Types of Movable Bridges

Movable bridges may be broadly categorized as swing, vertical lift, bascule and retractile, although some bridges do not fit into these categories. In the United States there are about 750 swing, 230 vertical lift, 720 bascule and five retractile bridges. Construction features of each type are described subsequently.

Swing bridges

In swing bridges the movable span, often termed the draw, rotates about a vertical axis. The type may be subdivided as to the manner of draw support when swung open (permitting navigation). If the dead load (self-weight) is supported on a pivot bearing at the axis of rotation it is termed 'center bearing'. The draw is balanced on this pivot. To keep the draw from tipping under unbalanced loads, such as wind, balance wheels are provided that roll on a circular track concentric with the pivot bearing.

Swing bridges when open, in which the dead load is supported by a nest of tapered wheels, are said to be rim bearing. In these bridges the superstructure is supported by a circular girder called a drum girder. A tapered plate (tread plate) is fastened to the underside of the drum girder. It bears on a nest of tapered wheels whose axes are oriented radially to the axis of span rotation. The wheels, in turn, roll on a tapered plate, called a track plate, that is fastened to a chair casting. The drum girder and the nest of tapered rollers are all held concentric by radial members connected to bearings rotating about a central pivot post.

In the closed position, the ends of the main longitudinal bending members (usually trusses) are lifted at the rest piers so that there will be an upward reaction at the truss ends for all combinations of live load and temperature.

Vertical lift bridge

The movable span of a vertical lift bridge moves vertically upward to provide clearance for passage of vessels. The ends of the lift span are attached to wire ropes passing over sheaves at the tops of the towers with the far ends connected to counterweights. The vertical lift type is subdivided into tower, span and tower-span drive.

In tower drive lift bridges there is drive machinery in each tower connected to the counterweight sheaves. The forces necessary to raise the span are transmitted to the counterweight ropes by friction. The action is identical to that of a traction drive office building elevator.

Span drive lift bridges have the machinery located on the lift span itself, usually above the roadway or tracks. There is one drive and the span is moved by a wire rope drive which hauls the lift span up or down. Alternatively, shafts extend longitudinally from the span drive machinery to the towers, where pinions engage a rack and the span crawls up or down the towers.

In tower-span drive lift bridges there is also only one drive. It is fixed on a structure spanning between the towers. Usually this drive is coupled to the counterweight sheaves and the action is that of a friction drive.

Bascule bridges

There are two principal categories of bascules: rolling lift bascules and trunnion bascules. Rolling lift bascules have the distinguishing feature that the ends of the main spanning members (bascule girders) are cylindrically curved and the movable span (leaf) rolls on these curved surfaces during opening and closing. As the curved ends of the girders roll shoreward the leaf tilts open to clear the channel. The leaf simultaneously rotates and translates. To close the bridge the leaf rolls toward the channel. This bridge type was developed and promoted by the Scherzer brothers in Chicago at the end of the 19th century.

The other, older, type of bascule is the trunnion bascule. Trunnion bascule leaves rotate about large shafts (trunnions) that are usually inserted through the webs of the bascule girders. The shafts may rotate with the girders, or the bascule girders may rotate about fixed trunnions. The leaves of trunnion bascules only rotate, they do not translate.

Retractile bridges

These bridges roll horizontally on tracks in order to clear a waterway. The movable spans of retractile bridges only translate, they do not rotate. The axes of the roadway and the navigation channel usually intersect at about 45° so that the movable leaf can be rolled clear of the channel without having to be rolled backwards over or under the approach.

Span Drive Machinery

The movable span can be driven by either a mechanical or a hydraulic drive. In mechanical drives reduction gearing is used to convert the high-speed, low-torque output of electrical motors to the low-speed, high-torque output required to drive the movable span. Two approaches are used for hydraulic operation. In one, hydraulic motors produce the high torque necessary to drive the movable span. An electric motor drives a hydraulic pump that supplies the high-torque hydraulic motor. This eliminates the need for all but the last or the last two stages of reduction gearing. In the other approach, the movable span is driven by the movement of pistons in the cylinders. Suitable articulation converts the linear action of the cylinder to motion rotating the movable span. Wire ropes are sometimes used to transmit drive forces in both mechanical and hydraulic systems.

Stabilizing Machinery

The machinery which stabilizes the movable span in the closed and open positions varies with the bridge type. For center bearing swing spans it comprises end lifts at the rest piers and center wedges at the pivot pier. These are used to lift the draw so that the trusses act as continuous spanning members when the draw is closed, from rest pier to rest pier with an intermediate support. Rim bearing swing spans also have end lifts but do not have center wedges. Both center and rim bearing swing bridges have centering latches or centering mechanisms to aid in positioning the draw in the closed position and prevent it from rotating while the end lifts are extended or retracted.

Vertical lift bridges have span locks to lock the lift span in the closed (lowered) position, guide wheels to guide the lift span as it rolls up the tower and, sometimes, buffers to decelerate the moving span as it approaches the upper and lower limits of travel.

The curved treads fastened to the bottom of the bascule girders of Scherzer bascules and the tracks on which they roll are usually considered stabilizing machinery. Other stabilizing machinery of bascules, both rolling lift and trunnion, are midspan locks, tail locks and live load reactions.

Electrical Power Equipment and Controls

The electrical equipment may be considered power equipment and control equipment, although it is sometimes difficult to distinguish between the power and control functions. Electrical power equipment comprises transformers, circuit breakers, motors, overload relays, resistor banks, thyristors, electrically-released brakes, panelboards, electromagnetic switches, conduit, boxes, submarine cables and wiring. The control desk, control switches, pushbuttons, indicating lights, meters, limit switches, navigation lights and air horns are considered control devices. There is obvious overlap of function, as for example with submarine cables which may contain control wires in addition to power wires.

Most movable bridges are powered by electric motors that are either directly coupled to gearing in order to drive a mechanism or coupled to hydraulic pumps which perform the same basic function, that of multiplying the torque output of the motor at the expense of speed. Few bridges are now powered by internal combustion engines or are hand-operated. Alternating current drive systems are most common, either as original equipment on newer bridges or as part of rehabilitation of older bridges. AC drive motors are usually of the wound-rotor type to permit the connection of external secondary resistance so as to obtain the desired speed–torque motor characteristics and, often, for speed control. In older AC systems speed is usually varied stepwise by switching resistance into or out of the secondary circuit using a drum switch and relays. In the state-of-the-art controls the secondary resistance is fixed in order to obtain the desired motor characteristic and speed is controlled by varying the input power to the motor by means of thyristors responding to a feedback signal.

Span drives of some larger lift bridges and a few other smaller movable bridges are powered by direct current. The direct current is usually supplied from a motor-generator set, the motor of which is usually an AC motor. DC drives were installed in the larger lift bridges in order to take advantage of the superior speed control capability provided by the DC generator–DC drive motor arrangement. This was the only practical means of stepless motor speed control commercially available before the development of the thyristor.

The motors of stabilizing machinery are usually AC squirrel cage motors because speed control is not needed. The operation of these motors is interlocked with the traffic control devices and the span drive by means of limit switches.

Traffic Control Devices

The traffic control devices on highway bridges comprise traffic signals, warning gates, warning gongs, resistance gates and warning signs. Warning gates are usually of the semaphore type, similar to electrically-operated railroad crossing gates. Resistance gates are intended to absorb impact of a moving vehicle which has passed beyond the lowered warning gates. There are three varieties of resistance gates: pintel gates, semaphore gates or vertical lift. Modes of operation also vary: hand, direct-geared electric or hydraulic.

INSPECTIONS

Types of Inspections

The machinery and controls of movable bridges are inspected in connection with bridge rehabilitation projects, to satisfy insurance requirements, and as part of the biennial bridge inspection program mandated by the US Department of Transportation. The thoroughness of the inspection varies, depending on its purpose. We have classified them as Types I, I-R, II, II-A and III.

Type I

A cursory visual inspection of the drive and stabilizing machinery, the traffic control devices and the electrical controls for this equipment. No parts hidden by guards or housings that require tools for removal are inspected. No mechanical or electrical measurements are made.

Type I-R

This is a visual reinspection along the lines of the Type I inspection. It is made some years after a higher-order inspection and has two main objectives. First, to ascertain the condition of components or systems found to be deficient in the prior inspection. Second, to discover other defects which may have developed since the last inspection.

Type II

A visual inspection during which access covers are opened for the inspection of hidden components. An electrical performance test is made for at least one opening/closing cycle, during which the power consumed by each normal span drive motor is measured and recorded on a strip chart.

Type II-A

An intermediate inspection which is essentially a Type II but includes more measurements. It is in accordance with the intent of the *Bridge Inspector's Manual for Moveable Bridges* published by the Federal Highway Administration (FHWA), hereinafter termed the *Manual.* The scope of the Type II-A inspection conforms essentially with the *Manual* but the extent of machinery disassembly and measurements is less and the field data are presented in a condensed form. On the other hand, electrical measurements that are not referred to in the *Manual*, but which have been found valuable in assessing movable span behavior, are included.

Type III

An in-depth inspection in which virtually every mechanical and electrical component is inspected visually and measurements are made to determine wear of mechanical parts, and some electrical equipment is tested. It is an expansion of the Type II-A inspection to meet the recommendations of the *Manual*, especially as to reporting field data.

The intermediate inspection, Type II-A, is being considered by New York State as a standard for biennial inspections. The scope of this inspection is described in more detail subsequently.

Scope of Type II-A Inspection and Report

The Type II-A inspection is designed to conform to the intent of the FHWA *Manual* but the amount of machinery disassembly is less than shown in the *Manual* and the field data are presented in a more concise format. The work items included in the scope are described subsequently.

Preparation

Prior to actual field inspection the team leader visits the bridge in order to assess inspection difficulties. The available mechanical and electrical drawings are studied in planning the inspection and to determine the probable original tooth thicknesses of all open gearing.

Bridge operation

Bridge operation is observed in the normal mode so as to assess condition of the normal mode drive, functioning and effectiveness of traffic control devices, interferences between movable and stationary parts of the bridge, controllability of the moving span and the effectiveness of stabilizing machinery. During bridge operation observers monitor the machinery for abnormal noises and vibration. The manner in which the bridge operator on duty handles the controls is also observed.

Performance tests

The power consumed by the normal span drive motors is measured and recorded on a strip chart during at least one opening/closing cycle for each movable span. The object of the test is to detect excessive self-weight imbalance of bascule and vertical lift spans or severe binding of the machinery. Electrical control deficiencies are sometimes found from these tests.

Span drive machinery

Every moving and stationary component of the span drive is visually inspected. The types of components inspected depend on the type of movable bridge but, generally, the inspection includes brakes, speed reducers, open gearing, drive shafts and couplings, bearings, mounting bolts, sheaves and wire ropes. The following measurements are made: gear tooth thicknesses, gearset backlash and clearance, and sleeve bearing clearances.

Shafts are visually inspected for cracking at keyways and shoulders, and for movement between coupling hubs and shafts.

The mechanical linkages of brakes are inspected, as are the brake wheels. Clearances between shoes and wheels are measured with the brakes set and hand released.

Samples of lubricating oil are recovered from speed reducers and sent to a laboratory for chemical analysis to determine the amount and sources of contaminants.

Stabilizing machinery

The mechanical components that stabilize the movable span when it is in motion and at rest are inspected. Depending on the type of movable bridge, these include treads and tracks, span locks and drives, centering devices, buffers, live load supports, trunnions, wheels and axles, end lifts, span guides, and wire ropes and adjusting devices.

The counterweight sheave trunnions of vertical lift bridges are inspected visually and ultrasonically. Ultrasonic testing is performed by a recognized testing laboratory experienced in this work.

As part of the machinery inspection movements at midspan locks and tail locks and live load shoe clearances are measured.

Electrical system

The major electrical components are visually inspected on the load side of the service disconnect. Span drive motors and motor-generators are observed while running to check bearings and for excessive noise and vibration. Brushes, commutators and slip rings are inspected. The installation and components of contactors, circuit breakers and drum switches are examined. The brakes are inspected for thrustor/solenoid operation, limit switch operation and deterioration of wiring. The extension of springs of spring-set thrustor brakes is measured. Span lock motors, end lift motors, navigation lights and traffic control devices are inspected and their operation observed.

Operability of accessible limit switches is determined and interlock limit switches are subjected to a simulated test where appropriate.

The exposed portion of the submarine cable is inspected for corroded armor wires and other damage. The electrical insulation resistance of submarine cable conductors is measured.

The control desk is examined to assess operability of control functions and safety with respect to misoperation and unauthorized use.

Traffic control devices

Traffic lights, warning gates and resistance gates are inspected for structural condition, mechanical operation and effectiveness in stopping vehicular traffic. Layout of the traffic devices is recorded for comparison with the relevant *Manual of Uniform Traffic Control Devices.*

The field measurements and condition observations are recorded on field inspection forms which were developed for this purpose and are included in an appendix to the report. Where appropriate deficiencies are photographed. The typical report contains the following:

- Descriptions of the existing mechanical, electrical and traffic control systems illustrated with a schematic span drive machinery diagram, a stabilizing machinery layout and a traffic control device layout.
- Photographs of the general arrangement of machinery, electrical equipment and traffic control devices, and of defects.
- Reduced size photocopies of annotated performance test strip charts.
- Discussion of the condition of the mechanical and electrical components emphasizing defects and corrective measures.
- Recommendations for in-depth inspections of inadequate components and a listing of defects and items that require attention.
- Evaluation of the traffic control devices with respect to the relevant *Manual of Uniform Traffic Control Devices.*
- Ratings of the major components of the mechanical and electrical systems and the traffic control devices according to the state numerical condition rating system. The ratings are based solely on visual observations and measurements. No numerical analyses are made.

The Type II-A inspections were developed to satisfy the requirements of the federal biennial bridge inspection program. They are primarily intended to discover defects that may compromise the safety of the bridge. Dangerous conditions are 'flagged' in writing to the state. However, local agencies are using these reports as the basis for improvements in maintenance and for minor rehabilitation programs.

REHABILITATION

Many of the older movable bridges have been rehabilitated to some degree. Usually the electrical systems have been revised. We found 90-year-old bridges with the basic original span drive and stabilizing system in place but with successive changes to the motive power—from reciprocating steam engines to direct current, to two-phase alternating current and then to three-phase alternating current. We have inspected some fifty different movable bridges from 5 to 100 years old and not found two alike, even though some were originally constructed from a single set of plans.

The usual criterion for movable bridge machinery and control rehabilitation in the northeast United States is to design for a 30-year extension of bridge life. For the older bridges the span drives may have to be completely replaced. For bridges less than 50 years old the original span drive is usually retained and specific components found to be worn or defective are replaced. For bridges whose electrical systems have not been rehabilitated in the last 40 years it is usual to replace the complete electrical system with a thyristor-controlled alternating current system. The unavailability of replacements for electrical components older than 40 years and the fact that the older panelboards have live fronts, which are now considered dangerous, play a part in such decisions. If the traffic controls are more than 30 years old they are usually replaced because they are often damaged and deteriorated, and because the existing traffic control device layouts of that age seldom conform to the current State DOT requirements.

CONCLUSION

The major types of movable bridges and their mechanical and electrical components were described. Because of aging, repeated stressing from traffic and the deleterious effect of marine environments, coupled with inadequate maintenance, the machinery deteriorates. In order to identify dangerous conditions and quantify the deterioration for repair and rehabilitation programs, field inspection and reporting procedures at various levels of thoroughness have been developed. Six of these were briefly described and one, the Type II-A, was outlined in detail. The report contents for this type of inspection was also included. Finally, current approaches to machinery and control rehabilitation in New York State were described.

27

Application of Radar and Thermography to Bridge Deck Condition Surveys

D. G. MANNING and T. MASLIWEC

Research and Development Branch, Ontario Ministry of Transportation, 1201 Wilson Avenue, Downsview, Ontario, Canada, M3M 1J8

ABSTRACT

This paper summarizes the theoretical and practical aspects of the application of radar and thermography to bridge deck condition surveys. A prototype vehicle was developed. The radar waveforms and the thermographic images were calibrated in the field to various forms of physical distress. It is shown that radar and thermography are useful tools for assessing the condition of bridge decks and that the two technologies are complementary; thermography is most useful for locating subsurface defects on exposed concrete surfaces and radar is most applicable to detecting defects in concrete deck slabs which have a bituminous surfacing.

INTRODUCTION

An effective bridge management system requires a comprehensive data base. At the network level, the data are needed in order to assess the overall condition of the bridge network and to predict the effect of alternative preservation and improvement actions. At the project level, more detailed data are required to select the most appropriate method of rehabilitating a particular structure and to prepare the contract documents.

Collecting data on the condition of bridges by traditional methods is expensive and the results are not always dependable. In the late 1970s, research studies were initiated by the Ontario Ministry of Transportation to develop rapid and reliable methods for collecting data on the condition of bridge decks. The studies were part of a larger program to improve the technology for rehabilitating bridges.[1]

The first studies were conducted on exposed concrete decks and were designed specifically to investigate methods of detecting delamination. Infrared thermography was found to be capable of identifying delaminated areas over a wide range of ambient temperatures.[2] During the period 1980–82, the studies were extended to investigate methods capable of detecting deterioration in concrete deck slabs which have a surfacing of bituminous concrete. Over 90% of the bridges in Ontario now have a waterproofing membrane and two 40 mm thick lifts of bituminous surfacing. A test site was constructed by selecting a typical bridge which was exhibiting corrosion-induced distress, surveying the bridge and then paving it without first making repairs. Areas of scaling and debonding were simulated prior to paving. The capabilities of eight techniques to detect the locations and types of deterioration in the test deck were evaluated. The most promising techniques were found to be radar and thermography.[3]

Further research concentrated on developing automated processing techniques to analyse the thermograms and radar waveforms and present the results in a form usable by bridge engineers. The product of this work was a self-contained vehicle named DART (deck assessment by radar and thermography) which, as the acronym implies, is equipped with both radar and thermography.[4]

MAJOR TYPES OF DETERIORATION IN BRIDGE DECKS

The most serious form of deterioration in bridge decks in North America is that caused by corrosion of embedded reinforcement. As the reinforcement steel corrodes, it expands and creates a crack or subsurface fracture plane in the concrete at, or just above, the level of the reinforcement. The fracture plane, or delamination, may be localized or may extend over a substantial area.

Scaling, which is a breakdown of the cement–paste matrix, is also a serious problem wherever it occurs. The disintegration of the concrete, which is caused by the freezing of concrete critically saturated with water, begins at the surface and gradually progresses so that the full depth of deck slab may be affected. It is most common in older bridges which do not have an effective air void system.

On asphalt-covered decks, bond failure may occur between the concrete deck and bituminous surfacing. Debonding can result in moisture being trapped on the surface of the concrete and, where thin surfacings are used, can lead to failure of the bituminous surfacing. Although the consequences of debonding are not as serious as either delamination or scaling, it can be confused with these two phenomena in surveys. Consequently, it is important to be able to identify and define debonded areas.

HOW THERMOGRAPHY AND RADAR DETECT DETERIORATION

Thermography

Infrared thermography is a remote method of sensing the energy emitted from the surface of an object. Consequently, the detection of deterioration by infrared thermography is based on the difference in surface temperature that exists between deteriorated and sound concrete under certain atmospheric conditions. Concrete is a poor conductor of heat and the differences in surface temperature develop on a bridge deck as a result of different rates of heat transfer occurring inward from the surface. Discontinuities which impede the conductive heat flow from the surface, such as a delamination or debonding, will result in a higher surface temperature during periods of heating. The reverse is true during periods of cooling. Because of the breakdown of the matrix, scaled concrete has a different coefficient of thermal conductivity than sound concrete, which also results in a difference in surface temperature.

The differences in temperature have been confirmed by implanting thermocouples in deck slabs and by the direct measurement of surface temperatures over known areas of sound and deteriorated concrete.[2,3] Measured temperature differences were also compared with a theoretical heat transfer model as a means of establishing operational constraints on the use of thermography for bridge deck surveys. An approximate solution of the general equation for transient heat flow[5] indicated that the most critical parameters affecting temperature differentials are solar irradiance and emissivity. However, wind speed is also critical to the theoretical calculation of emissivity.

Delaminations appear as well defined white areas in an infrared thermogram which are hotter (during daylight hours) than the surrounding areas of solid deck which are cooler and appear dark.

Scaling of concrete beneath a bituminous surfacing produces a characteristic thermal image which is a mottled grey-white tone. The more severe and extensive the scaling, the more mottled the thermal image appears.

Debonding is not easily identified by infrared thermography. Depending upon the depth of any delaminations, debonding may appear hotter or identical to the delaminations and, at other times, may go undetected.

Radar

The detection of deterioration by radar is based on reflections of a high frequency electromagnetic wave caused by changes in the electromagnetic properties of the material being probed. Whenever a transmitted wave encounters an electromagnetic discontinuity (deterioration) or change of material dielectric, it is partially reflected. The patterns created by the reflected waves are received by the radar antenna. The radar receiver takes a measurement of the time required for the transmitted pulse to travel to a target discontinuity and for the reflected pulse to return.[4] The time delay of the radar echo from a structural discontinuity is directly related to the depth of the fault.

The impulse radar used in the DART system has been described in detail elsewhere.[6] The radar is monostatic with the antenna design being based on a constant flare angle, variable-width open horn. The transmitted signal is a pulse which has a duration of approximately 1 ns and a repetition rate of 5 million pulses per second.

The reflected signal is sampled using the sliding gate technique, and averaged, producing an output voltage pulse waveform of approximately 1 ms duration. The actual radar waveform reflected from a metal plate on the deck surface is shown in Fig. 1. The signal has a duration of approximately 0·9 ns between negative peaks P and Q with the main positive peak at A. The peak at A represents the surface reflection and is used to calibrate the system. The transients at the beginning and end of the main pulse restrict the resolution in measuring the thickness of a material. The response after point R is very small, indicating the variations due to discontinuities can be detected in this region without too much interference.

Figure 2 represents, from a theoretical standpoint, the principal reflections which occur in a bridge deck. The peak at A represents the reflection from asphalt, the peak at C is produced by the asphalt–concrete interface and the reflection at D is from the rebars. Although the reflections are shown as being quite distinct, there may be considerable overlap of the transient portions of the waveforms such as at point B. However, this is not a serious problem provided that the main peaks are well defined. The thickness of the asphalt is determined by measuring the time between point

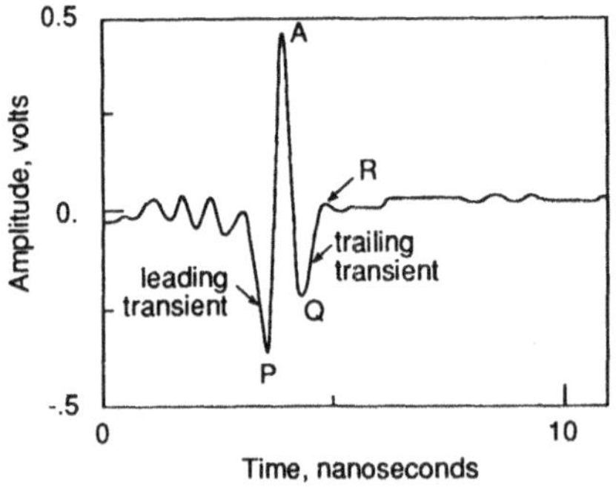

FIG. 1. Waveform reflected from a metal plate on the deck surface.

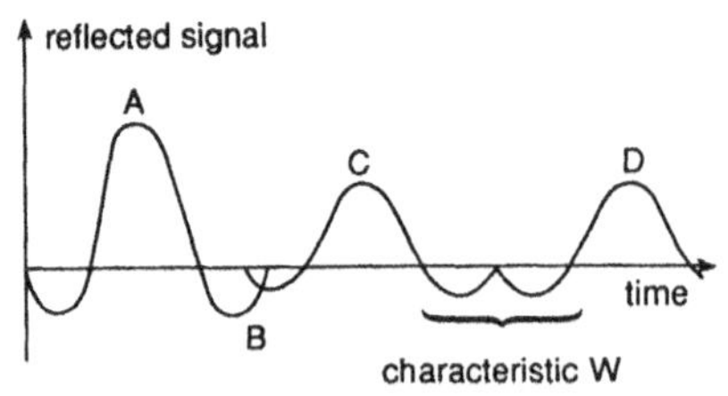

FIG. 2. Theoretical reflection from a sound bridge deck.

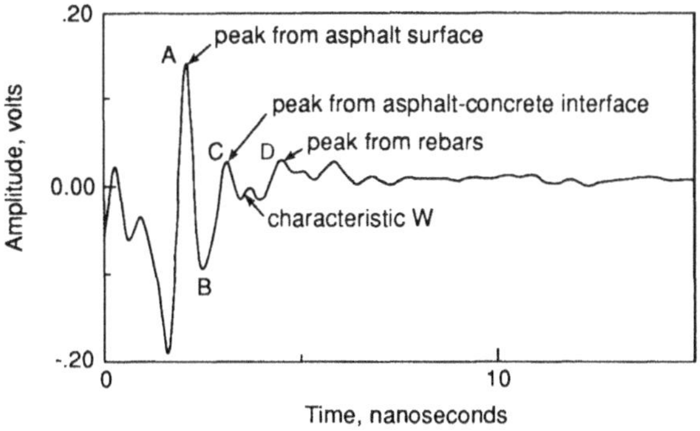

FIG. 3. Actual reflection from a sound bridge deck.

A and point C, taking account of the dielectric constant of asphalt which is calculated from field measurements.[7] The reflections from the asphalt–concrete interface and the rebars produce a *W*-shaped waveform which is called the characteristic *W*. This characteristic *W* is useful in identifying the difference between sound and delaminated concrete.

The radar reflection from a sound bridge deck is illustrated in Fig. 3. The peak at A represents the surface reflection from the asphalt, the peak at C indicates the reflection from the asphalt–concrete interface and the peak at D is the reflection from the rebars.

For analysis, two ratios are defined, the first being R1, the ratio of the magnitude at point B to the magnitude at point A, where point B is defined to be the first negative peak after peak A. The second ratio, R2, is the ratio of the magnitude at point C to the magnitude at point A, where point C is defined to be the first positive peak after point B. These two ratios have different uses.

Debonding

Since debonding affects the interface between the asphalt and the concrete, the reflection from the asphalt surface will remain essentially unchanged. However, the reflection from the asphalt–concrete interface will be altered, producing a reflected waveform with either increased or decreased magnitude at point C depending on whether the gap is filled with water or air. This gap will effectively change the R1 and R2 ratios of the radar waveform.

Scaling

Scaling breaks down the concrete matrix which may contain either air or water. In this case, the gap is much larger than occurs in debonding. For the purposes of modelling, it is assumed that the reflection from the gap is due to a single reflection from the asphalt–gap interface. With air in the gap, the reflection not only changes polarity, compared with a good asphalt–concrete interface, but the magnitude of the R1 ratio is increased substantially. With water in the gap, the polarity remains the same, but the R2 ratio is much larger. Since it is not known whether the scaled concrete contains air or water, both the R1 and R2 ratios are calculated and a search made for values larger than normal.

Delamination

Delamination occurs in the region between the surface of the concrete and the top mat of steel. There may be one or more fracture planes in the concrete containing air or water, resulting in at least one discontinuity between the concrete surface and the rebars. The additional reflections from the delamination mean that the characteristic *W* present in a good structure is no longer present.

The range of values used for evaluation are as follows:

	R1	R2
Good	$<0{\cdot}4$	$<0{\cdot}28$
Debonded	$<0{\cdot}4$	$>0{\cdot}28$
Scaled	$>0{\cdot}4$	$>0{\cdot}28$

THE DART PROTOTYPE VEHICLE

The exterior and interior of the DART vehicle are shown in Figs 4 and 5 respectively. The essential features of the unit are as follows:

(i) A hydraulically-operated telescopic mast mounted on the front of the vehicle. The mast is lowered for attaching the infrared scanner and refilling the liquid nitrogen. It is raised to its operating height of 5 m.

(ii) The infrared scanner. When equipped with a 20° lens and operated at an angle of 45° and height of 5 m, the scanner is capable of viewing the width of one complete traffic lane at a single pass. The scanner is connected through a control unit and filter to a high resolution video tape recorder.

(iii) The radar antenna is attached to a rail on the front of the vehicle using a sliding bracket so that it can be positioned anywhere within the width of the vehicle. The mounting height of 150 mm results in an elliptical footprint of 300 × 175 mm on the deck surface. The transmitter and receiver are also mounted on the sliding bracket. Returning radar echoes are received by the antenna and transmitted to the control unit inside the vehicle. The waveform is simultaneously

FIG. 4. Exterior of DART vehicle.

FIG. 5. Interior of DART vehicle.

displayed on an oscilloscope and recorded on a seven-channel FM tape recorder. The operator can monitor the signal from the control unit and from the tape recorder, thus permitting a check of the quality of the signal before and after recording.

(iv) A fifth wheel is attached to the rear of the vehicle for speed and distance measurements. Custom interface devices enable distance pulses to be stored on both the video and FM recorders.

(v) A microcomputer for processing the data.

(vi) The vehicle is equipped with heavy duty air conditioning, a generator to supply electrical power and racks with vibration-isolation mounts for the electronic equipment.

A schematic of the components of the thermography and the radar is given in Figs 6 and 7 respectively.

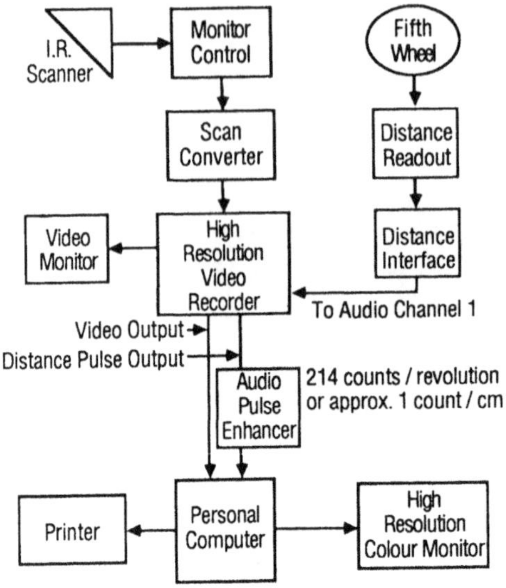

FIG. 6. Infrared system schematic.

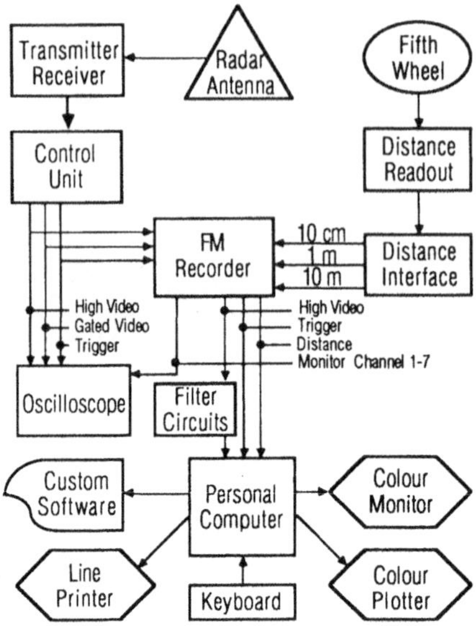

FIG. 7. Radar system schematic.

Data Acquisition/Processing

Radar

A central problem in the system software development is in reducing the volume of radar data while preserving all desired information. Data reduction in the data acquisition programs is accomplished in two different ways. First, for all waveforms only 10 ms of data before and after the surface reflection is considered relevant. After internal noise subtraction, only the second 10 ms of data are permitted to enter data processing subroutines. The second 10-ms slot is also used to compute the average level of the trace and adjust the trace with respect to zero. Secondly, for data that are processed using fifth wheel information, the 10-cm fifth wheel waveform is used to ensure that waveforms are sampled only every 10 cm.

The program MTO88, written in standard Fortran 77 language, performs all processing functions such as calculating the depth of asphalt, determining the R1/R2 ratios for scaling and debonding, and providing the zero crossing count for delamination detection. The software is designed to handle analysis for a bridge deck of any length with 12 grid lines. The processed results are stored as formatted data files for future reference and are also output, in tabular form, to the monitor screen and the line printer.

Thermography

For processing the data, the video record is fed into an on-board computer through an image digitizer interface. Using custom software, the infrared image is sampled in such a way that the oblique angle of view and other distortions of the infrared image are eliminated. Deterioration is identified through a combination of computer- and operator-assisted machine interpretation.[5] The computer produces a scaled graphics image of each lane of the deck, showing areas of delamination and scaling.

FIELD OPERATIONS AND EXPERIENCE

The DART vehicle is normally operated between May and November, with the infrared scanner being rented only for the months of June, July and August. However, the scanner has also been used in February to investigate its application in cold weather.

The vehicle is operated at a speed of between 3 and 4·5 km/h, making traffic protection necessary. At faster speeds the digitized version of the infrared image is distorted and the radar surface coverage capability is impeded.

Whereas one pass with the equipment permits an infrared scan of one complete lane width, the radar collects information only along a grid line approximately 300 mm wide. The number of grid lines required with the radar is a function of the condition of the deck (the worse the condition of the deck, the more grid lines are needed). In general, a minimum of two passes are made in each traffic lane. If the data indicate significant deterioration (as determined by the operator reviewing the waveforms on an oscilloscope), additional passes are made and, on badly deteriorated decks, a grid-line spacing of as little as 1 m may be used.

On most bridges the time taken to set up the traffic protection usually exceeds that taken to complete the infrared and radar surveys. Infrared surveys should always be performed before radar surveys. This is because any obstructions to solar radiation on the bridge, such as people or stationary vehicles, can cause false readings in the infrared image.

Thermography

In general, areas of delamination on exposed concrete decks can be identified at almost any time of day except near sunrise and sunset. Near sunrise the surface temperature of the delamination changes from being cooler to hotter than the average deck temperature. The reverse situation occurs shortly after sunset. The 'window' in which delaminations can be detected through an 80-mm thickness of bituminous concrete is quite narrow. The optimum time of day (in southern Ontario) is between 12 noon and 1 pm, when the delaminations are not only most easily distinguished but the area of the 'hot spots' approximates the actual size of the delaminated areas. Although the maximum difference in temperature usually occurs later in the day, the outline of the delaminations becomes less distinct because of heat transfer within the surfacing. After about 3 pm, the deck begins to cool and the delaminations fade and are no longer detectable by the scanner.

The optimum time of year for thermography will depend upon local climate and latitude. In southern Ontario, maximum temperature differentials were observed in early or late summer when clear skies prevailed, rather than in midsummer when overcast conditions were more prevalent. Ambient temperature in the range of −33°C (in February) to 32°C were investigated and were found not to have a significant effect on the detectability of deterioration. Differentials were, however, greater in summer than winter because of the greater intensity of solar radiation.

Although the equipment has a sensitivity of about 0·1°C, a temperature differential of 1·5–2°C is a reasonable practical minimum to counteract the effects of differences in emissivity caused by such factors as polishing in the wheel tracks, staining and patching.

A summary of the factors affecting the quality of the infrared image is contained in Table 1. More specific information is given in Ref. 8.

Radar

The major advantage of radar is the ability to propagate through layered media, thereby giving information on bituminous surfacings, concrete deck slabs, the interface between the two and any discontinuities. In fact, a bituminous surfacing more than 25 mm thick enhances waveform interpretation (compared with an exposed concrete deck) because interference between the radar echo reflected from the deck surface and the radar echo from deterioration below the surface is reduced substantially. On exposed concrete bridge decks, reflected signals from defects interact with surface returns. The ministry's existing software is applicable only to asphalt-covered bridge decks.

Investigations have been made to assess the radar's operating characteristics under a range of climatic conditions on a variety of structures. A comparative evaluation of the relative strengths and weaknesses of the DART's radar and infrared systems is provided in Table 1.

It is apparent from Table 1 that the radar and infrared systems used in the DART prototype complement each other very well. Taken separately, the radar system provides only the dimension of depth, while the infrared thermography provides a two-dimensional surface image. By using both techniques, it is possible to obtain a three-dimensional perspective of the condition of a bridge deck.

TABLE 1
Operational characteristics of radar and infrared thermography

Operational characteristic	Effect	
	Radar	Thermography
Meteorological limitations	Adversely affected by standing water only	Needs sun, little cloud and wind; inoperable with standing water
When can system be used?	Year round	Year round but must have sun and no rain
What hours of the day can system be used?	All day	Usually 9.00 to 15.00; optimum 12.00 to 13.00
Influence of external parameters		
Paint markings	None	Easily compensated for
Isolated debris	None	Easily compensated for
Oil stains	None	Easily compensated for
Skid marks	None	None
Snow/ice/water	Adverse	Adverse
Traffic	None	Adverse if solar irradiance affected
Signal noise level	Waveforms need careful monitoring and calibration	Hot spots may not be detected due to meteorological factors or secondary reflections
Signal reproducibility	Independent of time, excellent	Dependent on solar heating, variable
Selectivity		
Asphalt-covered deck	Excellent results with surfacing >25 mm	Good if surfacing <80 mm
Exposed concrete decks	Signal processing software not applicable	Excellent results
Penetration	Detects deep-seated effects; difficulty with surface effects	Detects only surface effects; no penetration
Inspection speed (limiting factors)	Maximum 8 km/h (pulse repetition rate)	Maximum 4·5 km/h (video digitization rate/camera scan rate)
Number of scans required	Many, depending on bridge width and coverage required	One scan per lane
Expertise required	High level of technical competence	Basic training plus experience on bridges

COSTS

The DART system represents a significant investment. The approximate value of the major items (in 1989 Canadian dollars) is summarized below:

Radar	\$75 000
Thermal scanner and converter	\$54 000 (or \$2400 per month rental)
Microcomputer and peripherals (including extra circuit boards)	\$27 000
FM recorder	\$13 000
Industrial VCR	\$5 000
Fifth wheel	\$7 000
Oscilloscope	\$2 400
Generator	\$1 500

Adding the cost of the vehicle, generator, external and interior racks, and numerous smaller items such as cables, filters and interface devices would make the replacement cost of the fully equipped DART vehicle approximately \$180 000. It should be noted that the ministry did not purchase the thermal scanner and converter but rents the equipment during the summer months for \$2400 per month. In addition, the cost of developing the software was \$96 000 for the radar and \$30 000 for processing the thermal images.

CONCLUDING REMARKS

The operating experience with the DART system has shown that, within known limitations, the prototype constitutes a viable, rapid, non-contact and non-destructive method for performing bridge deck condition surveys. It is envisaged that the system will play a major role in providing data for the ministry's bridge management system, initially at the project level with possible future expansion to the network level. It is anticipated that more reliable data will result in improvements in both the decision-making process and in fewer cost overruns in rehabilitation contracts.

Future work will involve upgrading the microcomputer. This will not only increase the processing speed, but the enhanced memory will eliminate the need for the FM recorder. Any other changes are likely to focus on making the system simpler to operate.

Other studies will be undertaken to determine the feasibility of applying radar to other components of bridge structures and to pavements.

REFERENCES

1. MANNING, D. G. and RYELL, J., Decision criteria for the rehabilitation of concrete bridge decks. Transportation Research Record No. 762, 1980, pp. 1–9.
2. MANNING, D. G. and HOLT, F. B., Detecting delamination in concrete bridge decks. *Concrete International*, **2**(11) (1980) 34–42.
3. MANNING, D. G. and HOLT, F. B., Detecting deterioration in asphalt-covered bridge decks. Transportation Research Record No. 899, 1983, pp. 10–20.
4. MANNING, D. G. and HOLT, F. B., Deck assessment by radar and thermography. Transportation Research Record No. 1083, 1986, pp. 13–20.
5. MASLIWEC, T., An experimental and theoretical evaluation of IR thermography for surveying the condition of bridge decks. *SPIE*, **934**, *Thermosense X* (1988) 19–27.
6. ALONGI, A. V., CANTOR, T. R., KNEETER, C. P. and ALONGI, A. JR, Concrete evaluation by radar theoretical analysis. Transportation Research Record No. 853, 1981, pp. 31–7.
7. CARTER, C. R., CHUNG, T., HOLT, F. B. and MANNING, D. G., An automated signal processing system for the signature analysis of radar waveforms from bridge decks. *Canadian Electrical Engineering Journal*, **11**(3) (1986) 128–37.
8. MASLIWEC, R. and MANNING, D. G., Bridge deck condition surveys using the DART prototype vehicle. Annual Conference Proceedings, Roads and Transportation Association of Canada, 1987, pp. C3–C36.

28

Inspection Based Reliability Updating for Fatigue of Steel Bridges

ANDREW G. TALLIN and MARK CESARE
Polytechnic University of Brooklyn, 333 Jay Street, New York 11201, USA

ABSTRACT

This paper reports on a study of the fatigue reliability of standard AASHTO bridge girder details. In this study, the linear elastic fracture mechanics (LEFM) model of fatigue crack growth was used. Reliabilities were found using techniques based on first-order reliability methods (FORM) modified to calculate inspection updated estimates of reliability. Three example analyses of two different details are given. Analysis results of these details indicate that even at inspection qualities significantly better than might be expected in the field, that effect of inspection which detects no damage, has only a limited effect on the estimated reliability.

INTRODUCTION

About one-third of steel bridges in the US are 50 years old, or older, and many more are nearing that age.[1] As the number of bridges nearing old age increases the need for inspection and maintenance becomes increasingly important. At the same time the resources which can be allocated to the proper maintenance of bridges is shrinking. Older bridges are more susceptible to problems of aging such as corrosion and fatigue. This paper demonstrates the use of first-order reliability methods (FORM) and linear elastic fracture mechanics (LEFM) to update the estimated probability of fatigue failure of steel bridge details based on the results of inspections.

CRACK GROWTH

There are two common approaches to the fatigue of steel, S–N analysis and fracture mechanics crack growth analysis. The S–N approach relates the amplitude of the stress range at a fatigue sensitive point to the life time in stress cycles. This method has been used extensively in bridge fatigue studies.[2,3] Because S–N analysis does not relate to a measurable indicator of damage, it is difficult to incorporate inspection observations into the fatigue analysis. The use of an LEFM model for fatigue crack growth allows information on the presence or size of observed cracks to be incorporated into descriptions of both failure and inspection events. The LEFM approach to the fatigue of steel bridges has been described by a number of researchers.[4,5] The LEFM approach to fatigue crack growth relates the range in the stress intensity ΔK at the crack tip to the rate of crack growth $\mathrm{d}a/\mathrm{d}N$ by the Paris–Erdogan equation:[6]

$$\mathrm{d}a/\mathrm{d}N = C\,\Delta K^m \tag{1a}$$

$$= C(SY(a, y)\sqrt{\pi a})^m \tag{1b}$$

where C and m are material constants. The factor $Y(a, y)\sqrt{\pi a}$ is the stress intensity factor and depends on the crack size a and the detail geometry y. Equation (1) can be rearranged and integrated to result in

$$C \sum_{i=1}^{N} S_i^m = \int_{a_0}^{a_{cr}} \frac{\mathrm{d}a}{(\sqrt{\pi a})^m\, Y(a, y)^m} \tag{2a}$$

$$= \Psi(a_{cr}) \tag{2b}$$

where S_i is the stress range of the ith cycle, a_0 is the initial crack size and a_{cr} is the crack size in the Nth load cycle. The function $\Psi(a_{cr})$ is the damage accumulated during the growth of a crack from the initial size a_0 to a crack of size a_{cr}. The sum $C \sum S_i^m$ is the cumulative load effect on the component causing the damage. Because of the complicated forms of $Y(a, y)$ which occur for practical geometries, integration of eqn (2) must be performed numerically.

In the case of bridge girder details a number of stress intensity functions (SIF) have been compiled by Albrecht and Yazdani[7] for a number of AASHTO fatigue sensitive details. These SIF were used in this study. These SIF take into account a number of factors.

- Surface or edge crack effects. Because cracks in components usually initiate at surface flaws, cracks begin as surface cracks during the first stage of growth. Cracks which initiate at points such as the end of a flange are edge cracks even after the crack has become a through crack.
- Elliptical crack effects. Surface cracks during the first stage of growth are elliptical and, due to three-dimensional effects, have a varying stress intensity along the tip of the crack.
- Width effects which adjust for the finite thickness of the webs and flanges.

- Stress gradient effects which account for non-uniform stress distribution near the crack tip.
- Evolution of the crack as it grows. A crack in a girder detail is assumed to begin as a surface crack, develop into a through crack and continue to grow as a through crack until the critical crack size is exceeded. As a result different geometric factors $Y(a, y)$ are used during each stage of crack growth.

RELIABILITY ANALYSIS

The reliability analyses done in this study were performed using first-order reliability methods (FORM). Reliability analysis using FORM begins by defining the failure event of each element or component using a limit state function $g_{\text{fail}}(\mathbf{Z})$. Whenever $g_{\text{fail}}(\mathbf{Z}) \leq 0$ failure occurs. The limit state is a function of a vector of physical random variables $\mathbf{Z}$ such as loads, material and geometric properties. The surface $g_{\text{fail}}(\mathbf{Z}) = 0$, called the limit state surface, divides the space of random variables into the failure set and the safe set. The probability of failure is the probability content of the failure set $p_f = P(g_{\text{fail}}(\mathbf{Z}) \leq 0)$.

In FORM the vector $\mathbf{Z}$ is transformed into a vector of uncorrelated zero mean unit normal variables $\mathbf{U}$, called u-space. The minimum distance from the origin to the limit state surface in u-space is equal to the first-order approximation of the safety index $\boldsymbol{\beta} = -\Phi^{-1}(p_f)$. The point on the limit state surface which is closest to the origin is called the design point and is denoted U^*. FORM can be extended to the reliability analysis of parallel systems $(g_1 \leq 0 \cap g_2 \leq 0 \cap \cdots \cap g_n \leq 0)$, where g_i is the limit state function for the ith event. For parallel systems U^* is the point closest to the origin which satisfies all the constraints. The probability of the joint event is estimated by linearizing each of the limit states which are zero at the design point U^*, called active constraints. The resulting estimate of the probability of failure is $p_f = \Phi[-\boldsymbol{\beta}; R]$, where $\boldsymbol{\beta}$ and R are the vector of $\boldsymbol{\beta}$ values and the correlation matrix of the active limit states, respectively. More detailed descriptions of FORM can be seen in a number of references, for example Ref. 8. A computer program, PROINSP,[9] based on the general purpose reliability program PROBAN was used to perform reliability and updated reliability calculations.

FATIGUE RELIABILITY

The fatigue failure of a component is defined as the event that a crack exceeds some critical crack length a_{cr}. However, since the damage $\Psi(a)$, defined in eqn (2b), is monotonically increasing with crack length, the capacity of the component can be defined as the amount of damage absorbed by the component as the crack grows from a_0 to a_{cr}, $\Phi(a_{cr})$. The accumulated load $C\sum S_i$ exceeds the damage evaluated at a_{cr}:

$$g_{fail}(\mathbf{Z}) = \Psi(a_{cr}) - \sum_{i=1}^{N} S_i^m \leq 0 \tag{3}$$

where $\mathbf{Z}$ is a vector of random physical properties, including material properties, loads and geometric parameters.

Inspections are similar to failure events and depend on the result of the inspection. Two inspection results considered here are (1) a crack is not detected and (2) a crack is detected and its size is measured. The first event is described by the limit state

$$g_{nf}(\mathbf{Z}) = C\sum_{i=1}^{N} S_i^m - \Psi(a_D) \leq 0 \tag{4}$$

where a_D is the minimum detectable crack. The event $g_{nf}(\mathbf{Z}) \leq 0$ occurs when the accumulated load $C\sum S_i^m$ is smaller than the amount of damage which must be accumulated to grow a crack from a_0 to the detectable crack size a_D. The second event is described by the limit state

$$g_{fd}(\mathbf{Z}) = \Psi(a_I) - C\sum_{i=1}^{N} S_i^m = 0 \tag{5}$$

where a_I is the crack size measured during the inspection. The event $g_{fd}(\mathbf{Z}) = 0$ occurs whenever the damage accumulated in growing a crack from a_0 to a_I equals the accumulated load $C\sum S_i^m$.

The result of an inspection can be used to update the probability of failure. If I and F denote the inspection event and the failure events, respectively, then the probability of failure given that the inspection event occurs is from the definition of conditional probability:

$$p_{f|I} = \frac{P[F \cap I]}{P[I]} \tag{6}$$

The probabilities of the joint events $F \cap I$ and I can be found using FORM,[10] where the event $F \cap I$ is modeled as the parallel system consisting of the failure event (eqn (3)) and inspection events (eqns (5) and (6)).

EXAMPLES

Cover Plate

A 1·25-in welded cover plate terminus (AASHTO category E)[11] on a plate girder was analysed. This cover plate is similar to the cover plates which were

TABLE 1
Yellow Mill Pond Bridge

Quantity	Distribution
Miner's stress	$N(1{\cdot}4, V=0{\cdot}2)$
log C (m)	$N2(\mu=\langle -9{\cdot}17, 3{\cdot}2\rangle; \sigma=\langle 0{\cdot}069, 0{\cdot}055\rangle; \rho=-0{\cdot}971\,48)$
Initial crack size	$LN(0{\cdot}018\,43, 0{\cdot}002\,08)$
Average daily truck traffic	$N(5\,669{\cdot}0, V=0{\cdot}1)$
	Fixed values
Final crack size	4·3 in
Crack aspect ratio	0·25
Thickness of cover plate	1·25 in
Thickness of flange	1·26 in
Width of flange	16·47 in
Thickness of web	0·76 in
Weld size	0·5 in

observed to develop cracks after only 12 years of service on the Yellow Mill Pond Bridge in Connecticut.[7] Table 1 shows the distributions for each of the random variables used in both the failure and inspection limit states. The failure criterion was the development of a through crack greater than 8·6 in in length. The inspection times were selected at the time when the reliability index $\boldsymbol{\beta}$ fell below 2·0 or when the failure probability p_f rose above 0·023.

Figure 1 shows reliabilities of the uninspected detail along with the updated reliabilities due to inspections resulting in no cracks found at 24, 33 and 40 years. The inspection intervals decrease from 24 to 7 years after three

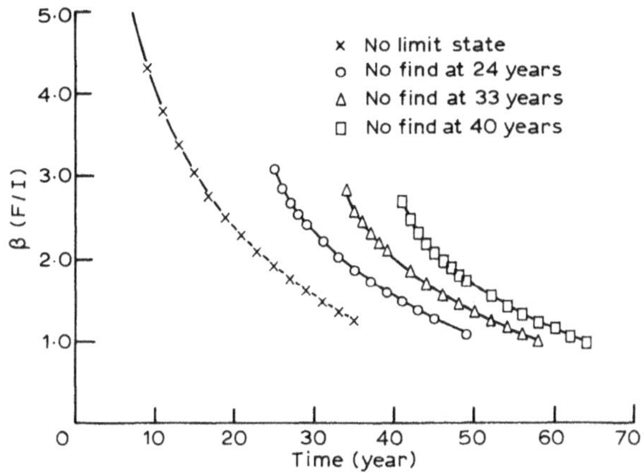

FIG. 1. Updated reliabilities for the Yellow Mill Pond Bridge.

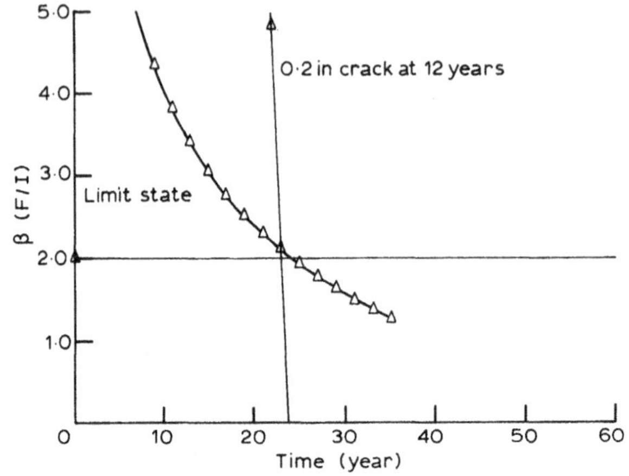

FIG. 2. Updated reliabilities for the Yellow Mill Pond Bridge where 0·2-in crack is found at 12 years.

inspections. This does not show the characteristic increase in inspection times for the fatigue crack growth limit states investigated by Madsen *et al.*[10] These results show that the failure event F and the no-crack detected event I have a strong enough degree of stochastic dependence for only a short time after the inspection.

The detectable crack criterion used here is of especially high quality for bridge inspection, having a mean minimum detectable crack of only 0·35 in. With such an inspection precision the probability of detecting a crack at 24 years is over 30%. Because the crack growth rate is much higher for larger cracks, a decrease in the inspection quality (i.e. an increase in the detectable

TABLE 2
Rolled beam (W30 × 360)

Quantity	Distribution
Miner's stress	$N(10{\cdot}0, V = 0{\cdot}2)$
log C (m)	$N2(\mu = \langle -9{\cdot}17, 3{\cdot}2\rangle;\ \sigma = \langle 0{\cdot}069, 0{\cdot}055\rangle;$ $\rho = -0{\cdot}971\,48)$
Initial crack size	$LN(0{\cdot}001\,2, 0{\cdot}000\,58)$
Average daily truck traffic	$N(500{\cdot}0, V = 0{\cdot}1)$
	Fixed values
Final crack size	2·5 in
Crack aspect ratio	0·67
Thickness of flange	1·68 in
Width of flange	16·665 in
Thickness of web	0·945 in

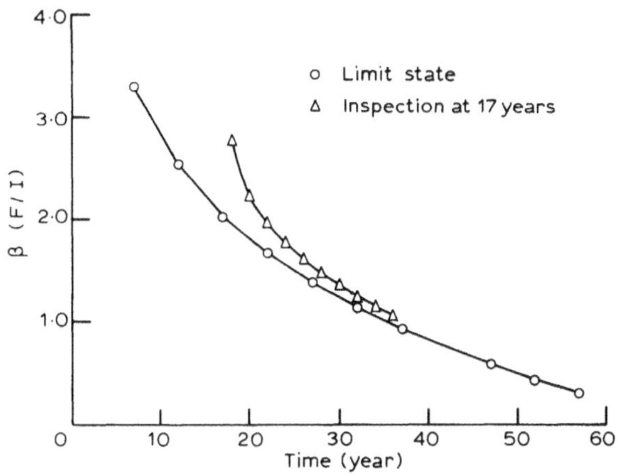

FIG. 3. Updated reliabilities for a W30 × 360 rolled section.

crack size) causes the time between inspections to decrease. For example, if the detectable crack size is equal to the through thickness the probability that there exists a crack with a short remaining life time is quite high.

Figure 2 shows the estimated reliability of the same detail after a crack of 0·2 in was found after 12 years of service. The reliability immediately following the inspection is elevated, $\beta \approx 20$ at 15 years. However, the updated reliability falls to below $\beta = 0$ by 24 years.

Rolled Beam

A W30 × 360 rolled section (Table 2) was analysed at several levels of applied stress range and for a single inspection where no crack was detected. The failure criterion was the development of an edge crack of 2·5 in in the flange. As in the case of the cover plate the inspection time was selected at the point where the reliability falls below $\beta = 3{\cdot}0$.

Figure 3 shows reliability index for both the inspected and uninspected detail due to a Miner's stress range equal to 10 ksi. The inspection at 17 years lifts the reliability immediately after the inspection; however, the reliability soon approaches the initial uninspected reliability.

CONCLUSIONS

The method of updating reliability described here can be used to estimate reliabilities of fatigue sensitive details conditioned on the results of inspections which result in either no detection of a crack or a measured size of a crack. Because the LEFM based fatigue analysis relates physical quantities such as crack size and stress range the method of updating estimated reliabilities using LEFM is straightforward.

The examples here show the limited effectiveness of even fairly precise inspections of details which had been observed to fatigue easily. Because the inspection qualities used for bridges can only detect cracks which are advanced in age, the effect of an inspection which detects no damage on the estimated reliability is limited to a short time past the inspection.

The example of the inspection which resulted in a detected crack gave updated reliabilities which quickly fell off following the discovery of the crack. In such cases the rate of deterioration of the estimated reliability can be used to determine the speed at which repairs must be made in order to maintain an acceptable level of safety.

ACKNOWLEDGEMENTS

The work described in this paper was supported by grants from the National Science Foundation (grant number MSM-8657854) and Det Norske Veritas.

REFERENCES

1. GALAMBOS, C. F., Bridge design, maintenance and management. *Public Roads*, **50**(4) (1987).
2. MOSES, F., Probabilistic load modeling for bridge fatigue studies. *IABSE Colloquium on Fatigue of Steel and Concrete Structures*, Lausanne, Switzerland, 1982.
3. NYMAN, W. and MOSES, F., Calibration of bridge fatigue design model. *Journal of Structural Engineering*, **111**(6) (1985).
4. FISHER, J. W., *Fatigue and Fracture in Steel Bridges—Case Studies*, John Wiley & Sons Inc., New York (1984).
5. YAZDANI, N. and ALBRECHT, P., Risk analysis of fatigue failure of highway steel bridges. *Journal of Structural Engineering*, **113**(3) (1987).
6. PARIS, P. and ERDOGAN, F., A critical analysis of crack propagation laws. *Journal of Basic Engineering, Trans. ASME*, **85** (1963).
7. ALBRECHT, P. and YAZDANI, N., Risk analysis of extending the service life of steel bridges. Maryland DoT Report No. FHWA/MD-84/01, 1986.
8. MADSEN, H. O., KRENK, S. and LIND, N. C., *Methods of Structural Safety*. Prentice-Hall, Englewood Cliffs, NJ, 1986.
9. TALLIN, A. G. and SKJONG, R., *PROINSP—User's Manual*, A.S. Veritas Research Report 87-2018, Hovik, Norway, 1987.
10. MADSEN, H. O., TALLIN, A. G., SKJONG, R. and KIRKEMO, F., Probabilistic fatigue crack growth analysis of offshore structures with reliability updating. *Proceedings Marine Structural Reliability Symposium*, Arlington, VA, 1987.
11. AASHTO, *Standard Specifications for Highway Bridges* (1977), 12th edn. American Association of State Highway and Transportation Officials, Washington, DC.

29

Diagnostic Dynamic Testing of Bridges on Brenner Motorway

RAINER FLESCH

Structural Dynamics/BVFA Arsenal, A-1030 Wien, Faradaygasse 3, Austria

and

KARL KERNBICHLER

Institute for Reinforced Concrete, Technical University Graz, A-8010 Graz, Technikerstrasse 4, Austria

ABSTRACT

A dynamic method for the damage evaluation of large bridges is developed by BVFA and Technical University Graz. The method is a combination of dynamic in-situ tests and dynamic calculations. The basic idea is to detect damage to bridges via changes of the dynamic parameters. The steps in the process are dynamic in-situ testing, mathematical modelling, fitting of the mathematical model to the test results and sensitivity investigations. The development of the method is presented by discussing the history of the process. Further, the steps are discussed, giving references to detailed information and some examples from past projects. The latest investigations were started in October 1988 on five bridges on the Brenner motorway. The tests were finished in April 1989 and analysis of measurements is still under way. More detailed results will soon be available.

OBJECT

Due to the drastic increase of the number of vehicles passing over bridges, especially heavy trucks, methods for bridge inspection become more and more important. The main aims are:

- safety inspection of the bridge, and
- detection of damage at an early stage for minimising the cost of repair.

At the moment mainly visual inspection techniques with limited scope for quantification of damage are used. This situation is unsatisfactory in our world of high technology. Hence in 1981 BVFA and Technical University Graz started the development of a global method for the inspection of structures using vibrations. The method has been greatly improved over the years using the increasing capabilities of computers and measuring equipment. Experience mainly with prestressed bridges has been obtained. The basic concept of the method is to formulate a dynamic model for the virgin state of the structure. In general, damage decreases the stiffness and increases damping, resulting in changes of dynamic properties being different for different modes of vibration. The dynamic properties are modal frequencies (eigenfrequencies), mode shapes and modal damping ratios. Hence damage will lead to a certain pattern of deviation of dynamic parameters and can be used for localisation and quantification of damage in a global manner.

The following steps are carried out:

- establishment of a structural dynamic model for the vertical direction (FE model),
- experimental determination of modal parameters by dynamic in-situ tests, and
- systematic system identification (fitting of the structural model to the modal test model).

The procedure described above is the baseline evaluation of the dynamic properties. It should be done when the construction of the bridge is finished and before the bridge is opened for traffic.

After a certain time of use (e.g. 2–5 years) the investigations should be repeated. If changes of modal frequencies, mode shapes and damping ratios are found in the experiment, the structural model must be fitted again to the test results. Comparing the old and new mathematical model, changes of stiffness in certain parts of the structure can be detected and quantified in a global sense (over the FE element, including the actual damage).

If changes exceed a certain limit, which has to be fixed for any particular case, the damage area found by the global method should be investigated in detail by a local method.

With the detailed information (e.g. exact location of cracks, crack width, depth of cracks, etc.) the fitting of the structural model to the experimental results can be further improved. With the resulting optimum model the redistribution of internal moments and forces for design loads due to the stiffness change can be calculated. Based on the results, in any particular case a decision must be made on whether the structure is still safe or not.

The concept described above can be strictly followed only for new bridges, when the baseline was evaluated after finishing construction and before opening of the bridge, so that no disturbance occurs as a result of the traffic. For most existing bridges there is no baseline available. For some bridges used for several years first measurements were carried out during the last few years, giving the baseline for that certain point in time. Often the evaluation was done after major rebuilding or visual inspection. It is argued that severe damage, currently invisible, will increase by the next routine measurement to such an extent that it can be detected via the change of dynamic parameters.

For older bridges it would be helpful to have a variant of the method where damage can be detected from a single in-situ test. At the moment only ideas exist (e.g. comparison of tests carried out with different levels of excitation force or with different locations of the excitation, or by the systematic placing of additional masses during the test to increase the opening of cracks, resulting in a further stiffness decrease and hence increasing the probability of detection).

The method consists of the following parts (see Fig. 1):

- structural (mathematical) modelling,
- in-situ testing,
- analysis of test results (first step of system identification: elaboration of modal test model), and
- systematic system identification (fitting of mathematical model to the modal test model).

Systematic system identification is still a developing field with many unsolved problems. Hence additional approaches which can help to localise and quantify damage are most desirable. One possibility is the use of structural dynamics modification software for sensitivity investigations. Using only the modal test model, a decrease of the bending stiffness is applied systematically to the structure, giving the change of the dynamic properties due to a local stiffness change. This fruitful approach will be discussed later in more detail.

Depending on the mode shapes, the single modes have a different local sensitivity to stiffness changes. The more modes with different shapes that are available, the easier will be the localisation and quantification of any damage. In the past investigations have been focused on the change of modal frequencies although this parameter only varies in proportion to the square root of the stiffness change. In the future the changes of the mode shapes, which are much more pronounced, will be used for interpretation. Before changes of damping ratios can be interpreted the damping mechanism of R/C structures must be better understood.

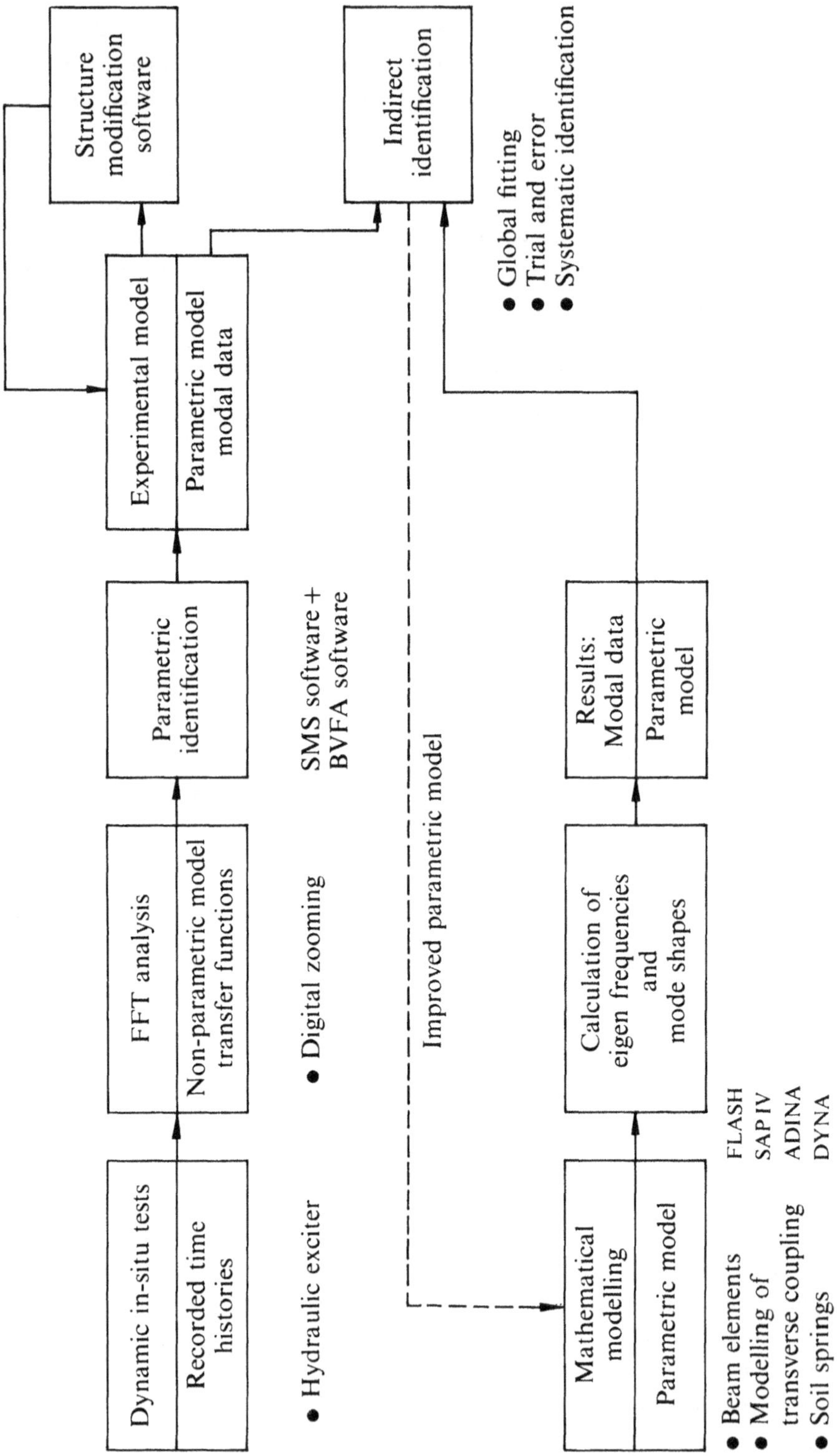

FIG. 1. Steps of the dynamic method for safety inspection.

THE HISTORY OF APPLICATION

The work was started in 1981. For the Raach Bridge dynamic and static tests were carried out before and after cutting of a number of tendons. The artificial damage was possible because the bridge had to be removed later for a new traffic scheme. Compared with today the test equipment was quite primitive, but it showed that the method works in principle. Much more information was obtained by the dynamic method than by the static approaches. The results are reported in Ref. 1.

During the first investigations of the Obernberg Bridge[2] the test technique was improved significantly. A maximum difference ($\Delta f = 0{\cdot}006$ Hz) for modes 1–7 was obtained between measured and calculated modal frequencies. Two of the higher modes were also closely fitted to the test results by a more detailed modelling of the open caisson foundation of pier 2.

For the Gänstorbrücke in Ulm (FRG) tests were carried out before and after major repair work. Measurements were carried out for three states. Changes of modal bending frequencies and modal torsional frequencies were detected. The increase of stiffness resulting from the repair work was established by comparing the structural models for the different states. The results are given in Refs 3–5.

In November 1985 investigations of the Lavant Bridge were started. Lavant Bridge is one of the greatest prestressed framed bridges in Europe. The maximum span is 160 m and the maximum shaft height is 130·35 m. The tests could be carried out in the virgin state before it was opened for traffic. For the analysis of the measurements a modal analysis software package from SMS was used for the first time. Further, sensitivity investigations were carried out using the structural dynamics modification software. The results are given in Refs 6 and 7.

Starting in October 1988 a baseline evaluation was carried out for five bridges on Brenner motorways. The bridge investigated first was again the Obernberg, this time after widening of the cross-section (adding one lane). In the latest test series a new reaction mass exciter driven by a hydraulic actuator was used. For all previous projects an eccentric mass exciter had to be used.

To obtain experience with other types of bridge a R/C arch bridge (Äussere Nösslachbrücke) and a composite bridge with steel girders and a concrete slab (Gschnitztalbrücke) were included in the programme. Further, for one mushroom slab bridge (Nösslachbrücke) tests were carried out to study the behaviour of gaps and bearings between the six substructures of the bridge. Finally, another R/C frame bridge was tested to provide a baseline after reconstruction in areas of sliding foundations due to hillside creep.

Hence the latest series could be carried out with excellent test equipment and analysis capability, but with disturbance by traffic. As routine tests have normally to be carried out under conditions of traffic disturbance, it is important from the latest investigations to find procedures to eliminate these disturbances. The problem will be discussed in more detail later. The analysis for all five bridges is still under way, with more results available soon.

For the next 2 years three new projects are planned. Projects one and two are baseline evaluations for a three-span prestressed bridge and a composite bridge in the virgin state in Budapest (Hungary). Project three is an investigation of a multi-span prestressed bridge at the border between Austria and Yugoslavia. In this project tests of substructures during the construction are also planned which will be very fruitful for improving the indirect system identification approaches.

DESCRIPTION OF THE METHOD

Focus and Limitations for the Application

All bridges tested before the latest series were large prestressed bridges. Investigations focused on that type, because the most frequent damage to prestressed bridges is bending cracks having areas of distorted bond at both sides of a crack, giving some kind of smeared stiffness decrease in keeping with the assumption in the structural model. Hence the method seems to be suitable for prestressed bridges. A more detailed discussion of different forms of damage and expected effects on dynamic properties are given in Refs 6 and 8.

In principle the method can also be used for R/C bridges, steel bridges or composite bridges, especially if changes in boundary areas (foundation of piers, bearings, etc.) is detected. In the literature some cases of loss of bearings are reported, damage which could be quickly identified by the dynamic method.

As it is not possible to close a bridge completely for testing, the measurements need to be carried out with traffic disturbance. Hence procedures to eliminate disturbance are necessary. Early results are discussed at the end of this paper. It is only possible to close one lane in each direction for about 2–5 days, the same time span used for visual inspection using the bridge inspection truck.

Next, influences from environmental conditions on the measured results must be considered. Experience has been gained relating to the influence of temperature. There is an influence if a stiffness change (e.g. due to an elongation in the longitudinal direction) occurs in the direction of a pronounced modal movement. The experience from Gänstorbrücke is reported in Refs 3–5 and 8. It can be concluded that temperature has an influence, but the single modes are influenced to a different extent. It is assumed that after exclusion of modes which are very sensitive to temperature changes that enough modes are still available to give the expected information.

Test Technique

As frequency changes due to damage are often small a precise test technique is necessary. In the past an eccentric mass exciter was used. Using a static frequency changer the frequency could be controlled with an accuracy of 0·003 Hz. The disadvantage of the eccentric mass exciter is the quadratic force and the very low force amplitude in the low frequency range. From experience the minimum force should be 1 kN to obtain good results. Recently a reaction mass exciter driven by a hydraulic actuator was developed by BVFA. The excitation force can be kept constant during the frequency sweep using a control program. The exciter, the hydraulic pump and the control equipment are mounted on a flat-bed lorry which is used for transportation of the heavy equipment to the excitation point. The excitation can be carried out in vertical, longitudinal or transverse directions. For inspection the bridges are mainly excited in the vertical direction.

The response is measured by velocity transducers (Hottinger SMU 30A). The signals are integrated to obtain the displacement response. As there is often a long distance between transducer and magnetic tape recorder the amplifiers are mounted in boxes together with the transducers. The frequency response function for each transducer was obtained on the BVFA shaking table. The function is approximated analytically and used for the correction of the measured transfer functions via a complex division. To have a mesh of measurement points adequate for sensitivity investigations the distance between measurement profiles should be about 5–10 m. Normally 4–6 transducers are distributed over the cross-section in each profile. First, 2–3 positions for the exciter, adequate to excite as many modes as possible, are selected. Then the transducers are placed in one profile after the other. For each configuration a frequency sweep is carried out in the frequency range 0–10 Hz lasting about 10 min. The time histories of excitation force and of the responses are recorded on magnetic tape.

From these records transfer functions are calculated using a FFT analyser. In the latest series about 1000 transfer functions were obtained per bridge within 1 week.

From 2–3 positions of the exciter 40–60 modes were obtained. More detailed information is given in Refs 6 and 8.

Analysis of Measurements

From the records of the response and of the excitation force, transfer functions are obtained by FFT analysis. Before the Lavant project, peak picking and circle fit algorithms were used to elaborate the eigen frequencies, mode shapes and damping ratios.

For the analysis of the Lavànt Bridge measurements the SMS modal analysis software MODAL 3·0 could be used for the first time. The software is very modular and provides a best fit algorithm and is very easy to use. The total concept seems to be aimed at mechanical engineering problems. Special features of tests of very large civil engineering structures had to be handled by additional software. Detailed information is given in Refs 6 and 8.

Dynamic Calculations

The dynamic calculations were carried out by Technical University Graz. SAP IV and FLASH were used for mathematical modelling. During every project attempts were made to find the method with an adequate level of accuracy. Only beam elements and additional spring elements were used. In the opinion of the authors, the inclusion of more complicated elements would not improve the quality of modelling since material parameters of concrete structũres are often quite uncertain (deviations from plan cross-sections, variance of mass density, etc.). These uncertainties must be allowed for using correction factors.

About 1200 DOFs were used for each model. In general, the bending stiffness can be modelled well but problems can arise for torsional stiffness and cross-section deformation for open cross-sections.

To get precise results important details must be modelled well. In some cases the transversal coupling of adjacent bridges via the carriageway slab had to be modelled. The problem was solved by eccentrically connected beam elements.[2,4] The influence of the foundation was sometimes modelled well by additional springs.[2]

Systematic System Identification

Systematic system identification means the fitting of structural models (FE models) to test results. The state of the art is given in Refs 9 and 10, and in the proceedings of the workshop held in Lambrecht (FRG) in 1987 (see Ref. 6).

In past projects systematic system identification was carried out in two relatively simple steps:

- In a global manner using the modulus of elasticity. Hence the modulus was used as central fit parameter, representing also uncertainties and deviations of cross-sections, mass densities and boundary conditions, etc.

- By trial and error in the areas of repair work, especially in the case of Gänstorbrücke.[3–5]

For routine inspection systematic methods are necessary. Methods are under development at Curt Risch Institute (University Hannover, FRG) but they are limited to models with 30 DOFs at the moment.

Sensitivity Investigations

It was shown before that systematic system identification can be quite difficult for large systems. Hence the application of the structural dynamics modification (SDM) software provides a powerful tool for interpretation of changes to modal properties. For the investigations only the modal test model is necessary, hence no FE model has to be fitted to the test results. The software is applied to the baseline test model. To investigate the influence of local cracks on the modal parameters the decrease of bending stiffness has to be modelled. A 3-DOF beam element (negative rib stiffener) is 'roved' over the bridge, being implemented in three adjacent DOFs at one time. In that way the influence of a local stiffness decrease (smeared over two adjacent distances between measurement profiles) on the modal parameters can be elaborated. The basic equations used for the software are given in Refs 6 and 8, together with all references. Changes of modal frequencies due to a local stiffness decrease are presented in Refs 6 and 8. Changes of mode shapes are given in Refs 7 and 8.

In the future a systematic approach using the values mentioned above for the interpretation of measured changes of dynamic parameters will be elaborated.

As an example some changes of modal frequencies obtained for the Lavant Bridge are shown in Fig. 2. For every position given by the horizontal axis you can find the decrease of modal frequencies on the vertical axis, the negative element being centred at the enumerated position.

In the future SDM will also be used for systematic system identification.

Disturbances by Traffic

To establish a practicable routine test technique the method must be applicable during traffic flow. The influence of traffic was studied in detail in the latest test series. The results are still being processed. The first lesson learnt is that at least one fixed transducer should be provided, which remains in the same position during all frequency sweeps (e.g. in the profile of excitation). Comparing the resulting transfer functions, disturbances can be detected. Probably a procedure for data correction can be established. The imaginary part of the transfer function in the driving point must always have positive values, hence negative values at certain frequencies are a criterion for disturbances during that sweep.

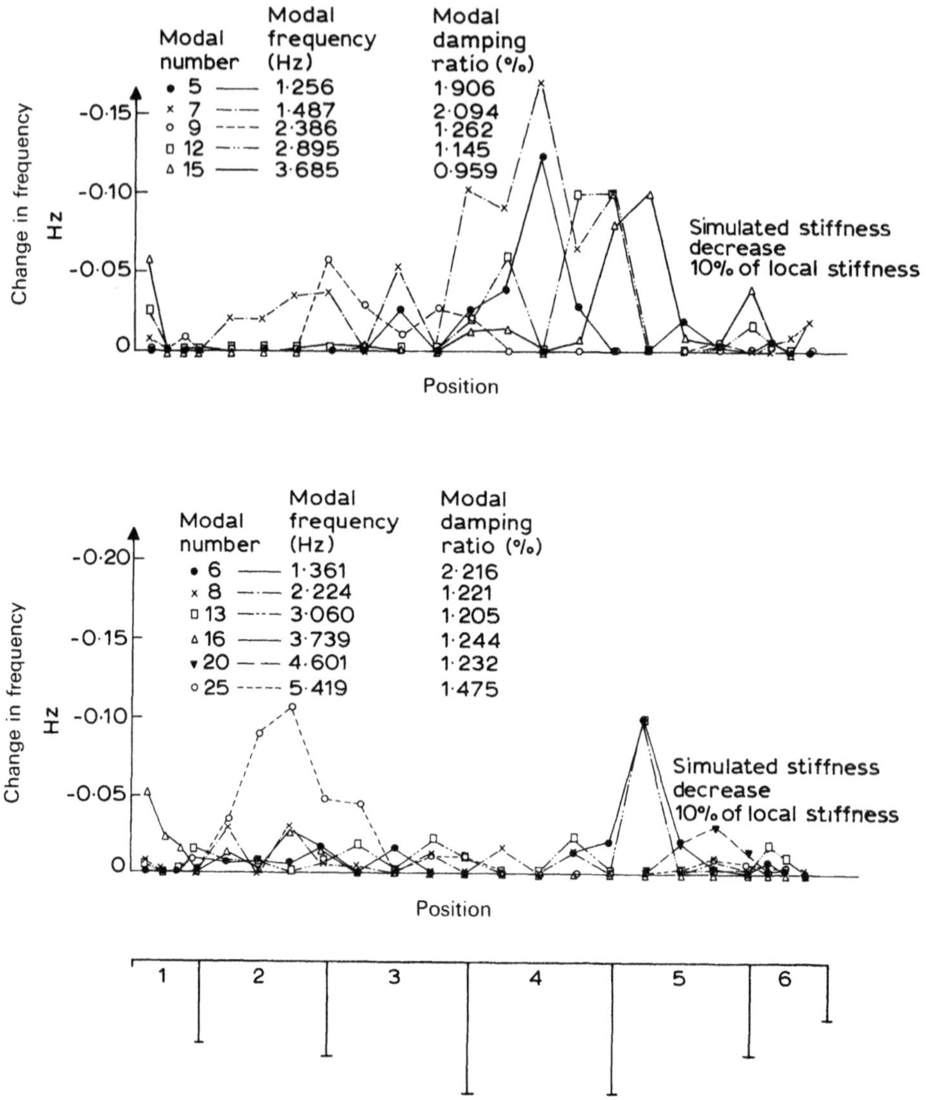

FIG. 2. Decrease of modal frequencies due to a local stiffness decrease, calculated for Lavant Bridge.

In general, the traffic has the following influences on the results:

- Additional excitation. As for the calculation of the transfer functions a cross-spectrum is used. The results are only distorted if the disturbance has a momentary frequency component equal to the actual sweep frequency. As the sweep has to be repeated for every measurement profile, there is a low probability that disturbances will occur at the same modal frequencies in each profile. Due to the large number of profiles distorted values can be recognised and eliminated, with enough information remaining to determine the mode shapes. In addition, if the baseline was evaluated for the virgin state without traffic, the original mode shapes are known, making the analysis of routine measurements much easier.

- As the additional excitation is not synchronous with the artificial excitation, traffic can also produce an additional damping mechanism.
- Additional moving mass. The influence from heavy trucks could be of special importance.

The influence of additional moving masses seems to be most severe but it can be assumed that modal frequencies are always decreased. Hence, using the large number of results, the most probable values for each modal frequency and damping ratio can be established. During the analysis of measurements additional modes (computational modes) are used during curve fitting to take into account the disturbances. These additional modes, often having negative damping values, are removed finally.

CONCLUSION

The history of application illustrates the development of the dynamic method. It seems to be possible to produce a technologically advanced approach with greater capability than visual inspection but with comparable time required for the inspection. It is necessary to carry out the routine tests under traffic. The first results of the latest series promise a way to establish the most probable values for modal frequencies and damping ratios.

At the moment analysis is very time consuming, but a further automation and reduction of the time necessary for analysis will be sought. At present the method is mainly applied to large prestressed bridges because their main damage mechanism fits quite well with the model assumptions.

The changes of modal frequencies due to structural damage are often very small but the changes of the mode shapes are often much more pronounced due to the shifting of modal nodes.

Before changes of damping ratios can be interpreted the damping mechanism of R/C structures must be better understood. The structural dynamics modification provides a powerful tool to study the influence of local stiffness changes on the modal properties. This information can be the basis for a localisation and quantification of damage at a later point in time.

The latest test results for bridges of the Brenner motorway will be presented in forthcoming papers.

REFERENCES

1. Kernbichler, K. and Flesch, R., Static and dynamic tests, their qualification for bridges inspection and long-term observations of bridge structures. RILEM Symposium, Budapest, 1984.
2. Flesch, R., Kernbichler, K. and Dürnsteiner, Ch., Dynamic testing and modelling of Obernberg Bridge. *Proc. Int. Conf. on Num. Meth. for Transient and Coupled Problems*, Venice, 1984.
3. Flesch, R., Kernbichler, K. and Grübl, P., Brückeninspektion mittels dynamischer Untersuchungen. 8. GESA-Symp., VDI-Berichte, Nr. 514, 1984.
4. Kernbichler, K., Flesch, R. and Rauscher, G., Dynamische Untersuchungen von Großbrücken (Massivbrücken), in-situ Versuche und Rechenmodelle. Tagung Dynamische Probleme, Universität Hannover, 1984.
5. Flesch, R., Kernbichler, K. and Rauscher, G., Dynamische in-situ Versuche und Rechenmodelle—Praktische Anwendung auf Großbrücken/Massivbrücken, ÖIAZ, 131. Jg., Heft 10, 1986.
6. Flesch, R. G. and Kernbichler, K., Bridge inspection by dynamic tests and calculations—dynamic investigations of Lavant Bridge. *Proc. Workshop on Struct. Safety Evaluation Based on System Identification Approaches*. Friedr. Vieweg & Sohn, Braunschweig, Wiesbaden, 1988.
7. Flesch, R. G. and Kernbichler, K., A dynamic method for the safety inspection of large prestressed bridges. *Proc. Int. Workshop on Nondestructive Evaluation for Performance in Civil Structures*. Department of Civil Engineering, University of Southern California, Los Angeles, 1988.
8. Flesch, R., Die Methoden der Baudynamik mit spezieller Berücksichtigung ihrer Anwendbarkeit zur Bauwerksinspektion. *Habilitationsschrift*, Teil 2, Technical University Graz, 1988.
9. Natke, H. G., Die systematische Anpassung von Rechenmodellen an Versuchswerte als Verfahren zum Nachweis des dynamischen Systemverhaltens. *Bauingenieur*, **57** (1982).
10. Natke, H. G., Einführung in Theorie und Praxis der Zeitreihen- und Modalanalyse. Vieweg & Sohn, Braunschweig, 1983.

30

Experience with the Management of Cable Stayed Bridges in Korea

HELMUT WENZEL

VCE, Vienna Consulting Engineers, Vienna, Austria

ABSTRACT

Cable stayed bridges are sophisticated structures and more attention has to be paid to their management, i.e. inspection, maintenance and repair. In Korea several bridges of this type were designed by foreign consultants in co-operation with local companies. This turned out to be a procedure that did not produce satisfactory results. Two different types of structures, steel structures versus concrete structures, and two methods of design and management are presented here and discussed.

INTRODUCTION

In the period from 1985 till 1989 the author's company had the chance to work on four cable stayed bridges in Korea. Two of them are steel structures, designed by British consultants and built by local contractors. The other two are concrete bridges, designed by VCE and also built by local contractors, but under the supervision of the designer.

There are different owners and so the method of managing the structures is different. In general, we found that the understanding of the requirements of bridge management is too small and it is difficult to succeed with new concepts. The following report shows that it is nearly impossible to introduce a standard close to the European state of the art.

After a description of the structures the paper will present how the bridges were designed, built and supervised, which inspections were carried out, which defects were found and which repair activities took place. The final section deals with a proposal for more effective management.

DESCRIPTION OF THE STRUCTURES

Jindo Bridge (see Fig. 1)

The spans are 70 + 346 + 70 m. The pylon is also a steel structure. The cables are of the locked coil type. The pavement consists of 5 cm asphalt only.

Dolsan Bridge

This bridge is similar to the Jindo Bridge, with smaller spans only (70 + 260 + 70 m). The structural system and appearance are the same.

Olympic Grand Bridge (see Figs 2 and 3)

The 30 m wide deck is supported by cable pairs anchored in the median. The concrete structure is very slender. The cables are made of single strands

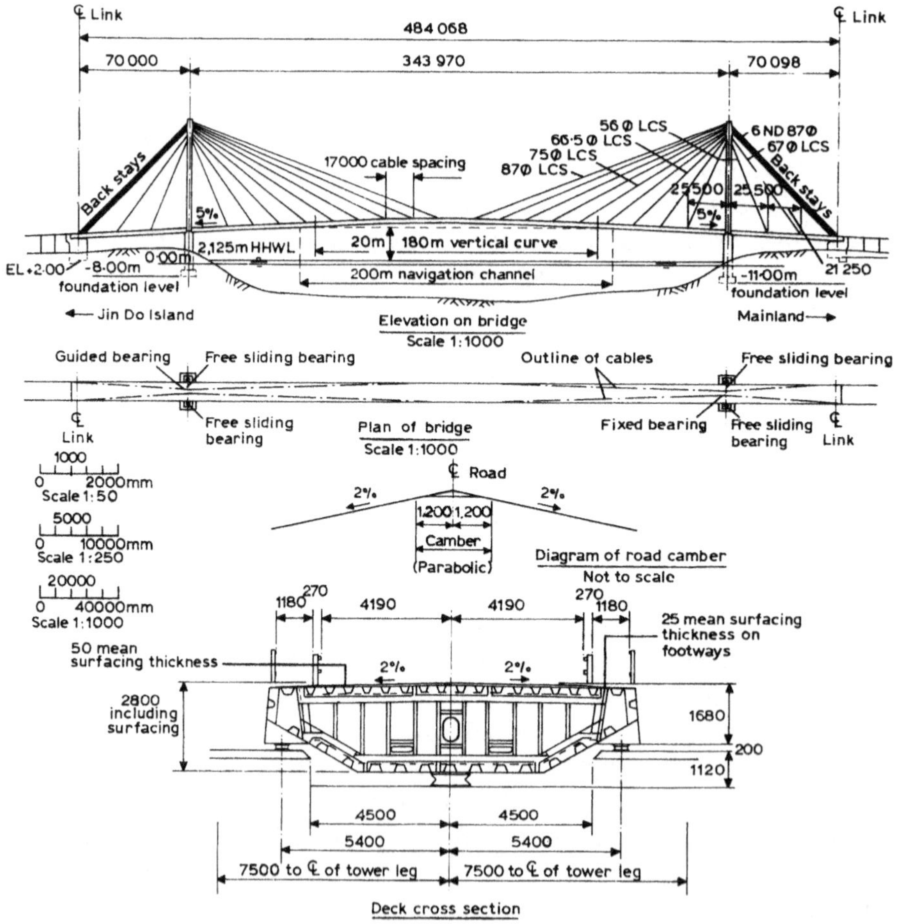

FIG. 1. Jindo Bridge, general arrangement.

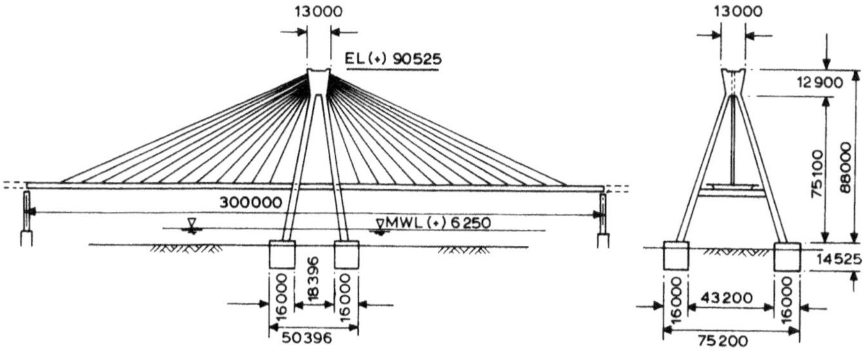

FIG. 2. Olympic Grand Bridge, front and side view.

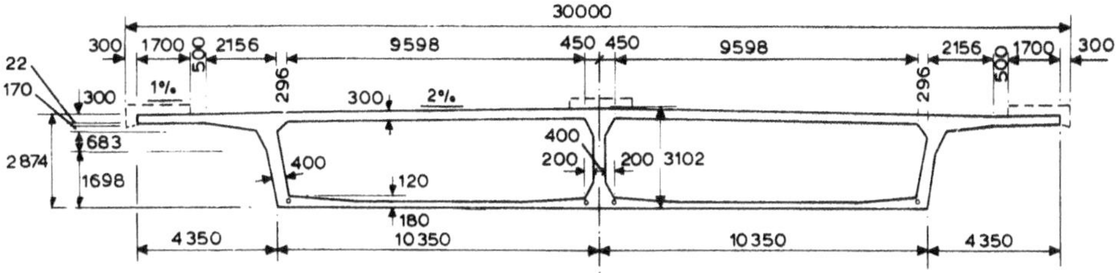

FIG. 3. Olympic Grand Bridge, typical cross-section.

0·6 in, 37–61 of them in one PE sheath tube, and were grouted after erection with cement mortar. The pylon is a concrete structure. The main span is 300 m. Construction was by the cast in-situ free cantilever method.

Haeng Ju Bridge (see Fig. 4)

This is a two-lane bridge built by the incremental launching method using auxiliary piers pushing 60 m spans. The main spans were converted into a cable stayed bridge using prestressed concrete members. The width is 16·5 m, the deck height 4·5 m.

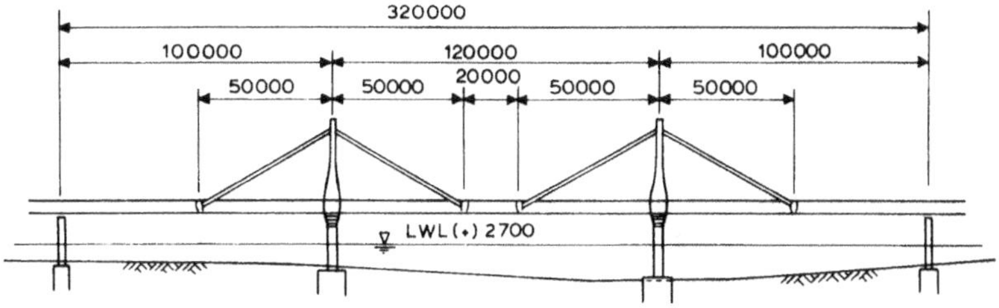

FIG. 4. Haeng Ju Bridge, side view.

DESIGN

The arrangements for designing bridges in Korea is typical of the new industrialised countries. There are some well trained engineers, but the amount of work is huge, and whoever is successful in one job will be promoted and all his skills are lost. In Korean society a great distance can be observed between generations and know-how is not transferred to potential rivals. This fact makes it impossible to build up a strong design capacity and it is very common to meet new engineers in the most demanding jobs.

Foreign consultants are engaged to design structures in co-operation with local firms. Training programmes are arranged and specialists educated. But, for the above reasons, these activities do not bring the required success. The specialists are promoted away from their jobs and the next one starts from the beginning.

This happened for all the bridges described in the previous section.

The two steel bridges were designed by a well-known British consultant in co-operation with a big local engineering firm. The intention was to create substantial local knowledge to deal with all future requirements for these structures. In fact the author found after only 4 years that even minor things could not be handled by the local firm. It was impossible to find the relevant drawings in their files. A good design got a bad reputation, due to lack of understanding.

The design of the Olympic Grand Bridge has a long history and was finally produced by VCE in co-operation with the contractor. Engineers were trained but disappeared after completion of the job. The design was handed over to the same local engineering company and site supervision was carried out together with VCE. Again the author's company was confronted with a number of engineers who had to be trained from the beginning and who disappeared after some time. The author's company therefore failed to install a strong engineering force capable of handling the requirements of inspection and maintenance of this bridge.

A similar story applied to the Haeng Ju Bridge.

The design of the Haeng Ju Bridge bearings were modified to save money. The new design did not consider local stresses and the bearings were deformed and had to be exchanged. The result was a 4-month delay in construction. Figure 5 shows the bearing design before and after modification.

Although there were different clients for all the structures, the same lack of understanding was found. A better procedure was recommended to them, but accusations were made that the local situation was being misunderstood. The argument of higher overall costs for the structures was ignored because different activities were covered by different departments from different

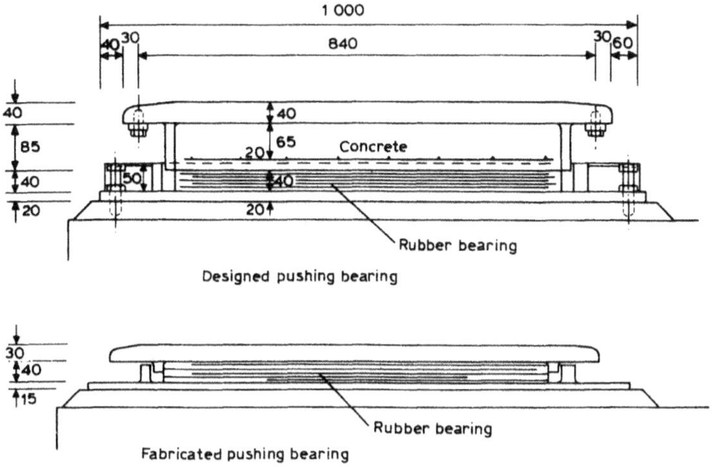

FIG. 5. Bridge bearing as originally designed and finally fabricated.

budgets. This is also common in Europe, but the understanding at least exists.

It is essential to put more emphasis on client education, rather than consultant training, to improve the situation.

CONSTRUCTION AND SUPERVISION

The construction industry in Korea is one of the strongest in the world. Accordingly, their self-confidence is disproportionately developed. This leads to problems when it comes to the building of structures not fully understood by the engineers. Plenty of 'good' ideas are born to reduce costs and essential parts of the structure are altered or completely deleted. In most cases there is no authority to set limits and important parts are wrongly produced.

Jindo Bridge (see Fig. 6)

The biggest Korean steel manufacturer received the order to build this bridge. Foreign supervision was not considered. Several items were executed differently to the original design. The cables were ordered in Japan. A proper stressing procedure was not available and no adjustment took place after the whole structure had settled. This, combined with lack of tolerance control, led to the problems described in the section titled 'Inspection'.

The problem of vibrations, which is normally handled by the designers after erection, was completely neglected. Rumours were spread that there was something wrong with the design. Only after 4 years, when the problems grew worse, were experts asked to reappraise the situation.

FIG. 6. Jindo Bridge.

Good design ideas disappeared before completion of the structures and much money now has to be invested to upgrade the bridges to the desired level. The engineering is also more expensive now than proper supervision would have cost.

Olympic Grand Bridge (see Fig. 7)

It was possible to convince the client that proper supervision is well worth the investment. The main reason was not related to the lifetime costs; it was the fear of problems with a structure important for the Olympic Games.

In this case the designer who had the basic ideas was appointed to supervise the construction works. A lack of information did not occur and many of the 'good' ideas could be avoided.

The contractor was one of the big bridge contractors in Korea, and due to the fact that the alternative design was backed by them good understanding made co-operation easy. All the new methods introduced by the author's company were understood and applied seriously. There was no major incident during the whole construction period.

Again education of the local engineering partner was attempted for him to take over the functions of bridge management, but again a satisfactory status was not reached.

Haeng Ju Bridge

Experience gained through proper supervision of the Olympic Grand

FIG. 7. Olympic Grand Bridge during construction.

Bridge was ignored at the Haeng Ju Bridge. Again the contractor pointed out that the material suppliers would provide supervision free of charge, which led to considerably lower quality at higher prices.

INSPECTION, MAINTENANCE AND REPAIR

The understanding of these items was completely lacking. Inspections are done only if something strange happens or damage is done by vehicles.

A proposal put forward by the author's company for a minor inspection every year and a major inspection every 6 years was not accepted.

The idea of inspecting before the warranty period expires was also new to them. The main argument put forward was that very seldom was any company held liable for damage to their work. That this requires better supervision during construction was not accepted.

Jindo and Dolsan Bridges

The call for an inspection came after complaints about vibrations during strong winds. A team of engineers was sent to Korea and a kind of routine inspection was performed. Three major defects were found.

Firstly, most of the cable bearings were loose, because the complicated fixing proposed by the designer was changed to a simple method that did not work. The final fabrication is shown in Fig. 8.

FIG. 8. Anchorage of the backstays.

The elastomeric blocks were inserted on top of the anchor structure and the fixing device was not installed. It is not shown in the contractor's shop drawings. Due to the vibrations the pads went down along the cables. At the first and last cable the client's workers installed a wooden box to support the pads, which was better than nothing. The slipped pads can be seen on the bottom of the anchor tube.

At the pylon the same situation was found by VCE. The pads had come down all the way to the deck anchor.

Due to this fact no damping was available to the cables and there is doubt about the number of load cycles endured to date and also whether there is any damage to the cables at the anchors, which cannot be inspected with simple methods.

The amplitude of the cable vibration was not excessive but too much for comfort.

The second fact was that the cables are stressed very differently and not according to the design. This can be seen particularly at the backstays, where only three out of six cables bear 80% of the loads. The loose cables showed unusual behaviour during strong winds. Their vibration was more horizontal than vertical and an uncomfortable horizontal impact was observed at the slender bridge deck. Due to the small width of the bridge and the long span, the horizontal excitation can be considerable.

The rubber cable bearings that were provided to damp the vibrations should have been fixed in position by a tiny steel structure to avoid

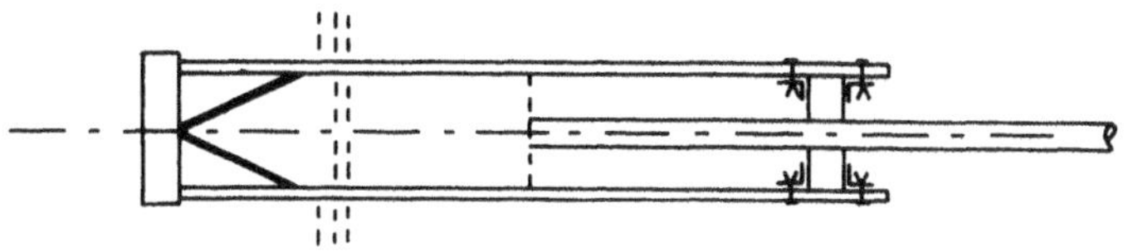

Designed cable bearing (fixing device)

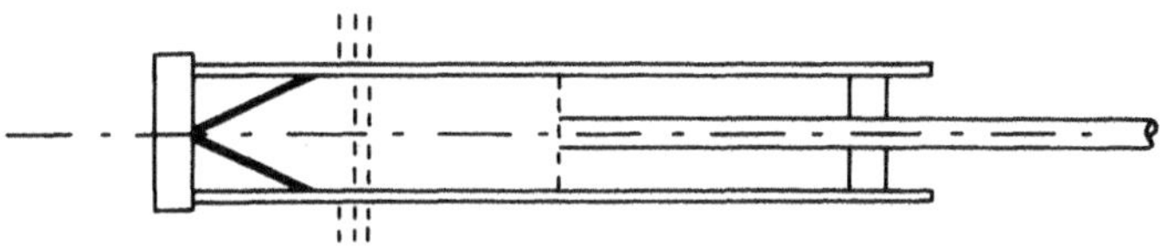

Fabricated cable bearing (friction only)

FIG. 9. Bearing as originally designed and finally fabricated.

movement during vibrations. These angles could be installed only after the erection and the work was difficult. On the other hand, the cable bearing has to be pushed in and the friction is high after installation. To save the difficult fixing device erection the simpler detail was adopted, neglecting the long-term behaviour of the bearing. Figure 9 shows the detail before and after modification.

The third item found was a cracked asphalt surface and other damage to lighting poles and hand rails. This can also be explained by the excessive vibrations during the typhoon season. One of the poles was buckled completely.

After the first visit a report was presented that pointed out what measures would be necessary to maintain and repair the bridge, together with a cost estimate and schedule. Since then nothing further has happened because there could be no budget allocated. The suggestion that the cables might break without warning after a certain number of load cycles did not help.

The proposed inspection procedure was not introduced. The bridge is still inspected by the relevant road master, who has no technical education related to bridge engineering.

Olympic Grand Bridge

Due to sufficient site supervision during construction and execution in line with the original concept, including an inspection after erection, it was possible to create a structure of excellent quality with low expected lifetime costs.

The design and supervision concept was a success.

During construction attempts were made to save money using inferior quality materials and alteration of difficult items, but subsequent supervision hindered all attempts to do so. The success was satisfactory for everyone. The client got an excellent structure and the contractor had no difficulties and finished well ahead of the schedule, which is very unusual in Korea.

This successful experience was not, however, adopted as a future standard.

Haeng Ju Bridge

Even the good example a few kilometres upstream, the Olympic Grand Bridge, did not result in the installation of a similar procedure for this bridge. Attempts were again made to save money by deleting consultancy services. Basic design, detail design and site supervision were carried out by three different parties without any communication.

The result was a 4-month delay at the beginning of the works, because the contractor did not understand the design and ordered wrong materials. Then came construction 150% slower than the concept. At present the construction has had to be stopped due to difficulties with the accuracy. One can imagine what kind of quality will be handed over to the client.

CONCLUSION

It is sad that bridge management does not exist in countries like Korea, because there would be the chance of avoiding all the mistakes that happened in Europe during the past 40 years. The hopeful beginnings introduced by foreign consultants and supported by a few local engineers have not been successful due to an unfortunate promotion system and split competences and budgets for activities that should be seen in a combined way.

It will be found that an average of 1·5% of the erection costs have to be spent for maintenance every year after the growth of the countries' highway network slows down. At that time costs for the work done now will be considerably higher than the average. But it is too much to expect understanding of these long-term problems.

The VCE proposal given to the clients in the Far East is:

- —Provide an inspection by reasonably educated engineers (locals) every year.
- —Provide a detailed high-class inspection every 6 years.
- —Allocate an average budget of 1·5% of the investment costs of all bridges for inspection and maintenance.

31

Performance Monitoring of Glued Segmental Box Girder Bridges

PETER WALDRON, MAHMOUD RAMEZANKHANI

Department of Civil Engineering, University of Bristol, Queen's Building, University Walk, Bristol BS8 1TR, UK

and

BEN BARR

University of Wales, School of Engineering, PO Box 917, Cardiff CF2 1XM, UK

ABSTRACT

Over the past 5 years three major glued segmental bridges have been instrumented for the measurement of strain and temperature. A comprehensive data base of results now exists containing strains and temperatures measured at a number of important cross-sections in each of the bridges. Since reference readings were recorded just a few days after casting, the results form a unique record for the assessment of time-dependent effects such as creep, shrinkage and loss of prestress at all stages before, during and after erection. Recent developments reported here include load testing of the structures, to provide confirmation of structural performance, and the installation of a telemetry system to enable the continuous monitoring of instrumentation to be conveniently managed from the office.

INTRODUCTION

Segmental methods for constructing concrete box girder bridges are now commonly used throughout the world. The term segmental construction refers to any concrete bridge structure that is cast in a number of discrete longitudinal segments. Several alternative techniques have evolved over the years.[1] These are usually categorised according to the method of erection, namely (i) balanced cantilever, (ii) progressive placing, (iii) span-by-span or

(iv) incremental launching, and by the method of casting, either (a) cast *in situ* or (b) precast. Of these different techniques, the balanced cantilever method is by far the most common, accounting to date for approximately 85% of segmental bridges built in the USA and two-thirds of those in the UK. Whereas the other three methods of erection are best suited for continuous viaducts with spans less than 60 m, the balanced cantilever approach has been used extensively for spans up to approximately 250 m, using in-situ concrete, or 140 m where precast segments have been employed.[1]

Structures Monitored

The research described here concentrates on the balanced cantilever method as the predominant erection technique. More specifically, the study is centred on glued segmental construction in which match-cast precast segments are prestressed together through a thin layer of epoxy resin (Fig. 1).

FIG. 1. Grangetown Viaduct under construction by the glued segmental method as a balanced cantilever.

This technique is becoming increasingly competitive for medium span bridges and has been employed in the construction of seven of the 16 segmental bridges completed in the UK in the last decade.[1]

The three bridges chosen for this study reflect the full range of structural configurations for which glued segmental construction is a viable option. The various features incorporated in the three bridges included straight or highly curved alignment, constant or variable section depth, rectangular or trapezoidal cross-section and differing methods of erection either by launching girder or crane.

(a) River Torridge Bridge

Located 1 km north of Bideford, North Devon, the structure forms part of the 8·4-km Bideford bypass.[2] Completed in May 1987, it carries two lanes of traffic 29 m above mean high water level over the Torridge estuary.

The bridge consists of eight continuous spans, each up to 90 m long, with a total length of 645 m. The superstructure, which is straight in plan, is formed from 251 segments each weighing up to 105 t, varying in depth from 6·1 m at the supports to 3·1 m at midspan. Segments were match-cast on site and erected using a purpose-built launching gantry, itself weighing 150 t. As each segment was positioned a thin layer of epoxy resin was applied to the matching surfaces immediately prior to prestressing, thus ensuring watertightness and a uniform transfer of stress across the joint.

(b) Grangetown Viaduct

The two remaining viaducts, which have been completed more recently, are both located on the peripheral distributor road currently under construction around Cardiff.

The Grangetown Viaduct is over 1 km long and is the longest glued segmental bridge in the UK. The twin trapezoidal box girder superstructure is made up from 641 match-cast segments each weighing up to 74 t with a near-constant depth of 2·8 m increased locally to 3·5 m at the supports. Both the eastbound and westbound carriageways consist of 17 spans up to 72 m in length.

(c) Cogan Viaduct

The Cogan Viaduct, which is the second structure to be instrumented at the Cardiff site, provides a 15 m navigable clearance to the River Ely at high water. With a 95 m main span and a radius of curvature of only 285 m for some of the 60 m approach spans, it is more complicated than the adjacent structure although only one-third of the total length. Unlike the Grangetown Viaduct, a rectangular box section was selected to accommodate the variable section depth made necessary by the large main span, which is the longest in the UK for this type of structure. In all some 300 segments were used to construct the twin carriageways, varying in weight from 43 to 117 t.

Segments for both of the Cardiff viaducts were match-cast on site by the short line method. Although deck erection of Grangetown Viaduct commenced using a launching girder, the majority of units in both structures were erected by crane.

PROGRAMME OBJECTIVES

The principal objectives of the research were threefold:

(a) the assessment of long-term time-dependent effects such as creep and shrinkage in the concrete and loss of prestress in the steel;
(b) verification of short-term structural performance due to the application of vehicular loading; and
(c) the measurement of differential temperature in concrete box girder bridges and an assessment of its structural significance.

All three objectives are best met by the development of generalised computational models validated against high quality field data. To this end an extensive programme of field monitoring was undertaken complemented by a parallel study in the laboratory to provide the necessary input data on the creep and shrinkage of the concretes used for construction.

INSTRUMENTATION

(a) Time-dependent Effects

Instrumentation for the long-term (20 years) measurement of time-dependent concrete strain was similar in all three bridges. In the Torridge Bridge four segments were selected for monitoring within a half-span of one of the central 90 m spans.[2] Three further segments at approximately midspan, quarter-span and support sections were instrumented within a single half-span in both the remaining bridges. A typical 72 m span was selected in the Grangetown Viaduct and a highly curved 60 m side span of constant depth in the Cogan Viaduct.[3]

Each of the ten segments selected was instrumented with embedment-type vibrating wire strain gauges cast in the concrete during construction. These were located at a number of discrete points on the median line of the walls forming the box section. In some segments gauges were deployed only in the axial direction for the measurement of longitudinal strain; in others the gauges were arranged as three-element rosettes to determine the component of shear strain as well. Nearly 300 gauges were used altogether. These proved to be very robust and reliable with a failure rate of only 1%.

Due to the lack of security on site and the remoteness of individual segments prior to erection, data were acquired and entered into a computer data base manually. Although inconvenient for the first few months after casting and during erection, when readings were required every few days, the procedure was the best that could be achieved and worked well.

Once in a data base the raw results were reduced automatically to units of strain and adjusted for the effects of temperature. They could then be considered in a variety of different formats on the computer screen before being produced as hard copy for the validation of long-term performance. The two most useful formats for presentation of data have proved to be the distribution of strain (axial, shear or principal) around the cross-section at any number of ages, and a strain/time plot for any number of gauges.

An example of the former is given in Fig. 2, which shows the distribution of axial strain around the instrumented segment adjacent to pier 4 in the Torridge Bridge immediately prior to erection (160 days), after erection of the half-span containing the instrumented segment (197 days), after continuity was established with the adjacent span (246 days) and at completion of the bridge (450 days). Figure 3 is an example of the complete strain/time plot for the four axial strain gauges cast in the corners of the same segment.

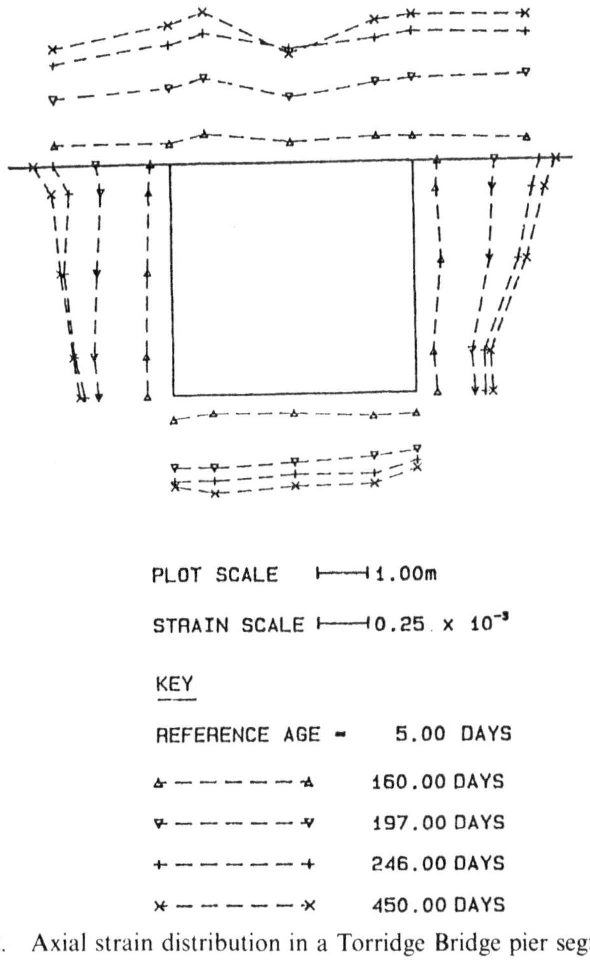

FIG. 2. Axial strain distribution in a Torridge Bridge pier segment.

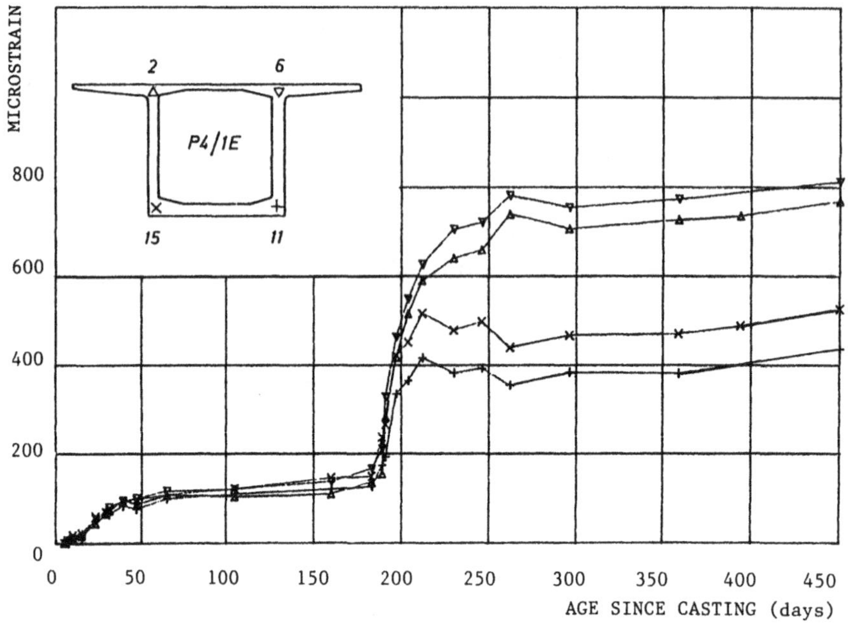

FIG. 3. Strain/time plot for the axial strain gauges in the four corners of the Torridge Bridge pier segment.

The basic time-dependent properties were obtained from numerous prism specimens, each with a single vibrating wire strain gauge located at its centre, manufactured from the batches of concrete used for the construction of the instrumented units. Some of the prisms were sealed fully soon after casting to prevent the loss of moisture; the remainder have either been left unsealed or partially sealed to represent more accurately the environmental conditions at the various bridge sites.

A number of these prism specimens are being used for the assessment of Young's modulus, Poisson's ratio and the coefficient of thermal expansion and their variations with age. The remainder are being used for the long-term measurement of creep and shrinkage effects. Once again data are collected manually and entered on a computer data base for later interpretation. In addition to straight strain/time plots, it is then a simple matter to obtain plots of other parameters such as creep coefficient, which are necessary for the validation of the time-dependent analysis.

By assessing all of the data, evidence is beginning to emerge regarding the accuracy of the available methods of analysis and the validity of the various creep and shrinkage algorithms used for design. A time-step analysis has been developed for the assessment of any segmental bridge constructed using concretes of different ages. Strain at any point and at any stage of construction can be calculated from a knowledge of the physical properties of the actual materials used for construction and the sequence and age at which certain erection events occurred. Results of this analysis can then be compared with those obtained from field measurement and from those derived by using algorithms recommended in different international codes of practice. Early results from these comparative analyses indicate that all major code recommendations have shortcomings but that the ACI-78 recommendations provide the best fit to experimental data.[4,5]

(b) Short-term Performance

In the two Cardiff bridges short-term structural performance was assessed by load testing. This has provided sufficient data against which the original design calculations and any alternative analytical approaches could be validated.

For each test, the transporter used for the delivery of segments during construction was loaded to 130 t and stopped at a number of positions along the viaduct on the instrumented and adjacent spans. In both load cases the tests were first performed on the bridge centreline, to provide information on flexural performance, and then at maximum eccentricity, for an assessment of torsional behaviour. Results were recorded manually as before and entered on to a separate data base for later consideration. Figure 4 shows an example of axial strain in the four corners at one of the instrumented sections within the Grangetown Viaduct due to the application of the load test vehicle at eight different longitudinal positions.

In this way, by using existing instrumentation installed for the monitoring of long-term performance, and by making use of existing site equipment and staff, it was possible to provide valuable confirmation on performance at very modest cost.

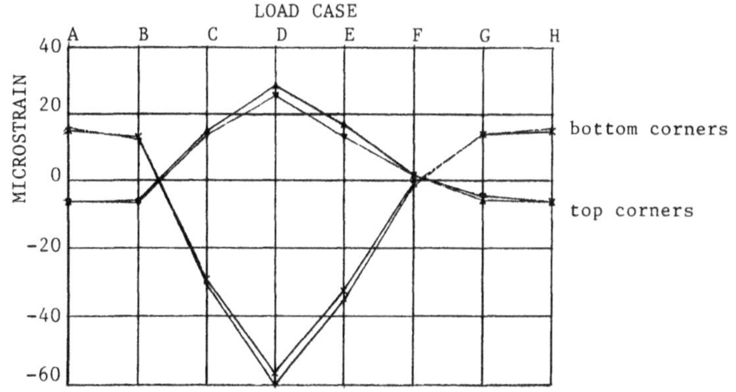

FIG. 4. Axial strains in the corners of a segment of the Grangetown Viaduct during load testing.

(c) Differential Temperature Effects

One of the four segments instrumented for strain in the Torridge Bridge was also used as a pilot study for a possible investigation of thermal effects. A number of thermocouples were installed within the concrete of one of the segments for the measurement of temperature. Early results indicated that, under certain climatic conditions, significant differentials occurred both through the thickness of the individual walls and over the entire depth of the bridge.

Since a more detailed investigation was warranted, a further segment was selected for monitoring temperature effects in the Cogan Viaduct. This segment was within the same half-span to be instrumented for strain but was in addition to the three segments containing vibrating wire gauges. The instrumentation consisted principally of arrays of thermocouples placed across the thickness of each wall element forming the box section, supplemented by a number of individual thermocouples located at the corners and near the outside surface of the concrete (Fig. 5). The segment was also heavily instrumented with electrical resistance strain gauges placed orthogonally in pairs adjacent to the thermocouples for the purpose of monitoring short-term longitudinal and transverse strains due to diurnal temperature variations.

Since the purpose of the thermal investigation was to monitor differential temperatures caused by rapid changes in the environmental conditions, a manual system of data acquisition was no longer practicable. As there was no particular merit in monitoring the bridge in the difficult site conditions prior to completion, this aspect of the monitoring has only recently started. A Solatron data acquisition system driven by a PC is now installed in the relatively secure closed cell of the bridge, powered from the mains circuit running through the box for lighting.

After running the system automatically for several weeks and collecting data from each of the 150 sensors every hour, it became apparent that

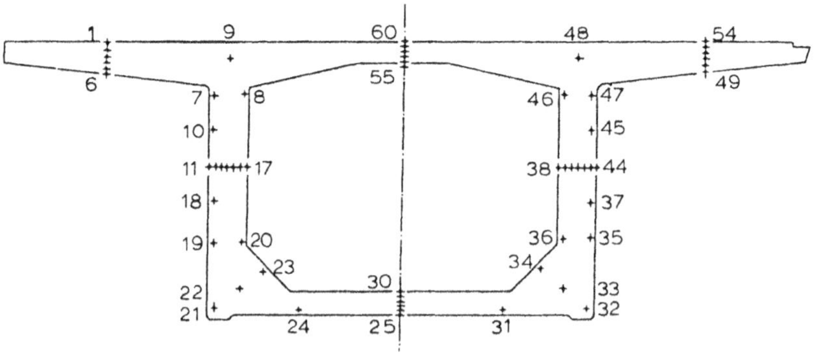

FIG. 5. The position of thermocouples in a segment of the Cogan Viaduct.

handling and interpreting the large amount of stored data would be a major task. Moreover, data would ideally need to be collected over several seasons in order to monitor the most onerous loading conditions due to differential temperature effects. The problem has recently been overcome by the installation of a telemetry system using a standard telephone line. This enables the remote computer on site to mimic the host computer back in the office. It is now possible to switch the data logging system on or off, to change the frequency of readings, and to transfer data back to the office automatically. This innovation, shown schematically in Fig. 6, has provided greatly increased flexibility and efficiency in data acquisition at very little additional cost.

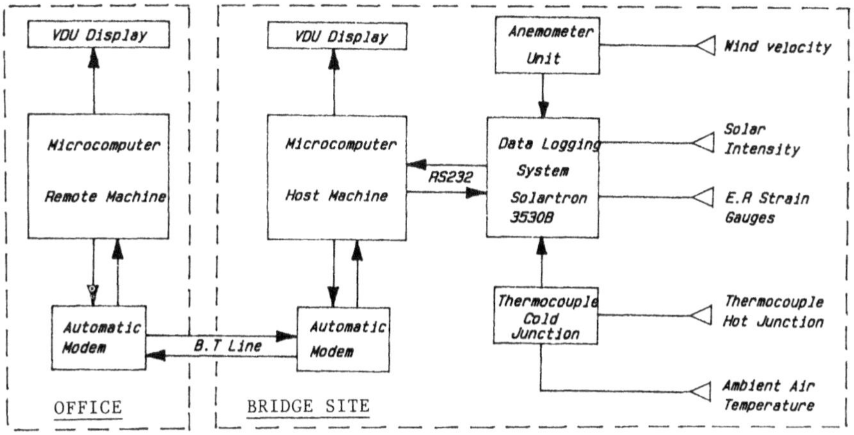

FIG. 6. Data acquisition and telemetry system employed in the Cogan Viaduct for monitoring differential temperatures.

Although the system has only been installed recently, interesting data are already being collected. Figure 7 shows the variation of temperature over the depth of the box during a relatively hot spell in June 1989. The surface of the 100 mm thick blacktop is considerably hotter than the ambient temperature; the differential between the top and bottom concrete surfaces is in line with the maximum design values given in BS 5400.[6]

Another point of interest is the effect of shading on the webs. Figure 8 shows the outside surface temperatures at the centre of both webs, one of which is always in the shade whilst the other is shaded for the greater part of the day by the overhanging side cantilever. The very rapid change in temperature caused by this effect is just one of the many factors complicating a detailed thermal analysis.

One such method of computer analysis currently under development is being validated against this field data. This can calculate and allow

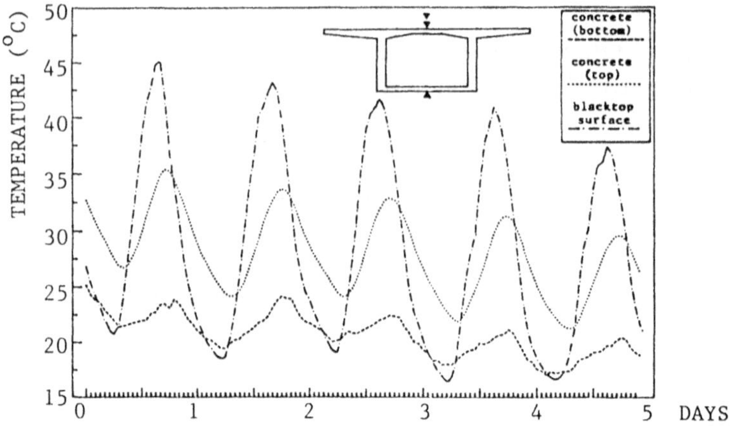

FIG. 7. Temperature differentials measured over the depth of the instrumented segment in the Cogan Viaduct.

automatically for shading due to bridge configuration as well as for the effects of bridge location and orientation, time of day and season, degree of cloud cover and turbidity, thermal properties of the constituent materials, and the nature and colour of each surface in terms of solar radiation absorptivity, emissivity and convectivity. Based on a finite element approach, the method requires design or measured values for these various quantities to be input. Then, in order to validate the analysis against measured differential temperatures, values of those parameters which vary

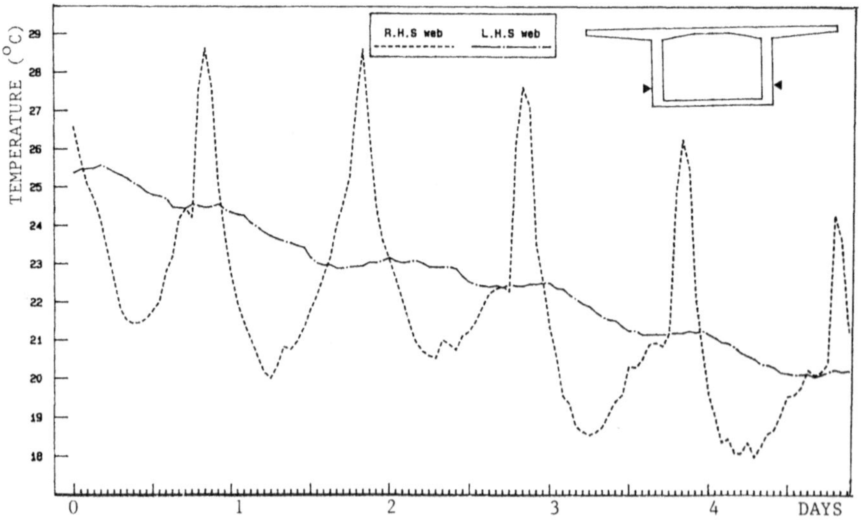

FIG. 8. The effects of full and partial shading on temperatures measured on the outside surfaces of both webs over a 5-day period (June 1989).

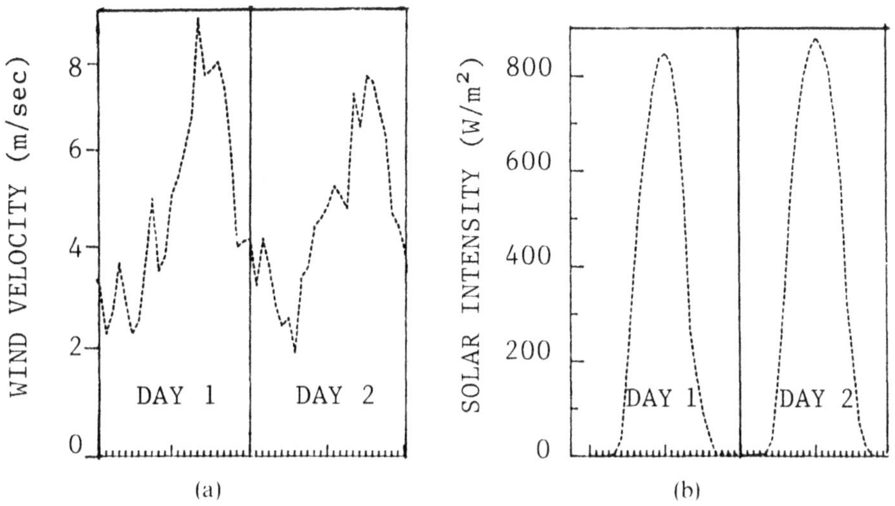

FIG. 9. Variations of (a) wind speed and (b) solar intensity measured on the Cogan Viaduct over a 2-day period (June 1989).

continuously, such as wind speed and solar intensity, must also be input. These are measured on site with an anemometer and solarimeter, and recorded automatically with the other data as continuous functions against time, as shown in Fig. 9.

Results from the analysis are encouraging and permit temperatures to be estimated within a few degrees anywhere in a concrete section. A conventional finite element stress analysis may then be used to compute the distribution of stress both longitudinally and transversely. Early results indicate that the transverse stresses due to the frame action around the closed cell of a box girder may be equally important as the longitudinal stresses, which have already received some attention.

CONCLUSIONS

Meaningful instrumentation has been installed in three segmental bridge decks at modest cost which may be used for studying both the long-term time-dependent behaviour and the short-term response due to applied load. With the introduction of inexpensive portable computers, it is timely to start creating data bases of actual performance during construction and service life to assist with the technical management of real structures. Such instrumentation systems should be installed as a matter of routine, not only in bridges of complex geometry or novel design but also in a number of typical bridges of standard configuration.

ACKNOWLEDGEMENTS

This research is being undertaken as part of the co-ordinated programme of large- and full-scale testing of structures funded by the Science and Engineering Research Council. Additional funding and support has been provided by South Glamorgan County Council.

REFERENCES

1. ZIADAT, G. S. and WALDRON, P., Segmental construction—a state-of-the-art report. Department of Civil Engineering, University of Bristol, Report No. UBCE/C/86/1, December 1986, p. 159.
2. ZIADAT, G. S. and WALDRON, P., Measurement of time-dependent behaviour in the River Torridge Bridge—instrumentation and early results. Department of Civil Engineering, University of Bristol, Report No. UBCE/C/87/4, October 1987.
3. BARR, B. I. G., WALDRON, P. and EVANS, H. R., Instrumentation of glued segmental box girder bridges. *Int. Assoc. Bridge Struct. Engng Colloquium*, Bergamo, Italy, 1987, pp. 175–89.
4. WALDRON, P. and ZIADAT, G. S., Assessment of the long-term behaviour of segmental bridges. *Proc. Inst. Struct. Engrs*, Building Res. Est., Seminar on the Life of Structures, Brighton, April 1989.
5. ZIADAT, G. S., Time-dependent analysis of prestressed concrete segmental bridges. PhD thesis, University of Bristol, September 1988.
6. BRITISH STANDARDS INSTITUTION, BS 5400: The design of highway bridges, Part 2, 1984.

32

Remote Computer-Aided Bridge Performance Monitoring

T. D. SLOAN

Civil Engineering Department, The Queen's University of Belfast, Belfast BT7 1NN, UK

J. KIRKPATRICK

Department of the Environment Roads Service, Commonwealth House, 35 Castle Street, Belfast BT1 1GU, UK

and

A. THOMPSON

Civil Engineering Department, The Queen's University of Belfast, Belfast BT7 1NN, UK

ABSTRACT

A project to monitor the in-service behaviour of the Foyle Bridge, near Londonderry, is described. Special instrumentation to measure the dynamic movements of very large structures has been developed, together with software to control it and to process the data obtained. The system can be run unattended and can be operated by remote control. Typical results are presented.

INTRODUCTION

In 1984 the Environment Committee of SERC announced a programme of structural research based on large- and full-scale testing of structures. As part of this overall programme a project to monitor the behaviour, in service, of the Foyle Bridge is being carried out. The work is funded jointly by the Science and Engineering Research Council and the Department of the Environment (NI) Roads Service, and is being run by the Department of Civil Engineering, The Queen's University of Belfast.

The new Foyle Bridge, near Londonderry, NI, opened to traffic in May

1984. It is a high level crossing of the River Foyle, some 3 km downstream from the Craigavon Bridge in the city centre. Designed by Freeman Fox and Partners and built by RDL–Graham Joint Venture, the bridge is 866 m long and comprises three main spans totalling 522 m constructed in steel, together with a 344 m approach viaduct of 13 spans in prestressed concrete. Full details of the design and construction have been published elsewhere,[1–4] but the principal features of the structure are summarised here.

The whole bridge, both steel and concrete, consists of two independent parallel structures, as can be seen in Fig. 1. The main steel structure consists of twin box girders which vary in depth from 3 m at midspan to 9 m at the main piers, over each of which there is a diaphragm. Each girder was fabricated, off site, in three sections, which were then floated to site and winched into position. The site splices are located at the sixth points of the centre span and are bolted, except for the deck which is site welded. To achieve the desired bending moment conditions under zero applied load conditions, the ends of the side spans were raised 6·5 m before the centre span was fixed and lowered to the final position only after splicing was complete.

Both girders are fixed at the west abutment and there are expansion joints at the junction of the steel and concrete sections of the structure. At the main piers the girders are carried on pinned bearings, the horizontal movements being accommodated by flexure of the piers.

As there is no interconnection between the two girders they move independently of each other and, at midspan, a vertical differential movement of up to 500 mm can be expected under severe storm conditions.

At 30 m above water level the site is very exposed, being completely open to both north and south and with almost no shelter from any other direction. The basic design wind speed given in CP3, Chapter 5, for the area is 53 m/s, compared with 50 m/s for the Forth Bridge and 43 m/s for the Severn Bridge. Occupying as it does the most exposed site of any major bridge in the British Isles, the Foyle Bridge offers an excellent opportunity to monitor the in-service performance of a box girder structure.

The project is concerned only with the steel section of the bridge. The objectives are:

- To investigate methods of continuously monitoring long-span bridges with the aim of developing a cost-effective solution which could have application in similar situations elsewhere.
- To establish a 'footprint' of the structural response to a range of test loads when the structure is in the new condition.
- To monitor the behaviour of the structure under a variety of wind and traffic loadings.
- To compare the measured and predicted behaviour of the structure.

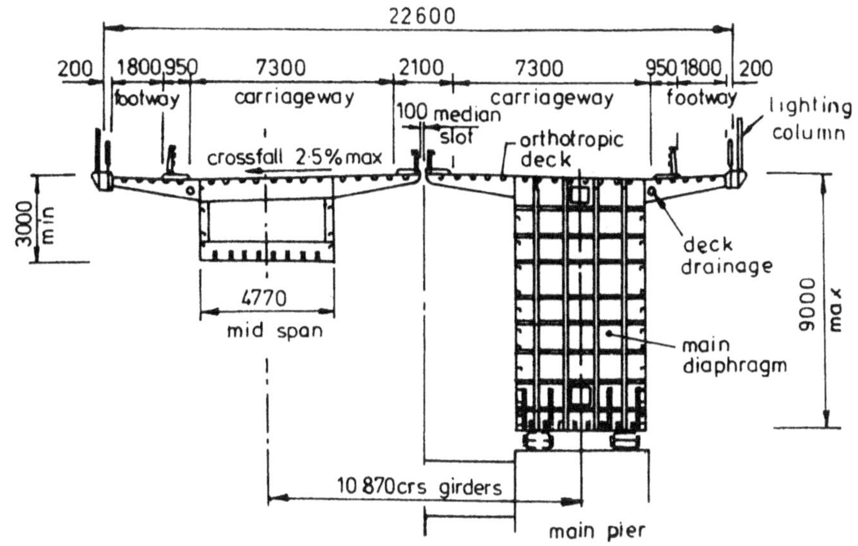

FIG. 1a. Cross-sections of steel girders.

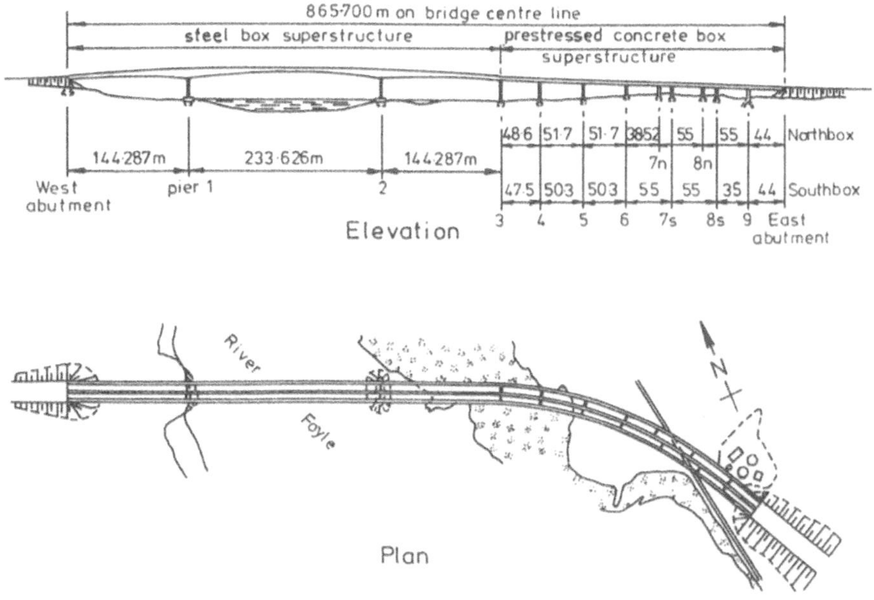

FIG. 1b. General arrangement of the Foyle Bridge.

REQUIREMENTS OF THE MONITORING SYSTEM

Preliminary studies indicated that, as an initial requirement, data would be needed on:

- deflection at the midspan point of the centre span;
- strains at midspan and at the intermediate supports;
- temperature changes in the structure;
- wind speed and direction; and
- movement of the expansion joint.

Several other requirements of a more general nature soon became apparent.

The bridge site is remote both from the university (130 km) and the DoE base in Coleraine (50 km). For security and safety reasons access to the interior of the structure must be restricted. Since, in addition, studies of response to wind, temperature and traffic have to be made on a continuous basis, the system had to be capable of operating unattended for long periods and of storing the resulting data for subsequent analysis.

Interruptions to the power supply, while infrequent, can occur at any time, but are more likely during storm conditions when it is particularly important to collect data. As funds were not available for an uninterruptible power supply, the system had to be capable of restarting automatically after a power failure. Furthermore, as transient voltages on a power supply are liable to produce unpredictable effects in a computerised system, it had to be possible, as a last resort, to reset and restart the system under manual remote control.

Automatic data collection generates very large volumes of data, up to 500 bytes/s in this case. Since, at this rate, even the largest storage devices would soon be filled, it was essential to have significant processing capacity on site so that a preliminary analysis of the raw data could be done immediately and only the significant parts retained for transmission to the university for further analysis.

In any system exceptional conditions of various types will occur from time to time, caused either by malfunctioning of equipment or external interference. In an unattended system such events must be logged so that corrective action can be taken.

Finally, the system had to be capable of expansion and refinement in the light of experience gained.

HARDWARE

A schematic diagram of the monitoring system is shown in Fig. 2. The core of the system is a DEC PDP-11/53 minicomputer fitted with 512 kb of

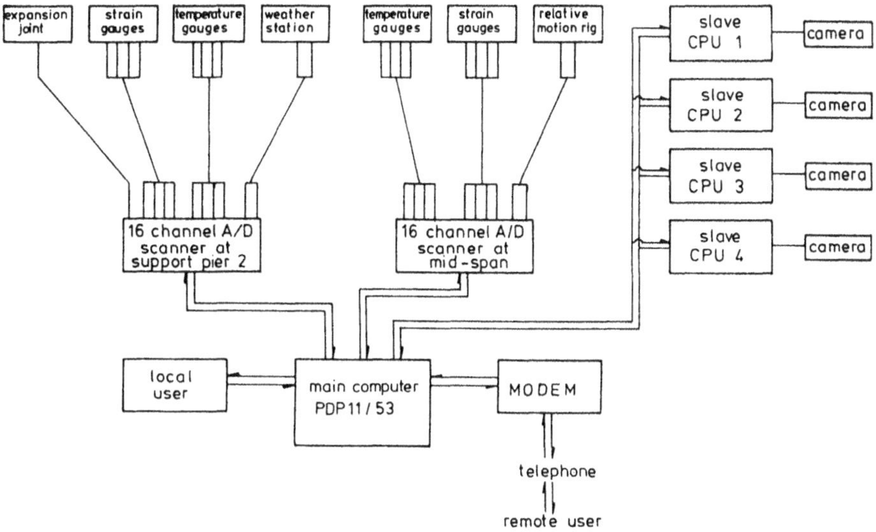

FIG. 2. Foyle Bridge equipment schematic.

memory, a 20 Mb Winchester disk drive and six RS232 serial line ports. The computer runs under the Micro/RSX-11M real-time, multi-tasking operating system and can be accessed either locally from a terminal in the bridge or remotely via a modem and dial-up telephone line from either the university or the DoE offices.

The modems are fitted with error correction logic which provides an error-free connection though at the cost of a slight reduction in effective operating speed. The hardware is configured so that a 'break' signal on the modem line will cause the processor to halt, in which state a re-boot can be initiated by the user.

Intermittent problems, caused by interference on the mains or the telephone line, can cause the modem to 'lock up' and fail to answer incoming calls. Purpose-built circuitry has been developed to reset the modem periodically and so ensure that such a condition cannot persist for more than 4 h. A more complete description of the system can be found in previous publications.[5,6]

Measuring the deflections of such a large and inaccessible structure presents considerable difficulty; the fundamental problem lies in establishing a frame of reference from which to make measurements. Conventional methods, such as levelling or the use of displacement transducers, were clearly impractical, due both to the distance to the nearest fixed point and to the need to record the response to dynamic loadings.

Initially consideration was given to the use of accelerometers but these had to be rejected for several reasons; they will not measure the very small accelerations experienced during the slow-moving deflections caused by

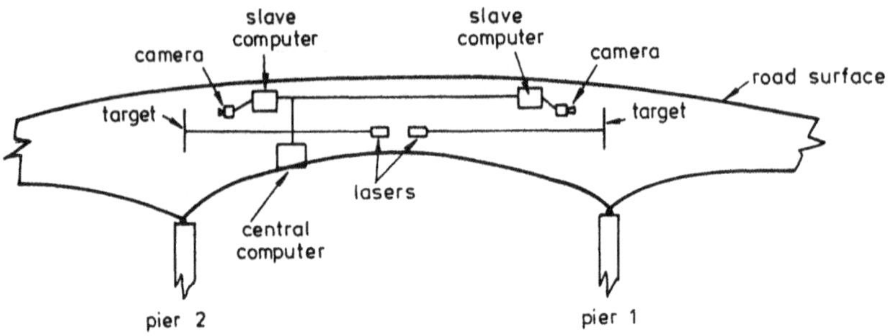

FIG. 3. Layout of equipment to measure deflection of main span.

heavy vehicles and even with fast-moving traffic or wind loading they would be operating at the limit of their resolution. The solution adopted is shown diagrammatically in Fig. 3. Helium neon lasers fitted with beam expander and focusing optics are mounted inside the box girders at midspan so as to project a spot of light of approximately 3 mm diameter on to targets which are fixed over the main piers. Any movement (either linear or rotational) of the lasers will thus cause the position of the light spot on the target to vary.

The movement of each spot is tracked using a solid-state camera in which the film carrier is replaced by a light-sensitive computer chip. The chip presents to the lens an array of 128×256 light-sensitive cells which are examined in turn by a specially developed program running in a BBC microcomputer. These computers, one for each camera, scan the complete arrays repeatedly and store the resulting coordinates of the light spot in memory for later transmission to the central computer.

All the other data are collected using transducers which generate electrical analogue signals. These are connected to a pair of free-standing 16-channel analogue-to-digital converters, each of which is connected, in turn, to the main computer via a serial line. This arrangement was chosen primarily to allow the A/D converters to be located as close as possible to the transducers and so minimise the length of the cables carrying the analogue signals. The alternative, which was to incorporate the A/D converter in the main computer, would have involved cable runs of up to 150 m with all the attendant problems of signal attenuation and noise pickup.

An anemometer was already available as part of an ice alert system which was installed when the bridge was built. It has been possible to tap into this system to record the wind conditions.

The movement of the expansion joint is monitored using a linear potentiometer which is linked to the A/D converter via a current loop line driver as the cable is 150 m long.

The relative movement between the two girders at midspan is also monitored using a rotational potentiometer mounted on one girder with the wire connected to the other. This has been included as a check on the performance of the laser deflection system.

SOFTWARE

The whole system operates under the control of the Micro/RSX operating system in the PDP-11 minicomputer. This is a real-time, multi-tasking, multi-user operating system in which the following features are specially useful:

- Tasks can be scheduled to run at specific times.
- One task can initiate and monitor the running of another.
- There are very good facilities for controlling the hardware without the need for programming in assembler language.
- The operating system can be configured so that, on start-up, a user-supplied command file is executed to allow the work schedule to be re-established after a power failure.

All data acquisition is under the control of a single program, SCNALL, which accepts requests from either the user or from another program, performs the scan and writes the results to a file in which full details of the operation—time, date, scan rate, channel identifiers—are recorded. Calibration factors are applied to all data as they are recorded so that readings taken at different times are directly comparable even though the equipment may have had to be reconfigured in the meantime. Data compression techniques are used to reduce the size of the data files to a minimum.

The maximum scanning rate of the cameras is 8·32 Hz and up to 8192 readings may be taken at one time. This allows the analysis of frequencies of up to 4 Hz, well above the structure's fundamental frequency of 0·4 Hz.

Application Programs

When studying the dynamic response of the structure it is obviously necessary that all the readings from a scan be available for analysis. However, for static or 'pseudo-static' (such as temperature effect) studies only a single value is normally needed. But because the structure is constantly in motion, even on the calmest day, such a single observation needs to be the average of a set of repeated observations. It is a simple matter to write an application program which will generate a request to SCNALL to perform a scan and then process the data file to extract the required information. Multiple applications can be concurrently scheduled for repeated runs; competing requests for the use of SCNALL are resolved via the operating system.

RESULTS

At the time of writing only a limited amount of data have been collected. Some preliminary results are shown in Figs 4–7.

Figure 4 shows the influence line for static midspan deflections due to a 75-t load moving across the bridge. This test was done on a calm day when the bridge was closed to traffic. Each reading is the mean of 1024 camera scans, to remove the effect of vibrations induced in the structure when the load was moved from one position to the next. The position of the load is measured from the east end of the steel structure.

Figure 5 shows the change in strain measured at the top and bottom of the box section at midspan during the same test.

Figure 6 shows the dynamic deflections recorded as a test load of 100 t was driven over the bridge from east to west at approximately 43 km/h. Initially the structure was almost motionless, what small movement there was being caused, probably, by a light breeze. As the load passed over the bridge, the midspan point first rose and then, as the load reached the centre span, deflected downwards by 220 mm before rising again as the load moved to the second side span. At the right-hand side of the trace, the structure can be seen vibrating at its natural frequency of 0·4 Hz, the amplitude reducing after the source of the disturbance was removed.

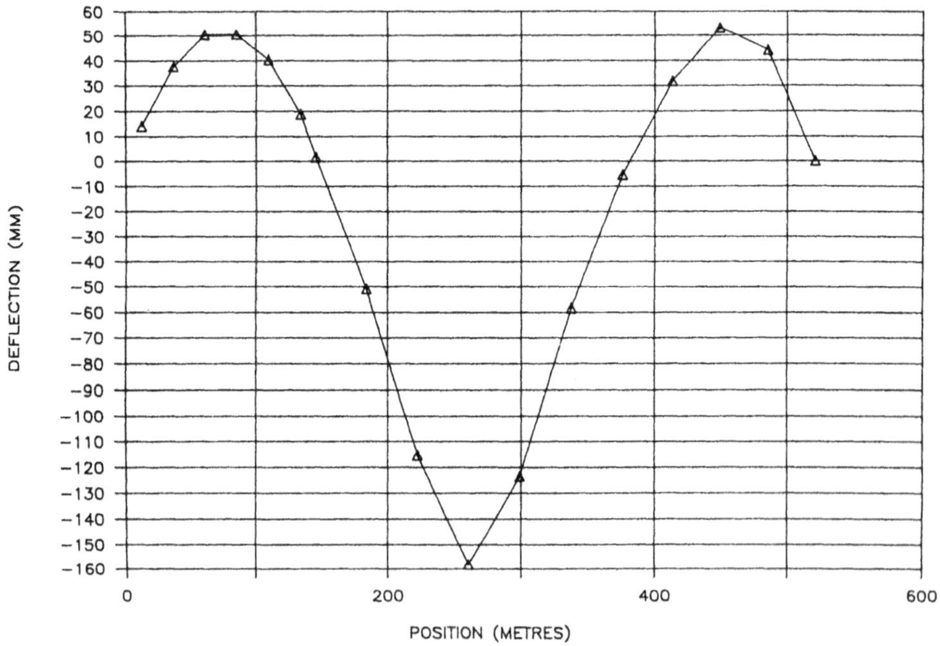

FIG. 4. Deflections at midspan.

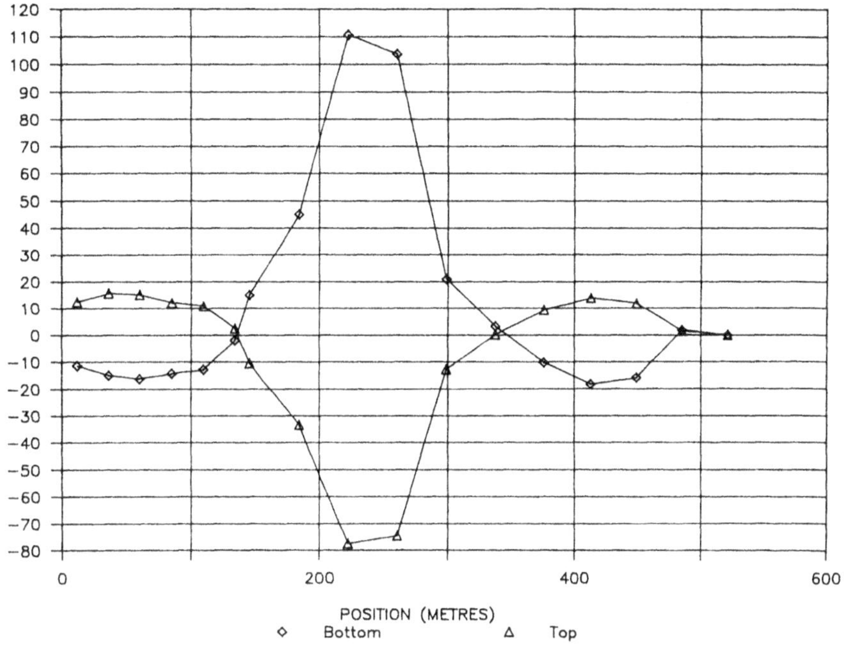

FIG. 5. Strain at midspan.

Figure 7 is a plot produced from data collected on the movement of the expansion joint caused by temperature changes in the structure. A clear trend can be seen, with the 'best fit' straight line having a gradient of 7·9 mm per degree change in steel temperature. This figure is about 2·0 mm larger than that to be expected from the expansion of the steel alone; the difference can be accounted for by the expansion of the concrete section of the bridge.

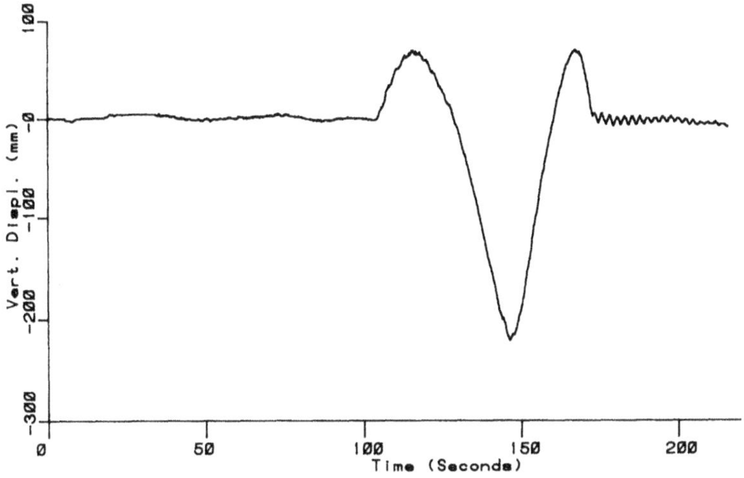

FIG. 6. Deflections due to 100-t moving load.

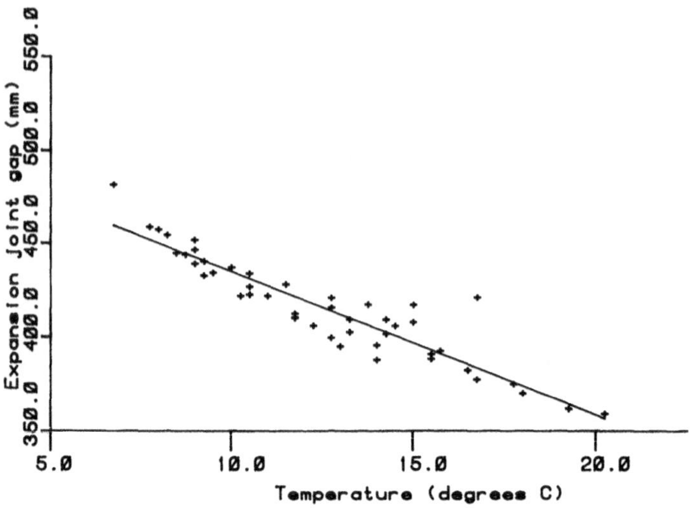

FIG. 7. Thermal movement.

A long-term study of this behaviour is currently under way and should allow confident predictions of the expansion joint movement with temperature to be developed. In the future this knowledge will form a useful check on the performance of the joint.

CONCLUSIONS

- A fully automatic monitoring system has been developed for a major structure. It can be run equally well under direct operator control or unattended.
- The system is flexible in that additional sensors may be added easily without the need for major reprogramming. The principles used could be adapted for use in other projects.
- The laser displacement system has been shown to be reliable in operation and to give consistent results.

ACKNOWLEDGEMENTS

The work described in this paper has been supported jointly by the Department of the Environment for Northern Ireland and the Science and Engineering Research Council. The authors also wish to thank Professor A. E. Long of the Department of Civil Engineering, The Queen's University of Belfast, for his support and encouragement throughout the project.

REFERENCES

1. PRESCOTT, T. A. N., STEVENSON, W. M. C. and NISSEN, J., Foyle Bridge: its history, and the strategy of the design and build concept. *Proc. Inst. Civ. Engrs*, **76**(1) (May 1984) 351–61.
2. WEX, B. P., GILLESPIE, N. M. and KINSELLA, J., Foyle Bridge: design and tender in a design and build competition. *Proc. Inst. Civ. Engrs*, **76**(1) (May 1984) 363–83.
3. QUINN, N. W., Foyle Bridge: construction of the foundations and viaduct. *Proc. Inst. Civ. Engrs*, **76**(1) (May 1984) 387–409.
4. HUNTER, I. E. and MCKEOWN, M. E., Foyle Bridge: fabrication and construction of the main spans. *Proc. Inst. Civ. Engrs*, **76**(1) (May 1984) 411–48.
5. LEITCH, J. G., THOMPSON, A. and SLOAN, T. D., A novel dynamic deflection measurement system for large structures. *Proc. Civil-Comp. 89*, Civil-Comp. Press, Edinburgh, September 1989 (to be published).
6. SLOAN, T. D. and THOMPSON, A., Development of an automatic data collection system for a major box girder bridge. *Proc. Civil-Comp. 89*, Civil-Comp. Press, Edinburgh, September 1989 (to be published).

33

Inspection and Repair of some Highway Bridges in Italy

MARIO P. PETRANGELI
Department of Structural and Geotechnical Engineering, 'La Sapienza' University, Via Eudossiana 18, Rome, Italy

ABSTRACT

This paper reports about the inspection and repair of a number of 25-year-old viaducts in service along the A3 highway linking central to southern Italy. Particular reference is made to dynamic tests that allowed useful information to be collected about the effective prestressing forces acting on the beams. The results obtained are compared with those derived from other in-situ tests. The criteria followed for the repair (or demolition) of the decks and for their seismic retrofitting are finally briefly presented.

INTRODUCTION

The 44-km long A3 Salerno–R. Calabria highway is the main road connection between the centre and the south of Italy, including Sicily. This highway was built about 25 years ago. It reaches the level of 1015 m above the sea, crossing the Apennines in a zone where deicing salt is often used.

The government agency (ANAS) responsible for the management of the road has invested about 200 million US dollars to inspect and rehabilitate the existing 264 main bridges, which have a total length of about 56 km, most of them with pc decks spanning over 35–40 m.

In this paper the results of the inspections carried out on seven viaducts as well as their retrofitting are illustrated.

The 32 simply supported pc decks were all of the same type (see Fig. 1) and showed severe damage, such as deterioration of the concrete surface and cracking of the beams.

First of all, an accurate visual inspection by means of a by-bridge was

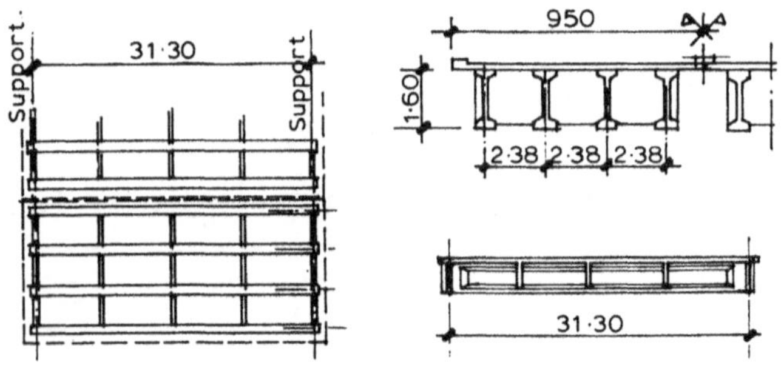

FIG. 1. Scheme of the decks.

performed; as a result of that three groups of decks, according to their deterioration, were defined:

(A) the pc beams had no visible cracks;
(B) the beams were cracked but the total width of the cracks measured on the four beams was less than 4 mm, moreover the total width of the cracks in a single beam was less than 2 mm; and
(C) the beams were cracked and one or both of the two previous limits were exceeded.

This preliminary classification has been kept as a reference for the instrumented tests.

METHODS OF INSPECTION

Materials

The following non- (or moderately) destructive tests on the decks were performed:

(a) Boring of 100-mm cores from the webs of the beams with measurements taken of the released stress; in the class C decks compression stresses as low as 30% of the theoretical values were found.
(b) Determination of the mechanical properties of the concrete of the cores as well as the carbonation depth, which ranged between 8 and 19 mm.
(c) Measurements of the ultrasonic pulse velocity in conjunction with the surface hardness.
(d) Windsor penetration tests.
(e) Systematic inspections of the state of the prestressing cables by means of an endoscope. The latter proved very useful since it showed when the injection of some cables was badly made or, in some cases, completely lacking.

Dynamic Tests

Fifteen spans of seven different viaducts underwent dynamic tests. The decks analysed had been previously classified (i) three in group A, (ii) eight in group B (moderate damage) and (iii) four in group C (severe cracks).

The investigation of the dynamic behaviour of each deck was carried out by means of low-frequency geophones (range of resolution 1·5–3000 Hz). Most of the tests were performed with two geophones, one on each side of the deck at the midspan, in order to establish the first modes of vibration (flexural and torsional); in a few cases eight geophones (four at the supports, two at the quarter-point and two at midspan) were employed to collect information about modes of higher order and about the behaviour of the neoprene bearing devices.

The data collecting system comprised an analogue-to-digital converter connected to a microcomputer capable of handling simultaneously up to 16 channels.

The maximum sampling frequency was 3·5 kHz when all the channels were operating, reaching 17 kHz if only one channel was active. To excite the decks vehicular traffic was utilised. The highway has never closed, at least one lane being always open. This procedure did not permit external actions to be precisely defined. Hence the analysis of the recorded signals has focused mainly on the free vibrations occurring after the passage of some heavy vehicles.

Figure 2 shows some typical time histories of the amplitude of the signals; the segments of these diagrams selected as 'meaningful' allowed the derivation of the Fourier spectra of the type shown in Fig. 3.

Table 1 presents the most important results for each deck: the vibration

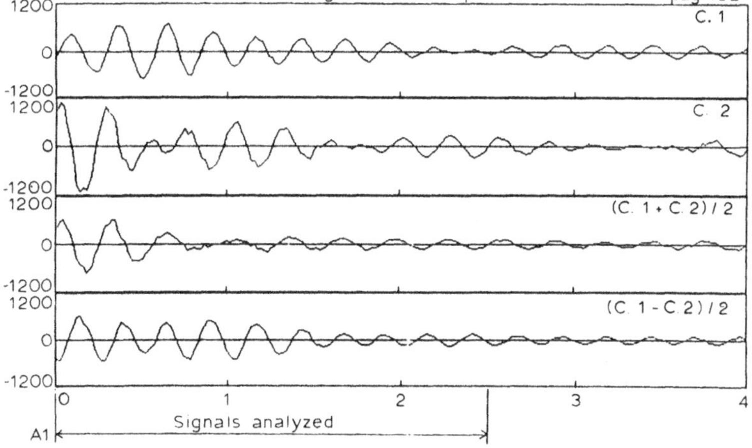

FIG. 2

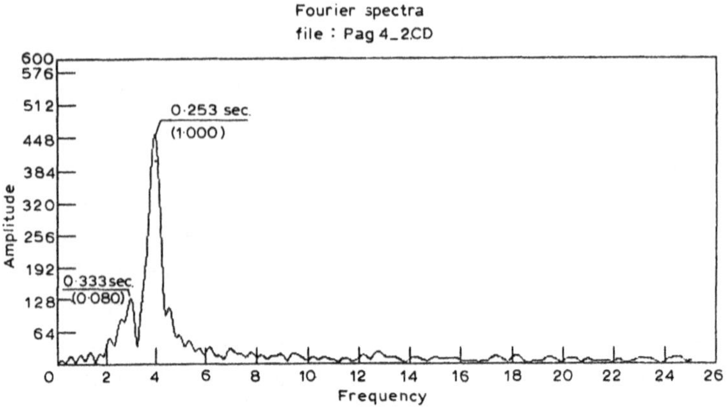

Fig. 3

period T_f for the flexural mode and T_t for the torsional mode. The three numbers in each column represent the extreme and average values found for all the analyses (generally four or five) carried out for each deck. In the table the ratio T_t/T_f and the classification previously mentioned are also reported.

One can see that the class A decks (light or no damage) have T_f values practically equal to the theoretical value expected of 0·27 s, while those showing severe cracks have T_f higher, up to 30% more.

It must be pointed out also that the ratio T_t/T_f could be useful in assessing the deck soundness since it seems to decrease as the damage increases.

TABLE 1

Deck		T_f (s)	T_t (s)	T_t/T_f	Initial class
1. S. Venere	3	0·31–0·31–0·31	0·27–0·27–0·27	0·87	C
2. S. Venere	4	0·30–0·31–0·31	0·27–0·28–0·27	0·87	B
3. S. Venere	5	0·31–0·32–0·32	0·27–0·28–0·28	0·88	C
4. Grotta I°	1	0·27–0·30–0·28	0·24–0·25–0·24	0·86	A
5. Grotta I°	4	0·30–0·32–0·31	0·25–0·26–0·25	0·81	C
6. Grotta I°	5	0·34–0·35–0·35	0·26–0·27–0·26	0·74	C
7. Grotta II°	1	0·30–0·31–0·31	0·24–0·25–0·24	0·77	C
8. Ranta	3	0·28–0·28–0·28	0·25–0·25–0·25	0·89	B
9. Ranta	4	0·33–0·34–0·34	0·28–0·29–0·28	0·82	C
10. Vomice	4	0·34–0·35–0·35	0·27–0·28–0·27	0·77	C
11. Vomice	5	0·30–0·31–0·31	0·26–0·27–0·26	0·84	B
12. Vomice	6	0·33–0·34–0·34	0·26–0·27–0·26	0·76	B
13. Bodetti	1	0·27–0·27–0·27	0·24–0·24–0·24	0·89	A
14. Spatolette	2	0·35–0·35–0·35	0·27–0·27–0·27	0·77	C
15. Spatolette	5	0·35–0·35–0·35	0·27–0·28–0·27	0·77	C

RESULTS OF THE INSPECTIONS

In order to decide the level of repair, knowledge of the effective prestressing force acting in the beams was of prime importance. Some data were derived from the stress releasing already mentioned. Further information was provided from the analysis of the dynamic tests. Since the natural frequencies of a grid frame depend on the stiffness *EJ* of the beams, it is possible to relate T_f to the flexural moment of inertia *J*, provided that the value of the *E* modulus is known with good accuracy by means of the non-destructive tests and from the cores.

Linking *J* to the height of the cracks and this last parameter to the actual prestressing force N_p, the measured T_f can be correlated with the unknown N_p. Figure 4 shows these diagrams for the decks referred to.

The tensile strength of the concrete has been assumed equal to zero because of the repeated live loads the decks have been subjected to in their life. The curve relating N_p to T_f shows how, for N_p lower than N_{pc} (value of N_p that combined with the dead load only gives zero compression at the bottom of the beam), T_f increases quickly with reduction of N_p. No information is available for values of N_p located between N_{pc} and N_{pt} (theoretical value).

Better results are possible if a known static load is located on the deck during the dynamic tests since a higher value of N_{pc} is achieved. In this case the problem is to have a mass rigidly connected to the deck or a vehicle whose dynamic characteristic is precisely known. The reliability of the dynamic tests has been confirmed from static load tests that were permitted on a few decks with the closure of one carriageway.

The results of these tests are represented in Fig. 5, giving the load–deflection curves in the midspan of a lateral beam. Line 't' gives the

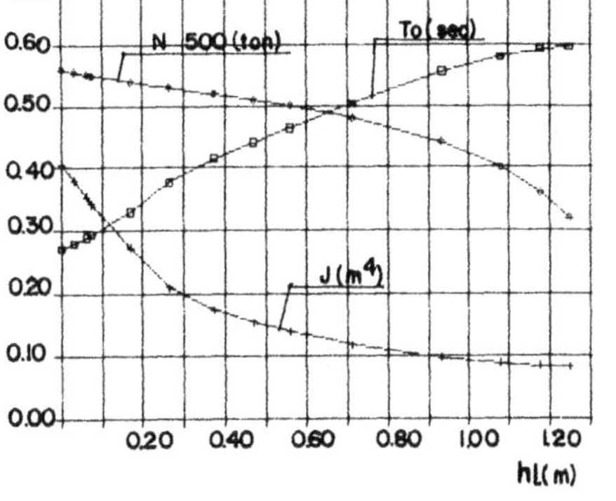

FIG. 4(a). Prestressing force *N*, fundamental period T_0 and moment of inertia *J* versus height of the cracks.

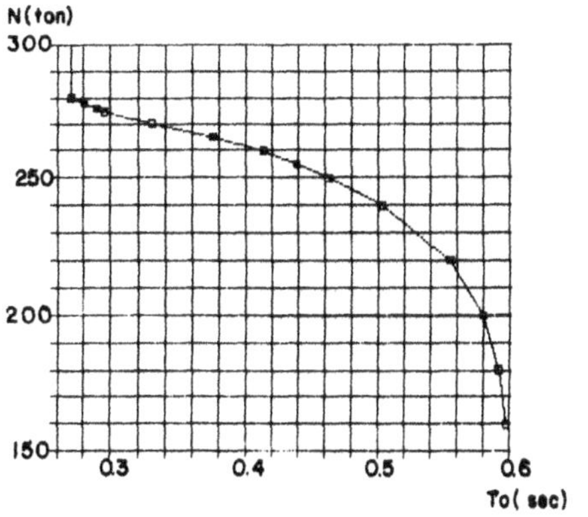

FIG. 4(b). Prestressing force *N* versus period T_0.

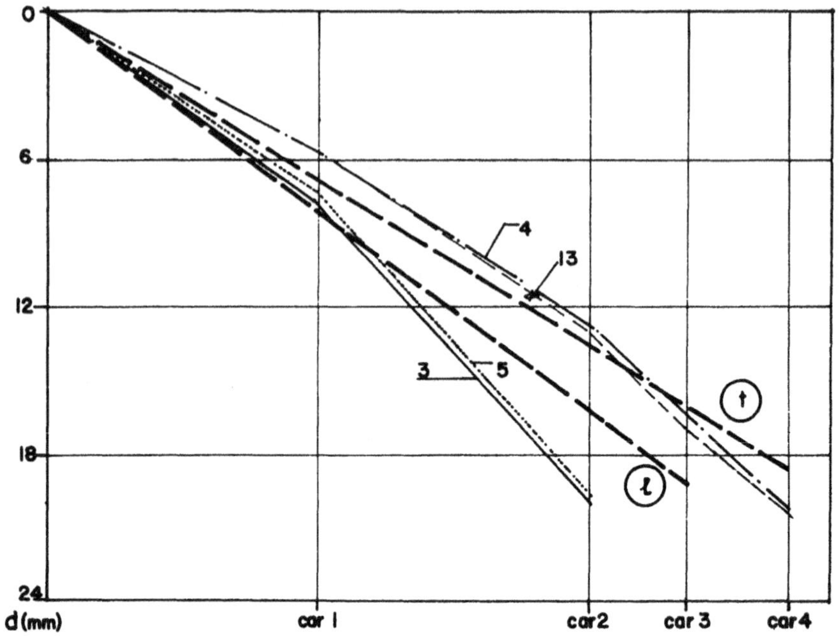

FIG. 5. Deflection d versus load in a lateral beam.

theoretical answer for the decks while line 'l' gives the same but with a value of EJ reduced by 20%; this was assumed to be a limit that, if passed, would dictate that the test be terminated.

These curves show how decks 4 and 13, with a natural frequency of 0·27 s (equal to the theoretical one), performed well during static tests.

Decks 3 and 5, which had T_f greater than 0·3 s, had the loading interrupted after two (of four) vehicles were located on the bridge.

METHOD OF REPAIR

On the basis of the experimental tests and of the subsequent computations carried out, three levels of repair were decided according to the residual prestressing force in the beams:

—Less than 40%; in this case the removal of the old deck and the construction of a new one was found to be cheaper than the repair.

—Between 40% and 90%; new prestressing cables were added to the beams by enlarging their bottom flange (see Fig. 5). Number and size of these new cables were based upon the existing prestressing force; their anchorage was produced by a steel frame or by a concrete block depending on the distance between the beams (Fig. 6).

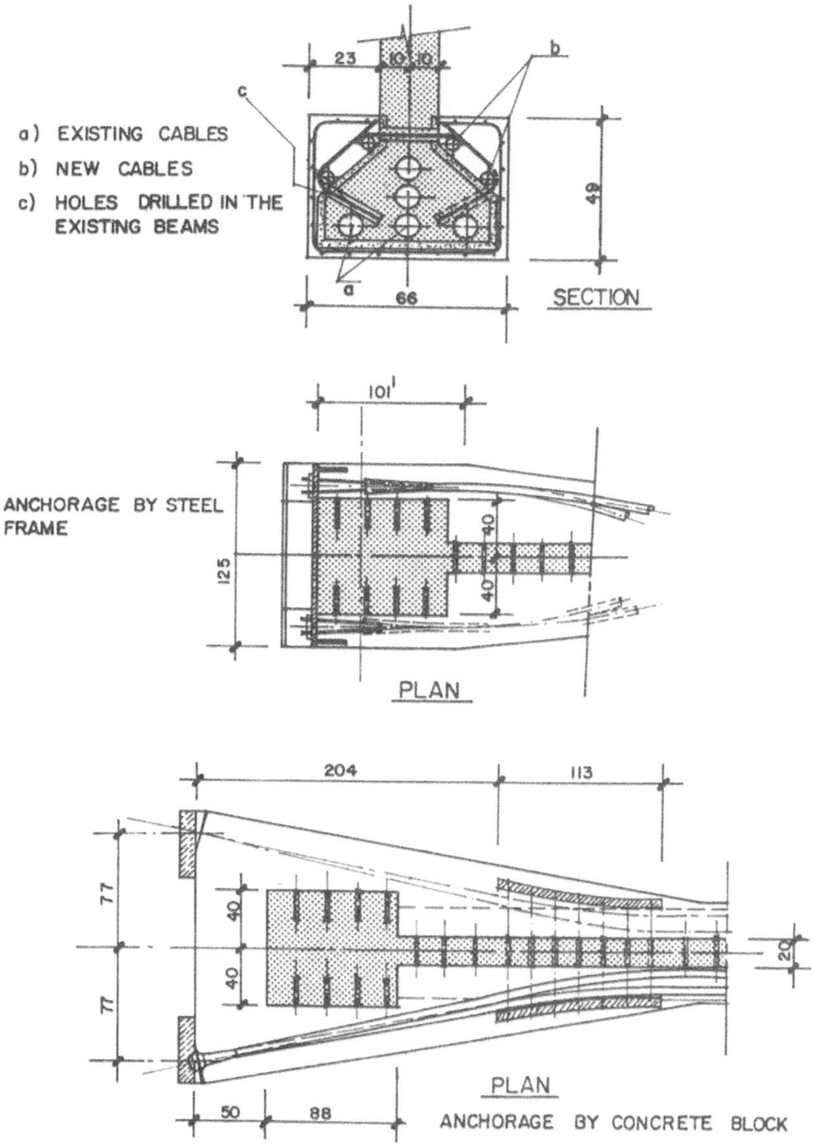

FIG. 6. Detail of the additional prestressing of the beams.

—More than 90%; no additional cables were provided but the repair of the beams' surfaces as well as reinjection of the existing cables was carried out.

In addition, a substantial variation in structure was adopted for all the viaducts in order to improve their response to seismic actions.

The decks, originally simply supported and separated on each pile by

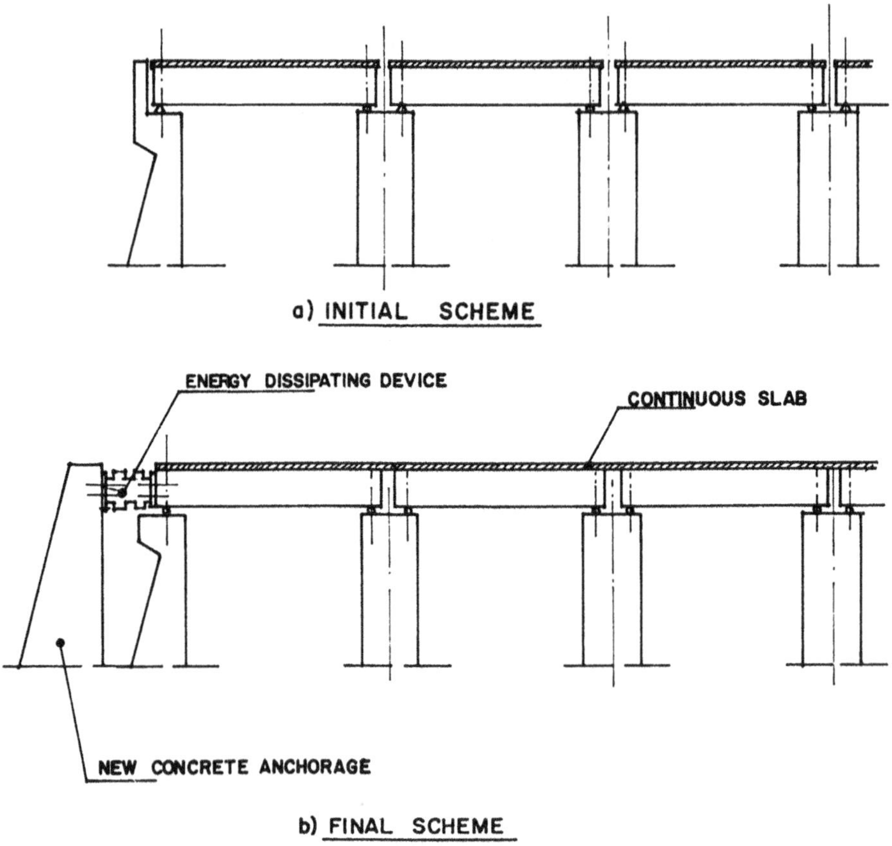

FIG. 7

joints, were connected together by the slab made continuous on the supports. All the bearings were changed, the new ones being free in the longitudinal direction. Fixed points only now occur on the abutments where energy-dissipating devices have been located.

These devices are of the elastoplastic type and have been designed to remain in the elastic range for earthquakes with a return period of 50 years; they will yield for stronger earthquakes and in this case they will be replaced.

Obviously the alterations to the abutments were necessary in order to improve their strength against the horizontal forces. This was done by adding new structures on the back of the existing ones, as shown in Fig. 7.

CONCLUSION

The 25-year-old bridges studied were found to be in a bad condition due to the extensive use of de-icing salt and to the absence of any kind of waterproofing between the deck slab and the surfacing. Besides that a dramatic reduction in the expected value of the prestressing force in the beams was found. To quantify this reduction the dynamic tests extensively used were of great importance since they allowed meaningful comparisons between decks of the same type.

The addition of new cables was the solution for the cases in which the loss of prestress was in the medium range. Economic reasons, i.e. the high cost of manpower and the high industrialisation reached in Italy in the construction of pc prefabricated bridge beams, mean that for the worst cases the replacement of the old deck with a new one was more economical.

Better behaviour under seismic action was obtained by making the slab continuous on the supports and by adding energy-dissipating devices on the abutments.

34

Inspection and Strength Evaluation of Concrete Highway Bridges in Czechoslovakia

KAREL DAHINTER

Pragoprojekt—Design, Engineering and Consulting Inc., K. Ryšánce 16, 147 54 Prague 4, Czechoslovakia

ABSTRACT

This paper describes some experiences and results of the investigation and evaluation of concrete highway bridges carried out under the state research programme. The complex diagnostic method is explained and illustrative examples of reinforced and prestressed concrete bridges are appended.

INTRODUCTION

There are about 46 000 highway bridges in Czechoslovakia with a span of more than 2·0 m, of which approximately two-thirds are plain concrete, reinforced concrete (rc) or prestressed concrete (pc) structures. The first of them were built at the beginning of this century, the main part of rc bridges between World Wars I and II, and bridges with pc superstructures after 1950.

The management of all highway bridges is controlled by two standards: ON 736220, 'Register of Bridges on Motorways, Highways and Urban Roads', and ON 736221, 'Maintenance of Bridges on Motorways, Highways and Urban Roads', and methodological guidelines appended to them.

According to these standards a physical condition is classified in seven rating levels:

—perfect, very good and good, with surface impairments only and non-reduced load-carrying capacity, when preventive maintenance is sufficient;
—satisfactory and bad, with structural deterioration and reduced load-carrying capacity, when rehabilitation is needed; and

—very bad and poor, with heavy deterioration and distress signs, when immediate traffic measures or closing-up must be done and extensive rehabilitation or replacement of the structure follows.

In the standards there are also basic specifications for three types of live loads for strength evaluation (legal weights of a vehicle in tonnes for class A/B bridges):

—the 'normal'—six two-axle vehicles in two lanes or three in one lane for narrow bridges (32/22);
—the 'exclusive'—one four-axle vehicle (80/40); and
—the 'exceptional'—special set for heavy loads with two three-axle tractors and one 14-axle trailer (280).

According to the decision of the central state authorities the load-carrying capacity of all existing bridges should have been checked by the end of 1993. This means that at first the condition survey and at second the strength evaluation should be done. For that purpose and for increasing the management level in general, special departments in Czech and Slovak republics were established as well as inspection teams in all regions.

As most of these bridges were designed according to former codes, with lower loading actions and material properties, their load-carrying capacity is not sufficient.

Due to this fact one of the state research programmes for the period 1986–90, directed by the Federal Ministry of Transport, has been aimed for 'Research of methods and technological measures for increasing the load-carrying capacity of highway bridges, both one-off and permanent'. The first of four partial programmes contains 'Methods of physical condition assessment' and was finished in 1988. The results are presented as follows.

COMPLEX DIAGNOSTIC METHOD

In general, the following investigations forming the complex diagnostics are used.

Visual inspections according to the national standards: routine checking, and main and extraordinary inspections carried out twice a year, once a year, every 4 years and for special purposes. Some devices are used, such as precision measuring equipment, crack microscope, telescope, etc.

Surveying of the shape, position, translations and deformations, when both classical measuring devices and a laser beam are used.

Geotechnical and geophysical surveys are always needed, at least as a reconnaissance in place with a geological data bank study. If necessary, explanatory boring is carried out to sample the core (core cutting), using pipe inspection TV camera with a video recorder and a device measuring strain characteristics of the earth body *in situ*, or the seismic methods of geophones or four-electrode resistivity measurement and stray current measurement.

Material testing includes the Schmidt hammer and CAPO test concrete surface strength techniques, and the ultrasonic pulse velocity test to check concrete quality. For checking the location and condition of reinforcement gamma radiography, cover-meter, half-cell potential and electrical resistance probes are used. Chemical effects of CO_2 carbonation and chloride ions for both concrete and reinforcement are measured by means of German instrument sets, 'Rainbow indicator' and 'rapid chloride test', respectively. Very rarely the techniques of concrete core cutting or determination of cement content are used, but always small holes are made to evaluate diameter and to check corrosion of rebars.

Strength evaluation. Tables presenting the load-carrying data on superstructures of precast units are available as well as on some in-situ concreted reinforced concrete slabs. All other bridges are to be reanalysed.

Dynamical analysis is provided to determine the characteristics of oscillation behaviour, natural frequencies and shapes, dynamical impact and logarithmic damping decrement.

Loading tests are used if there are any doubts as to the reliability of the structure. From the viewpoint of costs dynamical testing or response measurement are preferred to the statical if possible. However, the statical loading tests are in controversial cases irreplaceable.

The Results of the Complex Diagnostics

—Rating of physical condition with a list of damages and deteriorations.
—Determination of load-carrying capacity according to three types of highway live load.
—Recommendations for maintenance, repairs or rehabilitations if need be.

The final technical report for the bridge containing these conclusions is complemented by general arrangement drawings and photodocumentation.

CONCLUSION

The methods used for investigation were checked in field conditions, and their advantages and disadvantages were considered from the viewpoint of their reliability, speed, destructivity and costs. From the aspects of assessment of the structure behaviour and condition visual inspections, surveying and structural reanalysis with statical testing are the most important investigations. However, all other methods are important as well to create a whole picture of the structure, physical condition and load-bearing capacity.

ILLUSTRATIVE EXAMPLES

1 Reinforced Concrete Continuous Girder Bridge over the Sázava River near Kácov (Fig. 1) (completed 1914)

Deterioration and damage:

—water leakage and leaching due to damaged waterproofing membrane in the connection between the deck and main girders with cracking in all parts of superstructure;

—heavy damage in the bottom of main girders near the central piers, corroded rebars, spalling, scaling and delamination of concrete due to long-term carbonation, weathering, attacks by leaking water solution and, last but not least, the effect of heavy transport; and

—cast steel bearings on the abutments heavily corroded, damaged and one of them fully loose, not working as a support.

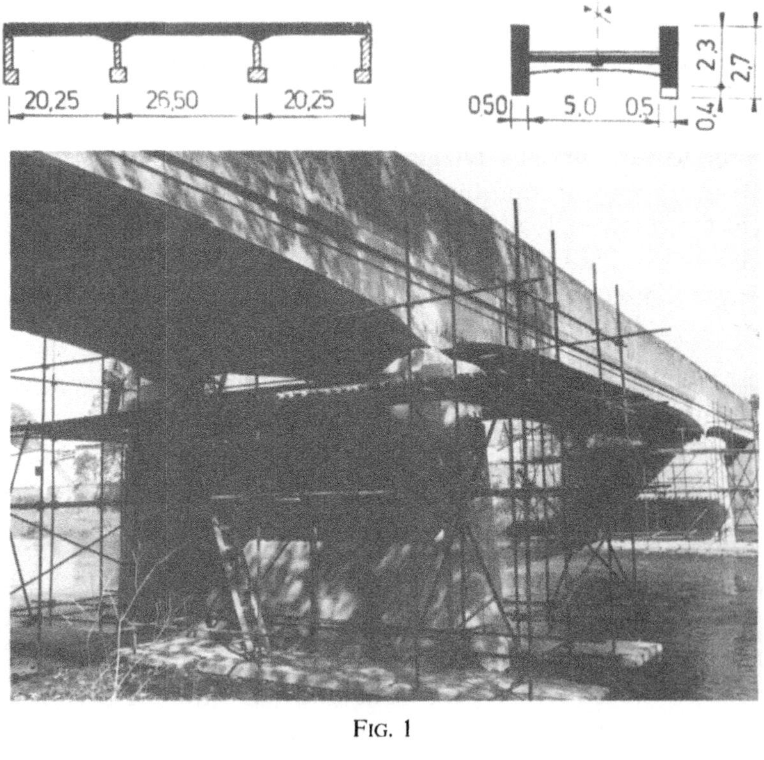

FIG. 1

Rehabilitation process:

—All deteriorated parts of concrete were removed, reinforcement cleaned and fully corroded stirrups complemented.
—Special epoxy solution was used for penetrating paint and injection of cracks.
—The missing parts of the concrete section were complemented by epoxy–mortar.
—The bridge deck was strengthened by an epoxy–mortar layer 10 cm thick, reinforced by welded mesh.

Complex diagnostics and rehabilitation results:

After finishing the rehabilitation work a reanalysis was done to determine the new load-carrying capacity of the bridge. The permissible vehicle weights of 20/40/80 t were checked by statical loading test and the bridge was opened to public traffic.

2 Three-hinged Arch Reinforced Concrete Bridge over the Ohře River in Karlovy Vary (Fig. 2) (completed 1932)

Deterioration and damage:

—deflection in the top hinge of about 10 cm;
—water leakage, leaching, cracks and spalling in the region of expansion joints, especially in the top hinge (Fig. 3(a));
—leaching in the construction joints of the deck and in the cable channel;
—corroded rebars, spalling and delamination of concrete of the deck supports (Fig. 3(b));
—local damage of the concrete surfaces of arches.

Complex diagnostic process and results:

—Repeated surveying of the shape determined the long-term and temperature deflection.
—Reanalysis of the structure on a space calculation model explained the long-term deflection as a consequence of creep and shrinkage of the arch after completion of the deck.
—Load-carrying capacity was calculated and performance of the structure was checked by nine positions of loading vehicles during the statical loading test; results were used for correction of calculation models by reanalysis.
—Dynamical analysis and measurements were carried out to determine dynamical performance.

FIG. 2

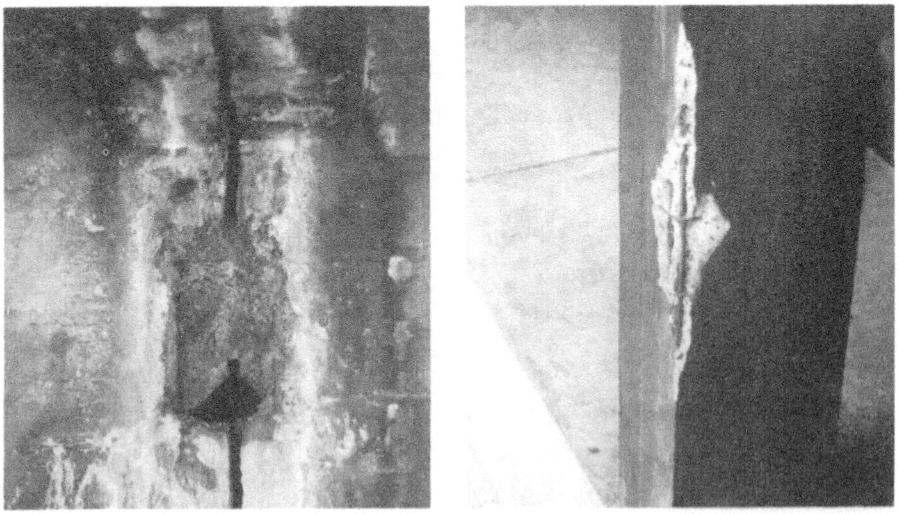

(a) (b)

FIG. 3. (a) Corroded top hinge; (b) damaged deck support.

—Principal geological and geophysical surveys were executed to check the foundation and corrosion conditions, including occurrence of stray currents.
—Schmidt hardness tests, ultrasonic pulse velocity tests, cover-meter and half-potential methods, and phenolphthalein carbonation tests with local probes were applied for material testing of the structure.
—Total rehabilitation of the bridge is indispensable and a preliminary design was prepared.
—Up to the rehabilitation a posted traffic (20/40/100 t) with reduced velocity (40 km/h) for a maximum of 2 years is permitted.

3 Precast Prestressed Concrete Composite I-Girder Bridge over the Berounka River in Beroun (Fig. 4) (completed 1954)

Complex diagnostic process and results:

—The old masonry piers and abutments from the year 1864 show only surface deterioration due to weathering and leakage of water from the bridge roadway.

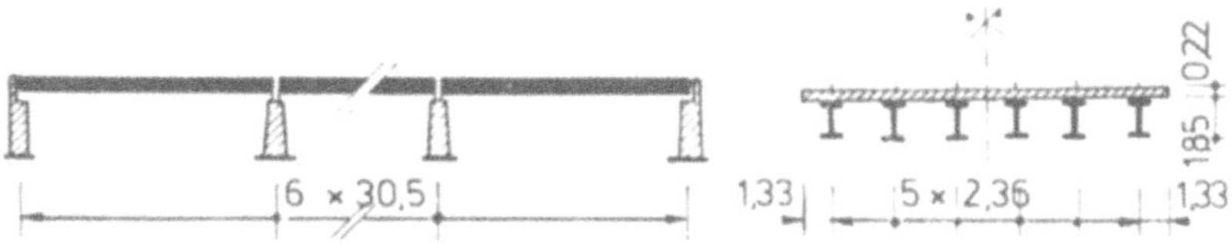

FIG. 4

—Most of the steel bearings, made of old rails, are slightly corroded, and need cleaning and protection paint.
—Superstructure is in a very good condition, concrete strength more than 50 MPa, statical and dynamical performance according to loading test fully corresponding with theoretical analysis.
—General rehabilitation of the roadway layers, waterproofing membrane and expansion joints are needed.
—Repair of cornice, railing and endpart surface of the girder under expansion joints.
—Load-carrying capacity needs no posting.

4 Precast Prestressed Concrete Flyover Crossing the DC Electrified Railway in Pardubice (Fig. 5) (completed 1964)

Complex diagnostic process and results:

—Very strong stray currents were measured.
—Special measures for application of half-cell potential methods were examined to determine the corrosion state of reinforcement.

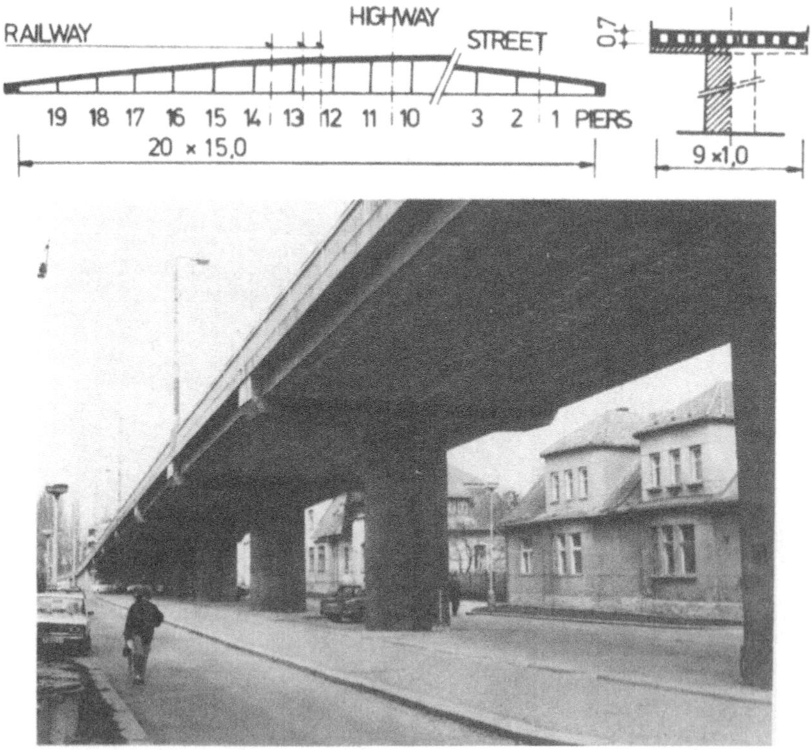

FIG. 5

—Cracks in connecting joints between the box beam due to long-term repeated live loading, especially in the area of quarter and three-quarter cross-sections, cause lowering of the distribution of loads.
—Statical and dynamical reanalysis and dynamical testing are to the date of compilation of this paper still in preparation.
—Special measures must be taken to protect the structure against corrosion.
—General rehabilitation of roadway and waterproofing is intended.
—Supplementary cast-in-place reinforced concrete slab on the box beams to impose the load distribution is recommended.

5 Cast-in-Place Prestressed Concrete Segmental Box Girder Bridge over the Ohře River in Drahovice (Fig. 6) (completed 1960)

Complex diagnostic process and results:

—Physical condition is very good.
—Long-term deformation due to creep and shrinkage led to excessive deflection of about 19 cm in the midspan which caused water draining difficulties on the roadway.

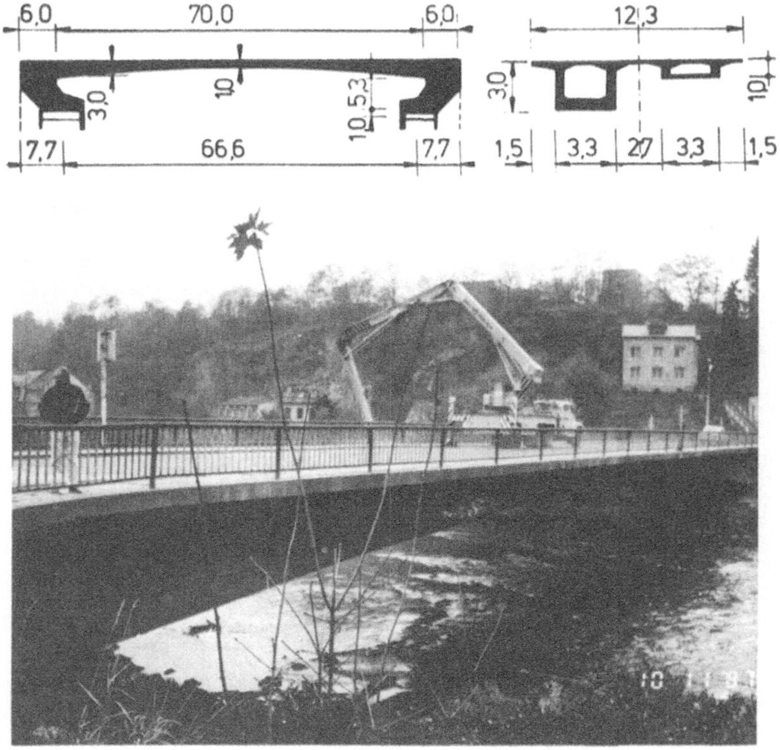

FIG. 6

—Unpleasant oscillation for users and considerably high live-load impact were observed. Dynamical response was measured in 1960, 1963 and 1987; the first natural frequency (2·40–2·33) and logarithmic damping decrement (0·055–0·036) remained principally unchanged. The observed dynamical impact for one vehicle is up to 75%.

—General rehabilitation of the roadway layers with waterproofing membranes and drainage system is intended.

6 Three-span Prestressed Concrete Frame Bridge with Cast in Place on Centring Box Girder (Fig. 7) (completed 1971)

With regard to the experience from bridge 5 one twin box girder with increased haunch height was designed, steel bearings were substituted by rigid connection and elastic walls respectively, and an increased coefficient of creep and shrinkage for the constructional superelevation in midspan was applied (total 280 mm). During construction one superstructure was damaged and several all-section cracks were caused in midspan.

FIG. 7

Preliminary results of complex diagnostics:

—cracks fully closed with leakage signs, otherwise physical condition very good;
—deflection to theoretical elevation of sound superstructure 80 mm (superelevation), the damaged one ± 0; and
—observed dynamical impact for sound superstructure up to 40%, damaged up to 20%; natural frequencies $f_1 \doteq 2{\cdot}5$ and $f_2 \doteq 4{\cdot}2$, logarithmic damping decrement $\doteq 0{\cdot}07$ (the same).

35

Prestressing with Fibre Composite Materials and Monitoring of Bridges with Sensors

REINHARD WOLFF and HANS-JOACHIM MIESSELER
Strabag Bau-AG, Siegburger Strasse 241, D-5000 Cologne 21, FRG

ABSTRACT

On several structures heavy-duty composite materials have proved their applicability as a corrosion-resistant alternative to conventional prestressing steel.

It is also necessary to monitor these concrete structures on a permanent basis in order to guarantee their durability over a long period of time. In prestressed structures cracks in the concrete and changes of the stress state in the tendons must be observed.

Bridges can be monitored in this way using sensor systems which are incorporated in the high-performance fibre composite prestressing elements or directly into the concrete.

INTRODUCTION

Progress in bridge construction is nowadays demonstrated by structures which are becoming increasingly audacious. This is particularly evident in the greater span widths of bridges, although the aesthetics of the actual structures and their integration in the natural setting of the landscape also play a decisive role.

At the same time, by making the optimal choice of the materials available, one should attempt to achieve a considerable increase in the useful life of these structures and to recognise any damage to the load-bearing structure at an early stage so that specific counter-measures can be initiated.

To meet these demands good results have already been achieved with the aid of optical fibre and capacitative sensors, both in the case of fibre

TABLE 1
Prestressing with fibre composites, material characteristics and comparisons

	Reinforcing steel BSt 500 S	Prestressing steel St 1470/1670	Polystal® (68% glass fibres)	Twaron® (aramid fibres)[a]	Carbon fibres[a]
Tensile strength (N/mm^2)	>500	>1 670	1 670	3 150	2 800
Yield strength (N/mm^2)	>500	>1 470	—	—	—
Ultimate strain (%)	10	6	3·3	2·0	0·7
Modulus of elasticity (N/mm^2)	210 000	210 000	51 000	125 000	400 000
Specific weight (g/cm^3)	7·85	7·85	2·0	1·45	1·75
Fields of application	Reinforced concrete structures	Prestressed structures			Stay cables, bracings

[a] Material characteristics corresponding to the fibres.

composite materials in prestressed structures and also in the monitoring of concrete load-bearing structures and prestressing tendons. According to the glass, carbon or aramide fibre composition, and with full exploitation of the individual material characteristics (Table 1), fibre composite materials open a possibility for the fulfilment of the most differing requirements.

Previous utilisation of fibre composite materials mainly concerned the glass fibre composite material which had till then been applied in the form of prestressing tendons for prestressed structures, tensioned mast bracing and as tie-rods for the rehabilitation of arches.

Further monitoring of the performance of the structure is made possible by integrating optical fibre and copper wire sensors in the tensioning bars.

THE PRESTRESSING TENDONS (HLV TENDONS)

Two companies—Strabag Bau-AG, Köln, and Bayer AG, Leverkusen—have been working on the development of glass fibre composite bars since 1978. These bars (trade name 'Polystal') have a diameter of 7·5 mm and comprise 60 000 E-glass fibres embedded in unsaturated polyester resin with a thickness of just a few microns strictly oriented along the direction of the bar. The fibre content is around 68%.

The tensile strength of the Polystal bars is comparable to that of high-grade prestressing steels, roughly 1670 N/mm^2, but in the other properties Polystal shows significant advantages. The composite prestressing elements are considerably lighter than their steel equivalents (Polystal has a density of 2 g/cm^3 versus 7·85 g/cm^3 for that of steel), are electromagnetically neutral and can be used in aggressive environments. Protection of the glass fibre against chemical attack and mechanical damage has been assured by a specially developed coating technique.

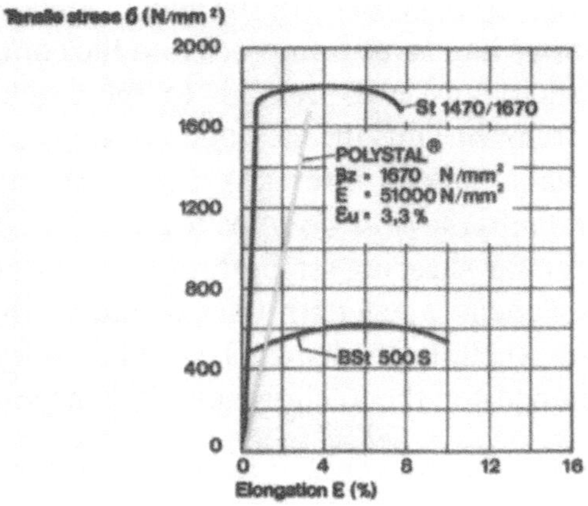

FIG. 1. Stress–strain diagram of a Polystal® bar in comparison with steel.

At 51 000 N/mm² the modulus of elasticity is only a quarter that of steel; tension force losses arising from creeping and contraction of the concrete drop correspondingly to one-quarter. In contrast to conventional steel, the composite glass fibre material has no plasticity limit; the stress/strain behaviour of Polystal is linear practically up to the point of failure (Fig. 1).

In the case of structural elements which are prestressed with steel, high degrees of plastic deformation occur when the plasticity limit is exceeded, providing the desired warning prior to the incidence of failure. Trials on structural elements have shown that, as a consequence of the low modulus of elasticity, a structural element prestressed with Polystal bars possesses a high degree of elastic expansion when permitted loads are exceeded. These deformations also give a prewarning of failure (Fig. 2).

A total of 19 of these glass fibre bars are bundled to form one prestressing

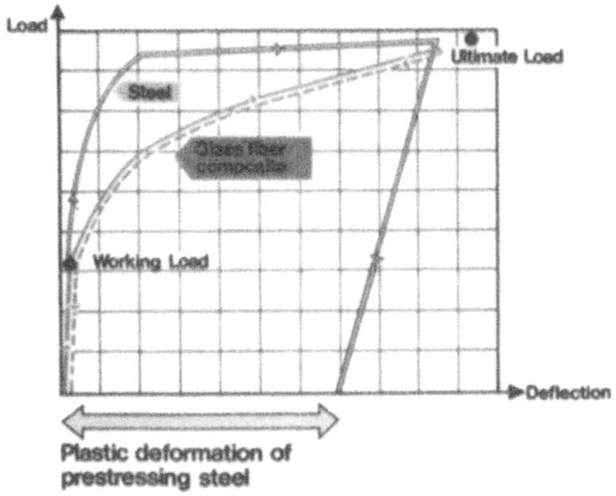

FIG. 2. Load–deformation relation of a prestressed concrete structure.

tendon with a working load of 600 kN. In the field of anchorage engineering completely new solutions had to be found because the composite glass fibre material will only bear transverse pressure to the extent of 10% of its longitudinal tensile strength. Thus the lack of cold workability rules out the utilisation of upset heads, rolled-on threads or even the utilization of steel wedges 'biting' directly into the 'soft' composite glass fibre bar material. The relatively low interlaminar shearing strength of the resin matrix requires a comparatively greater anchorage length than would be the case with steel prestressing tendons. Anchorage elements specific to the material had therefore to be developed for the anchorage of this new material.

FIG. 3. Anchor head of a composite glass fibre prestressing tendon.

The development of a tubular grouted anchorage (Fig. 3) by Strabag Bau-AG heralded a breakthrough for the anchorage of HLV composite bars. The composite bar is set inside a sectional steel tube in an artificial resin-based grouting mortar which has been specially developed for this purpose. The use of these grouted prestressing tendons covers the entire spectrum of the lightweight and mediumweight prestressing tendons (up to 1000 kN working load) and the complete field of soil and rock anchoring.

THE SENSORS

Nowadays increasingly high demands are made on monitoring methods for the assessment and control of the stress/strain behaviour of concrete structures. With the use of glass fibre composite bars it is possible, for the first time, to integrate sensors capable of indicating the stress state for the monitoring and control of the whole structure.

Today two different methods of monitoring (based on two different types of sensors) are carried out successfully.

Optical Fibre Sensors (Fig. 4)

This type of sensor is a specially coated optical fibre covered with a spiral of thin wire. The pitch of the spiral is so dimensioned that, if there is axial stretching of the sensor, the wire spiral will press on to the optical fibre and generate micro-bendings in it. These influence the permeability of light, which is directly proportional to the sensor's tensile strain. These circumstances also allow the localisation of a failure, which up to now was not possible with the employment of load cells. The sensor, which is approximately 2 mm thick, is either embedded in the centre of a glass fibre bar or directly in the concrete, or both.

Copper Wire Sensors (Fig. 5)

The individual bars of the prestressing tendons are permanently monitored by copper wire sensors incorporated in the cross-section of the composite bars, with four copper wires in each bar. As with a capacitor, an electric field is thereby generated between the respective wires located in the centre and one of the wires arranged externally. The glass fibre material between the wires (a dielectric) is defined by the dielectric constant. Assessment of the change in strain in the prestressing tendons is now based

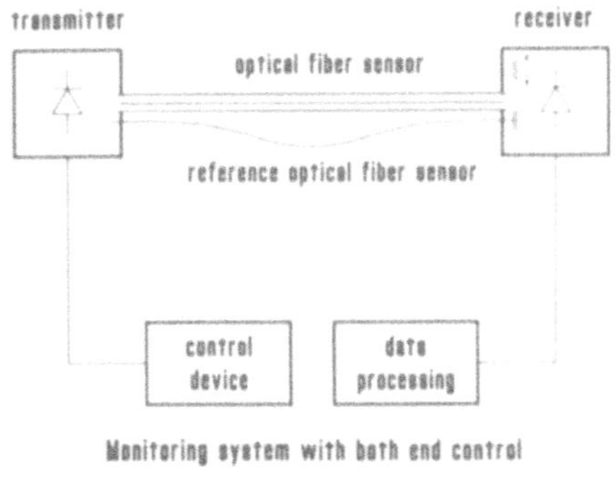

FIG. 4. Diagrammatic illustration of the measuring and monitoring system using optical fibre sensors.

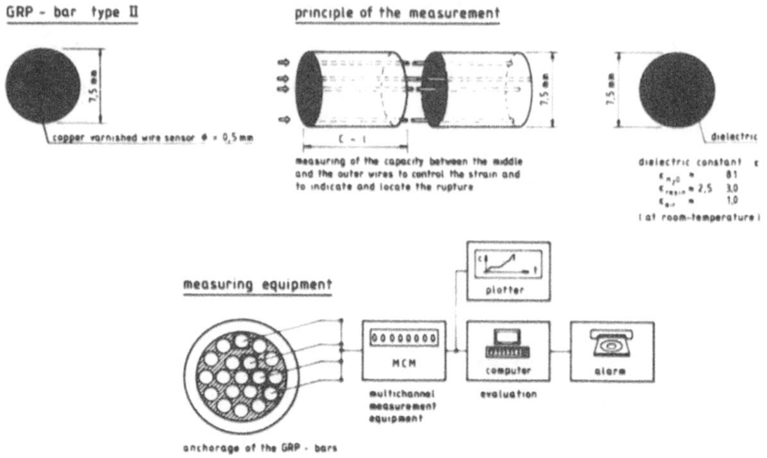

FIG. 5. Sketch showing the principle for the monitoring of prestressing tendons using copper wire sensors.

on the principle of the change in capacity produced by means of the central copper wire relative to the others; the change in capacity and the change in strain have a mutually proportional behaviour.

PRACTICAL EXAMPLES

The Ulenbergstrasse Bridge in Düsseldorf (Fig. 6), the first prestressed concrete bridge designed worldwide for extremely heavy road traffic loads (bridge class 60/30), is a demonstration construction project for the new material Polystal.

Its outline design provides for a 15 m wide two-span solid slab bridge with spans of 21·30 and 25·60 m, and a slab thickness of approximately 1·44 m.

In the longitudinal direction the bridge is prestressed by the use of 59 heavy-duty composite prestressing tendons (working load 660 kN each) with 19 heavy-duty composite bars per tendon. The bridge is prestressed by post-bonding achieved by means of an artificial resin-based grouting mortar specially converted for this purpose.

With regard to bridges, the next application was the construction of the pedestrian bridge to Marienfelde leisure park in Berlin in 1988 (Fig. 7). This is a two-span double-webbed slab/beam bridge with spans of 22·98 and 27·61 m; its superstructure has an overall height of 1·10 m and a slab width of 4·80 m. This bridge was executed with partial prestressing without bond for the first time in Germany. A total of seven prestressing tendons with a working load of 600 kN each run underneath the spans, around each of the

FIG. 6. Ulenbergstrasse Bridge, overall view.

two transverse beams in the bay section, and then upwards along the central column and over the transverse beam.

The most recent application to date has been the stabilisation of the arch of the 'Marie d'Ivry' metro-station in Paris (Fig. 8). Excavation in the vicinity of the tunnel tube resulted in ground settlement, which in turn caused the arch of the station's cross-section to crack. In order to improve the safety

FIG. 7. Berlin-Marienfelde Bridge, overall view.

FIG. 8. 'Marie d'Ivry' metro-station in Paris, tie-rod view.

required for the arch, the client decided on a tie-rod comprising glass fibre composite prestressing tendons. A total of 36 tendons with a working load of 650 kN each were installed in April 1989.

REFERENCES

1. WAASER, E. and WOLFF, R., Ein neuer Werkstoff für den Spannbeton, HLV-Hochleistungs-Verbundstab aus Glasfasern. 'Beton' 36 (1986).
2. MIESSELER, H.-J. and PREIS, L., Hochleistungsverbundwerkstäbe als Bewehrung im Beton- und Erdbau. *Bauen mit Kunststoffen*, **2** (1988) 4–14.
3. KÖNIG, G. and WOLFF, R., Hochleistungsverbundwerkstoff für die Vorspannung von Brückenbauwerken. IABSE Congress, Paris-Versailles, 1987.
4. LESSING, R. and MIESSELER, H.-J., Experiences in monitoring of load-bearing structures with optical fiber sensors. IABSE Congress, Lisbon, 1989.

ASSESSMENT AND EVALUATION

36

Bridge Capacity Assessment and Control of Posting, Permit and Legal Vehicle Loads

FRED MOSES

Department of Civil Engineering, Case Western Reserve University, Cleveland, Ohio 44106, USA

ABSTRACT

Several recent projects conducted by the author have concerned new methods for bridge evaluation, including strength capacity and safe life assessment. These projects have led to new AASHTO proposals for capacity evaluation and flexible methods for regulating safe loads on bridges. Economic pressure to revise truck weight regulations have also been considered. Proposals for evaluating permit trucks for different highway classifications have been reported. A comprehensive study of optimal truck weight regulations to balance vehicle productivity with increased costs for bridge repairs and replacement have been studied. These projects have been supported by the US Transportation Research Board and various state agencies.

INTRODUCTION

The emphasis for the highway industry in the United States has shifted to maintenance, rehabilitation and conserving the existing road network. Bridges are a vital link in the highway system and, in part, because of their conservative design, bridges have been allowed to deteriorate over many years because of deferred maintenance and repairs. There are some 600 000 bridges in the United States under a wide variety of ownerships and control. The Federal Highway Administration now estimates that more than 200 000 bridges are inadequate and lists over 125 000 as structurally deficient on the National Bridge Inventory System. There are some 5000–8000 replacements per year, so that for the foreseeable future the inventory of bridges in the United States will contain numerous structures incapable of carrying today's standards of truck weight.

In the present bridge inventory, about half the structures are more than 50 years old, which partly explains why so many are deficient. During recent decades, truck weights and volumes have grown enormously, while funds for inspection, maintenance, repair and rehabilitation were often not available. Despite this situation, bridges have maintained relatively high safety records because traditionally engineers used conservative methods of design which produced high levels of reserve strength. With increasing truck loadings, these reserves are eroded and many bridges must be replaced or else load posted for restricted traffic usage.

At the same time that budgetary pressures are restricting the upgrading of the bridge system, there are numerous proposals to allow increased demand on the system. Although the US Interstate System allows a legal load of 80 000 lb (356 kN), many states permit heavier loads. This is either done through a liberal permit system or else a 'grandfather' exemption clause which allows higher loading for certain classes of vehicles. A review of state regulations indicates over half utilize exemptions to allow vehicles on a routine basis up to 140 000 lb gross weight (623 kN). Requests for special permits for truck movements even above 300 000 lb (1330 kN) have also become relatively routine.

Two important aspects of bridge safety are impacted by these heavy vehicles, including the overall capacity of the structure to withstand the heaviest load combinations and also the reduced remaining life that may occur due to repetitive loadings which induce a cumulative fatigue damage. Because of the wide variety of structure type, span, material, geometry, age and design level the response of bridges to new truck demands will vary.

In recognition of both the increasing safety risk in bridges and the larger demand being placed on the system, there has been considerable research sponsored in recent years to promote better understanding of bridge loads, response analysis and strength capacity, and fatigue life. This research had led to improved methods of load capacity evaluations and bridge management policy to optimize utilization and conserve resources. This paper reviews recent work by the author in several of these areas. The topics to be covered include:

(a) guidelines for load capacity evaluation,
(b) guidelines for estimating safe remaining lives,
(c) evaluating heavy vehicle permits, and
(d) impact analysis of proposed truck weight regulations.

SAFETY METHODOLOGY

In the applications that follow, a consistent safety strategy is needed for analysis. This is necessary because traditional calculation methods by bridge engineers have widely varying levels of conservativeness contained within their design procedures. This variability in the safety level may be due to selection of load criteria, analysis techniques or assessment of safety factors for structure components and systems. Unlike pavement or roadway management programs which optimize only the economic components of the pavement, bridge management must relate to safety as well as economy. It is well recognized that traditional 'safety factors', whether in bridges or other structures, are not absolute criteria for safety. A major development for structural codes in recent years has been the introduction of structural reliability concepts to assist code writers to formulate safety checking models and derive appropriate safety factors. These reliability models recognize the uncertainties in modeling extreme load events, inexact analysis predictions, and the variability in material and system strength capacity. The goal is to establish a reliability or safety index which incorporates the actual safety margins and the uncertainties or randomness in the strength, load and analysis procedures. Safety margins are calibrated to ensure that target reliability indices are satisfied. To the practising engineer, such reliability based codes retain the traditional format of tabulated safety factors. Although instituted worldwide for design codes, reliability is also being used for evaluation procedures for existing structures. The uncertainties in evaluation are naturally different from those encountered in design situations.

Currently, guidelines for evaluating existing bridges are provided in the *AASHTO Manual for Maintenance Inspection of Bridges*.[1] Recommendations for introducing flexibility into the evaluation procedures are given in general qualitative terms. Quoting directly from the manual, '...a higher safety factor for a bridge carrying a large volume of traffic especially if it includes many heavy loads...'. Similarly, 'factors of safety used in rating must provide for reasonably possible overloads... and lack of knowledge as to the distribution of stresses'. Those quotes demonstrate the awareness of differences between design and rating, and that the latter should recognize that an existing structure should be influenced by observations and analysis that can directly be made. Nevertheless, the checking procedures in the manual are deterministic and do not take advantage of the apparent data base of the actual structure.

The intent of structural reliability theory is to characterize the uncertainties in load intensity, load effect analysis and strength capacity, and then allow for consistent and rational codified safety decisions. These methods have been adopted in many codes in the United States where it is known as load and resistance factor design (LRFD), in Canada and the United Kingdom where it is known as limit state design (LSD) and in other countries where it is called a partial safety factor format. To make it feasible for implementing reliability in rating a large number of structures, it is necessary (a) that the structures will be analysed by conventional methods and (b) that load and resistance factors be tabulated in a form that the engineer need not be concerned with probability or reliability theory in performing the checking equation.

The opportunity to introduce a comprehensive reliability approach to evaluating existing bridge structures is due to several circumstances. Recent code changes, such as the Ontario Highway Bridge Design Code introduced in 1979,[2] demonstrated that formal reliability methods can be used to calibrate safety factors based on uncertainty levels for all components in the design and evaluation process. Further, extensive field performance on bridges is available ranging from in-service weigh-in-motion (WIM)[3] to full-scale ultimate capacity determination.[4] These data show that current design and evaluation parameters represent idealizations and approximations that, while broadly applicable to design, do not necessarily apply to specific conditions of the bridge being evaluated. The flexibility of acquiring and using site-specific inspections, and statistical loading and analysis information, during the evaluation calculations become readily apparent.

LOAD CAPACITY EVALUATION

In the United States, all existing bridges must be load evaluated on a frequent periodic basis and recorded in the national inventory. This evaluation or rating follows a physical inspection of the bridge's condition. Following the evaluation, a bridge found deficient may be posted for reduced legal traffic loads or else scheduled for rehabilitation or replacement. The AASHTO manual allows evaluations at an inventory level corresponding to the design stress condition (55% of yield stress) or at a higher operating level, typically 75% of yield stress. The latter is used by many but not all jurisdictions in its posting decision. The selection of rating stress level is usually fixed for all structures by an agency and does not provide flexibility to account for the actual condition of the structure or to quantify other important site or engineering efforts that might be considered in the rating process.

As a consequence of these limitations, the author and his colleagues introduced a reliability based LRFD oriented rating procedure.[5,6] These have recently been approved by AASHTO as a guide specification.[7] The procedures utilized reliability theory and an existing data base to calibrate load and resistance factors which are intended to achieve a uniform and consistent evaluation procedure. Resistance factors ranging from 0·55 to 0·95 are affected by the bridge condition survey or level of deterioration noted, inspection and maintenance effort, and especially by the presence of redundant load paths. Dead load factors are influenced by direct field determination of overlay thickness. Live load factor selection ranging from 1·3 to 1·8 is based on traffic volume, sources of overload or control of heavy truck traffic and structural analysis method used. Impact values depend on deck roadway roughness and range from 0·1 to 0·3. Bridges with computed rating factors below 1·0 must be load posted.

With the factors tabulated in the guide specification, bridges may reach or even exceed previous operating ratings for those bridges which receive frequent qualified inspection and have adequate maintenance programs and loads corresponding to reasonable levels of traffic and enforcement. Conversely, bridges which do not maintain these conditions or have nonredundant components will find their ratings falling possibly to inventory levels or even lower. Evaluators will find options in these guidelines by which ratings can be improved by recommendations for more frequent and detailed inspection and maintenance, improved structure analysis and especially control of heavy overweight vehicles. In all instances, it is expected that the use of the tabulated factors will provide rating evaluations which meet the target reliability levels. The target levels used in establishing the factors were calibrated to reliability analysis with specific field performance experience.[8]

SAFE REMAINING BRIDGE LIVES

In most specifications, fatigue checking provisions are given in the form of allowable stresses which must be satisfied by a designated loading case. Such provisions do not indicate how to calculate the remaining life of an existing bridge which is needed to make cost effective decisions regarding inspection, repair, rehabilitation and replacement. Many agencies indicated in a survey that remaining life calculations are needed especially to assess rehabilitation options and in some cases to assist in making bridge posting decisions.

The author and his colleagues recently presented methods for assessing remaining life of steel bridges which were accepted by AASHTO as a guide specification.[9–11] The procedures in these guidelines used probabilistic concepts which provide realistic description of the fatigue conditions in a bridge. Appropriate target reliabilities were calibrated for both redundant and nonredundant categories. Further recommendations were given to quantify the effect of different levels of effort in reducing uncertainties (analysis or site data acquisition) and thereby improve the predictions of remaining life. To simplify implementation, definitions of detail categories were the same as present AASHTO rules. The options available to the evaluator if the computed remaining life is inadequate include;

(1) calculating fatigue life more accurately with recommended procedures;
(2) restricting traffic on the bridge;
(3) repairing the bridge; or
(4) instituting periodic detailed inspections.

The advantages of the procedures developed in the remaining life study include:

(1) Methods which realistically reflect actual fatigue stress conditions in bridges.
(2) Consistent procedures for both design and remaining life evaluation.
(3) Suitable flexible procedures for using site data.
(4) Allow record keeping of remaining life to be periodically updated.
(5) Extensive use of recent traffic and fatigue research.
(6) Reliability targets calibrated from current performance.
(7) Use of appropriate levels of analysis depending on application.
(8) Do not significantly depart from current methods of fatigue design.

Numerous examples and illustrations of the fatigue evaluation (as well as corresponding design provisions) were provided.

PERMIT REVIEW

Permits for overloaded vehicles are frequently issued by highway agencies in the United States and elsewhere. Most states face increasing pressures to allow heavier and greater numbers of overweight truck permits. This raises questions regarding bridge safety and management policy and especially the methods for reviewing bridges for permit loads. Typically, permit checks utilize the stress levels corresponding to operating levels (75% of yield) in the AASHTO specifications or, in some cases, agencies permit even higher stress levels. It is clear that such simple allowable stress checking methods may not produce uniform reliability due to considerations of bridge dead load, bridge geometry and normal truck traffic. Reliability based load factors offer advantages for more uniform permit checking rules. Presently, AASHTO is reviewing reliability based load and resistance factor methods as a possible alternative in its bridge design manual, and as stated above has reviewed favorably its use in bridge evaluation methods.

A project conducted by the author and his colleagues is under way to derive appropriate load factors for permit decisions.[12] Among the situations considered are permits for (a) single passage (no control of movement), (b) single passage with speed and lane position control, and (c) routine overloads with no restrictions on frequency of movements.

The maximum load effect may occur with a permit vehicle simultaneously on the structure with one or more adjacent random heavy vehicles. The occurrence probability of such multiple presence events depends on traffic volume, bridge span and geometry, speed and control of permit vehicles. Vehicle arrival simulation models have been developed to predict the maximum loading response and the corresponding distribution of load effects (fatigue life impact is considered separately in the permit study). Using target reliability levels calibrated from other bridge safety studies, it is possible to determine appropriate live load factors for permit checking. These permit review methods are being currently put in a specification format for appropriate review. The aim is to maintain within a bridge management system options for assisting in moving special loads without introducing unacceptable bridge risk levels.

TRUCK WEIGHT REGULATIONS

Many industrialized countries face increasing political pressures to raise current allowable legal vehicle loads. The economic incentives in terms of increased productivity appear to be large since for a small percentage increase in fuel consumption a much greater percentage increase in payload and profit is possible. The considerations for increased truck loads are based on safety, pavement damage and risk to bridges. One proposal in the United States which minimizes pavement damage is the so-called Turner proposal. This plan increases the number of vehicle axles which distributes the load and reduces pavement damage (since the latter is approximately proportional to axle weight raised to the fourth power). The effect on bridges of spreading the load is minor, and unless overall truck lengths are increased bridges will be adversely impacted by any heavier truck loadings. Length increases usually are restricted by problems of vehicle maneuverability and safety. It is possible, however, that Turner vehicles will provide enough savings in pavement damage to offset the larger bridge costs. In an effort to define the bridge costs, the author and his colleagues have participated in a recent TRB and FHWA study.[13] The main goal was estimating bridge costs that are likely to be incurred by state agencies following the introduction of new vehicle regulations. Sensitivity studies to compute costs if agencies adopted new bridge evaluation methods such as described above were also considered and are discussed below.

Bridge costs were divided into four categories. These include (a) overstress—new bridges, (b) overstress—existing bridges, (c) remaining lives—new bridges and (d) remaining lives—existing bridges. Over ten different proposed truck weight regulations (or scenarios as they are called) have been studied, including Turner configurations (increased weight with increased number of axles), as well as other variations of proposed weight rules such as the Canadian interprovincial proposal, the TTI model, extensions of the existing federal highway bridge formula and proposals suggested by various trucking associations. In each scenario, the costs were evaluated for each of the cost categories.

For new bridges, an estimate was made that $3 billion per year is now needed for new bridge replacements and this sum would be affected by any new regulations. That is, design loads would have to be increased which would raise costs. For example, we found that increases from present HS20 design load to HS25 (25% increase in loads) would raise average new bridge costs by about 4%. Applying such extrapolations to the load impact of each proposed regulation gave an estimate of increased bridge costs for new structures.

The major cost item, however, proved to be the impact on existing bridges. Using the FHWA computerized NBIS, a breakdown was performed of the almost 600 000 existing bridges to produce distributions by category of span, material, type (single or continuous) and by current operating load capacity. The present deficiencies under current weight laws were first identified. The next step in each scenario was to compare for each category the live load effect for the new scenario with that of existing weight regulations. Hence, the number of additional bridge deficiencies was estimated and the corresponding replacement costs computed. (In a survey, states indicated that replacement is often more economical than rehabilitation whenever strength upgrading is required.) The costs for the replacement of deficient existing bridges provide a major part of the overall cost impact. It was found that steel bridges incurred about two-thirds of the total cost impact (although they are a lower percentage of the population) with reinforced concrete and prestressed concrete the remainder. Continuous span bridges were a significant component of the cost, especially for truck vehicle proposals, which tend to cause large increased moments in the negative moment regions.

In general, it was found that the important factor in any new weight regulation was the proposed gross weight and the corresponding vehicle length (exclusive of distance from steering axle to drive axle). Some of the scenarios were found to more than double the existing number of structural deficiencies (currently about 125 000 out of the total population) with corresponding increases in bridge replacement costs. Breakdowns of impact by interstate and primary routes and also nonprimary routes was also done.

A sensitivity study also found that any deterministic changes in bridge evaluation procedures are not likely to cause significant reduction or mitigation of the impact of new weight regulations. For example, more liberal stress levels in evaluation rules will reduce the number of currently deficient bridges but many of these same bridges will then become deficient under the proposed new regulations.

The above-cited study used a distribution of current operating load capacity levels as a measure of evaluation. These are, in current practice, deterministic stress criteria. A study now under way by Dr Michel Ghosn of the City University of New York with the assistance of the author is looking at modeling the bridge population according to a distribution of reliability indices.[14] Hence, bridges would be replaced under either present truck regulations or the new regulations only if their reliability level was deficient. As described above, the bridge reliability index depends on traffic volume, methods of analysis, and bridge condition and maintenance. It appears possible to create a new truck weight formula which trades off increased reliability for shorter span bridges (where most deficiencies now occur) with lower reliability levels for longer span bridges. The latter structures have high dead/live load ratios and have been found in reliability studies to often exceed required target reliability levels.[8] Such a trade-off in formulating truck weight guidelines could produce a new truck weight regulation which has relatively small cost impact on bridges but does increase truck productivity, i.e. longer vehicles may be allowed heavier loads than presently allowed with some reduction in gross weight for shorter vehicles. These studies are still under way but some preliminary results support these conclusions.

The other two cost items mentioned above were in the fatigue life areas. A model of the distribution of steel span bridges was created which reflects their fatigue life. The model was calibrated to the present annual expenditures on fatigue related problems and associated damage costs in steel bridges. New weight regulations (both weight and volume changes are needed for fatigue) were used to find the expected corresponding future costs in fatigue life.

The impact of these fatigue costs was studied with different discount factors and other variables. The overall cost impacts, although significant, were generally small compared to the cost effect on the strength capacity of existing bridges. Similarly, the last cost item, namely cost impact of fatigue rules on new steel bridges, was small. That is, required strength increases to carry heavier vehicle loads also implicitly increases fatigue life so that this additional cost item appeared small. Further study, however, is under way of possible impacts on acceptable fatigue details under heavier loading conditions.

CONCLUSIONS

A variety of bridge studies are discussed above related to bridge safety, evaluation, fatigue life estimation, review of heavy overload permit requests and consideration of new truck weight regulations. Details of these studies are contained in the references and guideline specifications that have arisen out of this work. A common thread in these studies is a need for a systematic and defensible approach to bridge safety. Risk analysis has now become an accepted tool for code writers and reliability based design procedures a technique for implementing uniform safety criteria. Such work needs to be elaborated on so that confidence by users is increased. Risk criteria or corresponding reliability indices become an important basis for a rational bridge management policy. Otherwise, engineers are continually being asked to stretch the limits of the bridge system by permitting heavier permit loadings or even expanded truck weight regulations. Allowable stress procedures fail as criteria for bridge safety management since proponents of higher loads are aware of the conservative nature of these stress calculations. Bridge tests have shown that predicted stresses are often much lower than measured stresses. Also, proponents of higher loads claim that if a high stress is acceptable in one instance it should be acceptable in all cases. A reliability approach serves to incorporate all aspects of the load–capacity equation and identify the uncertainties which merit the requisite safety margins to avoid undue risk. Applications of reliability analysis described herein for load capacity evaluation, assessment of safe remaining life, permit review and recommendations for truck weight regulations have reinforced past engineering judgment but also allow recent research and test findings to be incorporated.

Recent developments in the United States and Canada in reviewing and adopting load and resistance factor formats for both bridge design and evaluation show the promise of such procedures being incorporated within a bridge management system.

ACKNOWLEDGEMENTS

The author wishes to acknowledge colleagues and associates who participated in these efforts. In particular, at Case Western Reserve University, Dr D. Verma assisted with the load capacity studies, K. S. Raju with the fatigue life assessment and Dr G. Fu with the permit studies and truck weight regulations. Dr Michel Ghosn also helped in the basic reliability modeling and analysis for several projects. C. Schilling, consulting engineer, helped with all aspects of fatigue modeling. The firm of A. G.

Lichtenstein made several important contributions in development and implementations, and Imbsen and Associates was a valued collaborator in the load capacity studies. Finally, the important help of Ian Friedland of the NCHRP staff, Harry Cohen of the TRB, and L. Talbert and V. Dalal of the Ohio Department of Transportation are appreciated.

REFERENCES

1. *Manual of Maintenance Inspection of Bridges.* American Association of State Highway Transportation Officials, Washington, DC, 1982.
2. Ontario Highway Bridge Design Code. Ontario Ministry of Transportation and Communication, Downsview, Ontario, Canada, 1983.
3. MOSES, F., GHOSN, M. and GOBIESKI, J., Evaluation of steel bridges using in-service testing. Presented at TRB Annual Meeting, Washington, DC, January 1986. Published in Transportation Research Record 1072, Washington, DC.
4. BAKHT, B. and CSAGOLY, P. F., Diagnostic testing of a bridge. *ASCE Journal of Structural Engineering*, **106** (7) (July 1980).
5. MOSES, F. and VERMA, D., Load capacity evaluation of existing bridges. NCHRP Report 301, NCHRP 12–28(1) Final Report, Transportation Research Board, Washington, DC, December 1987.
6. VERMA, D. and MOSES, F., Calibration of a bridge strength evaluation code. *ASCE Journal of Structural Engineering*, **115**(6) (June 1989).
7. Guide Specification for Strength Evaluation of Existing Steel and Concrete Bridges. AASHTO, Washington, DC, 1989.
8. GHOSN, M. and MOSES, F., A reliability calibration of a bridge design code. *ASCE Journal of Structural Engineering*, **112**(3) (1986).
9. MOSES, F., SCHILLING, C. G. and RAJU, K. S., Fatigue evaluation procedures for steel bridges. NCHRP Report 299, Transportation Research Board, Washington, DC, November 1987.
10. MOSES, F., SCHILLING, C. G. and RAJU, K. S., Reliability-based bridge life assessment. ICCOSAR '89, San Francisco, May 1989.
11. Guide Specification for Fatigue Evaluation of Existing Steel Bridges, approved by AASHTO, Washington, DC, 1989.
12. MOSES, F. and FU, G., A reliability analysis of permit loads on bridges. In progress, report to Ohio Department of Transportation, Case Western Reserve University.
13. MOSES, F., Effects on bridges of alternative truck configurations and weights. In progress, National Academy of Science, Transportation Research Board.
14. GHOSN, M., Bridge overstress criteria. In progress, sponsored by Federal Highway Administration at City University of New York.

37

The Use of Reliability Analysis in the Assessment of Existing Bridges

CAMPBELL MIDDLETON

Department of Engineering, Cambridge University, Trumpington Street, Cambridge CB2 1PZ, UK

and

ANGUS LOW

Ove Arup & Partners, 13 Fitzroy Street, London W1P 6BQ, UK

ABSTRACT

As part of a contract for the UK Transport and Road Research Laboratory the authors have developed procedures for assessing the reliability of some common types of concrete bridges.

Initially upper-bound plastic assessment techniques were developed which could be used as an alternative to lower-bound elastic or non-linear finite element methods to estimate the collapse strength of a bridge. However, this deterministic technique makes no provision for the many uncertainties facing an engineer when assessing an existing structure. These could concern the strengths of materials used, the deterioration of these materials, the load history and the form of hidden parts of the structure.

Structural reliability theory allows the effects of such uncertainties to be assessed rationally. An advanced level II reliability procedure following the method outlined in CIRIA Report No. 63 (The Rationalisation of Safety and Serviceability Factors in Structural Codes, 1977) was applied to the plastic collapse analysis to determine the notional probability of failure of a range of bridge types and dimensional configurations.

1 INTRODUCTION

Reliability theory has not as yet been widely used for the assessment of bridges. The complexity of the computations, limitations in the theory and general unfamiliarity have all been constraints on its use. With recent extensions in the theory[5] and readily available computing power it may now be possible to develop practical procedures to complement the more traditional methods used in the assessment of existing bridge structures.

In this work a number of possible plastic collapse modes were examined for each of the chosen bridge types. This provided a deterministic calculation of the strength of the bridge in relation to the specified loading and predicted the most likely collapse mode geometry. The probability of this failure occurring was then derived using reliability theory which examined the statistical variability of each of the basic parameters that contributed to the loading on the bridge and its strength to resist the applied loads.

The notional probability of failure for a bridge (typically 10^{-5}–10^{-25}) has little relevance to the real likelihood of collapse as most actual failures relate to gross errors or catastrophic events. However, the relative values can be used to rank the safety of a number of bridges and thus provide a basis for assessing the priorities for remedial work or bridge replacement in relation to different bridge configurations of deck type, span length and width. In addition, by using these methods an engineer assessing a complex bridge can explore the sensitivity of its reliability by making allowances for his knowledge or ignorance of the strengths and variability of individual components.

The methods described here are applicable for use in assessing both newly designed and existing structures, provided some details of the dimensions, material types and reinforcement are known.

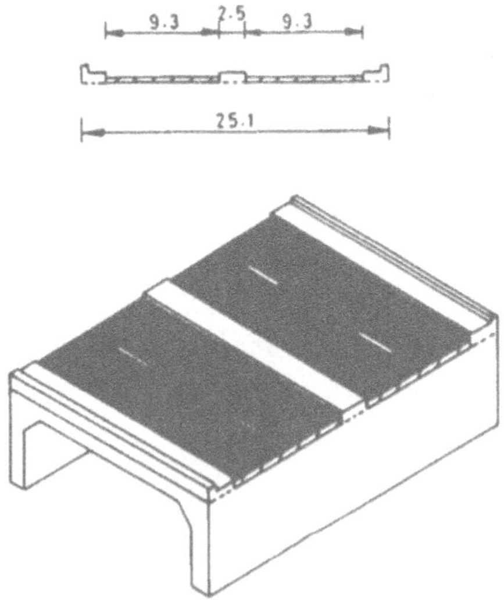

FIG. 1. Dual two-lane portal slab bridge. All dimensions in metres.

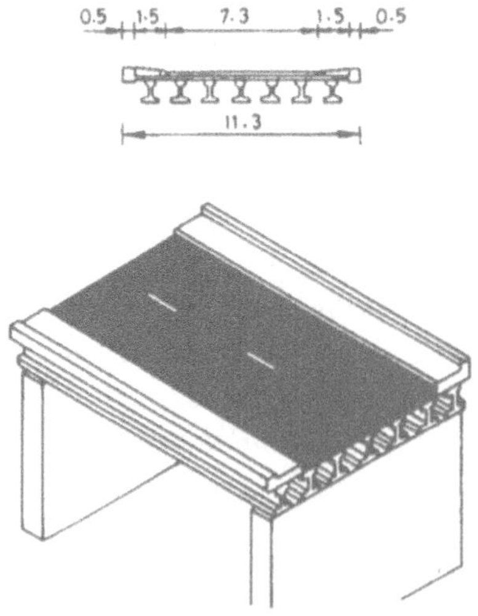

FIG. 2. Single two-lane M-beam bridge. All dimensions in metres.

2 BRIDGE TYPES AND DIMENSIONS

Two of the most common forms of highway bridge structures were selected for detailed examination. These were in-situ reinforced concrete portal slab bridges and simply supported, precast, prestressed M-beam bridges with in-situ reinforced concrete deck slabs (see Figs 1 and 2 for typical examples).

By reference to the Department of Transport's bridges database and departmental standards for cross-section design, a number of different span lengths, cross-section widths and span/depth ratios were selected as representative of the current and likely future stock of these types of structures in the UK. These cross-sections included single two-lane, dual two-lane and dual three-lane structures with span lengths ranging from 6 to 28 m and span/depth ratios from 11 to 30.

3 REINFORCEMENT DETAILS

The reinforcement details required for use in the collapse and reliability analysis were obtained by designing a new bridge corresponding to each of the selected dimensional configurations. Equally these details could have been obtained from existing bridges by reference to design plans or by site measurement.

Longitudinal and transverse reinforcement in the slabs, as well as prestress details and shear links in the M-beams, were designed using standard design practice based on elastic design moments and code provisions for shear reinforcement.

4 LOADING

In addition to dead and superimposed dead loads, two live loadcases from BS 5400: Part 2 were considered in the collapse analysis. These were the various lane loading combinations (HA with knife edge) and the heavy vehicle load (HB with 25 units). The most severe combination of HA lane configuration and HB vehicle location was found by the automated computer procedures in the collapse analysis and subsequently used in the reliability analysis.

5 COLLAPSE ANALYSIS

A number of possible plastic collapse mode shapes were selected as being appropriate for the bridge deck types examined. A set of geometric parameters describing these failure modes were optimised to give the lowest load factor against failure based on minimum energy assumptions:

Load factor LF

$$= \frac{\text{ED}}{\text{WD}} = \frac{\text{Energy dissipated in the plastic zones and/or yieldlines}}{\text{Work done by the applied loads}}$$

5.1 Portal Slab Failure Modes

Two geometric yieldline failure modes were considered for each of the portal slab bridges:

Mode 1 Full-width (FW) transverse yieldlines across midspan and both supports perpendicular to the traffic direction (Fig. 3).

Mode 2 Partial-width (PW) yieldline failure mode with variable geometric failure mechanism parameters to define the apex angle (AL) of the triangular plate and width of failure across the deck (CC) (Fig. 4).

5.2 M-beam Failure Modes

An M-beam deck comprises a number of discrete beams connected by an in-situ deck slab. At collapse each individual beam will be either intact or failed.

The method of analysis adopted in the collapse program considered a progressive failure of adjacent beams across the deck. It allowed the edge beam to collapse first and then continued calculating the load factor for the specific failure geometry after each successive beam failure until all beams had failed to form a full-width collapse mode. Two failure modes were considered for the M-beam deck bridges (see Figs 5 and 6).

Mode 3: Combined shear and bending failure (Fig. 5)

This failure resulted from propagation of an inclined crack, the location of which was defined by points A and B, as shown in Fig. 5.

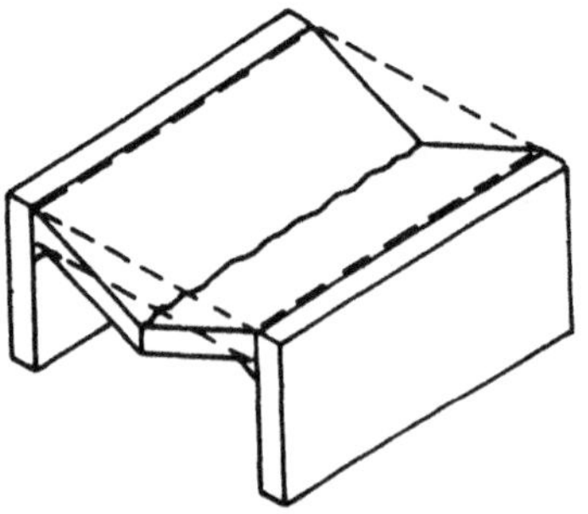

FIG. 3. Full-width portal slab failure mode.

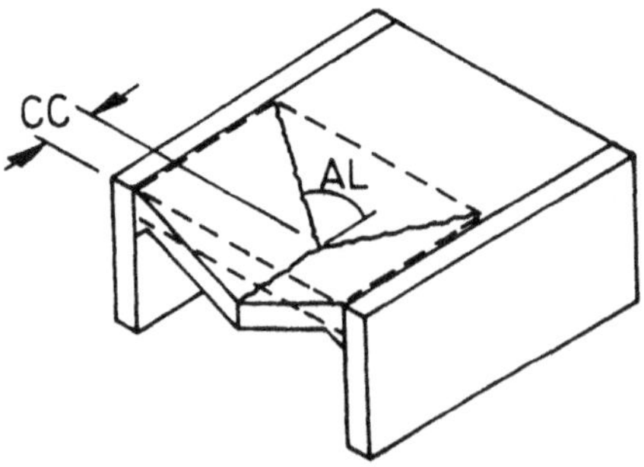

FIG. 4. Partial-width portal slab failure mode.

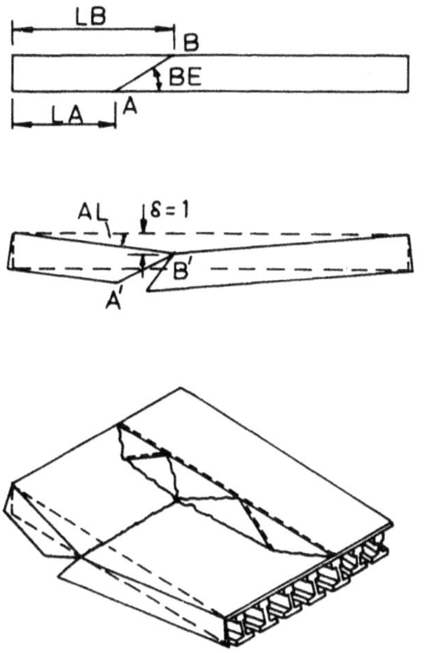

FIG. 5. M-beam failure mode 3, combined shear and bending.

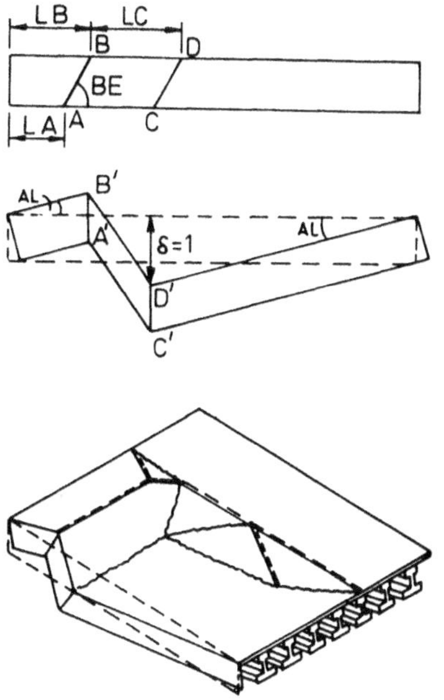

FIG. 6. M-beam failure mode 4, shear.

For the work/energy evaluation a unit virtual displacement ($\delta = 1$) from point B to B′ was assumed in the top of the failed beam.

Initially the program introduced a vertical crack on one edge beam, resulting in a pure flexural failure. The crack was then inclined by increasing dimension LB and the same progressive failure across the deck assessed. This procedure of optimising crack inclination and the number of failed M-beams was repeated for the crack initiating at the support, $\frac{1}{8}$ span, $\frac{1}{4}$ span and midspan in order to find the lowest load factor for this collapse mode. For modes with an inclined crack ($LB > LA$, $BE < 90°$) the shear steel intersected the failure plane and added to the energy dissipated in the section.

The deck slab between the failed and unfailed M-beams was assumed to fail in a series of triangular plates. This mode was found to produce a lower energy state than using a single triangular plate between beams.

Mode 4: Shear failure (Fig. 6)

This failure mode had a zone failing in shear (ABCD) in which energy was dissipated in the distortion of the concrete and in the straining of shear steel. The energy dissipated in the plastic shear zone of the beams was obtained by reference to Nielsen.[3]

The failure geometry was defined by assuming a unit virtual displacement downwards from point D to D′. This resulted in both points B′ and D′

rotating through the same angle, AL. A similar iterative procedure to that adopted for calculating the lowest factor for mode 3 failures was adopted here, except one extra geometric variable was introduced—the width of the shear zone parallelogram (LC).

The position of initiation of the shear zone, defined by a line from point A to B on Fig. 6, was examined at the support, $\frac{1}{8}$ span and $\frac{1}{4}$ span. In addition, the inclination of the parallelogram (defined by dimensions LA and LB), width of shear zone (LC) and number of failed beams across the deck were optimised until the lowest load factor defining the most likely shear failure mode was determined.

5.3 Governing Failure Mode

The governing or most likely overall failure mode was selected by comparing the load factors obtained for all the different failure modes under each of the specified loadcases and choosing the lowest for each bridge. This mode was then used in the reliability analysis to evaluate the probability of such a failure occurring.

6 RELIABILITY ANALYSIS

6.1 Basic Variables

Reliability theory was used to determine a 'notional' probability of failure for each of the bridges examined.

In normal practice the major parameters used in structural design and analysis are considered constant (e.g. dimensions, material strengths, applied loads). However, in reality many of these parameters are subject to statistical variation. For example, although a design may specify 40 MPa concrete cube strength, the concrete that is actually placed in the structure may well have a range of values between, say, 38 and 50 MPa. These values could be represented by a normal or log-normal distribution curve with a particular mean and standard deviation.

The level II method of reliability analysis used here revolves around selecting a set of the most important variable parameters, which are referred to as basic variables, and investigating the effect of this variability on the strength of the structure studied.

Allowance was made in the reliability analysis for uncertainties in the strength (or resistance) of the structure and applied ultimate collapse loads arising from possible variations in the values of yield stress of the concrete and steel, position of the main reinforcement, magnitude of the applied live loads and permanent loads, and the accuracy of the structural model.

Nine basic variables (X_i) were selected for the bridges examined:

X_1 = FY = characteristic yield stress of steel (kPa)
X_2 = DERR = error in placement of reinforcement bars from specified position (m)
X_3 = FCU = characteristic concrete compressive strength (kPa)
X_4 = WDLL = work done by live loads (kNm)
X_5 = WDSDL = work done by superimposed dead loads (kNm)
X_6 = GC = density of concrete (kN/m^3)
X_7 = H = thickness of deck slab (m)
X_8 = UNC = structural model uncertainty (non-dimensional)
X_9 = FPY = characteristic yield stress of prestressing strand (kPa)

6.2 Limit State Function, Z

Using the principles of plastic analysis, failure was defined to have occurred when the work done by the loads (WD) equalled the energy dissipated in the yieldlines or plastic zones (ED).

By defining the failure function Z such that

$$Z = \mathrm{ED} - \mathrm{WD}$$

then

$$Z \leq 0 \Rightarrow \text{failure} \qquad Z > 0 \Rightarrow \text{safe}$$

As in the collapse analysis, the expressions for energy dissipated and work done in the plastic zones were related to moment capacities, strain energies, virtual displacements of the loads and geometry. These parameters were themselves functions of the basic variables such as steel yield stress, concrete strength and reinforcement location which were described as probabilistic functions to account for their variability. Hence

$$Z = g(\underline{X}_i) = 0 \text{ at failure where } \underline{X}_i = \text{set of all basic variables}$$

A failure function corresponding to each of the four postulated failure mechanisms was incorporated into the program.

6.3 Advanced Level II Method of Reliability Analysis

In the level II method of reliability analysis the magnitude of each of the basic variables is changed from their original mean value until a combination of values is obtained for which the failure function Z equals zero and collapse has occurred. The manner in which these parameters are varied depends on their standard deviation and also their relative importance to the loading on, or strength of, the structure. The difference between the value of each variable at collapse and its original mean value can be assessed to determine the likelihood of the event occurring and hence the probability of failure of the bridge. The reliability index (β) gives a quantitative measure of this difference and can be directly related to the probability of failure of the bridge. Sensitivity factors (α) were derived in the reliability analysis which give an indication of the relative importance of each of the basic variables to the reliability of the bridge.

7 RESULTS

7.1 Portal Slab Bridges

7.1.1 Collapse analysis

The full-width (FW) failure mode was found to govern the most likely collapse mode of all the portal slab bridges examined except for the two shortest span structures ($L=7$ and 9 m) for the wider dual two- and three-lane bridges. In these cases the partial-width (PW) failure mode governed. Basically this is because the lengths of yieldline in which energy can be dissipated tend to be greater in the partial-width mode relative to the full-width mode for all but the shortest, widest bridges in which the partial-width failure occurs at the very edge of the bridge under a heavy HB vehicle.

7.1.2 Loadcase governing collapse

The HB loadcase was the dominant loadcase determining the most likely collapse mode in all the portal slab bridges examined (Table 1). With the wider, longer bridges it was less clear which would be the dominant loadcase. As width and span increased the effect of multiple lane loadings in the HA case became more significant and the relative importance of the concentrated HB load compared to the HA uniform lane load reduced. In these bridges it was found that the load factors for HA and HB loading were much closer together but the HB case still predominated (i.e. the LF for the HB loadcase was the lowest).

7.1.3 Reliability analysis

The failure probabilities of the portal slab bridges examined differed markedly for different span lengths, widths and span/depth ratios. For portal slab bridges with a span/depth ratio of 30 values ranging between 10^{-7} for a short (7 m), narrow (single two-lane) bridge and 10^{-24} for the longest (28 m), widest (dual three-lane) bridge were obtained.

Steel yield stress, live load and the structural model uncertainty were found to be the dominant basic variables governing the collapse of the portal slab bridges.

The reliability index (β) can be directly related to the probability of failure of the bridge with increasing β value corresponding to increasing reliability.

TABLE 1
Example load factors at collapse for portal slab bridges with span/depth ratio SDR = 30

Bridge type	Failure mode	Loadcase	Load factor at collapse
1. Dual three-lane (wide)	FW	HA	3·04
Span $L = 7$ m (short)		HB	2·93
	PW	HA	2·59
		HB	2·54[a]
2. Dual three-lane (wide)	FW	HA	1·93
Span $L = 28$ m (long)		HB	1·93[a]
	PW	HA	2·44
		HB	2·44
3. Single two-lane (narrow)	FW	HA	2·30
Span $L = 7$ m (short)		HB	1·93[a]
	PW	HA	2·74
		HB	2·18
4. Single two-lane (narrow)	FW	HA	1·73
Span $L = 28$ m (long)		HB	1·67[a]
	PW	HA	5·00
		HB	4·74

[a] Indicates governing failure mode and loadcase.

For example, the graph in Fig. 7 plots the reliability of some portal slab bridges against span length. The results for both partial- and full-width modes are shown. For structures failing by the full-width mode the reliability is relatively constant with span, whereas for the partial-width mode it greatly increases. It can be seen that the structures of most significance are the narrow, short span portal slabs with reliability indices near 5 corresponding to a probability of failure of approximately 10^{-7}.

7.2 M-beam Bridges

The highest nominal probability of failure obtained in the M-beam study (10^{-19}) was many orders of magnitude less than those obtained for the majority of portal slab bridges. The results indicated that the narrow, single two-lane M-beam bridges had a significantly higher probability of failure than the wider dual two- and three-lane bridges. This is because the heavy HB vehicle load is far more concentrated over a narrow structure than over a wider dual three-lane bridge, resulting in a higher likelihood of failure.

The most likely failure mode for the single two-lane bridges was found to be a full-width combined shear and bending mode with an inclined crack

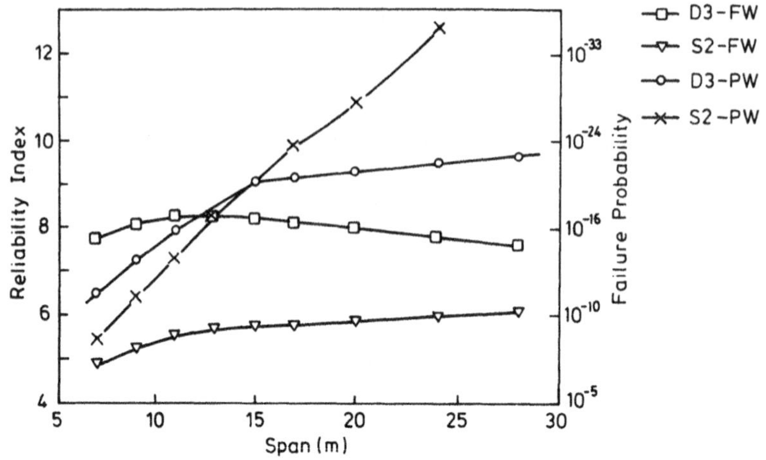

FIG. 7. Reliability index (β) versus span, both failure modes SDR = 30.

initiating from the $\frac{1}{8}$ span point. Some partial-width failures were observed in the wider dual three-lane bridges. The pure shear failure mode did not govern failure for any of the bridges examined.

The probability of failure of a given structure was not very sensitive to M-beam size, indicating that the adjustments made in prestress and shear link design to account for the different beam size adequately compensate for this change. The dominant variable in the M-beam analysis was the live load followed by the model uncertainty factor. It was interesting to note that the yield stress of the prestressing strand had a comparatively small effect on the results.

8 DISCUSSION

A significant finding in this study was the high probability of failure of short span, narrow portal slab bridges in comparison to the other bridges studied.

8.1 Sensitivity of Reliabilities to Basic Variable Parameters

A sensitivity analysis was performed for the portal slab bridges by separately doubling the standard deviation for steel yield stress and then for work done by the live loads. These changes resulted in significant increases in the nominal probability of failure of individual bridges.

8.2 Limitations on Assessment Technique

Clearly this sensitivity to basic variable parameters is unsatisfactory considering that their statistical properties are chosen in an imprecise manner from quite limited data and a degree of engineering judgement.

However, the methods outlined here are not intended for finding absolute reliabilities. It is proposed that they be used for ranking reliabilities of different bridges or identifying the most likely failure mode within one structure. The strength of the method is that it takes quantitative account of aspects such as brittleness, ductility and known variabilities which are not covered directly by more conventional methods of strength assessment.

Another difficulty in this project was developing realistic collapse models. There is very limited research available into the actual mechanisms of failure of concrete bridges and much of this relies on quite simplistic loading and failure models. In practice few engineers actually employ upper-bound methods when assessing bridge structures, possibly because they fear they cannot, from judgement, assess the lowest upper-bound failure mode. To a degree this can be overcome, using powerful modern computers, by selecting a large number of possible yielding/plastic collapse patterns and searching for the lowest upper bound.

It is also difficult to accurately model the ultimate strength of a bridge when factors such as ductility and membrane action can contribute significantly to the type of collapse mode and actual strength of the bridge.

9 FURTHER RESEARCH AND PRACTICAL APPLICATIONS

Currently a TRRL funded research programme at Cambridge University is working towards solving some of the problems outlined here and developing a more generalised computer program for assessing concrete bridges. This program will be calibrated against existing non-linear finite element programs with the aim of producing a program that can realistically model the lowest upper-bound failure loads for some common types of concrete highway bridges. It is intended to examine a broad range of collapse modes and refine the energy dissipation calculation to model the behaviour of the concrete more realistically. By generalising the loading subroutine any selected live load model could be examined in detail. Reliability theory will be applied to the derived collapse modes to allow for variations in the loading and strength parameters of the structure. Within this it is planned to make provision for the deterioration of reinforcement due to corrosion.

10 CONCLUSION

In this project procedures were developed for assessing the reliability of a number of common types of concrete bridges. Based on this work, a current research contract at Cambridge University is developing a suite of computer programs aimed at assessing the strength and reliability of a more extensive range of bridges subject to differing levels of physical deterioration under a variety of possible loadcases.

It is anticipated that the derived probabilities of failure may provide a basis for assessing the relative priorities of remedial work or bridge replacement in relation to different bridge types and spans, and thus contribute to the vital task facing our transport authorities in assessing the strength, safety and reliability of the existing population of concrete bridges. In this way bridges which fail crude assessment procedures may be allowed to safely continue serving the community.

ACKNOWLEDGEMENTS

The authors would like to thank the Transport and Road Research Laboratory and Ove Arup & Partners for permission to publish this paper. The work reported herein was carried out under a contract placed on Ove Arup & Partners by the Transport and Road Research Laboratory. Any views expressed are not necessarily those of the Transport and Road Research Laboratory nor the Department of Transport.

REFERENCES

1. JOHANSEN, K. W., *Yield-line Formulae for Slabs*. Cement and Concrete Association, Publication 12.044, 1972.
2. CONSTRUCTION INDUSTRY RESEARCH AND INFORMATION ASSOCIATION, The Rationalisation of Safety and Serviceability Factors in Structural Codes. CIRIA Report 63, July 1977.
3. NIELSEN, M. P., *Limit Analysis and Concrete Plasticity*. Prentice-Hall, New York, 1984.
4. THOFT-CHRISTENSEN, P. and BAKER, M. J., *Structural Reliability Theory and its Applications*. Springer-Verlag, Berlin, 1982.
5. THOFT-CHRISTENSEN, P. and MUROTSU, Y., *Application of Structural Systems Reliability Theory*. Springer-Verlag, Berlin, 1986.

38

Strength Assessment Methods for Concrete Bridges

P. A. JACKSON

Gifford and Partners, Southampton, UK

and

R. J. COPE

Polytechnic South West, Plymouth, UK

ABSTRACT

Two half-scale models of bridges with very lightly reinforced deck slabs have been tested to failure under both the British 'HB' design load and single wheel loads. The strength of the models is assessed using a variety of approaches: conventional elastic analysis using British and American practice, yield-line analysis, a semi-empirical approach developed in Northern Ireland which allows for membrane action in the slab, the Ontario Highway Bridge Design Code (which also uses membrane action) and finally non-linear analysis. The predicted strength is found to differ greatly between the approaches, the ratio of highest to lowest prediction approaching ten in one case.

Under single wheel loads the least conservative prediction, that obtained using non-linear analysis, proved to be most accurate although the predictions obtained using the other methods which allow for membrane action were also acceptable. Under the HB load the non-linear analysis again gave the best prediction and the other predictions which considered membrane action proved to be unsafe, apparently due to failure to model the interaction of global and local effects. The conventional methods based on elastic theory were conservative in all cases.

INTRODUCTION

In Britain the usual approach to the assessment of concrete bridges is the same as the design approach. The moments and forces are obtained from a linear elastic analysis and the sections are then checked against the requirements of a code of practice, normally BS 5400.[1] This approach, and the codes of practice used with it, has evolved over many years. Structures which comply with its requirements are always safe and serviceable but the reverse does not apply; structures which do not comply are not always either unsafe or unserviceable. This is relatively unimportant in design where ensuring that structures comply with code requirements is straightforward and the cost of providing additional load capacity is comparatively small. However, in assessment the situation is quite different. The cost of increasing the strength of an existing bridge by even 10% may be millions of pounds and similar effort may be required to make a structure comply with code requirements which are merely arbitrary rules defining what is now considered 'good practice'.

In the extreme case of the deck slabs of beam and slab bridges, the actual strength can be as much as ten times higher than is implied by conventional assessment methods. In this paper this is demonstrated using the results of tests on two half-scale concrete bridges and possible alternative assessment methods are considered.

DETAIL OF MODELS

Two half-scale models of M-beam[2] type bridges were constructed, using standard T2-beams[2] as approximate half-scale models of M4-beams. The beams were precast in a normal casting yard, transported to the laboratory and placed on elastomeric bearings. Formwork was then constructed for the deck slab, which was cast *in situ* using a half-size concrete mix with a 10 mm maximum aggregate size. The reinforcement and prestressing for the beams were designed in the normal way using BS 5400. However, because it was primarily the behaviour of the deck slab which was being investigated, 25% more than normal prestress was provided whilst the reinforcement in the deck slab was substantially lighter than would normally be provided.

The major reason why deck slabs are stronger than conventional design methods is that they are able to work by *compressive membrane action*, which is otherwise known as *arching action* or *dome effect*. In order to realise this effect some restraint is required and previous research has suggested that diaphragms are needed to develop this restraint. Analysis using the approach which will be considered in the section 'Non-linear Analysis' suggested, however, that this was not the case; the understressed concrete and steel surrounding the critical areas would provide adequate restraint. In order to investigate this, the first model, which is illustrated in Fig. 1, was deliberately made a worst case for restraint and had no diaphragms. The second model had five beams and it also had support diaphragms as well as parapet upstands. Both models had the full-scale equivalent of a 160 mm

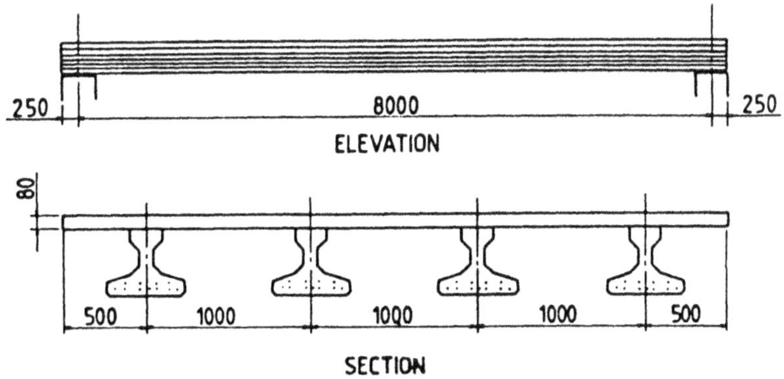

FIG. 1. Detail of first deck.

deck slab with the beams placed at 2 m centres, which is the widest spacing ever likely to be used for this type of deck.

ASSESSMENT

The strength of the bridges was assessed both for HB load and for single wheel loads. Four different assessment methods were tried and these will be considered in turn.

Conventional Approach

A conventional assessment of the first deck was performed. This was based on British design practice and used Westergaard's approach for the local analysis of the deck slab and a linear grillage for the global analysis of the bridge. The only departure from normal practice was that, in order to facilitate direct comparison with test results, measured rather than nominal material properties were used and all partial safety factors were set to one.

This procedure implied a failure load for the first deck of 14 kN per wheel when all 16 wheels of the HB rig were loaded and 21 kN when only one wheel was loaded. The difference was due to the influence of global transverse moments, the moments induced in the deck slab by its action in distributing load between the beams. These moments are not normally considered in North American practice. The normal AASHTO[3] approach to assessing deck slabs assumes a different shape patch load to that used in the tests. However, the analytical approach it is based on is very similar, apart from not considering global transverse moments. It would therefore predict a failure load of approximately 21 kN per wheel, both under single wheel loads and under the HB bogies.

The reason for the very low predicted failure load, only some 34% of the BS 5400 requirement for a 45-unit HB load, was that the reinforcement was lighter than normal. The second deck was provided with only one layer of reinforcement each way instead of two. The reinforcement was slightly heavier, however, with 8 mm high yield bars at 125 mm centres instead of 6 mm as in the first deck. Because this steel was located 10 mm below mid depth, the sagging moment capacity was greater than for the first deck. The strength assessed using Westergaard's approach and considering only sagging was therefore higher at approximately 16 kN per wheel with all 16 wheels loaded and 25 kN under a single wheel. However, this implied a hogging moment in excess of the slab's capacity and the strength assessed using Pucher's charts was similar to that quoted above for the first deck, hogging moments being critical.

Yield-Line Analysis

A normal yield-line analysis, which ignores membrane effects, implied a failure load under a single wheel of approximately 100 kN for the first deck and 85 kN for the second.

Plastic theory, upon which yield-line analysis is based, suggests that global moments and forces could redistribute away from the critical areas of slab and thus that the local strength of the slab would not be reduced when all 16 wheels of the HB vehicle were loaded. Indeed, with such a lightly reinforced deck slab, it might be argued that the global longitudinal compression induced in the deck slab by the full HB load would increase the local longitudinal moment capacity of the slab and hence increase its strength. However, this force makes the slab behave as though more heavily reinforced and thus reduces its ductility. Because of this it is debatable whether yield-line theory is a valid means of analysis for this case.

Punching Theory

Previous research into deck slab behaviour[4,5] has shown that slabs subjected to local loads fail by punching at loads which can be substantially above those predicted even by yield-line theory. The reason for this high strength is that the slabs are able to support load by compressive membrane action and the new draft assessment version of BS 5400[6] suggests that this effect should be considered in assessment.

Two approaches have been developed for predicting the failure loads, one by Kirkpatrick *et al.*[4] and the other by Hewitt and Batchelor.[5] Kirkpatrick's approach gave a failure load of approximately 150 kN for the second deck. Since Kirkpatrick recommended that diaphragms should be provided to develop the restraint, his approach is not strictly applicable to the first deck. However, ignoring this, the failure load it predicted was 185 kN. The reason for this being higher than for the second deck was that the concrete was stronger.

Although Kirkpatrick did perform some tests with two wheels loaded (and he found that the failure load per wheel could be reduced) his approach does not enable the effect of loading more than one wheel to be quantified.

Like Kirkpatrick, Hewitt and Batchelor recommended the provision of diaphragms and therefore their approach is not strictly applicable to the first deck. Their approach also requires an empirical restraint factor. Setting this to 0·6, as suggested for concrete bridges, gave failure loads which were very similar to those predicted by Kirkpatrick's approach.

Hewitt and Batchelor's approach has been used as the basis of an assessment method given in the Ontario Highway Bridge Design Code.[7] This recommends that the local strength of the slab should be assessed using charts which are based on Hewitt and Batchelor's approach using the 'conservative' restraint factor of 0·5. The global strength of the bridge is then assessed independently using a conventional linear elastic grillage analysis. This approach gave a failure load for the second deck of approximately 150 kN per wheel, the implication being that failure would take the form of one beam failing in flexure. For the first deck the prediction would have been approximately 140 kN per wheel but again, because of the lack of diaphragms, the use of the approach would not have been recommended.

For the first deck, the best interpretation of the Ontario code appears to be that yield-line methods should be used to assess the local strength of the slab.

Non-linear Analysis

Both models were analysed using a non-linear program which has been developed by the authors. This program uses comparatively simple line elements but, because of novel features of the elements and because it considers all six degrees of freedom at each node, it is still able to model in-plane forces reasonably realistically. This enables it to model membrane action. The program was developed from non-linear finite element programs which have proved capable of analysing deck slabs allowing for membrane action.[8]

Because the research project was originally aimed at developing design methods rather than assessment methods, and because serviceability is critical in design, the computer models were designed to give the best predictions for the serviceability behaviour. Two computer models were used for the first deck, one with a fine element mesh and the other with a coarse element mesh. The fine mesh analysis was used primarily for serviceability analysis although it was taken up to failure under a single wheel load. The coarse mesh was used for the failure analysis under all 16 wheels of the HB load.

The analysis predicted that at service load levels the slab would be most heavily stressed under the wheel nearest the centre of the deck. However, it predicted that failure would take the form of a local brittle bending compression failure under the wheel farthest from the centre of the deck. The predicted failure load was approximately 100 kN per wheel. Analysis using the finer mesh predicted that the failure load under a single wheel applied near to midspan would be approximately 220 kN.

Only one computer model was used for the analysis of the second deck and this was intermediate in detail between the two models used for the first deck. As for the first deck, it predicted that failure would take the form of a brittle bending compression failure under one wheel. Unlike for the first deck, the critical wheel was one of those nearest midspan of the deck. Concrete crushing first occurred on the soffit over the web of a beam rather than on top of the slab immediately under the wheel, as in the first deck. The predicted failure load was marginally higher than for the first deck.

Experience has shown that the program tends to be conservative when a realistic element mesh is used, as in the analysis of the second deck. Indeed it had been intended that the program should tend to err in the safe direction. However, the use of an over-coarse mesh, as in the analysis of the first deck, makes it less conservative.

TESTS

First Deck

The deck was first loaded with the design service HB load applied in a critical position; 120 cycles of this load, 5000 cycles of a lower load and three cycles of a 20% greater load were then applied in this and several other positions. The loading rig was then returned to its first position and the load reapplied. These tests were intended primarily to investigate the service load behaviour. However, they also served to ensure that any cracking which would be likely to occur in the real bridge would occur in the model.

On completion of these tests the bridge was still in good serviceable condition, despite having been loaded to over three times the 'ultimate strength' of the slab as given by conventional linear analysis. The load was then increased until failure occurred. At a load of approximately 103 kN per wheel the wheel farthest from the centre of the bridge punched through the deck. This failure looked like a classic 'punching shear' failure. However, a close study of the slab immediately before the final load increment had been applied had revealed a line of crushing concrete extending from the wheel which later failed towards the adjacent wheel. It appeared that this crushing was the cause of the failure.

As failure had approached, extensive flexural and shear cracks developed in the two most heavily loaded beams and the maximum beam deflection was approximately 60 mm. It did not appear, however, that beam failure was imminent. In a subsequent test of a single beam with the appropriate width of in-situ top flange, failure occurred from a deflection of 110 mm. The load was then over 20% greater than it had been at a deflection of 60 mm.

Since the failure had taken such a local form the bridge was still in good enough condition to enable two single wheel tests to be performed. In both cases failure took the form of the wheel punching through the deck with little warning. The failure loads were 204 and 226 kN, approximately twice the failure loads per wheel as when all 16 wheels had been loaded.

Second Deck

The second deck was subjected to a similar load history to the first. On completion of the service load tests the slab was more extensively cracked than the first had been. In particular, there were cracks on the top of the slab which failed to close fully when the bridge was unloaded. No cracks had appeared in the top of the slab of the first deck until the final loading to failure was well advanced.

As with the first deck, the eventual failure took the form of one wheel, this time a wheel near to midspan, punching through the deck. Again as with the first deck, concrete crushing had been visible a load stage before the failure occurred, this time on the soffit along the edge of a beam. The failure load was somewhat greater than for the first deck, at approximately 122 kN per wheel.

The sudden failure under a wheel reduced the total load on the bridge and hence reduced the deflection of the beams. There was insufficient time for the hydraulic pressure in the four jacks of the loading rig to equalise before three further wheels, one under each jack, punched through the deck. Despite this there was enough slab left in reasonably good condition to enable some local tests to be performed. These included two single wheel tests which were directly comparable with those which had been performed on the first deck. Punching failures occurred at loads of 176 and 184 kN.

DISCUSSION

As expected, the tests showed that conventional assessment methods using linear elastic theory are very conservative for this type of structure. The margin of conservatism is so great that it would be unreasonable to condemn a deck slab as unsafe on the basis of such an assessment. They should only be used to class a deck slab as safe. In the authors' view it is reasonable, even then, to smooth out the peaks of the elastic moment distribution over a finite width of slab. That is, to base the section check on the average moment intensity over a width equal to, say, the lesser of six times the effective depth of the slab or half the slab span.[9]

The yield-line analysis of the slab of the first deck gave an answer which was very close to the actual failure load when all 16 wheels of the HB vehicle were loaded. However, the failure mode was so different from that implied by yield-line analysis that this can be little more than coincidence; yield line theory depends on ductility whilst the failure mode was very brittle. The predictions of yield-line theory were conservative for all the other tests. However, this may not always be the case; there are indications (including a non-linear analysis of the first deck with 40% of the tendons removed)[9] that if the beams had been weaker the bridges would have failed in the same mode but at a lower load.

Kirkpatrick's approach gave good predictions for the failure loads under single wheel loads. The predictions were slightly conservative with a mean ratio of actual failure load to predicted failure load of approximately 1·2. This is very similar to that which he observed for his own tests. The ratios were similar for the two decks, indicating that, contrary to Kirkpatrick's suggestion, his method is applicable in bridges without diaphragms. However, it appears theoretically that this may not be the case if the wheel load is applied very near to the end of the bridge, that is near to the unsupported edge of the slab. The test performed with the wheel only 1·1 m from the end of the deck gave a 10% lower failure load than the test performed near the centre of the deck.

In both bridges, when all 16 wheels of the HB vehicle were loaded, failure of the deck slab occurred at a significantly lower load than predicted by the punching theories considered in the section titled 'Punching Theory'. The implication of the Ontario code that the high local strength of the slab would cause a global failure, that is a failure of a beam, to precede slab failure proved incorrect. What appeared to happen was that the global transverse moments induced by the differential beam deflections reduced the local strength of the slab.

The non-linear analysis gave good predictions for the failure load under full HB load. Also, unlike the other methods, it predicted the failure mode very well; it predicted precisely where concrete crushing would first occur and which wheel would punch through the slab first. Although the failure mode is relatively unimportant to someone assessing a bridge, this is significant; prediction methods which predict the correct failure load but the wrong mode must be considered highly suspect.

The non-linear analysis also gave a good prediction for the result of the one single wheel test for which it was investigated fully. However, analysis of other tests[9] suggests that the methods considered in 'Punching Theory' tend to be slightly better for such cases and, since they are also far simpler to obtain, they may be preferred. Punching theory could also be used for assessment under full HB load in cases where there are no significant global transverse moments, for example in a bridge with intermediate diaphragms. However, even then there are indications[4,9] that the interaction between adjacent wheels could reduce the strength and some allowance should be made for this.

CONCLUSIONS

Conventional analysis is very conservative for this type of structure. It should be used only to class a bridge as safe, not unsafe. Even then it is reasonable to make the assessment less conservative by evening out the peaks in the predicted moments.

Both punching theory and non-linear analysis gave acceptable predictions for the strength under single wheel loads whilst yield-line analysis proved to be conservative by a factor of up to two.

Only non-linear analysis gave good predictions for the behaviour under full HB load. Predictions based on assessing the global and local strength of the slab independently, as is recommended by the Ontario code, could be unsafe.

ACKNOWLEDGEMENTS

The test work considered in this paper was undertaken at the British Cement Association whilst P.A.J. was an employee of that organisation. He would like to thank the association for supporting the project and for allowing him to publish the paper. He would also like to thank the many staff of the association who assisted with the test work.

REFERENCES

1. BRITISH STANDARDS INSTITUTION. *Code of practice for the design of concrete bridges*. BS 5400, Part 4, BSI, London, 1984.
2. PRESTRESSED CONCRETE ASSOCIATION. *Prestressed concrete bridge beams*. Prestressed Concrete Association, Leicester, 2nd Edition, January 1984.
3. AMERICAN ASSOCIATION OF STATE HIGHWAY AND TRANSPORTATION OFFICIALS. *Standard specification for highway bridges*. Washington, DC, 13th Edition, 1983.
4. KIRKPATRICK, J., RANKIN, G. I. B. and LONG, A. E., Strength evaluation of M-beam bridge deck slabs. *The Structural Engineer*, **62B**(4) (1984) 86–8.
5. HEWITT, B. E. and BATCHELOR, B. DE V., Punching shear strength of restrained slabs. *Journal of the Structural Division, American Society of Civil Engineers*, **101**(ST9) (1975) 1837–52.

6. DEPARTMENT OF TRANSPORT, Draft BD/88. *The assessment of concrete highway bridges and structures.* HMSO, London, 1988.
7. ONTARIO MINISTRY OF TRANSPORTATION AND COMMUNICATIONS. *Ontario Highway Bridge Design Code.* Downsview, Ontario, Canada, 1983.
8. COPE, R. J. and EDWARDS, K. R., Non-linear finite element analysis of eccentrically stiffened bridge decks. *Proceedings of international conference on finite element analysis in computational mechanics.* Pergamon Press, Oxford, 1985, pp. 431–48.
9. JACKSON, P. A., *Compressive membrane action in bridge deck slabs.* PhD thesis, Polytechnic South West/CNAA/BCA, April 1989.

39

Assessment of Stresses in Post-Tensioned Concrete Bridges

C. L. BROOKES, S. H. BUCHNER and S. MEHRKAR-ASL
Gifford and Partners, Southampton, UK

ABSTRACT

This paper presents the results of trials carried out on a post-tensioned concrete six-cell box structure to determine the in-situ stress levels. The techniques used were based on the stress-relief which occurs during the removal of concrete cores and the re-establishment of the stress fields using a novel jacking technique. The estimated stresses were compared with those predicted using a finite element analysis and a good correlation was obtained. The measured stresses generally ranged between 6 and 9 N/mm² although local variations were observed. The elastic constants for the concrete were assessed using the results of the jacking tests and laboratory compression tests.

INTRODUCTION

The long-term condition of a number of prestressed concrete structures has been the subject of several research projects in the United Kingdom in recent years.[1–3] These projects have been aimed at determining the in-situ stresses in the structures and the residual levels of prestress. Much of the work has been carried out during the demolition of the structures, which provides an ideal testing ground and a unique loading regime.

The research has involved the development of methods for determining the in-situ state of stress in concrete structures based upon measurements of released strains obtained during the removal of standard cores. In addition, a unique jacking system has been evolved which can be used to determine the in-plane elastic properties of the concrete and provides a direct measure of the existing stresses. Although these techniques have been used for residual stress determination in steel structures and rock mechanics, they have not been previously applied with any great success to concrete structures.

In the current study these techniques have been primarily used for the investigation of structures suffering distress and in conjunction with computer assessments related to principal inspections. The site data has been supported by extensive laboratory calibration trials.[4,5]

IN-SITU STRESS ASSESSMENT

Measurements of the elastic response of a stressed medium to the removal of a core can be used to provide an estimate of the existing stresses. Once the core is removed a specialised jacking system is inserted within the remaining hole. The jacking tests use the application of load to the periphery of the remaining hole to assess the strain response of the material and hence the in-plane elastic constants for the concrete. In addition, re-establishing the strains released on the surface and the use of superposition for the different loading directions provides a direct measure of the existing stresses.

The selection of core size and gauge length is generally controlled by the size of the largest aggregate particle in the material under test, although other physical limitations may exist. For the laboratory and site tests, cores of 150 and 75 mm diameter have been used in conjunction with 140 and

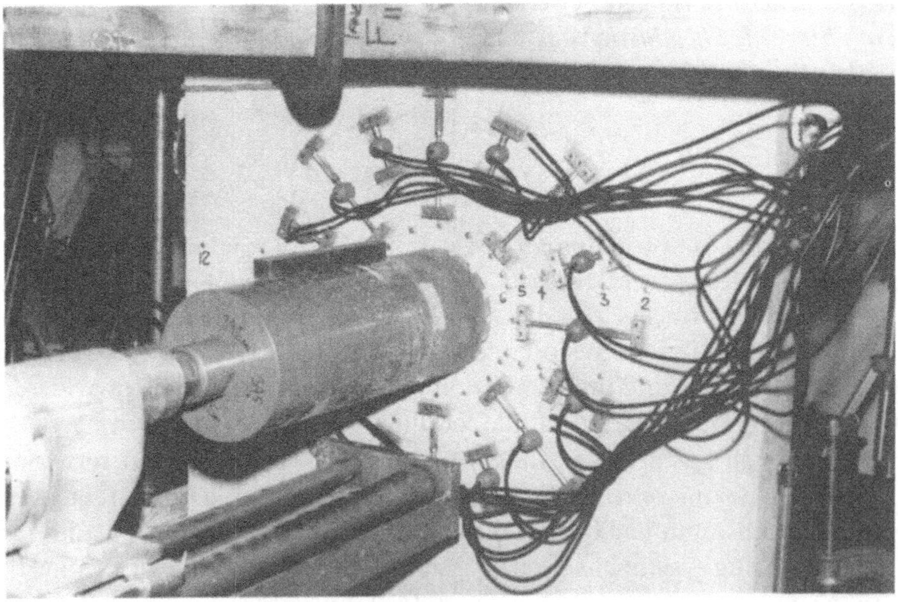

FIG. 1. Typical gauging arrangement.

64 mm long vibrating wire (VW) gauges, and 50, 100 and 200 mm long demec gauges. A typical gauging and coring arrangement for a 150 mm core position is shown in Fig. 1.

Analysis of the strains released after coring utilises the equations developed by Muskhelishvili[6] for the displacements around a hole in a plate under a known stress field. In order to analyse the results, the concrete is assumed to be an elastic homogeneous isotropic material under plane stress or plane strain loading conditions. This assumption cannot be considered as wholly valid for concrete, and as a result anomalies arise due to the presence of microcracks and the effects of heterogeneity and anisotropy caused by the aggregate particles. However, the laboratory test results have indicated that the methods can be used with a good degree of confidence, especially when the member is subject to an axial loading regime.

IN-SITU TESTS

An assessment of a 144 m long prestressed concrete bridge compared the stresses predicted using the in-situ stress-relief technique with those obtained from an in-depth computer study. The viaduct was constructed with three continuous in-situ spans of 43, 58 and 43 m set on a sagging vertical curve and supported on inclined abutment legs. The section of the bridge used in the computer model is shown in Fig. 2.

The main prestressing cables were applied externally within the box sections, passing beneath a number of diaphragms in the midspan section, thereby inducing longitudinal compressive stresses in the bottom flange of the deck. There are 12 cables, one on each face of the internal webs and one on the inner face of the edge webs. Each cable consists of 17/28 mm 19-wire strands and are Gifford–Burrow (CCL spiral strand) type. The strands were simultaneously tensioned from each end in two stages before being covered with concrete. In addition to the main cables, there are 168 PSC Freyssi

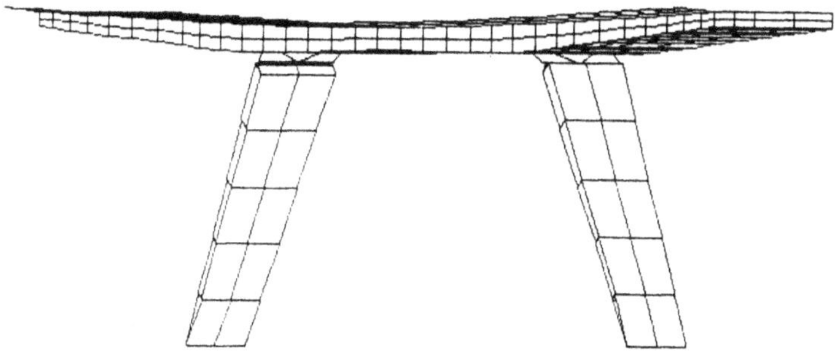

FIG. 2. Computer model showing section through bridge.

strand type cap cables, 84 over each leg, each cable consisting of 12/12 mm seven-wire strands. Two profiles were used for these strands, each stressed from either end alternately. A general loss of prestress of 16% was allowed for in the analysis. This low figure took into account the advanced age of the concrete at the time of stressing.

During the 25 years since the bridge was constructed water had seeped through manholes in the top slab and caused some corrosion of the external prestressing tendons. It was anticipated that any significant loss of prestress would be detected by use of the stress-relief technique.

Due to restrictions imposed by the depth of slab and the spacing of the main reinforcement, 75 mm diameter cores were removed. Care was taken to ensure that the stresses released were unaffected by the presence of the central diaphragm and so the coring positions were sited on both the north (N) and south (S) sides some 2–3 m away. Arrays of demec gauges were placed on and around the core positions, with additional rosette arrays of 64 mm long VW strain gauges placed around each position. As coring progressed strain release readings were taken on the VW arrays. Once coring was completed and a number of strain release readings had been taken, the core holes were used for jacking tests to determine the in-situ elastic properties and to obtain a second measure of the in-situ stresses.

COMPUTER ANALYSIS

The structure, excluding the abutments and foundations, was modelled elastically using the finite element method. The structure and permanent loading were symmetrically partitioned transversely to give a problem size of approximately 5000 degrees of freedom. First-order thin shell elements were used to represent all features of the deck and piers. Precast concrete hinges attach the legs to the deck and were incorporated in the model by coupling degrees of freedom to give similar articulation. The main and cap cables were included in their actual positions using 3D truss/spar elements attached to the shell mesh using rigid elements (multipoint constraint equations). Using a combination of rigid and truss elements to model the curved cable profile allowed the near uniform web mesh to be uninterrupted, thus maintaining good shell element aspect ratios.

Body force and pressure loading were used to apply dead and superimposed dead loads respectively, the pressure intensity being derived from a survey of the thicknesses of the finishes. Prestress loads were simulated directly by stressing the modelled cables. This was achieved by calculating the cable force profile immediately after locking off at the anchorages and estimating an equivalent axial temperature loading for each truss element to give the required force profile along the length of the cables. Once the temperature load vector had been applied to the model, the resulting cable forces were modified to account for secondary flexural effects. Second and subsequent sets of equivalent temperature loads are derived by introducing adjustment factors, resolving and checking cable forces. Generally cable forces converged within 0·5% of those required after four iterations. This procedure was automated with the aid of a spreadsheet computer program acting as a pre- and post-processor to the analysis.

RESULTS

During coring through the bottom slab it became evident that a reasonably high level of longitudinal compressive stress existed within this structure. The resulting stresses determined from the strains released during the incremental drilling test at a depth of 125 mm and those obtained from all the gauges after coring are shown in Table 1. These results indicated that the values of longitudinal stress predicted after only 125 mm of penetration were in good agreement with those obtained from all the gauges after coring had finished.

Once coring was completed, at a depth of approximately 175 mm, the major principal stresses on the north side of the central diaphragm were estimated to range between 6 and 9 N/mm^2. The results from the south side were generally higher than the north side. The estimated stresses ranged

TABLE 1
Estimated longitudinal stresses in viaduct

Core number	Estimated stresses (N/mm^2)				
	VW increment	Demec gauges		VW array	Jacking test
		Core	Hole		
1N	7·2	9·4	7·4	7·1	8·0
2N	7·1	7·5	5·1	6·3	5·4
2S	4·6	9·3	7·0	6·7	3·3
3N	4·5	—	6·6	7·0	—
3S	10·5	11·3	7·5	12·1	8·2
4N	5·3	8·8	7·5	8·7	—
4S	12·7	13·0	11·4	12·8	—
5N	5·6	9·0	4·9	5·5	3·8
5S	6·4	9·4	—	11·6	6·9
6N	7·6	—	8·5	11·5	8·3

TABLE 2
Comparison of jacking technique and standard compression tests for determining elastic modulus

Box number	In-situ jacking test (E_j)	Laboratory compression test (E_c)	Comparison (E_c/E_j)
1N	31·0	36·4	1·17
2N	35·7	34·4	0·96
2S	—	35·5	—
3N	—	—	—
3S	33·0	37·3	1·13
4N	—	—	—
4S	—	38·0	—
5N	35·6	37·0	1·04
5S	41·4	38·0	0·92
6N	33·0	35·3	1·07
Average	34·95	36·4	1·04

between 8 and 11 N/mm^2, and rose to a peak of approximately 13 N/mm^2 in box 4S. Peak values were seen to occur in the outer boxes of each carriageway with lower values in boxes 2 and 5. The principal stresses were generally estimated to be parallel to the longitudinal direction.

These results were obtained using the estimates for in-situ Young's modulus determined from the jacking tests. The results from these tests and subsequent laboratory compression tests are shown in Table 2. The average value of Young's modulus obtained from the jacking tests was within 4% of that obtained from the compression tests with maximum deviations of −8% and 17%. The results of the jacking test were also used to estimate the existing levels of stress in a number of the boxes. Table 2 shows that the estimates obtained using this technique were similar to those obtained from the other gauging arrangements and provided a further method of assessment.

The results of the computer analysis, based on a long-term reduction in prestress of 16%, are illustrated in Fig. 3 and compared with the average stresses determined from the in-situ tests. Using the reduced level of prestress, the predicted distribution of longitudinal stresses across the deck ranged between 7·2 and 7·9 N/mm^2. It was noticeable that the estimated stresses were in keeping with these predictions although they were slightly higher in places. A sensitivity study was carried out assuming constant loss of prestress across the section. The study indicated that in-situ stresses of 6 and 4 N/mm^2 would result from losses of prestress of 33% and 49% respectively. Therefore it appeared that there had been a localised loss of

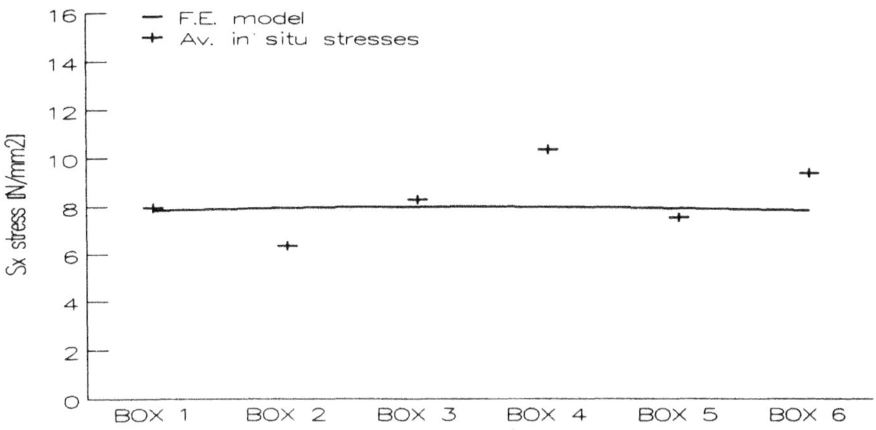

FIG. 3. Comparison of stresses obtained from the computer analysis and *in situ*.

stress of approximately 33% in the area of box 2, resulting in a redistribution of stresses into the adjacent cells.

There were a number of possible reasons why the concrete stresses determined from the analysis differed from those measured. Casting the deck *in situ* meant that there was a variation in the age of the concrete, both longitudinally and transversely across the deck, at the time of stressing. This would cause a variation in the short-term elastic behaviour of the concrete and in the long-term creep and shrinkage. In addition, the cable stressing sequence was such that elastic losses could accumulate on one side of the deck, causing lower stresses on that side. Other environmental factors could have contributed as the wind and rain direction tended to be from the west. Despite the apparent localised reduction in the levels of loss of prestress, the structure was assessed as being capable of carrying both HA loading and HA with 45 units of HB in accordance with the requirements of BS 5400 and BD 23/84.

CONCLUSIONS

The coring stress-relief technique provides a relatively quick and simple method of determining the in-situ stresses in prestressed concrete bridges. In conjunction with the jacking technique, this method can be used to determine the stresses in a deteriorating concrete structure or where the complexity of the problem requires the computer model to be verified. In addition, the jacking test provides a quick and reliable method of obtaining a measure of the in-situ modulus for the concrete.

The bridge structure tested was subject to high stresses due to the nature of the construction. The sagging vertical curvature of the deck and the high level of prestress resulted in estimates for the existing midspan stresses of 6–9 N/mm^2, with peak values of 11–13 N/mm^2 on the south side. These values were generally in keeping with the results obtained from a finite element analysis. However, localised anomalies were discovered which suggested that differential loss of stress had occurred between the boxes. Despite these localised effects, the structure was assessed to be able to carry full HA and HB type loadings as required by BS 5400. The use of the coring stress-relief technique allowed the assessment of the structure to be carried out on the basis of known facts rather than relying solely on a computer analysis. Thus the future management strategy for the maintenance of the bridge could be determined.

ACKNOWLEDGEMENT

The authors of this paper would like to thank the Science and Engineering Research Council for their support over a 5-year period during the development of the stress-relief technique.

REFERENCES

1. Buchner, S. H. and Lindsell, P., Demolition monitoring of the Taf Fawr Bridge. Final report, Gifford and Partners, Southampton, October 1986.
2. Buchner, S. H. and Lindsell, P., Testing of prestressed concrete structures during demolition. I.Struct.E./BRE Seminar on Structural Assessment—Based on Full and Large Scale Testing, Watford, UK, April 1987.
3. Lindsell, P. and Buchner, S. H., Prestressed concrete beams: controlled demolition and prestress loss assessment. CIRIA Technical Note 129, London, 1987.
4. Mehrkar-Asl, S., Direct measurement of stresses in concrete structures. PhD thesis, University of Surrey, September 1988.
5. Buchner, S. H., Full-scale testing of prestressed concrete structures. PhD thesis, University of Surrey, July 1989.
6. Muskhelishvili, N. I., *Some Basic Problems of the Mathematical Theory of Elasticity*. P. Noordhoff Ltd, Gröningen, Netherlands, 1963, pp. 202–10.

40

Assessment of Prestressed Bridge Beams

DAVID CULLINGTON

Bridges Division, Transport and Road Research Laboratory, Crowthorne, Berkshire RG11 6AU, UK

ABSTRACT

A number of pretensioned beams have been recovered from an M63 underbridge during demolition. They have no link reinforcement in the web. For this reason, if assessed using BS 5400: Part 4, they would not comply in shear. This paper contains a description of shear failure tests carried out on the 16 m long beams. The results indicate that the beams possess reserves of strength because of conservative assumptions in the method of calculation and the high strength of the concrete. Whereas it may be prudent to provide a minimum amount of shear reinforcement in the form of links for new design, there is a strong case for dispensing with this requirement in some assessments.

INTRODUCTION

Four motorway underbridges have been replaced at the Peel Green roundabout, to the west of Manchester, as part of the M63 widening scheme. Consideration was given to retaining the existing structures and widening them, but the pretensioned beams forming the decks contained no shear reinforcement, which is required by BS 5400: Part 4: 1984.[1] It was decided to replace the bridge decks primarily for this reason.

The consultant responsible for the scheme gave some thought to retaining the beams in spite of the absence of shear reinforcement. Calculations showed that the shear forces under factored HB loading were, at maximum,

Any views expressed in this paper are not necessarily those of the Department of Transport.

about 70% of the factored shear capacity as found using BS 5400: Part 4. Had the proportion been lower (perhaps 50%) the bridges might have been retained simply by means of an agreed departure from standard.

It was decided to test the beams when the decks were demolished for the following reasons:

—They were 30 years old and likely targets for assessment.
—They would contain the effects, if any, of time and heavy traffic.
—They might exhibit a shape or composite action effect which could cause them to depart from code predictions.
—Their strength, the load at onset of visible distress and the manner of failure would be of interest for future assessments.

The Shear Strength of Prestressed Beams

The behaviour of prestressed concrete beams in shear is complex but approximate methods are available for finding their strength for design and assessment purposes. BS 5400 requires two types of shear to be considered. These are indicated by the symbols V_{co} and V_{cr}.

The first of these, V_{co}, generally applies to short shear spans. It relates to a type of shear failure that initiates when the principal tensile stress in the web of a beam exceeds the ultimate tensile strength of the concrete. Its calculation is based on a number of simplifying assumptions. For instance, an I-beam is taken to be rectangular and the maximum principal tensile stress is assumed to occur at the centroid.

The second, V_{cr}, generally applies to larger shear spans and involves the combined action of bending and shear. It has been reported that the shear strength of a section of beam is reduced to below V_{co} when bending cracks are present. An empirically-based method of calculating the reduced strength appears in BS 5400, the basic data for which were obtained by tests on 190 prestressed beams.[2]

Failure in the V_{co} mode is expected to follow the appearance of inclined cracks in the web and to occur when a crack path is formed between the load and a point at or near the support. Failure in the V_{cr} mode is expected to follow the extension of a flexural crack in the bottom flange into an inclined crack in the web, and to occur when a path is formed between the load point and the bottom flange.[3]

Assessment Using BS 5400: Part 4

BS 5400: Part 4: 1984 is a design code, but at present it has to be used for assessment calculations as well. One important difference between design and assessment concerns the provision, or avoidance, of specific details. A detail being assessed may not be desirable according to current design practice, but the strength of the component may nevertheless be sufficient to resist the imposed loads.

The provision of web reinforcement in prestressed beams is a case in point. The current version of BS 5400: Part 4 requires web reinforcement to be provided in all cases. The version of BS 5400: Part 4 issued in 1978 was less rigid in its requirements. It permitted the use of beams with no reinforcement when the applied shear was less than 50% of the resistance and where tests had shown that reinforcement was not required. It would be helpful to relax the current design requirements for use in assessment provided that reliability did not suffer.

SPECIMENS AND TESTING METHOD

The Bridges

The M63 crosses the roundabout carriageway twice and there are thus four bridges on the site. Each bridge consists of three simply supported spans of 8, 16 and 8 m. Twenty-two beams form each span, placed side by side and transversely post-stressed through in-situ diaphragms. The top slab, cast on permanent formwork, is largely unreinforced. A 'triangular saw-tooth' shaped reinforcing bar provides interfacial shear connection.

The Beams

The beams (see Fig. 1) have 11 holes along their length for transverse prestressing. Apart from a small amount of reinforcement in the end blocks, the webs are free from shear reinforcement. For the tests, the transverse wires and diaphragms were removed, and the transverse holes refilled with

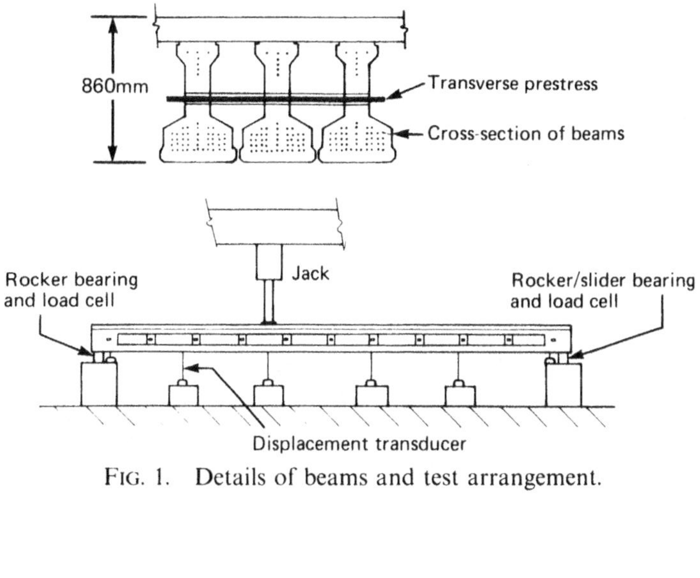

FIG. 1. Details of beams and test arrangement.

flowing mortar. Load positions were selected for the shear tests so that the principal bending cracks avoided the hole positions in areas of high shear.

Obtaining the Beams for Testing

The method of demolition was to break out the beams from the bridge in units of three joined together by the top slab. To recover the middle beam intact for testing required the slab to be cut through mid-way between the beams and the outer beams to be prised off.

In practice considerable difficulty was experienced. Cracking took place in the webs of the middle beams, in the vicinity of the transverse prestress, as the outer beams were removed. This was thought to be a result of the combined action of the release of prestress and the forces imposed by the separation process. Separation was finally achieved successfully by coring out the prestressing wires prior to prising the beams apart. Four beams suitable for testing were recovered.

Test Configurations

The configurations for the first four shear tests are summarised below:

Beam No. 1 end 1: point load at 2·9 m from support; beam supported over full span (15·7 m).
Beam No. 1 end 2: point load at 4·1 m from support; beam supported over span of 12·3 m, beam full length.
Beam No. 2 end 1: point load at 5·7 m from support; beam supported over full span (15·7 m).
Beam No. 2 end 2: point load at 1·6 m from support; beam supported over span of 8·1 m, beam 11 m long.

The beams were 0·86 m deep, and consequently the shear span to depth ratios for the tests were 1·9, 3·3, 4·8 and 6·6.

Method of Testing

The test arrangement is shown in Fig. 1. Loading was provided by a hydraulic jack of 1000 kN capacity operated through a servo-control panel. All tests were carried out under displacement control. There are a number of advantages in this: measured strains and displacements form a consistent set, failure is more controlled, and it is safer for observers who approach the beam to plot crack positions.

Instrumentation and Methods of Observation

The instrumentation consisted of load cells under the jack and at each support, transducers registering vertical displacement of the bottom flange on the beam centreline, and demec studs to determine bending and shear strains. Loads and displacements were recorded on a data logger. In addition, an *X–Y* plotter was connected to the jack load cell and a displacement transducer beneath the load point to give an independent indication of load–displacement behaviour.

Acoustic emission microphones were attached to the web of the beam in two of the tests: to collect basic data and give an indication of prolonged cracking after the application of each increment was completed—a sign of impending failure. Cracks were marked on the beam at various increments. Video cameras were used to record the failures.

RESULTS

Description of Failures

The crack patterns at failure are shown in Fig. 2. In all cases, in their final state, the beams contain two systems of cracking. Bending cracks are present

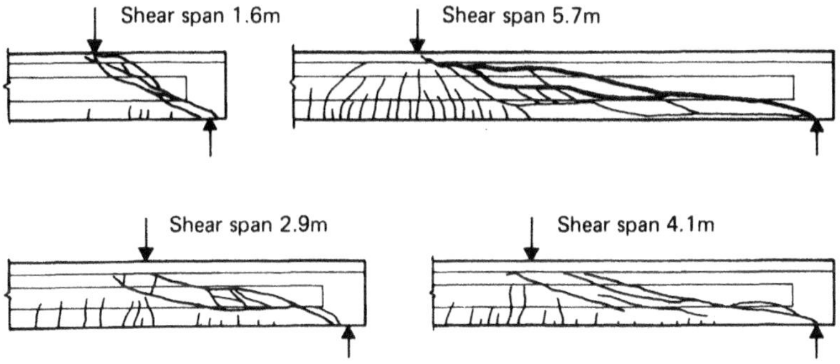

FIG. 2. Beam cracks at failure.

beneath the applied load, emanating from the bottom flange, and a number of pronounced cracks are present in the web extending between the load point and the near support. The two systems of cracking are independent and did not combine. It was the web cracks that caused failure. This is consistent with behaviour in the V_{co} mode. As expected the bending cracks are more developed for the larger shear spans.

Shear spans 2·9 and 4·1 m

The progression to failure was uncomplicated. Bending cracks appeared first and extended as displacement was increased. After further increments, web cracks formed suddenly as maximum load was reached and developed into their final shape under increasing displacement and falling load. It can be seen from Fig. 3 for the 2·9 m test that there was a considerable residual

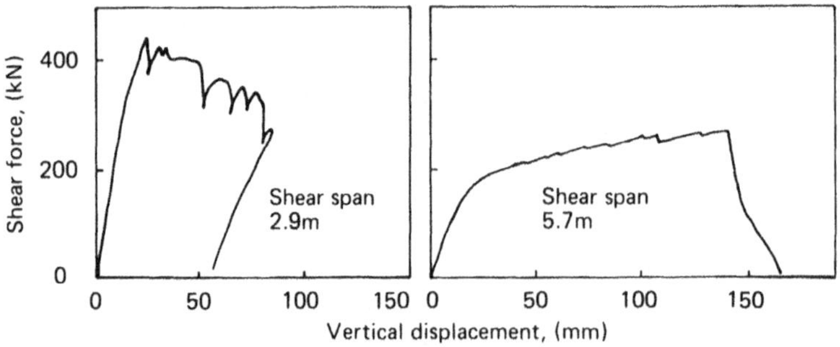

FIG. 3. Graphs of shear force against displacement.

load capacity after the point of maximum load had been passed. Figure 2 shows the cracking after the retraction of the jack. Most of the bending cracks had closed up and the web cracks had closed up partially.

Shear span 5·7 m

Cracking due to bending was more pronounced and, although a minor shear crack formed in the web, bending failure appeared to be a possibility. Before this could happen, however, new severe cracks appeared in the web at a shallow angle to the horizontal. A further small increase in load and a moderate increase in displacement took place as testing proceeded and the cracks widened. At this point the load–displacement curve was almost flat, as can be seen in Fig. 3. Finally, the top flange buckled upwards and split the beam from the load point to the support.

Figure 2 shows the position of the cracks in test 3 as they would have appeared if the split had closed up. Final failure was not in bending because there was no significant crushing of the top flange nor yielding in the prestressing wires. The bending cracks closed partially when the jack was retracted.

Failure was not considered to be of the V_{cr} type because the failure planes extended directly to the support and did not follow a flexural crack to the soffit of the beam. The shear crack nearest to the load point initiated in the web and its role in the final collapse was secondary to that of the two main failure planes indicated in Fig. 2.

Shear span 1·6 m

Failure was initiated by web cracking, as was the case for the other tests. The main difference was that the failure load was much greater than the web cracking load. Maximum load coincided with the development of severe web cracks. Subsequently the beam was able to sustain a modest increase in displacement without significant loss of resistance. Minor flexural cracking occurred before failure.

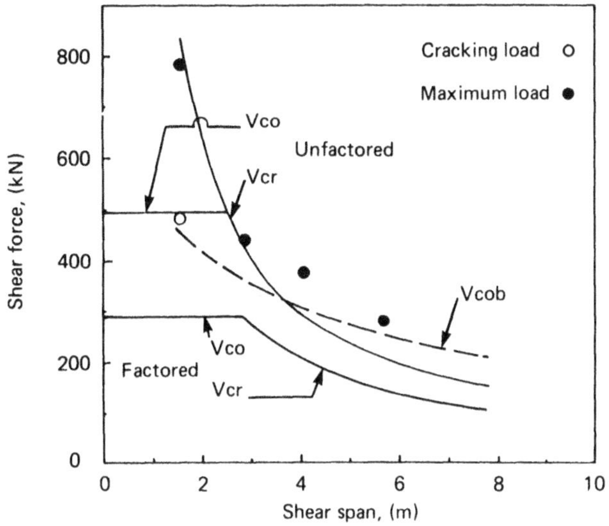

FIG. 4. Graph of shear resistance against shear span.

Measured and Calculated Shear Strengths

Figure 4 contains a comparison between the measured shear forces at failure and the shear resistance calculated by a number of different methods. In calculating the shear resistance it has been assumed that the composite section resists the total load, i.e. the dead load and the applied load. Values have been obtained in accordance with BS 5400 with partial factors included and assumed concrete characteristic strengths of 45 N/mm^2 for the slab and 60 N/mm^2 for the beams. These curves are termed 'factored' in Fig. 4. The strength values were deduced from the original design calculations with an allowance for ageing. Assumptions have also been made about the level of prestress. Further calculations have produced the unfactored curves. These use mean concrete strengths of 80 N/mm^2 for the slab and 90 N/mm^2 for the beams obtained from core tests. Partial factors for concrete strength and prestressing force, and some other conservative assumptions implicit in BS 5400, have been removed. The latter comprise allowances for shrinkage, for the effective depth adjustment in V_{cr} and for the use of a rectangular section in V_{co} (Ref. 3).

DISCUSSION

Performance of the Beams

The measured values of shear resistance lie between 1·6 and 2·7 times the factored calculated resistance (see Fig. 4). At critical points in the bridge deck the measured resistance is at least 2·7 times the required ultimate limit state shear force. On this basis the in-situ resistance after the application of partial factors exceeds twice the applied forces. The beams complied with BS 5400: Part 4: 1978 for use in the Peel Green bridges.

It was encouraging that the resistance of the beams did not fall to zero on the appearance of web cracks. Beyond the point of cracking the beams were able to sustain a considerable load as displacement was increased. This ability was least apparent at shear span of 1·6 m, but was compensated for by the high strength in this test. In all cases the appearance and behaviour of the beams after the formation of shear cracking suggested that there might have been arching between the load and the support balanced by tension in the wires. This is described by Neilsen[4] in his work on limit methods of analysis. It could have been the source of resistance in the later stages of the tests.

This behaviour confers a measure of ductility to the beams. Ductile failure is generally preferred to sudden brittle failure in structures. If there is an unexpected structural inadequacy or overload, displacements become large, cracking occurs and warning of a problem is given. It also enables redistribution of loads to take place. The presence of web reinforcement in a prestressed beam is known to enable higher loads to be carried beyond the onset of web cracking. The absence of web reinforcement in the beams tested may have reduced ductility but the behaviour gave no cause for concern in this respect. There is no evidence in the results to suggest that the absence of reinforcement led to premature failure.

Comparisons with Theory

According to BS 5400 the resistance of the beams to web failure, V_{co}, should be independent of shear span. This is because the prestress and the shear reinforcement (none) are constant along the length and the concrete section is practically constant. Consequently, if the beams had failed in a simple V_{co} mode, the shear forces at failure should have been roughly the same in all cases. Some allowance must be made for scatter in the results and the fact that only four tests are reported. However, Fig. 4 clearly indicates that the shear force at failure reduces with increase in shear span.

The V_{cr} calculation does not correctly identify the failure mode for these beams as this is supposed to depend on flexural cracks developing into shear failure. In practice the V_{cr} calculation serves a purpose because it correctly indicates a lower shear resistance at higher shear spans.

With allowance for experimental scatter the measured strengths follow the V_{cr} curve tolerably throughout the range of shear spans, including the region where V_{co} mode is expected to be critical. It is not known to what extent web modes of failure of the kind reported here were present in Hawkins data.[2] For magnitude alone the data fit moderately well (see Fig. 4). The arching/limit analysis method also leads to the prediction of a progressive reduction in shear resistance as shear span increases. In this case, however, there is no assumption about the interaction of flexural and shear cracking systems as implied by the V_{cr} calculation.

An alternative interpretation for the observed behaviour is offered in Fig. 4. The curve labelled V_{cob} indicates the shear resistance calculated on the basis that failure occurs when the principal tensile stress exceeds the tensile strength of concrete at the bottom of the web rather than at the centroid of the section. This possibility is discounted in the calculation for V_{co} given in BS 5400: Part 4. It is not particularly significant in practice because V_{cr} as calculated according to BS 5400 provides a lower, more conservative, value of resistance.

As the shear span increases, flexural stresses increase. The highest flexural stresses occur below the load and so it is at this point that the curve for V_{cob} has been calculated. This provides a simple basis for the calculations, but in practice cracking is likely to initiate closer to the support. Figure 4 also indicates the load at which a major crack formed in the web for the test at 1·6 m shear span. Cracking is evidence of a tensile stress exceeding the tensile strength of concrete and is the criterion for web shear failure of a beam. The V_{cob} curve lies below this and the other experimental points but follows the trend observed in the tests.

Performance of Bridges

To date the tests have shown that the factored shear resistance derived from the test is more than twice the design shear force at relevant points in the Peel Green bridges. There appeared to be no adverse effects from the absence of shear reinforcement, including any influence there might have been from traffic loading, time-dependent changes or deterioration. On this basis the Peel Green bridges could have been allowed to pass an assessment for further service. This conclusion could be applied to similar structures provided that a reliable estimate of concrete strength was available. However, the particular case of the Peel Green bridges was complicated by the fact that they were to be widened using BS 5400 design rules rather than simply being assessed for further service as they stood.

To allow the beams to continue in service after widening would have produced hybrid structures. In addition, the need to widen the structures necessitated the presence of civil engineering works on the site and the disturbance of traffic. The cost and effect of this presence may have been reduced by retaining the existing decks, but it would not have been eliminated. The balance of advantage was therefore shifted towards replacing the complete decks.

Had the bridges been subjected to service overloads, the tests indicate that cracking would have been visible on the bottom flanges before failure occurred. However, the cracks would not have been readily visible from the ground, only from a close inspection. Web cracks would not have been visible because the beams were placed side by side and the webs could not be seen. As the tests were observed to terminate in a different form of shear failure to that expected, some care is needed in inspections. It illustrates a problem of using a design code for assessment. If destructive tests are carried out on single elements to verify an assessment, design codes that do not identify the correct method of failure can be confusing.

The failure of a bridge deck is likely to be more ductile than the failure of a single beam. This is because a deck generally has a capacity to redistribute load effects transversely and some of these paths may have ductility. Provided that there is sufficient strength in the alternative load paths the failure load would also be higher. A non-linear analysis would be necessary to demonstrate load redistribution effects theoretically.

CONCLUSIONS

1. The Peel Green bridges would not have passed an assessment using BS 5400: Part 4: 1984 because they have no shear links in the webs. Tests on the beams indicated that they would have been satisfactory for further service.
2. The beams showed no ill effects from the absence of links, nor from 30 years of motorway service. Measured shear resistances, after allowing for partial factors of safety, were at least twice the required resistances. The beams possessed a measure of ductility.
3. The measured resistance reduced with increase in shear span and failure occurred in the webs. Bending cracks were visible on the bottom flanges in all cases but did not participate in the failures. The combination of characteristics is not consistent with V_{co} or V_{cr} behaviour as normally understood.
4. The decrease of resistance with increasing shear span, accompanied by failure in the webs, can be explained by considering the principal tensile stress at the bottom of the web rather than at the centroid of the section. Arching action also offers an explanation for the decrease of resistance with increasing shear span and may also explain the high resistance sustained after cracking.
5. Other bridges that contain prestressed beams without shear reinforcement could be assessed as satisfactory provided they were similar to those tested and the factored calculated resistance was sufficient. Unless the beams can be inspected thoroughly and show no signs of distress in spite of previous heavy loading, and the structures contain alternative load

paths, it may be preferable to ensure that the factored resistance is twice the required resistance. This may be the case if allowance is made for the increase in concrete strength with age.

ACKNOWLEDGEMENTS

The work described in this paper forms part of the programme of the Transport and Road Research Laboratory and is published by permission of the director. Help is gratefully acknowledged from Martin Crowe and Charlie Parkinson of Parkman Consulting Engineers, Les Clark, members of TRRL Bridges Division who have helped with the programme and staff from the Department of Transport, North West Regional Office.

REFERENCES

1. BS 5400: Part 4: Steel, concrete and composite bridges, Part 4. Code of practice for the design of concrete bridges. British Standards Institution, London, 1984.
2. HAWKINS, N. M., The shear provision of AS CA35-SAA Code for prestressed concrete. *Civil Engineering Transactions, Institute of Engineers, Australia*, **CE6** (Sept. 1964) 103–16.
3. CLARK, L. A., *Concrete Bridge Design to BS 5400*. Construction Press, London, 1983.
4. NEILSEN, M. P., *Limit Analysis and Concrete Plasticity*. Prentice-Hall, New Jersey, 1984.

41

Fatigue Assessment of Orthotropic Steel Bridge Decks

C. BEALES and J. R. CUNINGHAME

Transport and Road Research Laboratory, Crowthorne, Berkshire RG11 6AU, UK

ABSTRACT

Fatigue failures have occurred in the steel orthotropic decks of several bridges in Europe, some after little more than a decade in service. This has led not only to expensive inspection and repair procedures but has also highlighted the difficulty of assessing these decks, which lie outside the scope of most design codes.

Test procedures have been developed at TRRL which have been used to determine the fatigue lives of welded connections on particular bridges. The data collected and experience gained have led to a greater understanding of the behaviour of orthotropic decks which can be used for a more general assessment of fatigue lives. In particular, the influence of factors which are not easy to model mathematically, such as the effect of the surfacing, can be evaluated.

This paper describes the experimental techniques and the application of these procedures to the deck of the Severn Crossing.

Load tests, with a vehicle of known axle weights, are used to determine the influence surface of stress at the welded connections. Tests carried out with the bridge deck surfacing removed are compared with similar tests on the surfaced bridge with the asphalt at summer and winter temperatures. Where such trials on actual bridges are impractical, data may be obtained from laboratory tests on a full-scale deck panel. The merits of the different methods are discussed.

Fatigue tests on full-scale specimens representing sections of the bridge deck are used to determine the weld classification of the connection.

Any views expressed in this paper are not necessarily those of the Department of Transport.

A computer program developed at TRRL applies the procedures of BS 5400 (Part 10) to the measured stresses and the S–N data to estimate fatigue lives. Fatigue lives have also been estimated from stress spectra obtained under traffic loading.

The research highlights the sensitivity of the calculated fatigue lives to the position of the wheel tracks and to the asphalt surfacing. It is concluded that, for 'worst case' conditions, few connections in the decks of existing long span bridges meet the 120-year UK design life for fatigue.

INTRODUCTION

Steel orthotropic bridge decks are used where lightweight structures are required, that is for long span river crossings and for lifting bridges. Although few in number the long span bridges usually form vital road links and represent major capital investments. The Severn Crossing (the adjacent structures of Severn Bridge, Beachley Viaduct and Wye Bridge), for example, carries around 2 000 000 heavy goods vehicles per year between England and Wales, and cost £11m to build in 1966. The second Severn crossing is expected to cost around £150m plus £100m for approach roads.

The earlier long span bridges, such as the Severn Crossing and Forth Bridge (mid-1960s), were built to British Standard 153.[1] This standard was a specification for steel girder bridges and the fatigue clauses made no special allowance for the complexities of orthotropic decks. The current standard, BS 5400: Part 10 (code of practice for fatigue),[2] issued in 1980, specifically excludes orthotropic decks because of the complex stress analysis and classification of details. Against this background TRRL developed experimental techniques to assess the decks of the Severn Crossing for fatigue prior to decisions being made about the requirements for strengthening this part of the structure. These techniques have since been used to assess welded connections on other bridges.

THE ORTHOTROPIC BRIDGE DECK

A typical orthotropic bridge deck is illustrated in Fig. 1; the main welded connections are identified. For most long span bridges the longitudinal stiffeners (troughs) are either trapezoidal or 'V' shaped. Bridge decks with open stiffeners (bulb flats or angled plates) are not considered here. Transverse stiffeners (crossbeams) are welded to the deck plate and troughs during the fabrication of the deck panel. The crossbeams are subsequently welded or bolted to cross-girders or diaphragms.

Several different designs of trough to crossbeam connection have been used. For the Severn Crossing and Forth Bridge the troughs are in short lengths, around 3–4·5 m long, which are butted up to the crossbeams. In later bridges the troughs are around 14 m long and pass through cut-outs in the crossbeams, the troughs being 'spliced' together some distance away from the crossbeams. In the UK this later type also has cope holes around the apex of the trough and around the trough to deck plate weld. In other European countries the troughs are often welded all round. These differences are illustrated in Fig. 1.

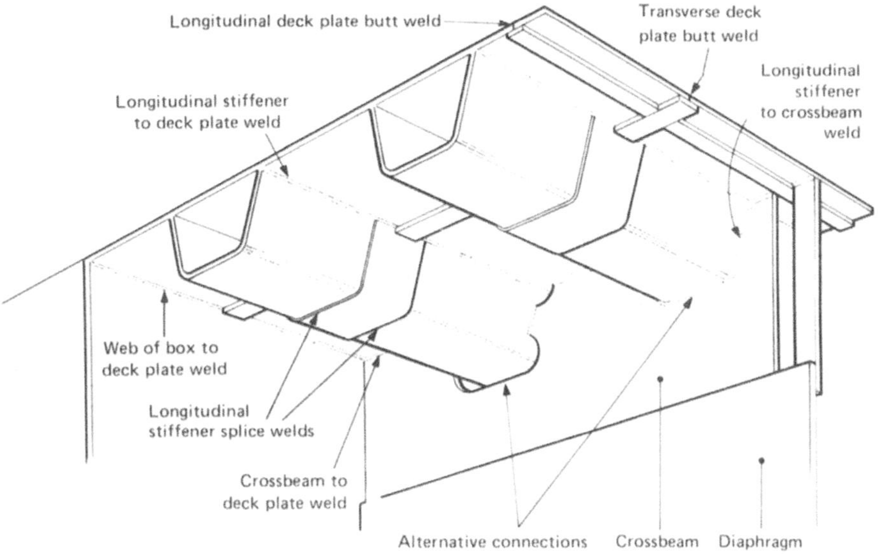

FIG. 1. Main welded connections in a typical orthotropic bridge deck.

Two forms of longitudinal stiffener splice joints are commonly used, either an infill plate, butt welded to the troughs on a backing bar, or an overlapping plate, fillet welded to the troughs.

Single-pass manual metal arc (MMA) welds were typically used for fillet welded connections though automatic or semi-automatic welding is used for long weld runs in modern bridges. Multi-pass butt welds are used to join the sections of deck plate.

Temporary attachments employed during construction can cause fatigue problems. One such attachment on the Severn Bridge was between a temporary diaphragm and the soffit of the trapezoidal troughs. This temporary 'flotation' diaphragm was used to seal the bridge box sections so that they could be floated down river from a temporary storage site to a position beneath the bridge from where they were lifted into place.

FATIGUE ASSESSMENT OF ORTHOTROPIC BRIDGE DECKS

The fatigue design of UK bridges is governed by Part 10 of BS 5400.[2] While the highway loading included in Part 10 is applicable to orthotropic decks, clause 1.5.1 of the code states that 'the stress analysis and classification of details in such a deck is very complex and is beyond the scope of this Part of this British Standard'.

Analytical methods can be used to calculate stresses at welded connections due to traffic loading but such calculations often require verification from 'measured' stresses (stresses calculated from measured strains). The accuracy with which the stress is determined is of great importance since fatigue life is inversely proportional to the third or fifth power of the stress range. Consequently, experimental stress measurements by load tests may be the simplest and most accurate method of assessing existing structures.

Some of the experimental methods have the advantage that the stress reduction (and fatigue life enhancement) of the surfacing may also be considered. This effect is extremely difficult to model mathematically and as such 'this effect should only be taken into account on the evidence of special tests or specialist advice' (BS 5400: Part 10, clause 6.1.4.2). It will be shown later that if this effect is not taken into account most connections close to the wheel tracks would fail to meet the required 120-year fatigue life.

Fatigue tests on full-scale specimens representing sections of the deck have been carried out to determine the classification of the connections. These tests are described later in this paper.

With the loading defined in the code and experimental methods used to determine the stress spectra and weld classifications, fatigue lives may be calculated for details on a bridge.

A choice of fatigue assessment procedures are given in the code. The first two methods are relatively simple and conservative, and involve the calculation of stress at the detail under the loading of a defined 'standard fatigue vehicle'. In the first method calculated stresses are compared with values of limiting stress for each class of detail. In the second method stress spectra are calculated and a simplified damage calculation performed. The most rigorous and least conservative method involves the calculation of stress influence lines for a specified set of commercial vehicles followed by a damage summation using the Palmgren–Miner rule.

The third procedure has been used as the basis for the assessment of details on the Severn Crossing using a range of techniques to determine stress spectra. These techniques and procedures can be applied to any bridge of this type.

DETERMINING STRESS SPECTRA

Stress spectra were obtained in a number of ways from measured strains using gauges installed close to the welded connections.

1 Static Tests in the Laboratory

A full-scale deck panel, 14·3 m long by 3·7 m wide, was tested at TRRL under a single static wheel load. The panel was supported in a reaction frame carrying a single wheel and axle assembly which could be moved to any position over the panel. The wheel and tyre were typical of those used on heavy goods vehicles in the UK. The wheel was loaded to 20 kN though loads up to 100 kN were possible.

The load was applied at over 500 locations on the panel to determine the full influence surface of stress for the gauge positions. Typical longitudinal and transverse influence lines obtained in the tests are shown in Fig. 2(a) for the crossbeam to deck plate connection. They show that the effect of the wheel load is very localised, the effective length of the influence lines being around 1 m; this is the case for most connections close to the deck plate.

Influence lines for the vehicle types of BS 5400 (Part 10, Table 11) were calculated by superposition of the single wheel stresses and stress spectra calculated using the 'rainflow' cycle counting method.

The laboratory method enables precise control of the wheel load and position to be obtained. Environmental effects are eliminated. The method does not allow the effect of surfacing to be assessed. Asphaltic materials are viscoelastic, that is the properties depend on the rate of loading. Consequently, the composite effect of the surfacing on the steel deck is quite different under a static load from that under a moving vehicle. For this reason laboratory tests were carried out with the panel unsurfaced.

Connections near the centre of the panel were tested to avoid problems with edge effects. Accurate modelling of support conditions is important, especially for connections at the crossbeam. The web of box and diaphragm plates are difficult to accommodate in a test panel of this type and were not incorporated in this particular panel.

2 Static Tests on the Bridge

These tests were carried out with a two-axle 16-t test vehicle of known wheel loads and axle spacings. An area of the bridge deck surfacing (20 m long by 3·7 m wide), over the instrumented details, was removed. Although the influence of the surfacing is small under a static load, strain readings take many minutes to stabilise because of creep of the asphalt.[3]

Strains were recorded with the vehicle stationary at a number of locations

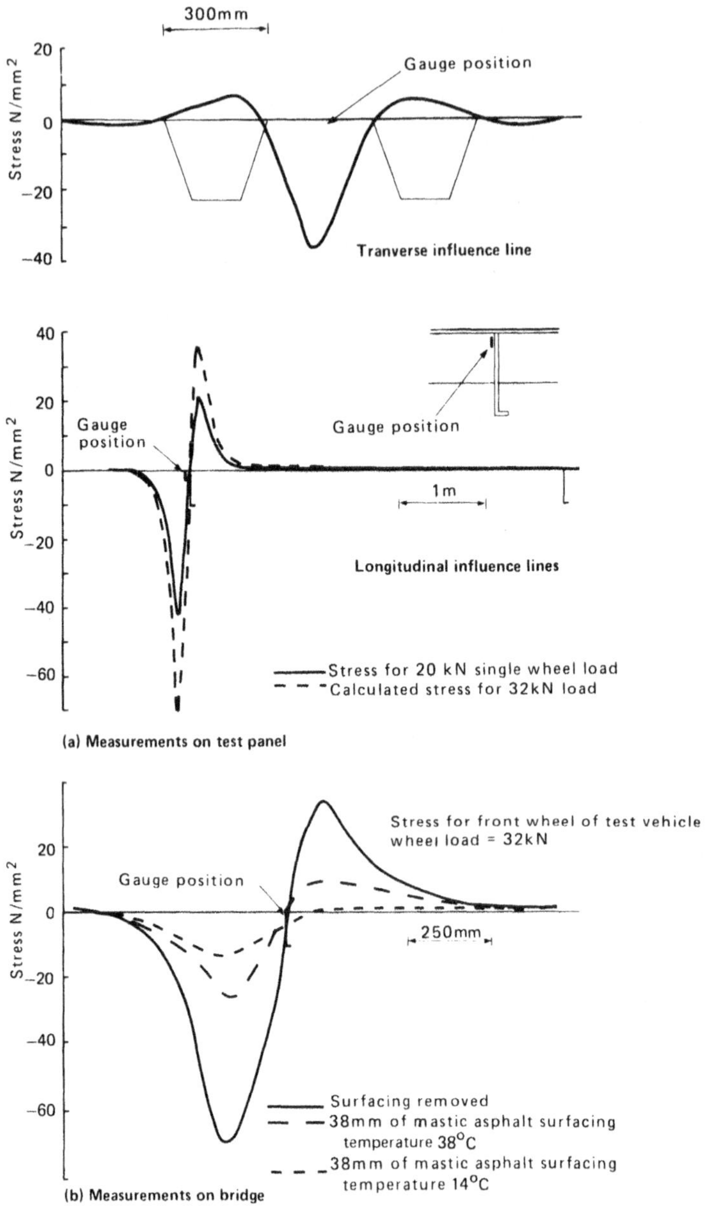

FIG. 2. Typical influence lines—crossbeam to deck plate connection.

on the deck, the load position being referenced to one of the front wheels. Interaction between the front and rear or nearside and offside wheels is negligible for most details because the influence of the wheel load is very localised. Consequently, single wheel influence line data can be obtained from these tests and the data treated in the same way as for laboratory static tests.

Exceptions are the connection between the trough and crossbeam and the stiffener splice joint, where the longitudinal influence line for a single wheel is very long. In these cases it is not possible to separate the strains for a single wheel from the vehicle loading.

Good agreement was generally found between the stresses measured in the laboratory and on the bridge. In the longitudinal influence line of Fig. 2(a) the 20 kN 'measured' stresses from the laboratory deck panel tests have been factored to the equivalent loading of the front wheel of the test vehicle (32 kN). These data can be compared with the influence line obtained from the test vehicle on the unsurfaced deck (Fig. 2(b)). The longitudinal scale of Fig. 2(b) has been increased but it can be seen that the peak stresses are in good agreement with the factored peak stresses from the laboratory tests.

The advantages of testing on the bridge are that there are no doubts about the accuracy of the model, and details which cannot easily be accommodated in a test panel can be tested. Bridge tests are also considerably cheaper and quicker to organise than laboratory tests, which may require the fabrication of a deck panel. The disadvantage of bridge tests is that lane closures are necessary to accommodate the test vehicle and the removal of the surfacing. In addition, loading from vehicles in adjacent lanes or carriageways, and changes in temperature, cause noise or drift of the strain gauge readings. Accurate positioning of the test vehicle is also difficult; errors in excess of 10 mm are normally noticeable on influence line plots.

3 Dynamic Tests on the Bridge

After the tests on the unsurfaced decks were complete the area was resurfaced. Hand-laid mastic asphalt was used for one half of the area and an epoxy asphalt on the other. In both cases a nominal 4 mm layer of an epoxy-based material was used as a waterproofing membrane.

To assess the effect of the bridge deck surfacings, strains were monitored using high-speed tape recording equipment as the test vehicle was driven over the strain gauged connections. A simple but effective method of recording the transverse position of the vehicle relative to the connection was by measuring from a reference line to the tyre imprint in a patch of sand. The vehicle was driven at constant speed and longitudinal scaling was calculated from electrical pulses on the magnetic tape generated as the vehicle passed over detector strips on the road located before and after the connection. Skilful driving was needed to run consistently close to the target position. Around 20–30 'runs' of the test vehicle were required to achieve a reasonable transverse distribution of points to describe accurately the full influence surface of stress.

Typical longitudinal influence lines are given in Fig. 2(b) for the mastic surfacing at 14 and 38°C. There is a considerable reduction in stress compared with the data from the unsurfaced deck, the greatest effect being when the surfacing was cold and stiff. Because of this temperature sensitivity load test data were required at a range of surfacing temperatures to assess fully the surfacing effect.

Previous tests[3] had shown that the effect of the surfacing was also dependent on vehicle speed, the greatest reduction in stress being obtained for fast-moving vehicles. In practice the change was found to be small for speeds in excess of 32 km/h (20 mph) and this speed was adopted for the test vehicle runs. Since most vehicles travel at speeds in excess of 32 km/h a calculation based on these data will tend to be conservative.

Single wheel influence line data were obtained following the analogue-to-digital conversion of the raw data. Once again this was not possible for details with long influence lines.

Each test requires a lane closure for only a few hours. Compared to laboratory methods, the uncertainties associated with modelling the deck and surfacing are avoided. The temperature and traffic data are collected separately from the strain data so that revised estimates of fatigue life can be produced to reflect changes, for example in traffic composition, without the necessity of repeating load tests. The disadvantage of this method is the logistical one of planning tests at the right times to obtain data at a range of temperatures.

4 Direct Measurements under Traffic Loading

'Intelligent' electronic equipment was used to monitor the response of the strain gauges to dynamic loads from the moving traffic. Stress spectra were output directly.

The advantage of this method is that the effect of temperature and vehicle speed on the composite action of the surfacing are automatically taken into account. The only disadvantage is that data are recorded in real time and many months of recordings were needed to assess the behaviour of the deck over the full range of surfacing temperatures and to take account of variations in traffic. An estimate of long-term trends in traffic can be used to predict fatigue lives from these data.

FATIGUE TESTS AND JOINT CLASSIFICATIONS

The fatigue classes in BS 5400: Part 10[2] are based on data from small laboratory specimens tested under axial loading. Only tests under tensile loading were included and the statistical analysis of the data took account of the absence of residual stresses in the specimens.[4]

The classification of welded joints in orthotropic decks may differ from the code because of the complex stress distribution and high stress gradients around the joint. Also there is evidence that fatigue strength is different for a joint subject to bending stress,[5] as is often the case in an orthotropic deck, compared with axial stress.

In tests at TRRL fatigue specimens were made to full scale and large enough to retain high residual stresses after welding. Care was taken to obtain good fit-up and to follow the correct weld procedures so that the specimens represented joints on a bridge without significant defects.

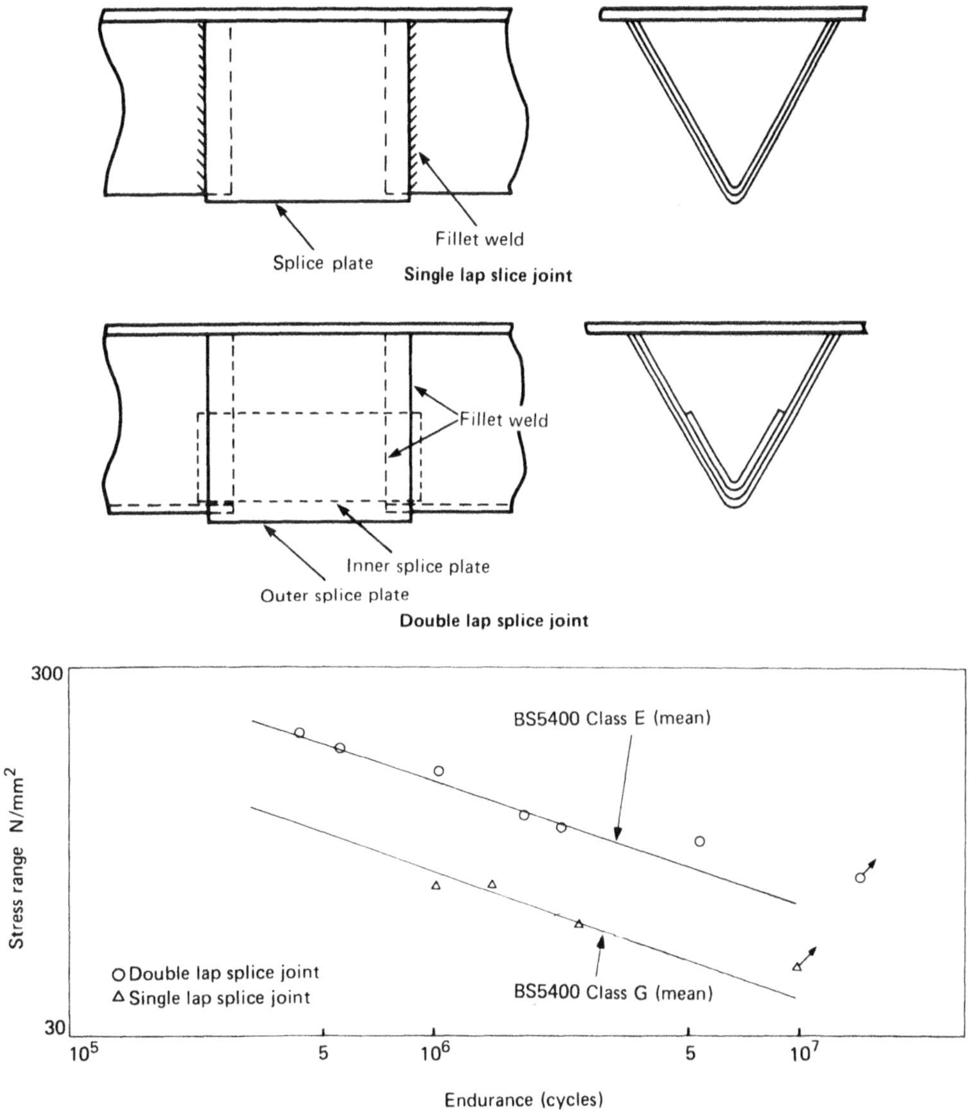

FIG. 3. Fatigue test results for the fillet welded trough splice joints.

Loading was arranged to replicate the stress distribution around the joint as measured on the bridge or deck panel. Strain gauges were installed at the same locations on fatigue test specimens as on the deck panel or bridge, so that the stress used to define the fatigue strength was directly comparable to that obtained from wheel loads.

Constant amplitude fatigue tests have been carried out on a number of deck plate joints. For example, tests on joints between lengths of longitudinal stiffener (trough splice joints) were reported by Cuninghame[6] for butt welded joints. Further tests have been carried out on fillet welded splice joints as these may be more economical to fabricate and less sensitive to misalignment of the troughs. Specimens consisting of 2 m long by 600 mm wide sections of deck containing a trough with a splice joint at mid-length were loaded in four-point bending. Results are shown in Fig. 3 in the form of *S–N* curves for longitudinal stress at the apex of the trough adjacent to the splice weld.

It will be seen that the fillet welded joints have lower fatigue strength than those with butt welds and so are less likely to be suitable for heavily trafficked bridges. In terms of BS 5400 classifications the butt weld is class D, the fillet welded joint with inner and outer splice plates falls within class E, and the joint with outer splice only in class G. Fatigue cracks in joints with outer splice only have been reported by Darnell *et al.*[7]

FATIGUE LIFE CALCULATION METHODS

Three calculation methods were developed to suit the methods used to obtain the stress data. Method 1 uses the single wheel influence line data from the laboratory or bridge tests to determine the fatigue life for the unsurfaced deck. Method 2 uses a similar procedure to calculate lives from the data from the surfaced deck but goes on to assess the overall life, taking into account the flow of vehicles across the bridge at different temperatures throughout the year. Method 3 uses the data obtained under traffic loading directly.

Method 1: Unsurfaced Deck—Static Test Data

A computer program has been developed at TRRL to calculate fatigue lives from experimentally determined stress influence surface data and the loading of BS 5400: Part 10. Data preparation mainly involves the entry of the singie wheel stress data and the coordinates of the loading positions. Influence lines for the vehicle types described in Table 11 of the code are calculated by superposition of the single wheel data. The vehicle influence lines are calculated at a number of transverse positions across the deck and due allowance is made for the transverse distribution of vehicles defined in the 'multiple paths' clause (C.1.4) of the code. For the assessment of existing structures the centreline of the transverse distribution of traffic can be determined by observation on the bridge.

A stress spectrum is calculated from the vehicle influence line data using the 'rainflow' cycle counting technique. Fatigue lives are calculated using appropriate *S–N* data and applying the Palmgren–Miner method of damage summation.

Once the data are assembled the program run time is short. It is therefore easy to repeat the calculations with traffic centred at a number of different transverse positions to determine the effect on fatigue life. This facility can be used to assess the effect of proposed changes to the position of traffic lanes on a bridge.

Method 2: Surfaced Deck—Dynamic Test Data

Fatigue lives may be calculated in a similar way from data obtained from load tests on the surfaced bridge. The data processing is more complex, analogue-to-digital conversion of the strains recorded on to magnetic tape first being required. It is also necessary to deal with the fact that the influence lines from the test vehicle are at random transverse positions.

Fatigue lives thus calculated relate to the surfacing temperature at the time of the load test. To determine an overall life relating to conditions throughout the year, allowance must be made for the variation of temperature throughout the day and year, and the variation in the flow of traffic at different times of the day.

Howells[8] recorded temperatures for a 2-year period (1969–71) on the Wye Bridge, presenting the data as the proportion of time, in 5°C bands, for each hour of the day. Data were also obtained giving the flow of HGVs over the Severn Bridge for each hour of the day, averaged over a year. These two sets of data were combined to obtain an estimate of the traffic flow at different temperatures. It was then possible to apportion the fatigue lives at specific temperatures to determine an overall value relating to the whole year.

Method 3: Surfaced Deck—Traffic Stress Data

Stresses at 12 welded joints in the deck of the Wye Bridge were monitored for 13 months. Stress spectra were stored by the recording equipment at 1 h intervals. Traffic flow was monitored during the recording period and the time and duration of interruptions to the flow over the instrumented area due, for example, to lane closures or traffic management schemes was noted. Fatigue damage was calculated from the stress spectra for the total recording period and factored to take account of interruptions to give an estimate of fatigue life for each joint.

CALCULATED FATIGUE LIVES

Table 1 gives fatigue lives for six connections on the Wye Bridge, calculated using the three methods described above. All lives relate to a 2·3% probability of failure (using mean—2 s.d. *S–N* data) and a traffic flow of 800 000 HGVs per annum. This is the observed traffic flow and is less than the 1·5 million vehicles required by BS 5400. The centre of the transverse distribution of traffic is assumed (in Methods 1 and 2) to be at the centreline of the carriageway. Where appropriate, surfaced lives relate to the hand-laid mastic experimental surfacing. Weld classes are given in brackets.

TABLE 1
Estimated fatigue lives

Detail (weld class)	Fatigue life (years)		
	Unsurfaced deck	Surfaced deck	
	Method 1 Static test data	Method 2 Dynamic test data	Method 3 Traffic stress data
Trough/deck (F)	6·5	94	>120
Longitudinal butt weld (F)	5·9	>120[a]	>120
Web of box/deck (D)	41	>120	40[b]
Trough/crossbeam (G)	4·3	13	18
Crossbeam/deck (D)	94	>120	>120
Transverse butt weld (F)	35	>120[a]	>120

[a] Insufficient data for full assessment but expected value >120 years.
[b] Surfacing cracked over web.

The lives given in Table 1 do not take account of damage sustained prior to resurfacing but would apply to repaired joints or to similar connections on a new bridge. It can be seen that, for the unsurfaced deck, none of the details meet the 120-year design life required by the code.

Some of the assumptions in Method 2 were chosen to give a conservative (low) estimate of fatigue life. For two of the connections the data obtained were at an insufficient range of temperatures to enable overall fatigue lives to be calculated. Nevertheless, the data suggest that both these joints would have fatigue lives in excess of 120 years on the surfaced deck.

Data for the Method 2 calculation were obtained within 6 months of resurfacing the strain gauged area. Collection of data for the Method 3 calculation began 12 months after resurfacing, by which time a longitudinal crack had developed over the web of the box. This explains the loss of effectiveness of the surfacing for this connection and the estimated life similar to that for the unsurfaced deck in this case.

Effect of Surfacing

The influence lines in Fig. 2(b) illustrate the large reduction in stress in the steel deck due to the bridge deck surfacing. They also show the effect of the variability of the properties of the surfacing with temperature. Table 1 shows that, for most connections, very short fatigue lives calculated for the unsurfaced deck are increased to above the 120-year design life by the surfacing. One exception is the trough to crossbeam connection where the stiffening effect of the surfacing has only a small influence on the stresses at the apex of the trough. Failures of this connection have occurred in service.

The results suggest that failure of the trough to deck plate connection should not occur within the lifetime of the bridge (strictly there is a calculated 2·3% probability of failure within 94 years) surfaced with mastic asphalt on an epoxy waterproofing layer. It should be noted that this surfacing system is stiffer than that originally used on the Severn Crossing and this difference may account for the fact that cracks have occurred in this connection after less than 20 years in service.

The results also show that the effectiveness of the surfacing can be completely lost if cracks develop in the surfacing over the welded connection. On some bridges it is quite common for cracks to develop over the web of the box within weeks of resurfacing. These are normally controlled by sawcutting the surfacing and sealing the joint with a flexible bituminous material. Cracks can also develop over other hard spots, such as the troughs and crossbeams, that is at the very connections the surfacing should be helping. This partly explains the difficulty of incorporating a surfacing factor in the codes. The search continues for a surfacing material with high stiffness (preferably at high and low temperatures) and a long fatigue life at the cold/brittle and hot/high strain extremes.

Transverse Position of Traffic

The transverse influence line shown in Fig. 2(a) illustrates the very localised effect of the wheel load on the stresses at a typical connection near the deck plate. Consequently, the fatigue life is very sensitive to the transverse position of the traffic. Using the fatigue assessment program it can be shown that the fatigue life of the web of box to deck plate connection on Wye Bridge could be increased from 12 to 80 years (unsurfaced deck, weld class F) by moving the traffic lanes 300 mm towards the centre of the bridge.

Consideration may be given at the design stage to the relative positions of the traffic lanes and connections such as the web of box to deck and the longitudinal deck plate butt welds. For more frequently occurring connections, such as the trough to deck plate connection (300 mm intervals on Wye Bridge), there is little scope for improvement by this method.

CONCLUSIONS

The analysis of stresses and classification of details in orthotropic decks is complex and beyond the scope of current codes of practice for fatigue. Tests are TRRL and elsewhere have produced fatigue data for the classification of most of the welded connections. Test procedures have been used to obtain stress spectra for these connections on a number of bridges from which fatigue lives have been calculated.

The data have been used to quantify the effect of the bridge deck surfacing and to show the sensitivity of the calculated fatigue life to the transverse position of the traffic.

The results suggest that few connections close to the wheel tracks meet the 120-year design life if the effect of the surfacing is ignored. The difficulty of incorporating a surfacing factor in the codes is recognised.

ACKNOWLEDGEMENTS

The work described in this paper forms part of the programme of the Transport and Road Research Laboratory and the paper is published by permission of the director. The work was carried out in the Fatigue Section of the Bridges Division, TRRL. The leadership and guidance of Mr D. E. Nunn is gratefully acknowledged.

REFERENCES

1. BS 153, Specification for steel girder bridges. Part 3B: Stresses. British Standards Institution, London, 1958.
2. BS 5400, Steel, concrete and composite bridges. Part 10: Code of practice for fatigue. British Standards Institution, London, 1980.
3. Morris, S. A. H., Stresses under dynamic wheel loading in a surfaced steel orthotropic deck with V-stiffeners. TRRL Report SR237, Transport and Road Research Laboratory, Crowthorne, 1976.
4. Gurney, T. R. and Maddox, S. J., A re-analysis of fatigue data for welded joints in steel. *Welding Research International*, **3**(4) (1973) 1–54.
5. Maddox, S. J., Fatigue of welded joints loaded in bending. TRRL Report SR84UC, Transport and Road Research Laboratory, Crowthorne, 1974.

6. CUNINGHAME, J. R., Steel bridge decks: fatigue performance of joints between longitudinal stiffeners. TRRL Report LR1066, Transport and Road Research Laboratory, Crowthorne, 1982.
7. DARNELL, W. J., EDMONDS, F. D., RUTLEDGE, J. C. and YONG, P. M. F., Fatigue problems in the orthotropic deck of the Auckland Harbour Bridge. Pacific structural steel conference, Auckland, New Zealand, 1986.
8. HOWELLS, H., Temperature spectra recorded in surfacings on steel bridge decks. TRRL Report LR587, Transport and Road Research Laboratory, Crowthorne, 1973.

42

Assessment and Rehabilitation of Suspension Bridges

PETER G. BUCKLAND
Buckland and Taylor Ltd,
1591 Bowser Avenue, North Vancouver, BC, Canada V7P 2Y4

ABSTRACT

The upgrading of suspension bridges, if required, is an extremely costly procedure. It is therefore worth some considerable effort on the part of the engineer to determine if rehabilitation is in fact really needed for the safety of the bridge or whether it is only required in order to satisfy a code.

The importance is emphasized of taking a critical look at criteria, including traffic and wind loads, aerodynamic stability and seismic response. Recent developments in wind tunnel testing are presented, with observations on long-term cable stretch, reliability of temperature measurements and the use of limit states design. Three examples are taken from suspension bridges ranging in span from under 200 m to over 800 m.

INTRODUCTION

Suspension bridges are no more immune to the ravages of time and the need for upgrading than other bridges—but they are potentially far more expensive to renovate. Worse than that, suspension bridges are usually on major routes where any interruption of service can have a huge effect on the community. It is therefore necessary to be doubly sure about a suspension bridge before committing to repair it. Is repair really necessary? How best to do it?

Techniques have been evolved over the last few years to answer these two questions with more certainty than before.

LOADING

The first item might best be to take a hard look at the loading. Is it really what the code says? What conventional wisdom says? Is it more? Less? Since the only function of a bridge is to support safely and effectively the loads imposed on it, particularly traffic loads, this is a useful starting place.

When the Lions' Gate Bridge in Vancouver, BC, was examined in the mid-1970s it was noticed that although it was designed for two lanes it was in fact carrying three lanes of densely packed traffic—about 60 000 vehicles/day. At first sight this appeared to be a 50% overload.

Instead of closing the bridge it was decided to take a new look at the loading. The trick is in formulating and then solving the probabilistic equations which describe the maximum loading expected to occur during the lifetime of the bridge—or some other defined period—with a certain degree of probability. This work has been reported elsewhere.[1,2] The results for the Lions' Gate Bridge are shown in Fig. 1.

The interesting points here are that for the major components of the

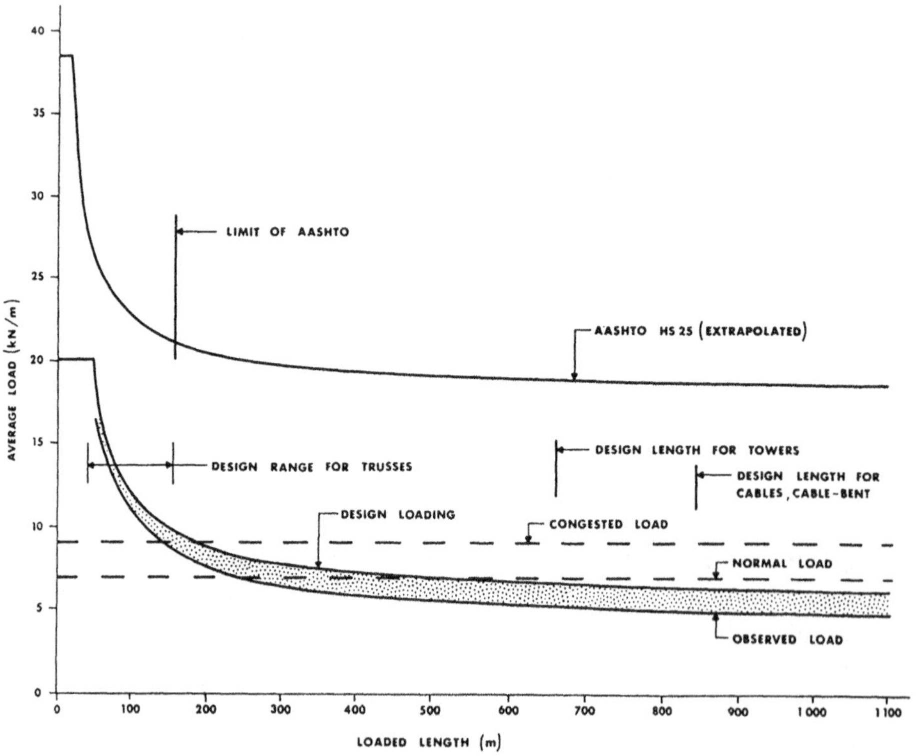

FIG. 1. Traffic loading for Lions' Gate Bridge. 'Design loading' based on 'observed load' was used for evaluation. The bridge was originally designed for 'normal' and 'congested' loads.

bridge—towers and cables—which are governed by long loaded lengths, the original designers' loading was in fact adequate. Such is not the case for shorter loaded lengths such as affect the design of the stiffening truss. Whether or not this situation is acceptable, it is at least the right way round; failure of the stiffening truss would be far less catastrophic than failure of the towers or cables (or anchorages).

Wind loading should not escape scrutiny either. The wind force is a function of several parameters, including expected wind speed, wind direction, turbulence of the wind, the shape of the bridge cross-section and the dynamic properties of the bridge.

The cost of determining these factors is tiny compared to the cost of modifying the bridge. Although the bridge engineer is trained to accept wind loadings from the codes, this is an instance where the engineer's involvement with criteria may pay handsome dividends to the client.

ANALYSIS AND SURVEY

Having decided with at least some confidence on the loads to be applied to the bridge, the next significant step is to check the condition of the bridge. Some aspects of this are obvious and relatively easy: looking for corrosion and fatigue cracking, measuring the tension in cable-band bolts, and even inspecting some of the cable itself.

Some other areas are virtually impossible to inspect with today's technology: embedded steel in the anchorages, for example, and the insides of spiral strands.

However, the real key to success comes in determining the dead load stresses in the structure. The technique is as follows, and it depends on the creation of a computer program that will accurately model the effects of changes on a suspension bridge.

Suspension bridges are difficult to analyse. They are non-linear in their behaviour and it is not possible, even theoretically, to have a condition in which all members are stressless at the same time. This makes the use of conventional non-linear analysis programs clumsy, at best.

For the examples herein the program SABER was developed. SABER not only calculates the effects of loading, it also allows the structure to be changed. Each loading or change is calculated directly from the preceding condition in a step-by-step manner. The steps are:

(a) Model the structure in the computer as it was built, which is not necessarily as it was designed.
(b) Impose on the model all known changes that have occurred to the

bridge—extra dead load, strengthened members, settlement of the foundations, drag of the anchorages, and so on.

(c) At this stage the computer will predict the expected geometry of the bridge.

(d) The bridge is also surveyed, at least for the key points such as elevations of midspans and tower-top deflections. The surveyed geometry, corrected for temperature effects, should be identical to that predicted by the computer. Experience with about a dozen suspension bridges suggests that such is almost never the case.

(e) The reasons for discrepancies between the predicted and observed geometries must now be sought. They will be caused by all those changes to the bridge that have occurred but which were not known. Apart from those suggested in (b), these could include long-term stretch of the cables or hangers, slippage of cables in saddles, errors in the construction of the bridge and incorrect information having been given to the evaluating engineer.

In some cases the causes of discrepancies cannot be pinned down exactly, but so far experience has shown that the effects of the causes can be bracketted so that there are upper and lower limits.

(f) Finally, when all permanent effects on the bridge are known, the resulting stresses can be calculated and added to the stresses from the various live and other loadings to obtain a complete picture of the real condition of the bridge.

The benefits of this powerful technique will be shown in the examples presented.

In some cases it is useful to supplement the information gained by site measurements. For example, if the polygon representing the cable geometry is known, all cable and hanger forces in a span can be determined by measuring the tension in one hanger.

TEMPERATURE EFFECTS

Surveying a suspension bridge only yields information if the temperature of the bridge is uniform, known and constant during the time of the survey. Conventional wisdom has been to survey the bridge during an overcast windless night, typically around 02:00 hours when the ambient temperature is fairly constant.

An interesting study on the Lions' Gate Bridge, however, revealed that the bridge continued to experience radiant cooling even though its temperature was 5°C below ambient. The measurements taken are shown in Fig. 2.

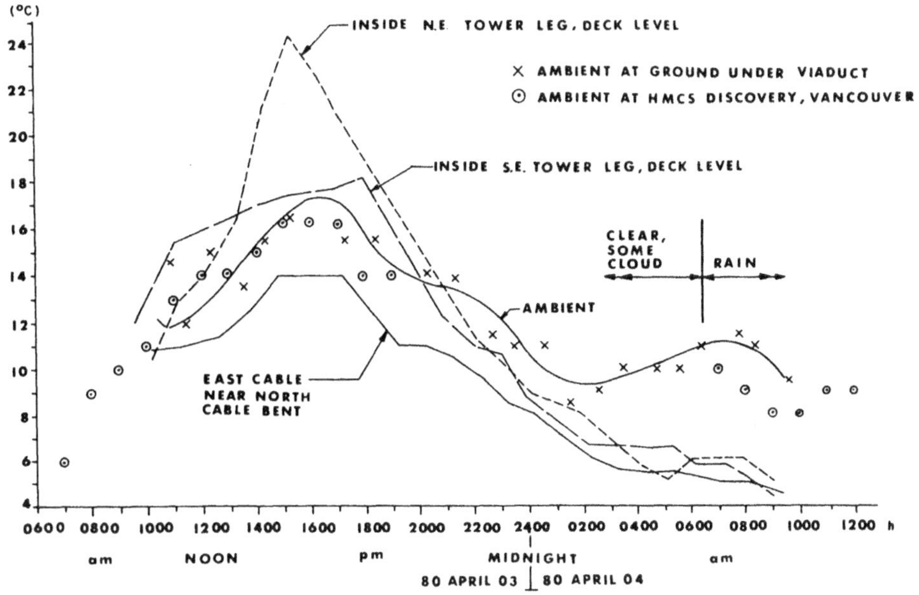

FIG. 2. Temperature variation with time, Lions' Gate Bridge.

SAFETY FACTORS

To this point, no mention has been made of safety factors. Since this question is addressed in another paper,[3] only two points will be made here.

Safety margins are provided to allow for all the unknowns in a situation, and limit states design reflects this philosophy by providing larger load factors for the loads that are least well known. Most major suspension bridges now in existence were not designed by limit states design, however, and it is possible to derive some benefit from this situation when considering the main members, which carry mostly dead load, and to which the smallest load factors usually apply. As a caveat, note that the safety factors should also reflect the importance of the members considered.

The second point is that if safety margins reflect ignorance then, when the ignorance is reduced by the methods described herein, perhaps the safety margins may also be reduced. Sensible judgement is required.

EXAMPLES

To illustrate the points made some examples of their use are given.

Lions' Gate Bridge, Vancouver, BC, Canada (Fig. 3)

In 1975 the concrete deck on the 670 m approach viaduct was replaced by a lighter, wider steel orthotropic deck in a series of $6\frac{1}{2}$-hour closures at night.

FIG. 3. Lions' Gate Bridge, Vancouver, BC, Canada.

By making the deck act compositely with the steel girders their ultimate capacity was enhanced. This technique was later used on the George Washington Bridge, the Golden Gate Bridge and others, and has even been patented (not by the originators of the idea) in the USA!

But the most interesting aspects of this study were the research into loading mentioned earlier, and the first use of the analysis-and-survey technique.

In this case the following known changes were modelled:

(i) unstressed lengths of members as per shop drawings;
(ii) settlement of the north tower foundation and the north cable bent (side tower) as surveyed;
(iii) some addition to the dead load; and
(iv) no movement of the anchorages as observed initially from a measurement of expansion joints and confirmed later by triangulated survey.

The geometry surveyed could only be attained in the computer model if the following changes were also added:

(v) long-term stretch of the main cables of about 0·038%;
(vi) long-term stretch of the hangers of about 0·075%; and
(vii) slippage of the cables through the north cable bent saddles of 60 mm.

Stretch of the main cables (a hexagonal arrangement of helical wound structural strand) induced deflections to the tower tops, which added to the P–δ effects. By using limit states design, even with conservative factors, the towers were deemed capable of carrying the induced moment.

Stretch of the hangers is most pronounced where the hangers are long, next to the towers, and throws extra load on to the bearings. This was later confirmed by measuring the bearing reactions, which were double what they were intended to be under dead load only. A solution is to lower the bearings when the bridge is next renovated, to throw more load back on to the hangers.

Slippage of the cables through the cable bent saddles was known to have occurred during construction in 1938, but was thought to have been stopped. Its continuance imposed some alarmingly high stresses on the cable bent legs. Those stresses could not have been revealed by any other method and in fact had been missed in an evaluation of the bridge 4 years earlier by a different firm of suspension bridge engineers.

The solution to the cable bent problem (Fig. 4) was to rotate the two foundations. This was mostly done under traffic by cutting through the mass concrete footings and inserting 3 m long rocker bearings (Fig. 5).

At the south end of the bridge a short rocker bisects the angle of the cables (Fig. 6). There was no slippage here, but the long-term lengthening of the cables must have caused the tops of the rockers to move towards the centre of the bridge. Close inspection, however, yielded no evidence that the pin had ever rotated. The 20-year-old paint film was still intact. The stresses in the

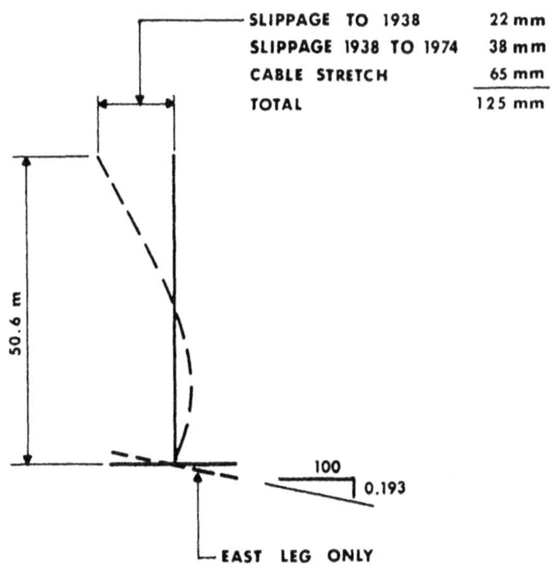

FIG. 4. Distortion of north cable bent Lions' Gate Bridge.

FIG. 5. The north cable bent footing, formerly of mass concrete, was cut and 'pinned' to allow straightening of the bent.

FIG. 6. South rocker posts with cover plates added.

rocker were thus indeterminate, depending on when the cables stretched and when the rocker last moved.

Jacking up the cable in order to attend to the rocker pin would have been very expensive. The pin had a half-shell bronze bushing but its bearing pressure was ten times the allowable pressure of the 1970s.

Instead it was decided to add plates to the flanges of the rocker so that its ultimate capacity would be greater than that of the bronze bushing. This way, if failure should occur, it would be a non-catastrophic failure of the bushing rather than catastrophic failure of the rocker itself.

Also on Lions' Gate Bridge, an exhaustive study was made of the natural frequencies, mode shapes and damping values of the bridge.[4] It was interesting that the initial measurements gave frequencies of up to 1·38 times the predicted values. Since this ratio is close to $\sqrt{2}$ and frequency is proportional to $\sqrt{(K/M)}$, where K is the stiffness and M the mass, it was thought that an error of 2 had been made.

Such was not the case, however. The main reason was that the breather joints in the deck were not 'breathing' at the small amplitudes measured. This significantly altered both the lateral and torsional frequencies, which were found to be coupled, and showed up the danger of relying on measurements alone, which in this case would predict, for example, a critical velocity for aerodynamic instability of about 38% greater than the true expected value.

Tacoma Narrows Bridge, USA

In professional collaboration with Arvid Grant and Associates, Buckland and Taylor Ltd used the analysis and survey technique to evaluate the Tacoma Narrows Bridge (Fig. 7) in 1984. The bridge has been well maintained, but what the evaluation showed was that the hangers have, as expected, elongated and thus increased shears in the stiffening trusses near the towers.

There was another learning experience in the survey. The bridge profile was surveyed from end to end and back during one night with no traffic on the bridge while readings were taken. The survey did not close by 1 ft and a simple error of one digit was suspected.

More careful analysis, however, revealed that the bridge temperature had been changing slightly while the survey was conducted. Thus the elevation of each point changed slightly between successive readings at that point. When a correction was made for this the survey closed.

Belgo Bridge, Quebec, Canada

This little suspension bridge, which carries logs on a conveyor over a span of 183 m, was built in 1917 and examined in 1982. By using the techniques

FIG. 7. Tacoma Narrows Bridge, WA, USA.

just described it was found that because of increases in dead load, seizing of rollers supporting cable saddles and differential settlement of foundations some tower members were carrying 2·5 times their calculated allowable loads under dead load only.

Because steel entering the anchorages, buried under a newsprint mill building, was known to be corroded but was not fully inspectable, it was decided to transform the suspension bridge into a cable-stayed bridge, apparently the first time this has been done.

The project took place during two closures of the bridge in successive summers and has been described in more detail elsewhere, but one of the critical aspects was the necessity of having the bridge partly carried by the old suspension system and partly by the new cable-stayed system with a central tower for almost a year (Fig. 8).

FIG. 8. Belgo Bridge in transition from a suspension bridge to a cable-stayed bridge.

During the hot summer months the suspension bridge sheds its load to the cable-stayed system, thus tending to overload the stiffening trusses in shear at the temporary end of the cable-stayed system. In the harsh Quebec winter the opposite occurs: the suspension bridge tries to take all the load, including that added for the cable-stayed system. This would overload the suspension system. Tuning the bridge to be always within these limits took some careful engineering!

Another comment on safety factors is appropriate here. If it had been decided that the stresses in the tower must be limited to about 0·6 of theoretical capacity, as is normal, the bridge would have had to be closed.

As the bridge is the only feed of logs to the mill for 8 months of the year, this would have had a disastrous effect on the economy of the mill and the town.

In fact some stresses were at least 50% greater than the theoretical capacities (i.e. 2·5 times normal allowable) under dead load alone. A temporary prop was installed at midspan which reduced the critical stresses by 10%, and the bridge operated with the stresses at least 35% above theoretical capacity (i.e. 2·25 times normal allowable stress) for 2 years. The argument was that the bridge had, in essence, survived a load test, and that reducing the load by 10% would be satisfactory for the short term. The client was made aware of the risks and the consequences of the alternatives. It was an informed client who made the decision as to which alternative to accept. The benefits of continued operation were the client's. It was not desirable for the risks to belong to the engineer.

AERODYNAMICS

Several existing suspension bridges have been deemed to be inadequate for aerodynamic stability. Some bridges, such as Golden Gate and the Bronx–Whitestone Bridge, have been modified to improve performance, and for one (the original Tacoma Narrows) modifications came too late.

It has been found that by using the most sophisticated wind tunnel techniques and by testing in turbulent wind flow with full aeroelastic models, taut strip or taut tube models, the expense of solving a problem that does not really exist can be avoided.

Once again, Lions' Gate Bridge provides a good example. Following movement during a hurricane in 1962, section model wind tunnel tests were performed the following year. The bridge was given a clean bill of health, but interpretation of the results was open to doubt. Based on the test results, this author estimated a critical velocity of only 55 mph (25 m/s). Such a number was consistent with incipient motion recorded in winds gusting to 72 mph (32 m/s) as these gusts would coincide with a mean wind of 50–60 mph (22–27 m/s).

What became apparent from the full model test in turbulent flow—believed to be the largest model tested in turbulent flow—was that the observed motion was in fact buffeting response and not the onset of instability.[5]

SUMMARY

Suspension bridges are expensive and disruptive to repair. Before embarking on a repair programme it pays to be very sure about the condition of the bridge, including those conditions that cannot be determined by routine inspection. For this purpose the analysis-and-survey technique is indispensable.

It also pays to take a careful look at all design criteria, including loads and safety levels. These conclusions are based on experience gained on a dozen existing suspension bridges. The examples in this paper represent a variety of sizes and degrees of rehabilitation required.

ACKNOWLEDGEMENTS

The figures relating to Lions' Gate Bridge originally appeared in two papers published in 1981 by the *Canadian Journal of Civil Engineering*. Permission to reproduce them here is gratefully acknowledged.

REFERENCES

1. BUCKLAND, P. G., MCBRYDE, J. P., NAVIN, F. P. D. and ZIDEK, J. V., Traffic loading of long span bridges. *Proceedings of Conference on Bridge Engineering*, St Louis. TRB Transportation Research Record 665, Vol. 2, Washington, DC, 1978.
2. BUCKLAND, P. G., NAVIN, F. P. D., ZIDEK, J. V. and MCBRYDE, J. P., Proposed vehicle loading of long span bridges. *J. Struct. Div. ASCE* (April 1980).
3. BUCKLAND, P. G., Canada's advanced national standard on bridge evaluation. In *Proceedings International Conference on Bridge Management*, ed. J. E. Harding *et al.*, Elsevier Applied Science Publishers, Barking, 1990, pp. 575–583.
4. BUCKLAND, P. G., HOOLEY, R., MORGENSTERN, B. D., RAINER, J. H. and VAN SELST, A. M., Suspension bridge vibrations: computer and measured. *J. Struct. Div. ASCE* (May 1979).
5. IRWIN, H. P. A. H. and SCHUYLER, G. D., Experiments on a full aeroelastic model of Lions' Gate Bridge in smooth and turbulent flow. National Research Council of Canada Laboratory Technical Report LTR-LA-206, Ottawa, 1977.

43

Structural Assessment of a Bridge with Transversal Cracks

CHARLES ABDUNUR

Laboratoire Central des Ponts et Chaussées, Paris, France

and

JEAN-LOUIS DUCHÊNE

Laboratoire Régional des Ponts et Chaussées, Le Bourget, France

ABSTRACT

To estimate the influence of cracks on the mechanical behaviour of a bridge, these discontinuities are assimilated to a series of elastic or plastic hinges, alternating with sound beam segments and jointly setting up a new system in equilibrium. The main difficulty resides in the calculation of the hinge residual flexural stiffness, which is a function subject to various parameters and assumptions. To offer a practical solution and facilitate realistic modelling, we developed an experimental method based on the relationship between bending moment and resulting curvature. Under a test load the measured curvature redistribution reflects the new mechanical response of the structure and leads to the actual stiffness functions of the hinges. These are introduced into a program of structural analysis. The new statical system, thus defined, enables the real stresses and residual strength to be estimated. In the technological field, prototype inclinometry instruments of high accuracy had to be developed. This method was tested in the laboratory then successfully applied on site.

INTRODUCTION

Flexure cracks are among the main structural defects observed while inspecting reinforced or prestressed concrete bridges. To assess the *relative* residual strength and optimise the strengthening of such defective structures, it is first necessary to explore their new statical systems. The actual stresses can thus be predicted under a given loading.

BASIC ASSUMPTION

The statical system may be obtained by the following basic assumption illustrated in Fig. 1: the cracked sections and their disturbed vicinities are assimilated to a series of elastic or plastic hinges H, alternating with sound beam segments B and jointly setting up a new system in equilibrium.

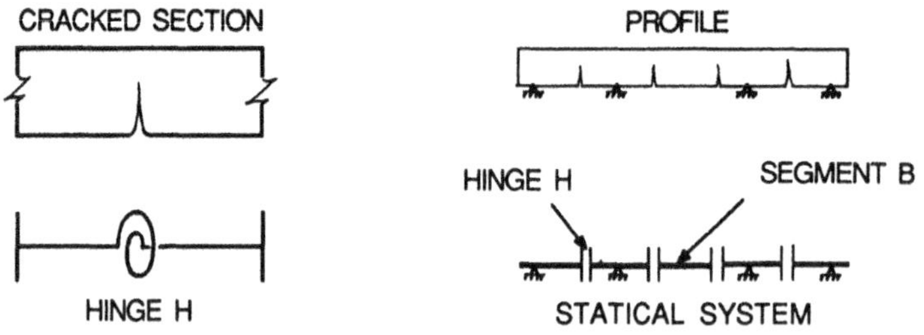

FIG. 1. Modelling assumptions of cracked sections of a bridge.

The resulting stress redistribution depends on the reduction in the flexural stiffnesses of the cracked sections and on the positions of these discontinuities with respect to the initial moment diagram.

EVALUATION OF FLEXURAL STIFFNESS

The sound beam segments B can generally be assumed to retain their initial flexural rigidities $[EI]_0$, which are theoretically given. For concrete bridges the calculation of $[EI]_0$ sometimes remains approximate, because of the varying modulus and the random contribution of non-structural elements to the deck moment of inertia. However, as the *relative* effect of cracking is under discussion, a reasonable value for intact sections can be adopted for both the initial and new mechanical systems.

The main difficulty obviously resides in the realistic calculation of the residual stiffness $[EI]_H$ for the cracked sections or hinges. In reinforced or prestressed concrete bridges, this quantity greatly depends on the extent and geometry of cracks, the constitutive laws of materials under cyclic loads, the redistribution of bond stresses, and the sense and 'viscosity' of crack movement.

The Experimental Option

Prior to theoretical modelling, it is essential to find an accurate experimental method for evaluating the flexural stiffness of a cracked section. Three main reasons can be given:

(a) Several variables of the complex function of residual stiffness are subject to assumptions which can only be verified by experiment.
(b) Rapid applications are often required to assess, in pragmatic terms, the actual mechanical behaviour of a structure.
(c) A specific metrology is at present being successfully developed for this purpose.

The Assessment Principle

The proposed experimental evaluation of stiffness, at a point x along the bridge, is based on the simple relationship between the bending moment $M(x)$, corresponding curvature $\theta'(x)$ and flexural rigidity $EI(x)$.

If, under a convenient test loading, the measured curvature diagram of the bridge is established, and if the stiffnesses $[EI]_0$ of the sound beam segments are given, then the actual bending moments and the cracked section stiffnesses $[EI]_H$ can be deduced through a chain application of the classical beam equation. The procedure is explained below.

Adopted Rotation Measurements

Curvature variation is usually obtained from the strain profile, measured with strain gauges in a sound section and with additional displacement sensors in a cracked section. This extensometry technique is obviously inconceivable for most bridges concerned, with multiple cracks and inaccessible residual sections.

To cope with the nature of the problem, we turned to broader aspects of measurement as provided by inclinometry, where substantial technological progress has recently been achieved: 10^{-6} rad accuracy, miniaturisation, robustness, simplicity and movability. These rotation measurements enable the detection of discontinuities in the bridge profile, thereby obtaining main angular deformations and sieving out local disturbances.

From a theoretical viewpoint, the following expressions of curvature $\theta'(x)$ are equivalent for a sound beam section of height h and position x, with a bending moment $M(x)$ and a strain difference $\Delta\varepsilon$ between the extreme fibres:

$$\theta'(x) = \underbrace{\frac{d\theta}{dx}}_{\text{Inclinometry}} = \underbrace{\frac{\Delta\varepsilon}{h}}_{\text{Extensometry}} = \frac{M(x)}{EI(x)}$$

For a cracked section the extensometry expression is very difficult to apply, owing to stress concentration and complex strain redistribution. The inclinometry expression remains reasonably valid.

Procedure and Interpretation

A test load cycle is applied and the resulting rotations are measured, at sufficiently close paces, to detect discontinuities and deduce curvature throughout the bridge spans.

Figure 2 outlines the interpretation of results at a given discontinuity or hinge. For a convenient load configuration producing a moment $M(x)$, and at every rotation discontinuity $\Delta\theta$, the curvature variation $\theta'(x)$ is plotted over a length covering the hinge H and the near parts of both adjacent sound beam segments B_1 and B_2, conserving their initial stiffnesses $[EI]_0$.

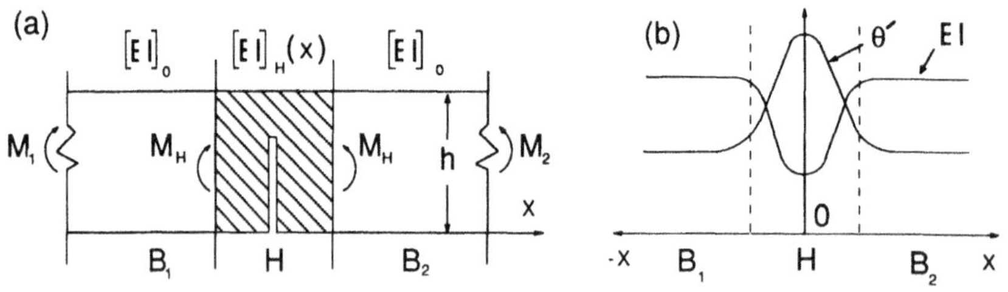

FIG. 2. (a) Moments M and stiffnesses EI in cracked sections H and sound segments B. (b) Redistribution of curvature θ' and stiffness EI.

The curvature response, in its redistributed form, reflects the new mechanical behaviour of the structure.

Applying the beam equation in B_1, H and B_2 leads easily to the relative residual stiffness of the hinge or cracked section:

$$\frac{[EI]_H(x)}{[EI]_0} = \frac{\theta'_B}{\theta'_H(x)}$$

the right-hand expression being deduced from the plotted curve.

This quantity is not only a function of position within the hinge but may also depend on the applied moment $M(x)$ if the hinge is plasticised.

With a reliable value of $[EI]_0$ the residual stiffness can be expressed in absolute form, $[EI]_H(x)$.

THE NEW STATICAL SYSTEM

The relative or absolute stiffness function of each cracked or plasticised section is scrutinised from a durability viewpoint then inserted into a computer program of structural analysis. The new statical system, thus defined, enables

—the prediction of the real stresses in the structure under any given loading, and
—the assessment of the load-bearing capacity and the optimal needs for strengthening.

The procedure can be repeated at a later stage to verify the effectiveness of eventual repairs or simply to follow up a time-dependent mechanical change.

PRELIMINARY FIELD INVESTIGATION

Two identically constructed bridges, one with transversal cracks and one apparently sound, were scrutinised under loading to test the feasibility and sensitivity of the proposed method as regards instrumentation and adopted criteria.

Under the same load system of a convoy, the observed rotations and deflections were respectively 25% and 30% greater in the damaged bridge (Fig. 3). Extensometry measurements, simultaneously carried out for mere comparison, confirmed the same percentage difference between the strain profiles of corresponding sound sections of the two structures.

A third damaged bridge of a similar three-span configuration was tested by inducing a controlled vertical fluctuation at one end support. Section rotations θ and crack gap δ remained linear with level change, except for a short interval around zero (Fig. 4).

The instantaneous crack width and strain responses to flexure are more pronounced on opening the crack than on closing it (Figs 4 and 5).

These general observations show the feasibility of consistent rotation measurements on site and their aptitude to detect crack effects.

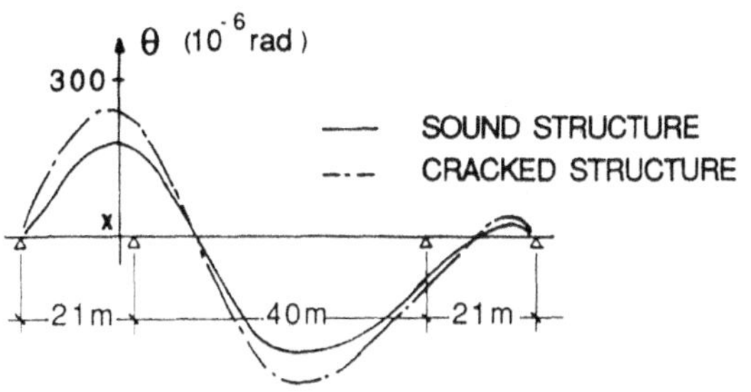

FIG. 3. Effect of cracks on the rotation influence line for a bridge section X under a passing convoy.

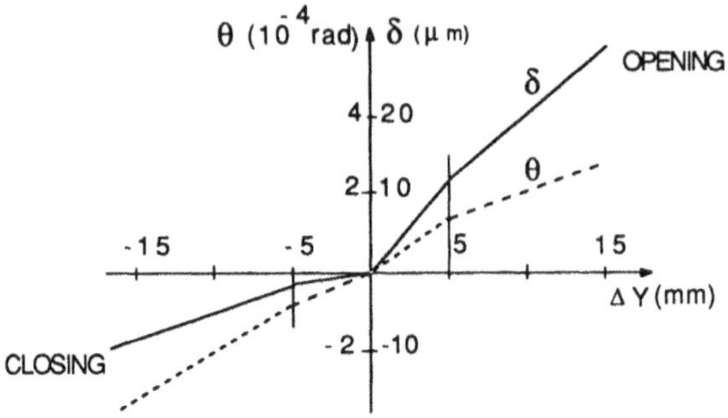

FIG. 4. Gap δ and rotation θ of a cracked section due to level change $\pm\Delta y$ at an end support.

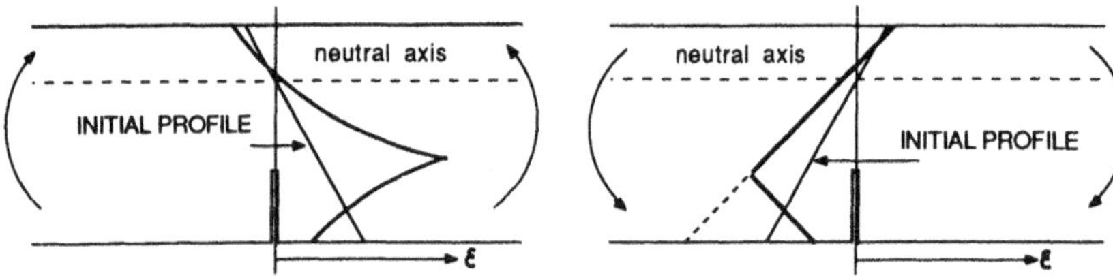

FIG. 5. Influence of the sense of flexure on strain redistribution in a cracked section.

LABORATORY TESTS

The influence of cracks on curvature and stiffness redistribution is examined in the laboratory, under a gradually increasing number of parameters, to explore further possibilities of rotation measurements and to verify or modify certain assumptions in view of a more realistic modelling. Strains and displacements were also measured but strictly for confirmation purposes of test results.

Research began on the preliminary model of a slotted steel section and continued on a prestressed concrete beam.

The Preliminary Model

A simply supported steel I-beam was subjected to a four-point flexure producing a constant moment over the span centre. Rotations θ_0 and deflections y_0 were measured.

A vertical slot was then cut at midspan, starting from the tension flange (Fig. 6). Under the same loading, the same measurements were repeated and completed by rotations θ_i and θ_s, respectively along the intact and slotted flanges, taken at close intervals throughout the span.

The results, summarised in Fig. 7(a), show the aptitude of rotation

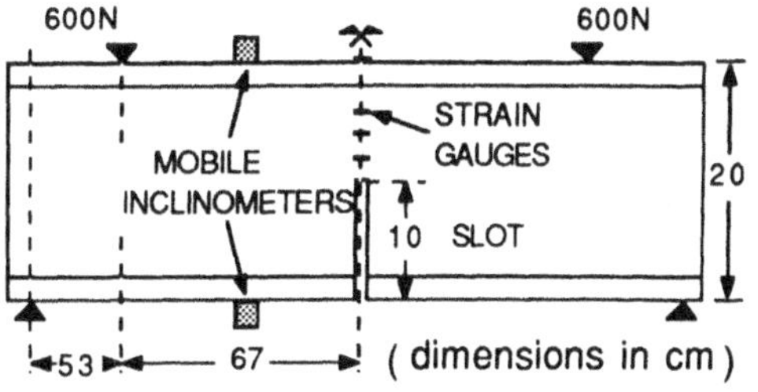

FIG. 6. Slotted steel rolled section under four-point bending.

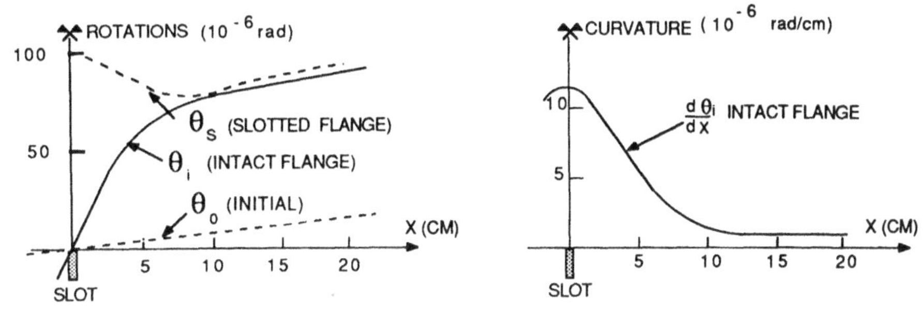

(a) (b)

FIG. 7. (a) Measured rotation redistribution around the slot in a loaded steel I-beam. (b) Deduced curvature redistribution.

measurements to define the stiffness redistribution zone around the slot. Rotations θ_i, at the intact flange, can be differentiated (Fig. 7(b)) to obtain the curvature or the inverse stiffness function (since the moment is constant). Rotations θ_s, at the slotted flange, evidently have a load-dependent discontinuity.

The theoretical model of the slotted beam, using the new stiffness function, yielded a mechanical behaviour in perfect agreement with direct measurements such as deflections and strain profiles of sound sections.

The Concrete Beam

We proceeded in the same way to test a simply supported micro-concrete T-beam prestressed with an unbonded tendon and lightly reinforced (Fig. 8).

Before cracking convenient four-point bending was applied. Rotations and deflections were measured to determine the actual initial stiffness $[EI]_0$ of the beam.

Three flexure cracks were induced under a temporary, cautiously increasing, point load at midspan. The cracked sections were instrumented with strain gauges and displacement sensors.[1] At one crack, after a minimum removal of the concrete cover, strain gauges were also fixed on both the unbonded tendon and the reinforcement bars to estimate the tensile forces actually developed.

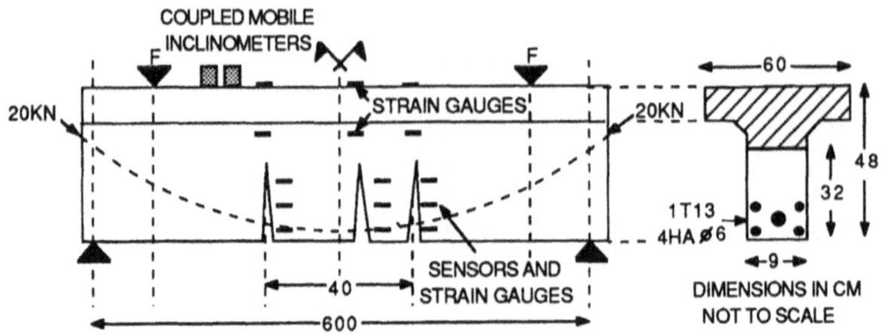

FIG. 8. Instrumented cracks of a prestressed beam.

Direct determination of curvature

The initial four-point bending was resumed. Corresponding differential rotations were recorded by two coupled inclinometers at close intervals along the intact upper flange of the cracked sections and their vicinities. Other measurements were maintained.

The curvature redistribution thus obtained, under a constant bending moment, leads to the inverse relative stiffness $[EI]_0/[EI]_H$ as a function of position x over the influence length of the three cracks. The experimental diagram is shown in Fig. 9.

Comparison with other data

As in the previous steel beam, the theoretical model based on the present stiffness redistribution was used to calculate the *increase* in midspan deflection, support rotations and normal fibre strain. These computed increments, owing to cracking, fully agree with direct parallel measurements.

Moreover, the curvature-deduced residual stiffness (Fig. 9) coincides with that estimated from the combined response of strain gauges and displacement sensors at each crack.

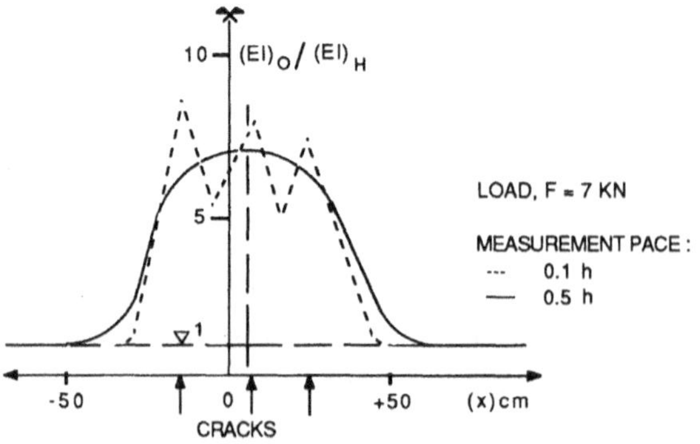

FIG. 9. Redistribution curve of the inverse relative flexural stiffness over a beam segment with three cracks.

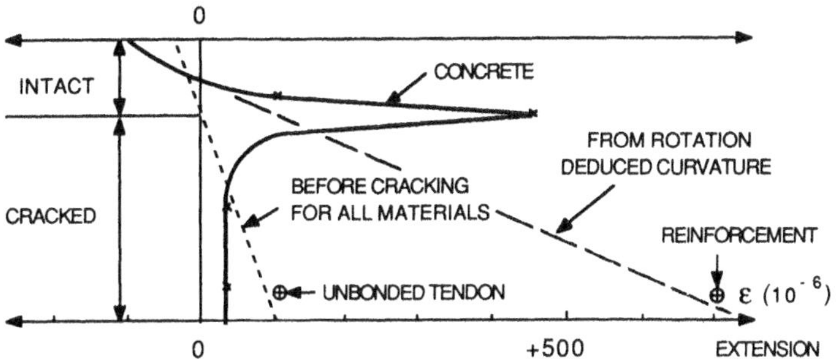

FIG. 10. Strain profiles and strain point values for concrete and steel at a cracked section. The rotation-deduced curvature coincides with strain data.

Figure 10 shows the logical yet complicated concrete strain profile obtained at a cracked section, hence the impracticality of classical extensometry for operational use in similar cases. It can be seen, however, that strain measurements on the upper remaining concrete section, combined with those on reinforcement bars, do confirm the curvature deduced from rotations.

Other Parameters

To study the effects of statical indeterminancy and bond redistribution, research is being extended respectively to continuous spans with a similar concrete section and to simply supported beams with various reinforcement patterns.

FIRST APPLICATIONS ON SITE

The rotation method is now being used on reinforced and prestressed concrete bridges with transversal cracks. At damaged sections the deduced relative residual stiffnesses confirm the bell-shaped distribution obtained in the laboratory and shown in Fig. 9. For the same relative height of cracks, field and laboratory stiffness values are very close at corresponding positions.

Over damaged segments the optimal spacing of rotation readings varies from 0·1 to 0·5 of the beam height.

In some cases a simplified version of the method may be sufficient, as in the following example. At a flexure crack of an inspected bridge, the actual rotation discontinuity $\Delta\theta$, measured by coupled inclinometers on either side, is represented in Fig. 11 as a function of bending moment, applied by a test load. This relationship is generally non-linear but can be linear for cracks already open under dead load. Hence the non-linear deviation, at the

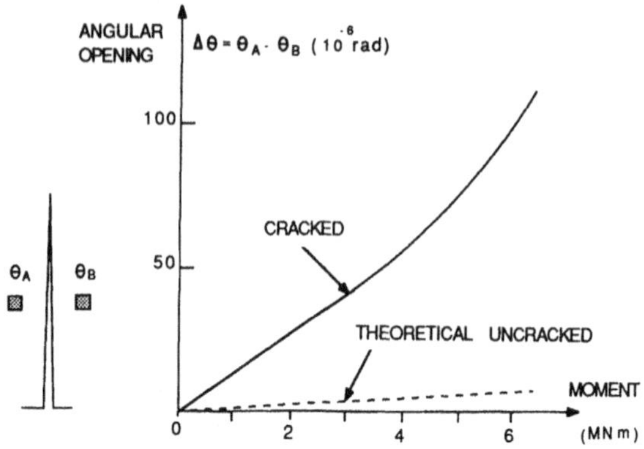

FIG. 11. Crack angular opening (measured on a bridge by two inclinometers) versus applied moment.

extremity of the curve, marks the advance of the crack tip under critical loading.

We can therefore replace the residual flexural stiffness $[EI]_H(x)$ by another experimental coefficient, $k = M/\Delta\theta$, which in this case is constant.

As already explained, the coefficients thus obtained for similar cracks enable the bridge to be recalculated on a more realistic basis.

CONCLUSION

An experimental method was developed to determine, under a test load, the curvature redistribution throughout the spans of a damaged bridge. This redistribution leads to the evaluation of the actual residual stiffnesses of the cracked sections.

The new statical system of the structure can thus be defined to enable

—the prediction of the real stresses in the bridge under any given loading, and

—the assessment of the load-bearing capacity and the optimal needs for strengthening.

The whole procedure can be repeated at a later stage to verify the effectiveness of repairs or simply to follow up a time-dependent mechanical change.

Supported by numerical techniques, this experimental approach is suitable not only for operational ends but also in research as a means to tackle the various parameters involved and to establish a more realistic theoretical model.

REFERENCES

1. Chatelain, J., Bruneau, J. and Duchêne, J.-L., Estimation par des essais de chargement du défaut de résistance à la flexion de certains tabliers en béton précontraint. International Conference on Inspection, Maintenance and Repair of Road and Railway Bridges, Brussels–Paris, 1981.
2. Godart, B. and Duchêne, J-L., Intervention sur le Pont de Champigny/Yonne. Conference on the Inspection and Testing of Structures, EPNC, Paris, 1987.
3. Godart, B., Approche par l'auscultation et le calcul du fonctionnement de ponts en béton précontraint fissurés. Euro-American Conference on the Rehabilitation of Structures, CEBTP, Saint-Rémy-lès-Chevreuse, France, 1987.
4. Abdunur, C. and Duchêne, J.-L., Mesures de rotations pour le schéma statique d'un ouvrage fissuré. International Conference on Measurements and Testing on Civil Engineering, Lyon, 1988.
5. Chatelain, J. and Godart, B., Evaluation de l'état mécanique réel de ponts en béton précontraint. IABSE Symposium, Helsinki, 1988.

44

Reliability Analysis Applied to Deteriorating Bridge Structures

JONATHAN G. M. WOOD, R. ASHLEY JOHNSON

Mott MacDonald Special Services Division,
20–26 Wellesley Road, Croydon, Surrey CR9 2UL, UK

and

CHARLES ELLINAS

Advanced Mechanics and Engineering, 4 Frederick Sanger Road,
Surrey Research Park, Guildford, Surrey GU2 5YT, UK

ABSTRACT

Reliability analysis techniques developed for steel bridges and offshore structures provide a methodology which can also be used to evaluate the safety of substandard or deteriorating reinforced concrete structures and arch bridges. This is illustrated by reference to cases of structures with alkali aggregate reaction and chloride-induced corrosion damage. The shortage of data on the statistics and deformation characteristics of concrete failure limits the quantitative application and highlights the necessity for more testing of deteriorated reinforced concrete. Consideration of overall reliability must include assessment of risks inherent in road closures for inspection and remedial work, the uncertainties of inspection procedures, and the reliability of repair and corrosion control.

INTRODUCTION

In the UK the CIRIA Report 63[1] has provided the framework for probabilistic methods for calibrating partial factors for loads and strength to achieve more consistent reliability in design. This partial factor approach, in a somewhat simplified form, has been incorporated in British Standards like BS 8110[2] for reinforced concrete and BS 5400[3] for bridges with widespread, but not universal, support in the engineering profession. Neither code provides a margin for deterioration.

This approach was used more rigorously in the aftermath of the steel box girder failures in the UK, Germany and Australia for design of remedial works, and to guide the major Merrison programme[4,5] of research and testing on steel box girders which led to the simplified BS 5400: Part 3 steel bridge code.

The methodology has since been further developed and refined in the UK offshore industry, predominantly for steel structures in the North Sea.[6] The logical framework provided by a probabilistic approach to determining the reliability of structures is tailor-made for resolving the major problems which arise when deterioration in bridges creates uncertainty about their current and future safety. In this paper we outline how techniques used in the design and appraisal of steel bridge structures and offshore structures are being applied to the particular problems of deterioration in reinforced concrete bridges with substandard details, alkali aggregate reaction and/or chloride-induced corrosion, and the potential for their application to other structures like arch bridges.

It is our view that reliability analysis must be based on a thorough knowledge and understanding of those factors which actually lead to

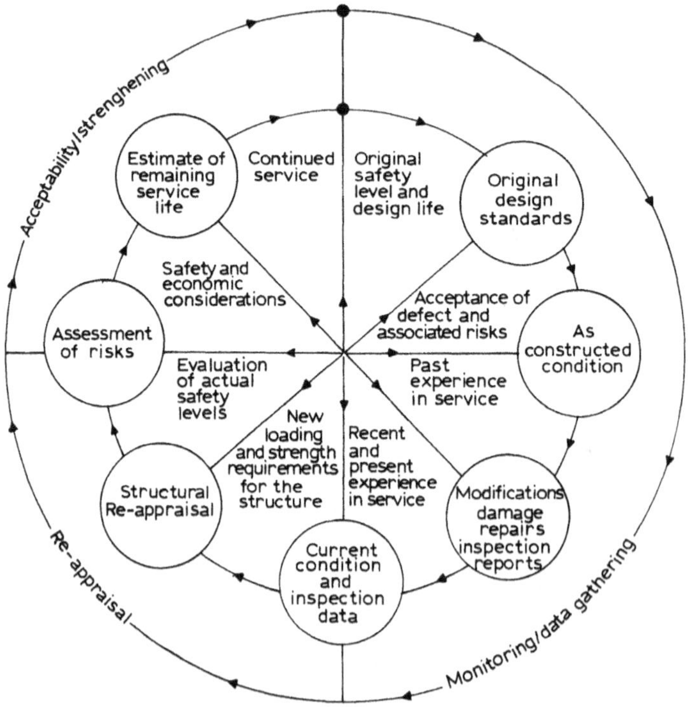

FIG. 1. The cycle of appraisal and reliability assessment.

structural failure as distinct from the convenient oversimplifications of structural behaviour in design codes. The analysis of well-recorded cases of structural failure[7–9] provides the first source of data. However, it soon becomes clear that actual failures are fortunately too infrequent to provide statistical data on loading, strength or the reliability of inspection.

As there are not enough case studies we must use site measurements and laboratory tests to provide us with the statistical data on the variability of loading and strength that is needed for proper analysis. Data on the reliability of inspection and monitoring techniques in detecting the severity and rate of structural deterioration will also be required. While general corrosion produces spalling and AAR produces cracking as warning signs, pitting corrosion is not easily detectable. Many of the most structurally sensitive parts of structures are not readily accessible for inspection. Determination of actual construction details and ground conditions provides another area of substantial uncertainty, especially for arch bridges.

Because the methodology of reliability analysis has been well established in other disciplines[10,11] it is not necessary to reinvent the framework of analysis when applying it to concrete structures (Fig. 1).

RELIABILITY IN DETERIORATING STRUCTURES

The majority of deteriorating reinforced concrete structures are old enough to have been analysed for design without benefit of a computer, often for smaller loads and temperature ranges than currently required. Some of their designers used robust over design as a substitute for precision. Others, without benefit of current knowledge, provided inadequate designs, particularly in shear, corbel design and in reinforcement detailing. Inadequate provision for articulation and thermal effects is a frequent source of massive overstress in appraisals but with a small serviceability risk and no collapse hazard.

The application of reliability approach to determining the rising risks associated with a deteriorating structure, which may also have been built to a lower design standard, identifies these types of uncertainty:

(a) Inherent uncertainties of variability of loading and strength.
(b) Uncertainties in our knowledge of loads, structural response, ground conditions, initial strength and deterioration which can be reduced by control analysis or testing.
(c) Uncertainties in the risks to the public arising indirectly from highway restrictions during inspection or remedial works and directly from falling spalled concrete or collapse.

Loading

Dead load can be determined more accurately for a built structure. Increases in the weight of services and surfacing can be controlled to enable loads and partial factors to be reduced.

The most severe traffic loadings on short span structures often arise from short special vehicles (e.g. cranes) in the 50–100 ton range, not from the variability inherent in normal C&U commercial lorry traffic. Traffic control of these 'over 50 ton' vehicles and lane restrictions, evaluated from load effects from the most adverse actual vehicle combinations, enable more accurate and reduced loadings to be adopted for substandard structures. However, reliability becomes very sensitive to the ability of police and highway authorities to enforce controls. For reliability analysis HA and HB loading are completely inappropriate. Garages of onerous real vehicles with associated frequencies of occurrence must be used.

In normal design a simplified approach to environmental loads (wind, overall temperature and differential temperature) is possible. For reliability analysis these must be considered explicitly with the refined traffic load analysis, particularly where probability of coincidence is high (e.g. high winds and/or cold increase the risk of accidents, which cause high load conditions of close-packed slow-moving vehicles on long span bridges). Impact loading from overheight or crashing vehicles, which is poorly represented in standards, gives a substandard risk of damage from overloading and must be considered.

When considering members in which AAR is developing the effects of restrained expansion must be considered as 'loads' in a similar way to thermal expansions.

Analysis of Overall Structural Response

Because most deteriorating structures predate computer analysis a detailed elastic appraisal inevitably shows areas of substantial over-provision of strength and areas of underprovision. Typically in bridge decks longitudinal flexural steel is overprovided except at the edge, while there is underprovision of transverse flexural steel. However, concrete is not as elastic as computer elements. Creep and cracking permit redistribution, provided ductility is available from suitably detailed reinforcement which has not been degraded by deterioration. The initial reserves of strength and ductility enable many deteriorating structures to continue in service without undue risks, despite severe local deterioration away from critical structural elements. The risks arising from localised deterioration in a structure are very sensitive to the availability of alternative load paths due to structural redundancy. The non-linear analysis of the structure from the local distress 'serviceability' stage through to the 'collapse' limit state enables this to be evaluated. It also provides an indication of potential cracking conditions which inspections may detect as a record of overloading and/or distress.

When comparing the result of detailed computer analysis with simplified code analysis the implicit assumptions of concrete design codes must be considered. The writers of codes carefully exclude all difficult analytical problems by placing limits on the detailing and proportioning of structures to ensure that

(a) strength calculations are based on simplified rules which do not relate to the actual mechanism of failure;
(b) structures are sufficiently ductile so that even if the idealised distributions of forces and stresses that come from structural analysis do not occur in the real structure the ductility will enable them to be redistributed before the ultimate limit state is reached; and
(c) many second-order effects like differential temperature and locked-in stresses from early thermal effects can be discounted due to the ductility provided.

This characteristic of concrete structural codes mirrors the detailing requirements for proportioning stiffeners and plates in steel codes to prevent the initiation of buckling before yield has occurred. The lessons learned from

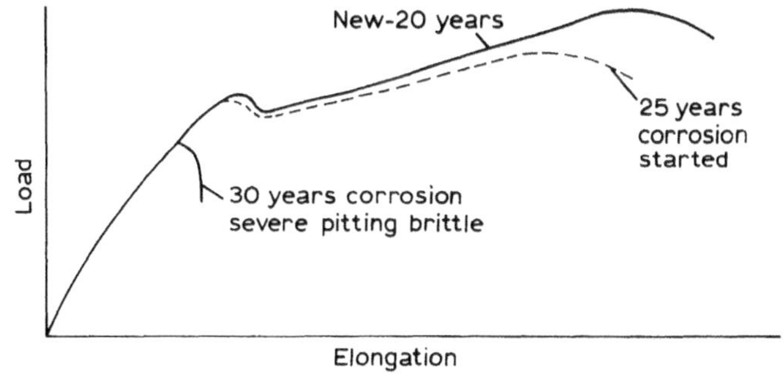

FIG. 2. Loss of ductility and strength in reinforced concrete with chloride ingress.

steel structures (in particular box girders) is that when buckling is initiated collapses occur suddenly without warning. The embrittlement of concrete structures (Fig. 2) by corrosion of steel or concrete deterioration produces the same type of rapid, catastrophic failure mode as buckling. Inspection and/or proof loading provide no warning of impending failures. It is the embrittlement from deterioration in concrete structures which provides the greatest increase in risk, not just change in strength.

EFFECTS OF ALKALI AGGREGATE REACTION ON MATERIAL PROPERTIES AND CONSTITUTIVE MODELLING

Our knowledge of the structural effects of alkali aggregate reaction has increased substantially since the early 1980s. AAR is probably the most difficult deterioration phenomenon to be analysed in concrete. In 1983 initial results of the effects of simple uniform AAR expansion in the finite element analysis of a pile cap were presented.[12] Following a review of those results and the research literature we stopped finite element analysis of alkali aggregate reaction until sufficient data on the actual stiffness, expansion and strength characteristics of concrete with AAR were available for a valid constitutive model to be evolved for finite element analysis. Over the last two years we have restarted finite element analysis work on AAR using data now available[13] This detailed evaluation is only appropriate to the most severely expansive and sensitively detailed parts of structures as identified using the Institution of Structural Engineers' approach.[14]

Firstly, it has been necessary to evaluate the expansive characteristics of the concrete and the extent to which this is restrained by the reinforcement. It is now clear that the forces generated by restraint of a severe reaction can be sufficient to yield the steel in most reinforced concrete members and it can produce compressive failures at expansion joints. Therefore these forces must be evaluated and considered as explicit forces in the analysis. In some structures they serve to prestress the concrete and have a beneficial effect. In other structures it is the magnitude of these forces combined with the reduction in strength which damage the structure. The magnitude of forces developed are very sensitive to the reactive aggregate and gel characteristics and to the conditions of exposure.

It has become clear that there is substantial variability in expansion at all levels, ranging from that around individual reactive particles up to the variation between adjacent pours. Even between nominally identical concrete pours this variability is high. The expansion is randomly generated by the variations in reactive aggregate concentration and alkali concentration within the materials. The local rate of expansion is very sensitive to moisture supply. It is this variability of expansion which gives rise to the characteristic map cracking in AAR. At its smallest scale it gives rise to the microcracking which changes the Young's modulus, tensile strength and compressive strength of concrete.

The stiffness changes in concrete with AAR are probably the most sensitive measure of microcracking. It is a simpler and more quantitative test and more representative of a structural volume of material than the qualitative techniques of petrography which help us understand the

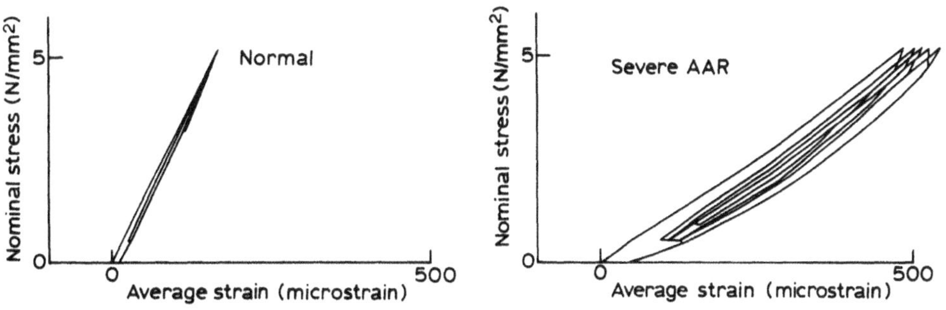

FIG. 3. Effect of AAR on stiffness of concrete.

deterioration. Figure 3 shows the change in load deformation characteristics of concretes with alkali aggregate reaction for a normal concrete and a severely deteriorated concrete. It can be seen that there is a reduction in the Young's modulus, a development of increased hysteresis and a progressive packing down of the core. This approach and its correlation with strength changes is described in Ref. 15.

In evaluating the tensile strength of concrete our understanding has been hampered by the difficulties of sample preparation and testing. The Brazilian splitting test is simple but unrepresentative of normal structural behaviour. It is sensitive to the inaccuracy of normal coring from existing structures. Its only merit seems to be that lots of people have done it before. Gas pressure testing is an interesting measure of tensile strength but difficult to relate to normal structural behaviour. Because the mode of failure of most concern in structures with alkali aggregate reaction is shear the torsion testing of cores[15] as illustrated in Fig. 4 gives us a much better measure of the changed

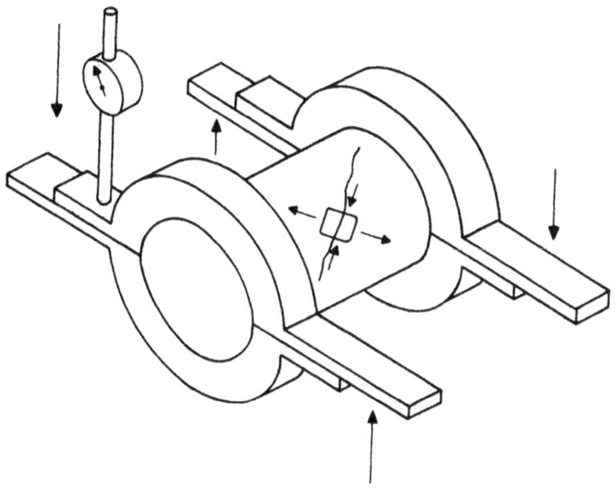

FIG. 4. Torsion testing of concrete cores.

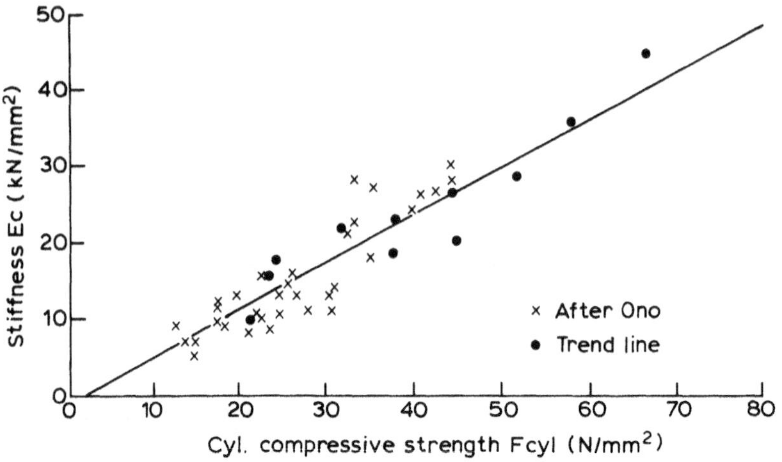

FIG. 5. Compressive strength test comparison, southwest concrete (5% reactive chart).

strength for evaluation of shear and bond behaviour. Results related to the microcracking as indicated by reduced Young's modulus are shown in Fig. 5. The growing train of strength below the design characteristic substantially increases failure risk.

Having identified the changes in concrete properties from AAR we can consider structural forms which will be sensitive to these changes.

PUNCHING SHEAR

An example is a flat slab bridge deck, with developing AAR, reinforced only in its top and bottom faces and supported on discrete columns. The rules for the evaluation of punching shear strength of normal concretes are largely empirically based on a limited series of tests. We were surprised to find in a review of the literature, and in discussions with those who carried out the work, that until recently none of these tests determined the load displacement characteristics to failure. Testing we have recently carried out shows that

(a) this punching shear failure mode is extremely brittle, as shown in Fig. 6; there is no inspectable warning of failure;
(b) the softening and unloading part of the curve is influenced by detailing of the flexural reinforcement but not sufficiently to make it ductile; and
(c) compressive membrane action can have a beneficial effect on punching strength in tests but in structures the high flexural stresses in adjacent concrete will limit this.

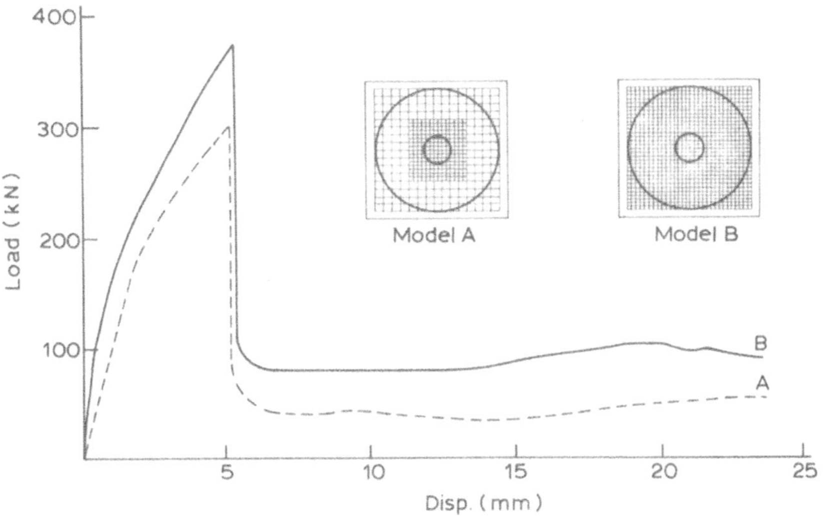

FIG. 6. Effects of reinforcement slabs on punching shear response.

Testing to determine changes in behaviour in slabs with AAR is in progress.[16]

Overall analysis shows that the deformation prior to fracture at a column head is not sufficient to significantly redistribute load to neighbouring columns. Once the ultimate strength is exceeded the residual strength capacity is well below the dead load component for concrete structures. This is the classic situation for progressive collapse. The normal ductility inherent in concrete frame design is no longer present. This type of progressive collapse from punching shear failures has occurred in the USA. One instance was precipitated by corrosion damage to a slab in a car park.

CORRELATION OF DETERIORATION WITH STRESS STATE

Another aspect of reliability analysis that needs to be considered is the correlation of deterioration rate with areas of high stress. In a bridge slab the reliability of waterproofing is sensitive to any cracking in the concrete deck. Cracking is most likely to occur at the most highly stressed locations and waterproofing breakdown can then preferentially supply water and salt to accelerate the AAR or corrosion. The basic reaction of alkali aggregate reaction continues in dry conditions and then can swell very rapidly with this subsequent introduction of water. The crosslinking of highly stressed areas with cracking leads to water ingress which then generates further cracking and creates a substantially more serious reduction in reliability than a random development of damage.

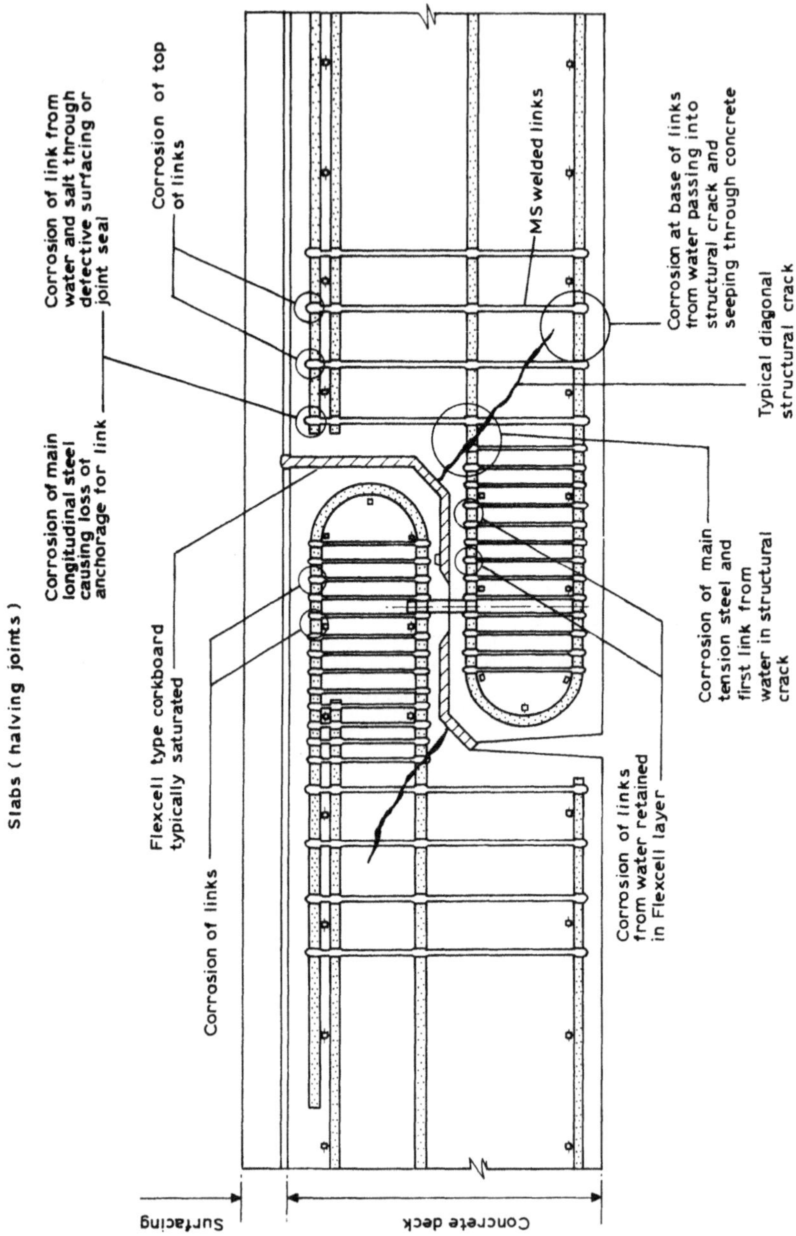

FIG. 7. Half-joint corrosion.

Similar phenomena occur with chlorides where cracking in highly stressed components permits preferential channelling of chlorides to the most highly stressed steel. This aspect needs particular consideration in assessment for reliability. A classic example of this is the half-joint shown in Fig. 7. Here the structural effects of the corrosion are further aggravated by damp, low oxygen conditions which favour the development of localised pitting corrosion. Recent research into corrosion in cracks suggests general rules on acceptable crack widths may underestimate their effect on the corrosion rates where chlorides are present.

DETERIORATION RATE MODELLING

The modelling of the reliability of deteriorating structures must introduce the third element, which is the predictive modelling of the deterioration phenomena and their rate of development.[17] That brings us round the circle to new design where the failure to consider durability and deterioration phenomena explicitly, and to provide a margin on initial strength to offset for deterioration, has been a major factor contributing to the problems we face in many structures today. The corrosion rate is strongly influenced by environment (oxygen supply, humidity, pH, chloride concentration and temperature). Similar AAR damage rates can be related to temperature and moisture conditions for a specific aggregate type. Reliability analysis shows that the control of the environment to limit deterioration rates, where this is possible, is one of the most effective methods of maintaining safety. This attention to detailing and maintenance of expansion joints and waterproofing is essential for safety.

CONCLUSION

The challenge for the coming years must be

(i) to build on our experience of reliability methods for steel structures;
(ii) to carry out testing to evaluate changes in the physical strength properties and the way in which deterioration phenomena change them;
(iii) to model the mechanisms of deterioration in terms of chloride ingress rates, corrosion rates or rates of development of alkali aggregate reaction; and
(iv) to ensure that a proper quantification of durability phenomena is introduced into the design process.

If one compares the number of concrete structures which are being demolished because of deterioration and lack of durability and those which have given rise to problems because they are under strength, it is clear that the main emphasis of education in the last 20 years has been too preoccupied with matters of structural analysis and too little attention has been given to durability.

To answer the client's question 'How long will it safely last?' we must adopt reliability analysis methods, extend our body of test and case study data, and hope we can learn faster than the corrosion and alkali aggregate reaction progresses.

REFERENCES

1. CIRIA Report 63, *Rationalisation of Safety and Serviceability Factors in Structural Codes.* CIRIA, London, 1977.
2. BS 8110, *Structural Use of Concrete.* British Standards Institution, London, 1985.
3. BS 5400, *Steel, Concrete and Composite Bridges. Part 1: General Statement 1988; Part 2: Loads 1978; Part 3: Steel Bridges 1982; Part 4: Concrete Bridges 1984.* British Standards Institution, London.
4. Institution of Civil Engineers, Steel box girder bridges. In *Proc. Int. Conf. ICE*, London, 1972.
5. Rockey, K. C. and Evans, H. R., *The Design of Steel Bridges.* Granada Publishing, London, 1981.
6. Smith, D., Csenki, A. and Ellinas, C. P., Ultimate limit state of analysis of unstiffened and stiffened structural components. In *Integrity of Offshore Structures*, ed. D. Faulkner *et al.* Elsevier Applied Science, London, 1987, pp. 145–65.
7. Walker, A. C., Study and analysis of 120 cases of structural failures. In *Proc. of Symposium on Structural Failures in Buildings.* Institution of Structural Engineers, London, April 1980.
8. Report of the Royal Commission, *Failure of West Gate Bridge*, Melbourne, 1971.
9. Woodward, R. J. and Williams, F. W., Collapse of Ynys-Y-Gwas Bridge, West Glamorgan. In *Proc. Instn Civ. Engrs, Part 1*, 1988, pp. 635–69.
10. Thoft-Christensen, P. and Baker, M. J., *Structural Reliability Theory and Application.* Springer-Verlag, Berlin, 1982.
11. Smith, D. and Ellinas, C. P., Rationalisation of tolerances through reliability analysis. In *Rational Fabrication Specifications for Steel Structures.* Institution of Mechanical Engineers, London, July 1988.
12. Wood, J. G. M. and Wickens, P. J., Structural effects of AAR and remedial actions. *Alkalis in Concrete 6th International Conference*, DBF, Copenhagen, 1983.
13. Wood, J. G. M., Wilson, J. R. and Leek, D. S., Physical behaviour of AAR damaged concrete in structure and in test conditions. In *Proc. of 8th Int. Conf. on Alkali–Aggregate Reaction*, ed. K. Okada *et al.* Kyoto, Japan, July 1989.

14. INSTITUTION OF STRUCTURAL ENGINEERS, *Structural Effects of Alkali–Silica Reaction—Interim Technical Guidance on Appraisal of Existing Structures*. ISE, London, 1988.
15. CHRISP, T. M., WOOD, J. G. M. and NORRIS, P., Towards quantification of microstructural damage in AAR deteriorated concrete. In *Fracture of Concrete and Rock, Recent Developments*, ed. S. Shah *et al.* Elsevier Applied Science, 1989.
16. CLARK, L. A. and NG, K. E., Some factors influencing expansion and strength of the SERC/BRE Standard ASR concrete mix. In *Conf. Proc. SERC–RMO*, 29 June 1989.
17. WOOD, J. G. M., WILSON, J. R. and LEEK, D. S., Improved testing for chloride ingress resistance of concrete and relation of results to calculated behaviour. In *3rd Int. Conf. on Deterioration and Repair of Reinforced Concrete in the Arabian Gulf*, Bahrain Soc. of Engrs/CIRIA, October 1989.

45

Computer-Aided Sketching of Load paths: An Approach to the Analysis of Multi-span Arch Bridges

W. J. HARVEY and F. W. SMITH
Department of Civil Engineering, The University, Dundee DD1 4HN, UK

ABSTRACT

Single-span masonry arches are notionally three times redundant. This neglects the fact that the abutments are not rigid, so the degree of redundancy is actually greater. In multi-span bridges the problem is compounded. Direct analytical techniques based on a study of deformation are of limited value in these circumstances.

Arches naturally release their redundancy by pushing the abutments outwards. They become three hinge structures. This approach does not allow closed solutions to the problem but rather indicates limits to the engineer's knowledge and understanding of the structure.

Some details of the approach are described, using Robert Stephenson's Royal Border Bridge at Berwick-upon-Tweed as an example.

INTRODUCTION

Thrust Lines and Elastic Analyses

In 1676 Robert Hooke[1] found a solution to the problem of the most important statically indeterminate structure then known. He kept his finding secret for fear of academic rivalry from those whose stature should have precluded such plagiarism. Of course we now know that his solution was of limited value because it provided only understanding, not a scheme of computation. This paper aims to show that understanding can be more useful than essentially meaningless calculated results.

In 1879 Castigliano[2] provided engineers with a computational scheme for indeterminate structures. In various guises his method has held engineers in thrall, and incidentally terrified students, ever since. Many of us recognise that elastic studies may obscure the true behaviour of a structure, but they do allow us to carry out closed-form calculations on structures which might otherwise be intractable.

Castigliano's theorems are particularly inappropriate for masonry structures. Masonry is hardly elastic, and arches in particular are very susceptible to small movements of their foundations, movements which must take place.

It is of course possible to use an elastic analysis in an exploratory way. That is to investigate the effect of movements of the supports and cracking, and different distributions of stiffness. Unfortunately this is rarely done because it requires too much effort on the part of the engineer.

Load Path Analysis

Engineers use load path analysis to assist in the conceptual design of complex structures. The relative stiffness of different paths from point of application to point of support is assessed on the basis of experience. The results are checked by more 'rigorous' methods as the design proceeds. The method proposed here allows the engineer to 'sketch' load paths very quickly on a computer screen. He receives quantitative feedback as to the implications of the load distribution he is assuming.

The Zone of Thrust

Hooke's line of thrust is of course a load path. The line of thrust ignores material strength, but it can be enhanced to show the material which would be required to support the thrust concerned.[3] In a two-dimensional structure, that is to say when studying the arch in elevation only, it is assumed that stress is uniform into the depth of the paper and so only the width of the thrust zone on the paper need be considered. This is obviously a very rapid process.

Graphical Indeterminacy

Barlow[4] gave a fine graphical representation of the indeterminacy of arch structures. He demonstrated with a physical model that there are many potential lines of thrust within an arch, any one of which may represent the actual performance of the structure. The freedom to investigate all the possible thrust lines in a structure may allow the engineer the flexibility he needs to understand its behaviour.

THE INDETERMINATE ARCH

Arches are notionally three times statically indeterminate. In modern structures engineers habitually release these indeterminacies by introducing hinges at the springings and the apex of the arch. Pippard and Chitty[5] showed that real arches will approach the three-hinged condition as they deform after the removal of their centring. Having demonstrated that masonry arches are not elastic rings they proceeded to use Castigliano to produce the MEXE analysis beloved of British bridge engineers. More recently considerable effort has been invested[6,7] in developing complex finite element programs for arch analysis. These still assume that the arch is supported on rigid foundations, which is manifestly untrue.

A force-based analytical technique is much more readily adapted to the problem of arches. The flexibility method consists in releasing indeterminate resultants until the structure becomes statically determinate, then analysing the effect of unit resultants. The stiffness method is the converse of this. All 'joints' are made rigid in space and the fixity reactions computed. Unit displacements are considered in order to find the 'true' deformed shape of the structure. Stress resultants (forces) are computed from the displacements.

A structure which tends to release its redundant reactions after construction may be best analysed by the flexibility method. This is especially true if a path can be opened to the exploration of the effect of reactions which cannot sensibly be computed. We will endeavour to demonstrate that such a technique is not just possible but actually easier than some others to implement.

THE RELEASED STRUCTURE

Natural Release: The Minimum Thrust View

A masonry arch deflects towards a three-hinge structure on release of the centring. This must be the case because the arch ring will shorten under stress and the abutments will deflect outwards. The shorter ring will no longer fit perfectly into the springings. As this deformation takes place, the horizontal component of thrust in the arch will reduce towards a minimum value. The minimum corresponds to the maximum possible rise of the zone of thrust within the arch ring (Fig. 1). Analysis of the three-hinged arch, which is the limit of this process, is very simple. It involves writing and solving two linear simultaneous moment equations. This structure may sensibly be regarded as the released form for the application of redundant actions.

Forced Release: The Maximum Thrust Case

It is of course conceivable, and was demonstrated in practice by Chettoe and Henderson,[8] that the springings may be forced together by external effects. This might be caused by the approach of a heavy vehicle. The limit

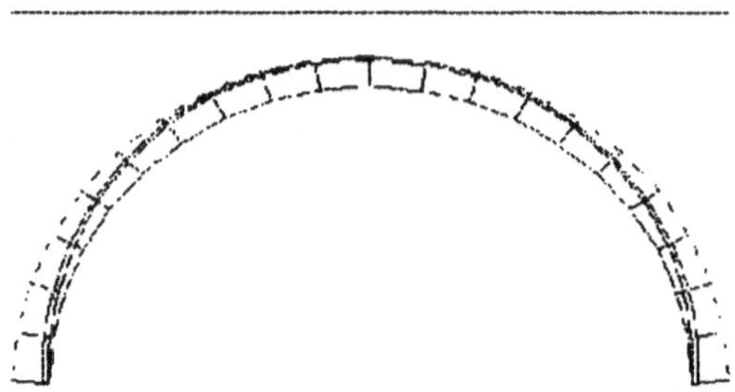

FIG. 1. Arch with minimum thrust.

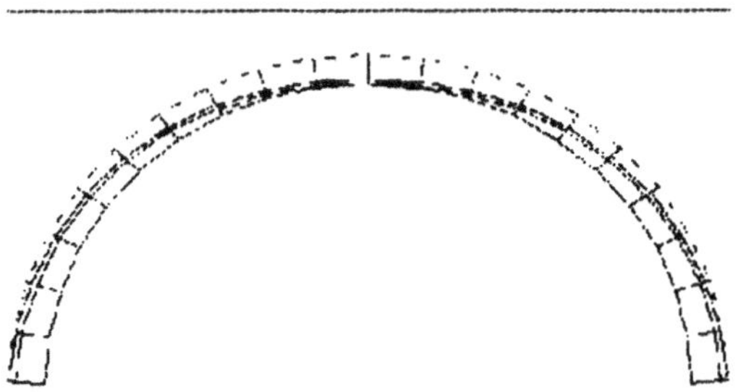

FIG. 2. Arch with maximum thrust.

case of this movement is when the arch begins to lift in a three-hinged jacking mechanism (Fig. 2). The result is still a stable structure which can be analysed as described above. In this case the rise of the zone of thrust is the minimum possible within the arch ring. The horizontal component of thrust in the arch is thus maximised.

The two cases described are exactly analogous to the limiting active and passive pressures exerted by soil on moving walls.

RESTORING REDUNDANCY

A Favoured Case: Minimum Stress

Having produced a released structure, we must now restore its redundant actions and reactions to model the real behaviour. Heyman[9] used the mechanism analysis to compute the line of thrust, which had a minimum deviation from the curve of the arch. Both Heyman and Harvey[3] chose to follow the curve of the intrados of the arch, but the arch centreline might equally well be used. This approach yields a line or zone of thrust which is wholly contained within the arch ring, thus implying the fully redundant structure (Fig. 3). A gross error is made in assuming that this is the correct solution. Heyman clearly did not fall into this trap. The implication is of minimum value of maximum compressive stress, which is not the form a structure naturally seeks.

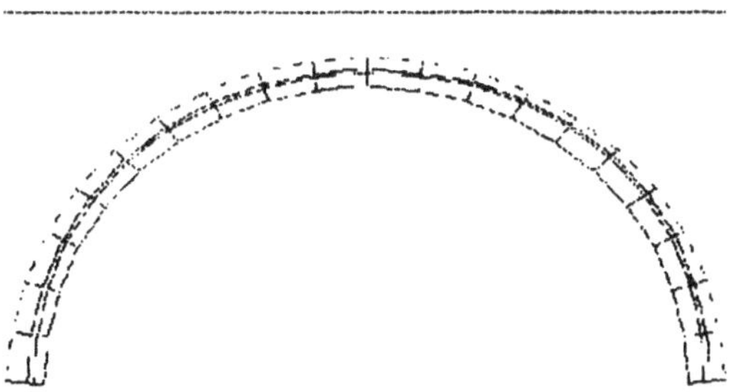

FIG. 3. Arch with minimum stress.

The Infinity of Real Solutions

The line of thrust defined by the above method lies at a point between the two extremes already discussed. Barlow[4] showed that these three solutions are only three of the infinite possible set which exist in an arch which is not on the point of failure.

The Importance of Reactions

The reactions which result at the supports of an arch are important because they control the behaviour of both arch and abutments. In some load cases a larger thrust will be needed, in others the interaction will lead to smaller values. If gross movements are excluded, the arch can deliver whatever reactions are required within the bounds set above. The two limits are reached with a range of abutment movements of less than 0·1% of the arch span.[10]

Exploratory Analysis

The illustrations presented thus far have all been produced using microcomputer software developed at the Wolfson Bridge Research Unit in Dundee. We have a program in commercial use based on the minimum stress theme. We have recently developed the maximum and minimum thrust models which were used for the diagrams. The final step has been to allow the user to alter the location, direction and value of the reactions for the arch. This encourages the engineer to explore the effect of redundancy by observing the relationship between reactions and the thrustline. Because the only computation involved is vector addition, results are presented as a relatively smooth animation even on an IBM PC AT.

MULTIPLE REDUNDANCY

Most real structures have rather more redundancy than the simple arch described above. It is perfectly practicable to use the exploratory technique on multiple spans.

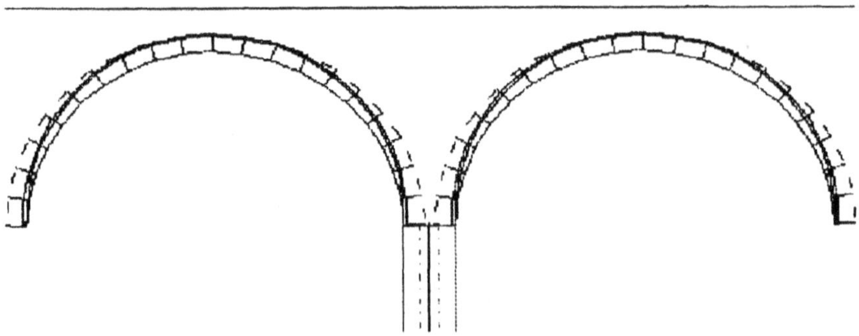

FIG. 4. Double span—both arches falling.

A two-span arch is the next step in complication. Figure 4 shows two adjacent spans of Stephenson's Royal Border Bridge at Berwick.

Here there are three reactions and six redundancies. The statically determinate release structure will have both arches descending (Fig. 4). It will be necessary to investigate the interaction of two resultants and so they must

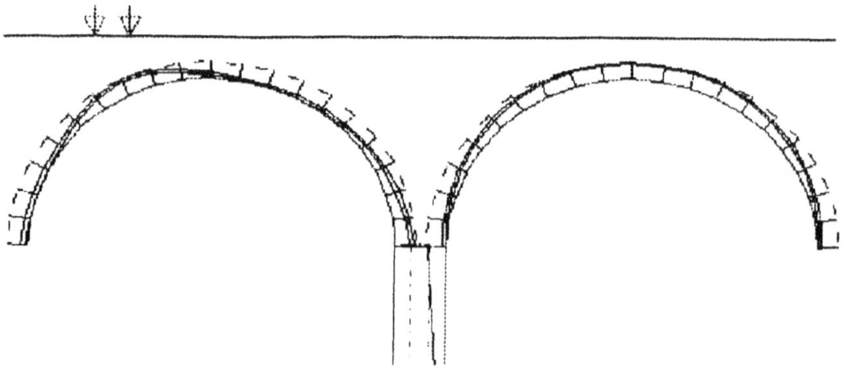

FIG. 5. Load to left—right span propping.

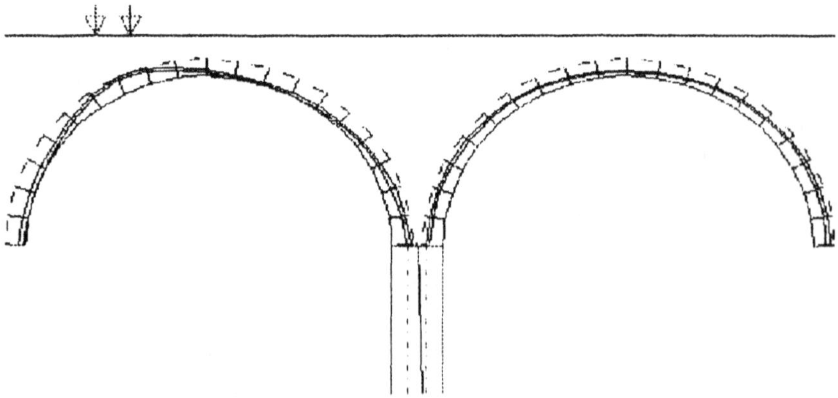

FIG. 6. Two spans with reactions adjusted.

be changed in turn. Each arch is dealt with separately. Reactions are adjusted to produce an acceptable zone of thrust. The two arches are then combined and the path of the resultant force down the pier is traced. If the two arches are analysed separately, the combined picture can be displayed adequately on a Hercules screen, and very attractively on VGA. The figures presented here are screen grabs from VGA using Word Perfect 5. Figure 5 shows the effect of a heavy load in the left-hand span of Fig. 4. In Fig. 6 the reactions in the right-hand arch have been adjusted to produce stability. Note that the maximum propping action has not been used.

Extension of the process

The graphical quality available on microcomputers is still somewhat limited. When two spans are being worked on it is usually advantageous to treat them individually before combining the graphics on the screen. A similar approach will of course be possible for a viaduct. Each arch can be treated as a separate entity and then the structure combined to show the effect at the piers. As screen resolution improves it will be possible to work on multiple spans on screen.

THE VALUE OF THE OUTPUT

No doubt many of the readers who have borne with us thus far will wish to ask the question 'so what?' What does a picture of the polygon of forces offer the engineer? Our view is that it allows him to explore the limits within which his structure is actually working. It will not give an absolute solution to the analytical problem, but the problem is not susceptible to analysis in that way. So many of the parameters concerned can only be defined in terms of limits. It is foolhardy to rely on even a few analytical solutions which do not clearly show what the implications are throughout the structure.

THE NEXT STAGES

This approach has been developed in a two-dimensional sense dealing with a very specific type of structure. The computation is based on a vectorial representation of the load path within the structure. As such it is definitely not limited to two dimensions. We hope to find the resources to develop the approach, first for complex three-dimensional masonry structures and then to structures where bending plays an important part in performance. The complexity in this latter stage is in representing the load path. The traditional approach of bending moment and shear force diagrams in different orthogonal axes is quite inappropriate and a major part of the development effort will be in finding more appropriate representations.

CONCLUSIONS

1. Slender gravity structures are fundamentally different from modern continuous structures and demand unique analytical tools.
2. The line of thrust view has long been popular and load path analysis is its modern descendent.
3. The flexibility method provides a basis for a semi-automatic form of load path analysis for masonry structures.
4. The method is capable of considerable development, especially as computers become faster and as display techniques improve.

REFERENCES

1. Hooke, R., *A Description of Helioscopes, and some other Instruments*. London, 1676.
2. Castigliano, C. A. P., *Théorie de l'Equilibré des Systèmes Elastiques et ses Applications*. Augustos Frederico Negro, Turin, 1879. Translated by E. S. Andrews, *Elastic Stresses in Structures*. Scott, Greenwood & Son, London, 1919. Also with an introduction by G. E. A. Oravas, *The Theory of Equilibrium of Elastic Systems and its Applications*. Dover, New York, 1966.
3. Harvey, W. J., The application of the mechanism analysis to arch bridges. *J. Inst. Struct. Engng*, **66**(5) (March 1988).
4. Barlow, W. H., On the existence (practically) of the line of equal horizontal thrust in arches and modes of determining it by geometric construction. *Min. Proc. Inst. Civ. Engng*, **5** (1846) 162.
5. Pippard, A. J. S. and Chitty, L., A study of the voussoir arch. National Building Studies Research Paper 11, HMSO, London, 1951.
6. Chrisfield, M. A., A finite element computer program for the analysis of masonry arches. Transport and Road Res. Lab. Report LR1115, Crowthorne, Berks, UK, 1984.
7. Chrisfield, M. A., Computer methods for the analysis of masonry arches. *Proc. 2nd Int. Conf. on Civil and Structural Engineering Computing*, Vol. 2. Civil-Comp. Press, Edinburgh, 1985, pp. 213–20.
8. Chettoe, C. S. and Henderson, W. H., Masonry arch bridges: a study. *Proc. Inst. Civ. Engng*, **8** (1957) 723–55.
9. Heyman, J., The estimation of the strength of masonry arches. *Proc. Inst. Civ. Engng*, **69** (Dec. 1980) 921–37.
10. Harvey, W. J. and Barthel, R., The relationship between thrust and springing movement in arches (to be published).

46

The Assessment of Masonry Arch Bridges—The Effects of Defects

CLIVE MELBOURNE

Department of Civil Engineering and Building, Bolton Institute of Higher Education, Bolton, Lancashire, UK

ABSTRACT

A large number of the existing stock of masonry arch bridges in the UK suffer from a variety of defects. The paper considers two of these defects—ring separation and spandrel wall separation. Ring separation in multi-ring arches is the loss of bond between successive rings. A series of laboratory tests on 1, 3 and 6 m span arch bridges is described. The significance of soil/structure interaction is discussed and the importance of defects quantified. The relative importance of the various parameters influencing the mode of behaviour and strength of the arch is discussed, thus allowing the engineer to formulate a mathematical model of the arch bridge to be assessed.

INTRODUCTION

Presently, in Britain, a programme of research is being undertaken to study the behaviour of masonry arch bridges under load. The programme is being funded primarily by SERC, British Rail and the Department of Transport. The aim of the research is to improve the methods of assessing the load-carrying capacity of this type of bridge as present methods appear overly conservative.[1,2]

The Bolton Institute Arch Research Team has been studying the behaviour of masonry arch bridges for some years as part of the national research programme. The work has encompassed small- and large-scale laboratory tests (up to 6 m span) as well as field tests.

Much of the earlier work related to the importance of the soil structure interaction and spandrel wall stiffening. However, an integral part of any assessment is an appraisal of the significance of defects which exist within the structure. Recent work at the Bolton Institute has concentrated on the significance of these defects on the load-carrying capacity. Initially two types of defect have been considered, ring separation in multi-ring brick arches and spandrel wall separation. Both types of defects have been identified by the industry as being commonplace and significant although, to date, engineers have only been able to take cognisance of them in the form of a subjective condition factor.

Ring separation occurs in multi-ring brick arches and is associated with the loss of bond between successive rings caused by weathering and/or stress cycling of the mortar.

MODEL TESTS

The models were of a parabolic profile with a span of 1000 mm, span/rise ratio of 3:1 and a width of 500 mm. The arch ring comprised two rings or courses of brickwork, either bonded or unbonded, around the full arc of the arch. In total the ring thickness was approximately 110 mm. The models were constructed using half-scale fletton bricks with an average compressive strength of $30\,N/mm^2$. They were constructed without spandrel walls. Because there were no spandrel walls the fill was contained by the perspex and wooden sides of the test rig. Prior to laying the bed face, each brick was coated with oil in order to minimise the effects of the bond strength on the model behaviour. A 1:1:6 (cement: lime: sand) mortar was used, it was mixed by volume and achieved an average 28-day compressive strength of $4\,N/mm^2$. Sand was used as backfill and compacted by vibration.

A knife edge load (KEL) was applied incrementally and monotonically up to failure at either the 1/4 or crown point.

Table 1 presents the results of the tests. Arches 1–3 were built such that

TABLE 1
Model test results

Arch number	Inter ring bed-joint material	Loading position	Experimental ultimate load (kN)
1	Mortar	1/4	13·2
2	Mortar	1/4	18·6
3	Mortar	Crown	31·0
4	Sand	1/4	6·9
5	Sand	1/4	8·1
6	Sand	Crown	8·8

the two rings of brickwork were fully bonded together using a mortar. (By bonding it is meant adhesion rather than brick bonding using 'headers'.)

Arches 4–6 were built with the mortar between the rings, being replaced by damp sand to simulate loss of adhesion.

All of the models which were loaded at the 1/4 span failed due to the formation of a four-hinge mechanism. For those models loaded at the crown, failure was due to the development of a 'classical' five-hinge mechanism. At failure in the 'bonded' arches no or little ring separation occurred and formation of the hinges was as a monolithic 'single' ring. In the 'unbonded' arches, other than in the region of the applied load, the unbonded rings were measured by embedment strain gauges as separating with increasing load. Development of the hinges in the models with initial ring separation was such that two thrust lines developed, one in each of the inner and outer rings. Hinges in the two rings were coincident and developed simultaneously. Collapse was sudden and caused by almost total physical separation of the two rings and full development of the hinges.

THREE-METRE SPAN BRIDGE TESTS

Two 3-m span bridges were built and loaded to failure in the institute's large-scale testing laboratory. The segmental arch barrel (radius 1875 mm) consisted of two rings of brickwork using class A engineering bricks. The brickwork was built in a 'stretcher' bond with no bonding between the rings other than through the mortar in the 'bonded' case and damp sand in the 'ring separation' case. The spandrel, wing and retaining walls were built in English bond using concrete commons. The walls were not attached to the arch ring. An average gap of 10 mm was provided between the spandrel walls and the arch ring. This ensured that both the effects of ring separation and spandrel wall separation could be studied.

Compaction of the 'graded 50 mm' limestone backfill was achieved using 100 mm layers and a vibrating compacting 'wacker' plate. The bridge was filled to 300 mm above the crown.

Both bridges were subjected to three loading conditions. Firstly, a 25·7 kN KEL was applied at 250 mm centres across the span to simulate a rolling load. Secondly, a 50 kN KEL was incrementally applied at the north quarter point, crown and south third point. Finally, a KEL was applied incrementally at the quarter point through to collapse. The elastic tests confirmed that the structure responded to the loading as a local effect, with soil pressure and brickwork strain changes being confined to the vicinity of the loading. This has been observed in field studies.[3]

Both arches failed by the formation of four-hinge mechanisms. In each

FIG. 1. Three-metre span arch barrel—ring separation.

case the spandrel walls cracked in the vicinity of the crown and rotated about the abutment remote from the KEL (Fig. 1). The sequence of hinge formation is given in Table 2.

In the bonded arch some ring separation occurred at the crown but the hinges formed at intrados and extrados. On the other hand, the unbonded arch produced extensive ring separation shortly after the formation of the second hinge. It is significant to note that the first hinge formed at approximately the same load in each test. As no ring separation cracking had occurred at this stage, it confirmed that the two arches were comparable. The unbonded arch deteriorated more rapidly after the formation of the first hinge and carried an ultimate load of 360 kN—a 33% reduction in carrying capacity compared with the bonded arch.

TABLE 2
Sequence of hinge formation

Hinge number	Position	Load (kN)	
		Bonded arch	Unbonded arch
1	Under load point, north quarter point	240	220
2	South quarter point	300	240
3	South abutment	400	320
4	North abutment	480	320
		(540 max.)	(360 max.)

As with the model tests, once ring separation occurred each ring formed its own pattern of hinges which interacted with each other.

An assessment of the bridge using the presently accepted 'MEXE' method[1,2] gave a modified axle load of 200 kN. This represents a load which is 85% of that required to cause the formation of the first hinge.

The soil/structure interaction was monitored using pressure cells, not only in the extrados but also in the spandrel walls and the backfill. The overall picture which emerged was one of a compacted backfill exerting pressure greater than earth pressures at rest and which restrained the arch and dispersed the applied KEL. Additionally, there was a frictional/cohesive resistance between the backfill and the spandrel walls and the extrados, which restrained the arch initially and which increased as the arch swayed into the backfill as hinges formed. This movement into the backfill not only increased the longitudinal horizontal soil pressure but also the deviatoric stress, as on the back of the spandrel walls and hence the resistance to movement.

A mechanism analysis using methods developed by Pippard and Baker[4] and Heyman,[5] and modified to incorporate the above factors, was used to predict an ultimate load-carrying capacity of 470 kN.

SIX-METRE SPAN BRIDGE TEST

The bridge was built and loaded to failure at the institute's large-scale testing laboratory and represented one of a series of full-scale tests coordinated by the TRRL. It had a span of 6 m with a rise of 1 m and an overall width of 6 m. The two-ring brickwork arch was constructed using solid engineering class concrete bricks. The brickwork was built in a stretcher bond with no bonding between the rings other than the mortar bed joint. The total thickness of the arch barrel was 220 mm. This represented quite a slender arch with a span:ring thickness ratio of 27. The spandrel walls were built in an English bond. A 50 mm graded limestone was used for the backfill material compacted in 100 mm layers to 200 mm above the crown; this was surfaced with a 100 mm thick bituminous layer. The overall length of the bridge was 14 m.

Throughout the construction process and subsequent testing instrumentation was provided to monitor all aspects of the bridge behaviour. Two separate series of load tests were conducted on the bridge. Full details of the tests are given elsewhere.[6]

The first comprised the application of a point load at various positions across the width and span of the bridge; these tests were carried out to simulate a wheel loading.

Initially it had been hoped to apply a point load of 100 kN; however, during the initial loading cycle adjacent to the spandrel wall at the quarter span, cracking was recorded at 70 kN. Loading caused separation of the spandrel wall and arch ring, and the cracking manifested itself inside the arch barrel as ring separation along the line of the quarter point, detected by strain gauges embedded in the brickwork. Further point loading cycles at different locations on the bridge caused the initial crack to open and close as the load increased and decreased respectively.

Following the point loading tests, the bridge was subjected to a KEL at the quarter span; loading was applied monotonically through to collapse. Separation of the spandrel/barrel interface commenced on the east side at 360 kN. Further loading caused the cracking to spread around the arch barrel. At 400 kN the first hinge in the barrel was observed underneath the load point. At 640 kN cracking of the spandrel walls was observed.

Failure of the bridge was due to the formation of a four-hinge mechanism at a total applied load of 1173 kN. The spandrel walls were lifted and rotated, as shown in Fig. 2. Extensive ring separation was coincidental with failure and the arch barrel/spandrel wall was almost completely separated around the full arc of the barrel. It is worth noting that the mode of failure was similar to that of model tests previously reported by the author and others.[7]

Typical graphs of deflection, soil pressure and brickwork strain are given in Fig. 3.

Formation of the first hinge under the KEL at 400 kN can be clearly deduced from the load/deflection graph, which shows a marked change in stiffness from the initial elastic behaviour. At the quarter point, remote from the loading, little movement took place until 800 kN, at which stage there was a rapid increase in deflection corresponding to extensive cracking of the arch and further hinge formation.

Surface-mounted vibrating wire strain gauges confirmed the stiffening

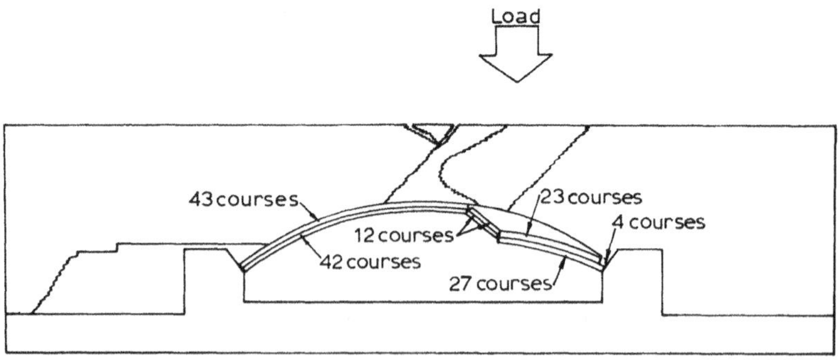

FIG. 2. Crack pattern at failure.

effect of the spandrel wall, which confined the thrust line to within the middle third at the three-quarter span.

Ring separation was monitored and propagation of the local ring separation (caused during the point load test) commenced at 400 kN.

Soil pressures were monitored not only on the extrados of the arch but also on the spandrel walls. Upon completion of construction the soil pressures were comparable with earth pressure at rest/unloading (i.e. $K_0\gamma h$ or $K_r\gamma h$, $K_r = 1/K_0$). The oil pressures under the KEL increased as the load was

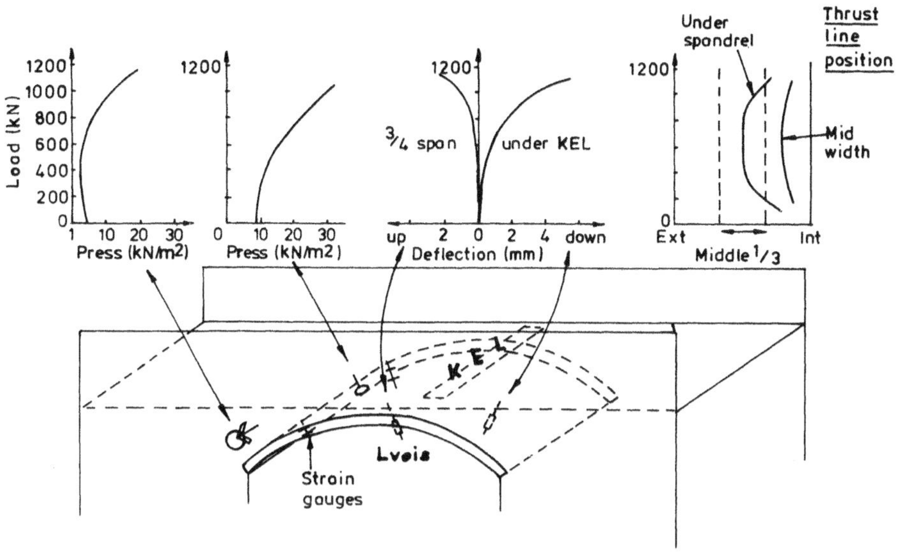

FIG. 3. Typical graphs.

applied and were compatible with a dispersal angle of 45°. There was a significant increase in the lateral pressure against the spandrel wall in the vicinity of the loading beam as the load increased. This was a contributory factor to the cracking in the wall. More significantly, there was little change in the soil pressure elsewhere in the bridge until the load exceeded 600 kN, at which stage there was a general increase in pressure on the side remote from the load caused by the arch barrel trying to 'sway' into the fill. It is important to note that even at failure the soil pressures did not reach passive pressures but did provide significant restraint to the arch.

Using a modified mechanism analysis adopting K_r pressures, together with the stiffening effects of the spandrel wall and cohesive/frictional resistance, an ultimate load capacity of 1010 kN was predicted and a load of 730 kN for the onset of mechanism behaviour.

CONCLUSIONS

(1) If free to do so, an arch bridge will fail due to the formation of a four-hinge mechanism.
(2) Using a modified mechanism analysis incorporating the lateral backfill pressures, spandrel wall stiffening and backfill cohesion/friction structural interaction, the onset of mechanism behaviour and the collapse load can be predicted. The prediction of the onset of mechanism behaviour could be used to set a serviceability limit state.
(3) At all stages of loading consideration must be given to the possibility of local failure (e.g. punching shear, ring separation, snap-through) and adequate factors of safety applied.
(4) Ring separation caused a reduction of between 56% and 33% in the ultimate load-carrying capacity of the model and full-scale two-ring brick arch bridges respectively.
(5) Passive soil pressures were not observed in any of the tests, even at gross deformation.
(6) Surface strain measurements indicated that the spandrel wall stiffening delayed hinge formation and influenced hinge positions.
(7) Where spandrel wall separation exists the cohesion/friction resistance on the back of the wall makes a significant contribution to stiffening the arch.

ACKNOWLEDGEMENTS

The author wishes to acknowledge the financial support of SERC, British Rail, TRRL and NAB, also the encouragement and support from the staff of the Department of Civil Engineering and Building at the Bolton Institute of Higher Education.

REFERENCES

1. *The Assessment of Highway Bridges and Structures.* Department of Transport Roads and Local Transport Directorate, Departmental Standard BD 21/84, Department of Transport, March 1984.
2. *The Assessment of Highway Bridges and Structures.* Department of Transport Roads and Local Transport Directorate, Advice Note BA 16/84, Department of Transport. HMSO, London, March 1984.
3. Melbourne, C., The construction of mass concrete arch bridges. Structural Faults and Repair Conference, London, June 1989, Engineering Technics, 1989.
4. Pippard, A. J. S. and Baker, J. F., *The Analysis of Engineering Structures*, 2nd edn. London, 1943.

5. HEYMAN, J., *The Masonry Arch*. Ellis Horwood Ltd, London, 1982.
6. MELBOURNE, C. and WALKER, P. J., Load test to collapse of a full-scale brickwork masonry arch. TRRL Contractor's Report (to be published).
7. MELBOURNE, C., QAZZAZ, A. and WALKER, P. J., Load testing to collapse of model and full-scale brickwork masonry arches. SERC, Repair Maintenance and Operation in Civil Engineering Conference, London, June 1989.

47

Theoretical and Experimental Investigations on Railway Bridges Dating from 1856 to 1895

F. MANG and Ö. BUCAK

Versuchsanstalt für Stahl, Holz und Steine (Testing Centre for Steel, Wood and Stone), Universität Karlsruhe, Kaiserstrasse 12, D-7500 Karlsruhe, FRG

ABSTRACT

When assessing the residual service life of old bridge structures, the question arises whether they can be still used after having exceeded the theoretical service life, or if they should be replaced by a new structure after having been newly classified, or if they are still usable after reinforcement or on the basis of shorter service intervals.

The determination of the residual service life of a bridge results in the difficulty of recording the exact history of the loads and the state of the bridge. Insufficient knowledge about the strength and fatigue behaviour of old steels and constructions applied aggravate the above decision.

This paper deals with experimental investigations on a complete bridge structure of the 'museum railway' of the community of Blumberg as well as with two developed bridge structures of the Federal Railways. The results of these component tests are compared with data known from literature as well as with other results of similar investigations performed at the institute.

INTRODUCTION

Today structures subjected to fatigue loads, for example railway bridges, are usually designed for a certain service life. The reason for this is the knowledge that after exceeding a critical number of load cycles with a sufficiently high loading level fatigue fractures occur. This calculated failure is covered with a corresponding 'safety factor'.

For older iron and steel structures, design 'for a set time' was not usual.

FIG. 1. General view of Koblenz/Waldshut railway bridge.

FIG. 2. Detail of the Koblenz/Waldshut railway bridge.

There are numerous structures which have been subjected to fatigue loading for more than a century.

After reaching and exceeding the 'standard service life', fatigue-loaded structures often cannot be reliably assessed with regard to possible residual life. One reason for this is insufficient knowledge about the behaviour of the static and fatigue strength of steels (wrought iron, puddled steel) and the structures used in the 19th century. The oldest railway bridge, the Koblenz/Waldshut between the Federal Republic of Germany and Switzerland, which is in the charge of the Versuchsanstalt für Stahl, Holz und Steine, dates from 1858 and is still in operation today. This bridge is shown in Figs 1 and 2.

FIG. 3. Local rust formation.

Extensive measurements under traffic load and investigations on the material allow the operation of this bridge to continue—at least for the next few years. Presently the gap in single elements due to rust formation (Fig. 3) seems to be a problem. After complete removal of the rust these areas are filled with a flexible lute (Fig. 4). The original values of the forces caused by the rust formation are being investigated and will be reported on later.

Since the design and strength of joints substantially influence the load-bearing behaviour of the total structure, such joints from old steel bridges have to be investigated more precisely. When assessing old existing bridges, material specimens have been taken from the braces or from less stressed

FIG. 4. Flexible lute.

areas of the construction for examination. The data obtained were and still are subjected to very high safety factors since a safe assessment is required.

The aim of recent investigations in Karlsruhe is to obtain real residual service life by studies on original joints and structural members or full structures, and comparison with data on simple specimens already tested in the institute or which are available from literature. Some characteristics may have to be newly determined.

GENERAL

Knowledge of the residual service life of old bridge structures is required by their operators, or by people who need to know how much longer these structures can be operated under present and projected loading conditions. The protection of historical monuments and economic considerations are the decisive factors.

One recent project was the so-called gun railway, also called 'pig-tail railway' (museum railway), which runs in the area of the Wutach Valley in the south of Germany. Figure 5 shows one of the bridge structures of the museum railway.

Until recently the question of residual service life has been tackled by gathering the following information:

(1) Survey of the bridge by visual inspection (state of corrosion). This was the determining factor for some of the bridges.
(2) If static calculations are available, determination of the peak stressed areas.
(3) Sampling of less critical but representative areas of the structure.
(4) Test investigations, for example chemical analysis, tensile tests and fatigue tests on original material.
(5) Additional calculations based on actual material properties.
(6) Determination of the previous loading on the bridge and expected future loading (load spectrum).
(7) Application of the Miner rule (linear damage accumulation hypothesis) for the determination of the residual service life as well as the safety factor for this particular bridge.
(8) Fracture mechanics investigations, for example COD tests.

Sufficient data for material properties have been established by preliminary surveying of various highly dynamically stressed structures which can be used for initial assessment. Significant investigations need only be done in the final phase for the purpose of confirming the assumptions made. Presently there is still a gap in knowledge about the behaviour of such

FIG. 5. A bridge structure of the museum railway.

bridges as a total system or on the bearing capacity of complete structural members in the original state. Questions about the rivet slip, the rivet initial load, the displacement under shear stress and load distribution with various elements of a structural member are still unknown. These parameters influence the fatigue behaviour. For this reason checks on full-scale structural members typical of an actual bridge are needed.

For this purpose a complete bridge with a length of 4·50 m dating from 1877 has been selected from various bridges of the museum railway together with the longitudinal girders and cross-girder joints of a further bridge system dating from 1895 with a span of 4·8 m which have been investigated in the Versuchanstalt für Stahl, Holz und Steine.

INVESTIGATIONS ON THE MATERIAL

Samples have been taken from five specimens of three bridges of the museum railway and three specimens have also been taken from the Stahringen Bridge.

With the exception of one specimen, which was taken from a base plate made of cast iron, investigations were carried out on semi-finished products made from puddled steel. The purpose of the investigations on these specimens was to find out whether the intensive investigations made on older materials from other bridges of the same period is possible.

The results of the tensile tests carried out on specimens from various parts of the bridge confirm the material data obtained from other similar structures.

PERFORMANCE OF FATIGUE TESTS

Tests on the Museum Railway Bridge

The bridge (Fig. 6) was installed in the 50 MN press of the institute. The photograph of Fig. 7 shows the installation corresponding to the critical load configuration LF 1.

Stress results of static tests under both load configuration 1 and load configuration 2 compared well with calculated values ($\sigma = 135\ \text{N/mm}^2$ calculated, $137\ \text{N/mm}^2$ measured).

Calculations carried out with the actual bridge dimensions before testing indicated that LF 1 was the critical load configuration.

For this reason a fatigue test was performed with this load configuration.

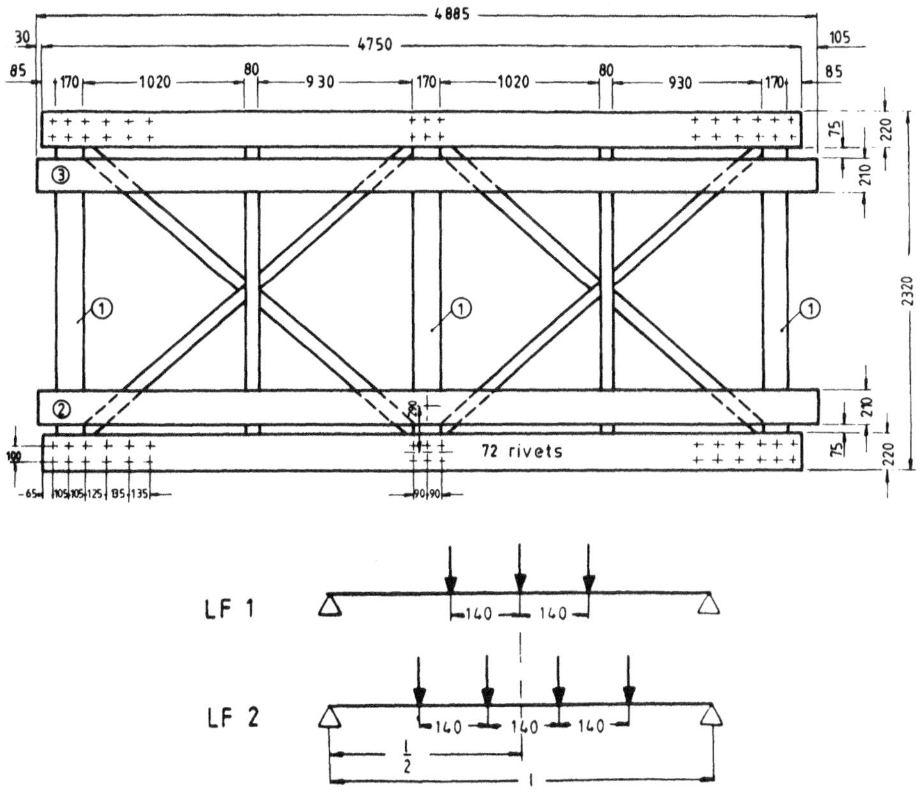

FIG. 6. Test specimen and loading arrangements.

FIG. 7. Bridge in the 50 MN testing machine.

The test load was *twice the maximum load expected* with a limiting stress ratio $R = +0{\cdot}35$. Under this load an edge stress of $\sigma = 137\ \mathrm{N/mm^2}$ existed on the main girder. The stresses have been calculated with the loads applied and the net iron section calculated according to the normal assumptions of static behaviour.

Flexible bearings were used in order to allow for possible irregularities of the bridge. To determine the stress state on the main girder important areas were monitored by strain gauges. With a load cycle of 108 070 the test was terminated since a correct load application was not possible because of the occurrence of a large crack. The cross-girder/main girder joint was where the failure occurred. This is shown in Fig. 8.

The bridge was cut into small pieces in order to investigate individually the main girders, longitudinal girders and the cross-girder/main girder joints (Fig. 9).

Tests on both Main Girders and Longitudinal Girders

The tests on the main girders with a span of 3300 mm were performed as a four-point bending test. The distance between points of load application was

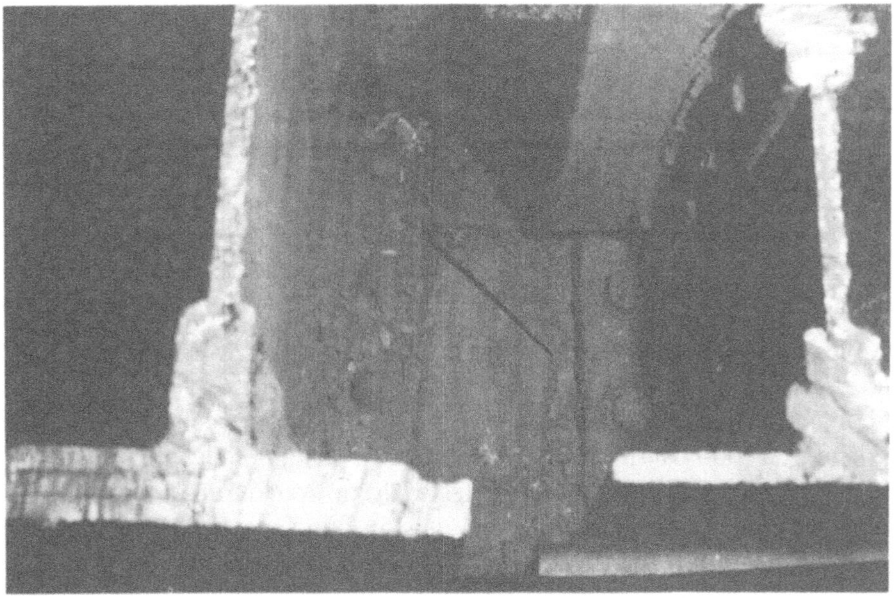

FIG. 8. Cross-girder/main girder joint—cracked web plate.

800 mm. With this arrangement a constant bending stress occurred in the central area of the main girder (Fig. 10).

The load for the first main girder was selected to induce a tensile stress of $\sigma = 137$ N/mm^2 and later 165 N/mm^2 at the extreme fibre. The limiting stress ratio was $R = 0{\cdot}2$. After 10 000 000 load cycles with an edge stress of 135

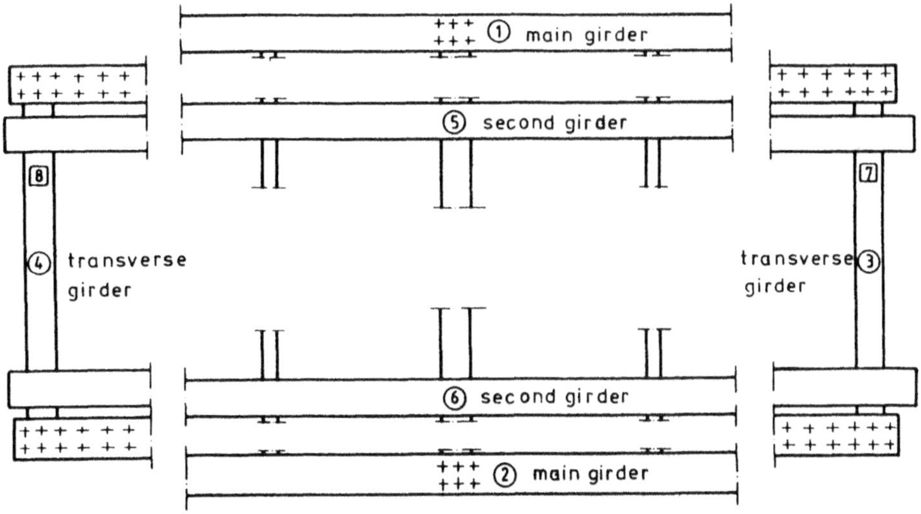

FIG. 9. Blumberg Bridge—classification of the test specimens after dismantling.

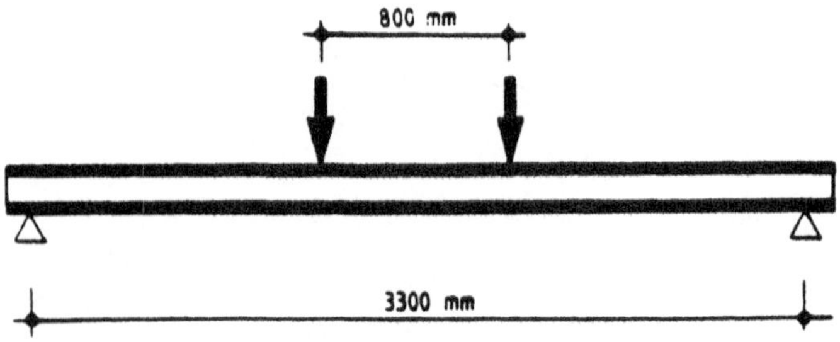

FIG. 10. The central area of the main girder.

N/mm^2 and without any discernible cracking, the load was increased to $\sigma = 165$ N/mm^2.

Under this load a crack in the web of the main girder occurred with a load cycle of 1 534 000. With this the load could be sustained despite the crack in the main girder. For this reason the test was continued until both angles and the chord plate of the main girder's tension flange were also broken. This occurred at a total load cycle of 1 572 600. The girder therefore took an additional 38 600 load cycles from the clearly visible incipient crack of the web up to the total failure of the main girder (Fig. 11).

The web was first to break followed by the angle section and last the chord plate. The cracks were random and not on a line.

FIG. 11. Rupture portions in main girder 1 after the test.

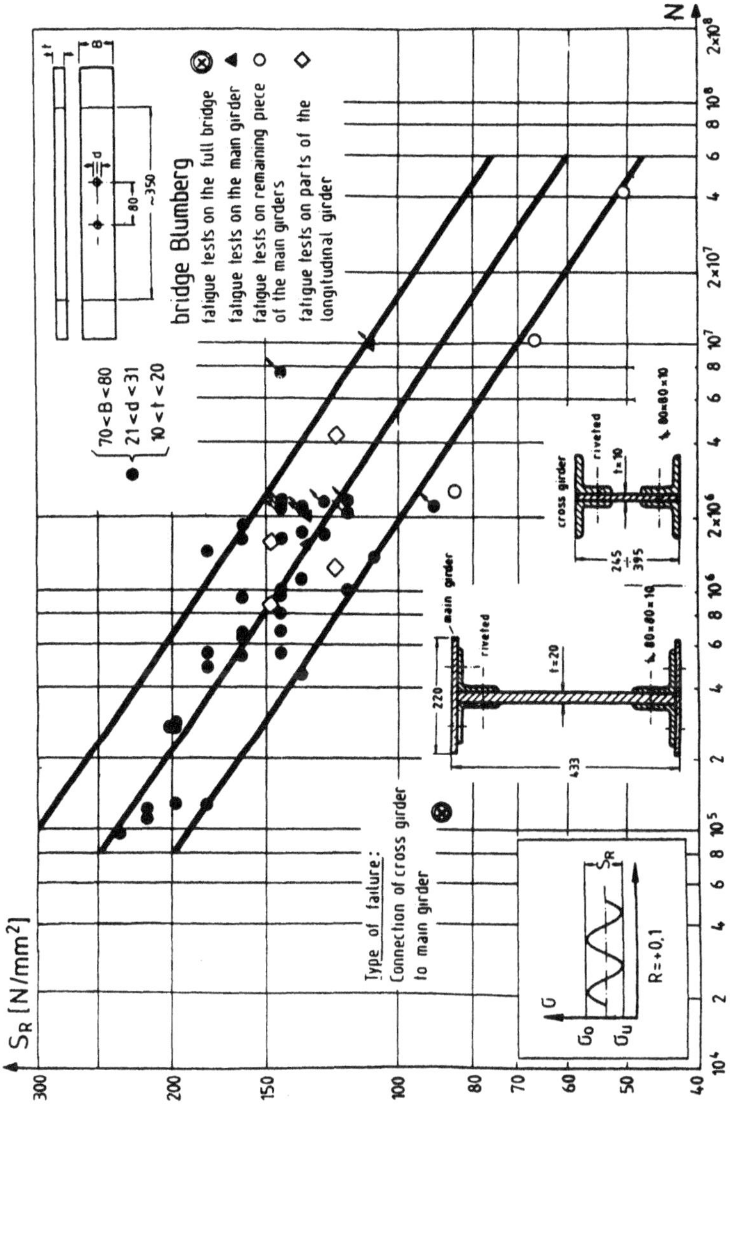

FIG. 12. Comparison of results of the tests on structural members of the Blumberg Bridge with former investigations on members with drilled holes.

With the second main girder, approximately 100 000 cycles were needed from the clearly visible incipient crack in the web up to total failure.

Three further test specimens were taken from the unbroken area of the main girder and tested in the three-point bending test with a span of 1500 mm. The upper loads were 450 or 600 kN, and the corresponding limiting stress ratio was $R = 0{\cdot}35$ or $+0{\cdot}1$.

Four test specimens were taken from the longitudinal girders and tested in a three-point bending test with a span of 1500 mm. With two specimens the upper stress was 165 N/mm^2 and with the others 135 N/mm^2. The tests were performed with a limiting stress ratio of $R = +0{\cdot}1$.

As with the tests on the main girder specimens, the cracks start from the bearing point and run diagonally through the web. Afterwards cracks occurred at the angle sections of some specimens.

TABLE 1
Results of the tests on dead-end transverse girder joints

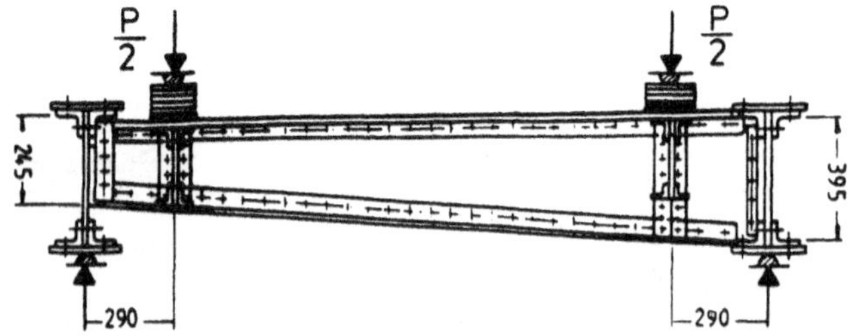

Test number	State	P_0 (kN)	P_u (kN)	Load cycles	Remarks
3	Dead-end transverse girder	450·0	157·5	2 560 400	Crack in the web between main girder 2 and longitudinal girder 7
4	Dead-end transverse girder	300·0 450·0	105·0 157·5	11 479 500 3 366 400	No crack Diagonal crack on the web plate between 1 and 5
8	Non-broken joints screwed together	450·0	157·5	27 237 400	Crack in the web between main girder 2 and longitudinal girder 7
7	Change of load arrangement so that broken area is not stressed	450·0	157·5	25 375 400	No crack

The results of these investigations were compared with older test data on puddle steel specimens with drilled holes existing in the institute. It is evident that the results from tests on the main girder and longitudinal girder are within the scatter range of the tests on small specimens (see Fig. 12).

Since the failure of the cross-girder joints occurred under test on the full bridge of the museum railway and investigations on such joints did not exist or rather were unknown from literature, emphasis has been placed on this type of failure in these investigations.

At first both ends of the bridge were tested with a limiting stress ratio of +0·35 (as with the test on the complete bridge) with various shearing forces, both as the three-point bending test and as the four-point bending test. The results with the corresponding test data are given in Table 1.

Afterwards the ends of the dead-end transverse girders were screwed together to a new specimen to enable joints with larger dimensions to be tested. Figure 13 shows the test specimen after failure.

From the second bridge a girder was subjected to bending with four cross-girder joints, and to a fatigue test. The limiting stress ratio was +0·1. The results are given in Table 2.

Tests on Wind Bracing Parts of the Bridge of the Museum Railway as well as on Small Specimens taken from the Bridge Structures

From both wind bracings of the bridge the central areas were kept in the

FIG. 13. Cracked cross-girder joint.

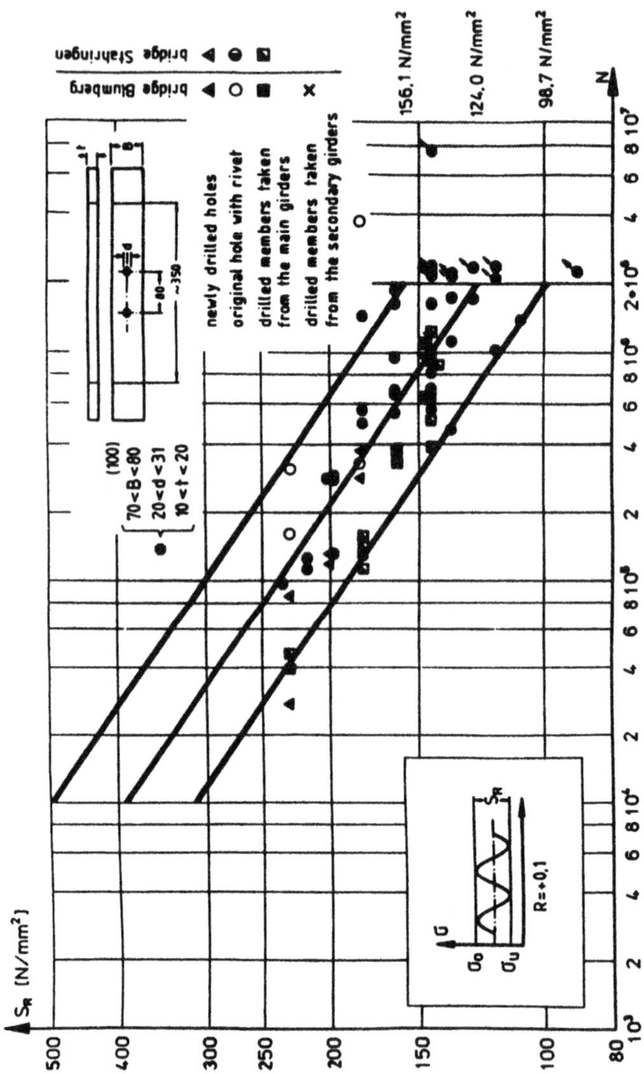

FIG. 14. Results of investigations on webs with drilled holes and on members with original rivets.

TABLE 2

Test number	State	P_0 (kN)	P_u (kN)	Load cycles	Remarks
1	Connection Cross-girder to main girder	350·0	35·0	15 083 712	No crack
		600·0	60·0	73 440	Crack under the load application

original state (with rivets) and subjected to a fatigue test with a pulsating machine. Two tests with a nominal upper stress of 250 and 200 N/mm^2 were performed. The results are plotted in Fig. 14 and compared with the results on punched webs of previous bridge investigations existing at the institute. It can be seen that the results are on the favourable side of the scatter range. At the same time it should be pointed out that the scatter is relatively high, as is usual with old structures.

Residual material from the bracings with newly drilled holes of 20 mm diameter was removed and subjected to fatigue tests under various stress levels. From the results plotted in Fig. 14 good correlation with old test data is evident. Several small specimens from the L-sections and chord plates of the main girders were taken and tested in fatigue. They also show good correlation with previous test values. With all specimens the incipient fatigue crack started from the rivet hole, as expected. Figure 15 shows the fracture surface of test specimens after failure.

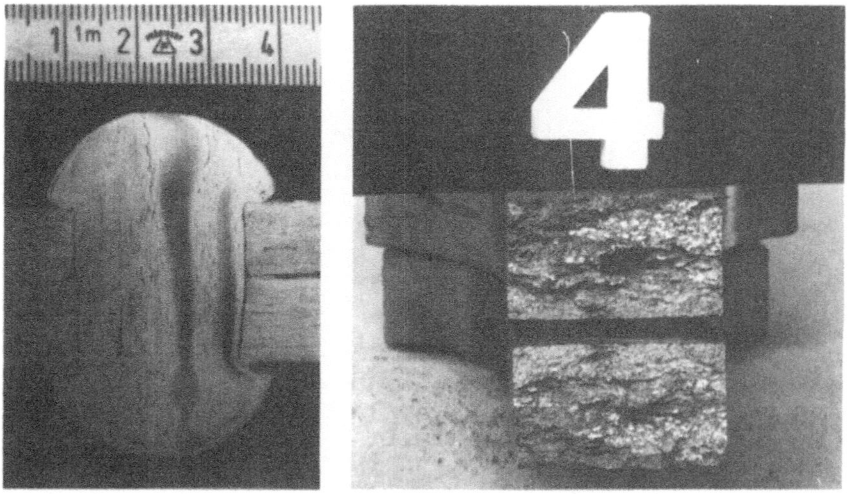

FIG. 15. Rupture surface of a solid rod—polished section on a rivet.

REFERENCES

1. MANG, F., Stahl im Altbau und Wohnungsbau. Abschlußbericht zur Forschungsstudie des Landes Nordrhein-Westfalen, VBI-72.02-92/77.
2. MANG, F., STEIDL, G. and BUCAK, Ö., Altstahl im Bauwesen. *Schweißen und Schneiden*, **1** (1985) 1–5.
3. BRÜHWILER, E. and HIRT, M. A., Das Ermüdungsverhalten genieteter Brückenbauteile. *Der Stahlbau*, **1** (1987) 1–8.
4. HERZOG, M., Erwiderung zur Zuschrift von Tschumi, M. auf Herzog, M., Abschätzung der Restlebensdauer älterer genieteter Eisenbahnbrücken. *Der Stahlbau*, **5** (1986) 159–60.
5. BAEHRE, R. and KOSTEAS, D., Einfluß der Vorbelastung auf die Restnutzungsdauer schweißeiserner Brücken. Bericht Nr. 7496 der Versuchsanstalt für Stahl, Holz und Steine der Universität Karlsruhe, January 1979 (unpublished).
6. STEINHARDT, O., Festigkeitsverhalten von Schweißeisen aus Brückenbauwerken des 19. Jahrhunderts. *ETR 26*, **6** (1977) 383–7.
7. N.N.: Unveröffentlichte Untersuchungen der Versuchsanstalt für Stahl, Holz und Steine der Universität Karlsruhe.
8. STIER, W., KOSTEAS, D. and GRAF, U., Ermüdungsverhalten von Brücken aus Schweißeisen. *Der Stahlbau*, **5** (1983) 136–42.
9. WENZEL, F., Erhalten historisch bedeutsamer Bauwerke. Jahrbuch 1987, SFB 315, und der Universität Karlsruhe (TH), Ernst & Sohn, 1988.

48

Structural and Material Damage to Concrete Highway Bridge Decks in Saudi Arabia

M. Y. AL-MANDIL, A. K. AZAD, M. H. BALUCH, A. M. SHARIF and D. PEARSON-KIRK

Department of Civil Engineering, King Fahd University of Petroleum and Minerals, Dhahran, Saudi Arabia

ABSTRACT

A seemingly large number of concrete highway bridges in Saudi Arabia are suffering signs of cracking, deterioration and, in some cases, localized failures to their deck slabs. Several reasons are believed to have contributed to these phenomena, including low quality materials, poor construction practices, lack of control over vehicular loads and the severe environmental factors.

A research programme has been initiated to categorize the types of damage to bridge decks and to identify the prevailing causal factors. Based on detailed in-situ and laboratory investigation of several defective bridge decks around the kingdom, damage may be broadly classified into (i) structural damage resulting from overloading and (ii) material damage resulting from environmental impact on the durability of concrete. This paper presents a typical bridge case study for each type of damage. Each case study involves a detailed description of damage inflicted on the bridge deck, along with the investigative approach adopted by the authors to determine the major causal factors. The paper is concluded by a 'damage likelihood chart' for the various types of bridge decks around the kingdom. The chart accounts for variations in the bridge geometry as well as its geographic location.

INTRODUCTION

Over the past two decades several hundred concrete highway bridges have been built in Saudi Arabia as part of the development of a modern highway network. Some of these bridges have suffered prematurely from excessive cracking, deterioration, loss of serviceability and failure. Several reasons are believed to have contributed to these phenomena, among which are low quality of concrete constituents, poor construction practices, lack of control over vehicular loads and the harsh environmental factors.

The impact of the harsh environmental factors on the durability of concrete structures in the Arabian Gulf region is well recognized in the literature.[1-3] Although these durability problems and the causal factors for deterioration which are applicable to a concrete structure are equally applicable to concrete bridges, the severity and unpredictability of bridge loadings and the direct exposure of the bridges to hostile environments have together paved the way for an aggravated assault on durability of bridge decks.

Realizing the seriousness and the scope of deterioration of concrete bridge decks, the Ministry of Communications (MOC) in association with King Fahd University of Petroleum and Minerals (KFUPM) have embarked on a national project to study the bridge deck cracking phenomenon in the kingdom of Saudi Arabia. The project was sponsored in 1984 by the King Abdulaziz City for Science and Technology (KACST) for a duration of 4 years.

As part of this national project, in-depth studies were conducted on 21 defective bridge decks around the kingdom. These bridges were selected as representative examples of the types of damage identified to be most prevalent in short-span bridge decks. Damage to the deck system has been broadly classified into (i) structural damage and (ii) material damage.

Structural damage to girder-slab bridges evinces itself as localized failure in the form of potholes resulting from punching shear or combined shear flexure failure. Damage in slab-type decks, with slab thickness appreciably greater than that in girder-slab decks, manifests itself in the form of a rectangular grid pattern of cracks in the soffit, with crack widths up to 2–3 mm.

Material damage in the form of corrosion of deck slab reinforcement occurred in the presence of a chloride and sulphate contaminated environment, hypothesized to have transpired either due to lack of control of mix ingredients and curing water or due to a subsequent ingress by the aggressive salts in the coastal regions.

This paper presents a representative bridge deck case study for each type of damage identified to be prevailing in short-span bridges in the kingdom, i.e. structural damage and material damage. The D2 bridge deck is a vivid example of localized failures inflicted on girder-slab bridge decks by the grossly overweight trucks. The EP2 bridge deck, on the other hand, presents an example of the concrete degradation in the environmentally hostile regions of eastern Saudi Arabia.

The paper concludes with a damage likelihood chart for the various types of existing bridge decks in Saudi Arabia. The chart is derived from the extensive surveys and inspections of damaged bridges around the kingdom.

STRUCTURALLY DAMAGED BRIDGE DECKS: AL-DARB BRIDGE CASE STUDY

General Description

This bridge is located on the Al-Darb to Abha road, 12 km northwest of Al-Darb Village. The MOC reference number is 081-0035-0211 and is code named D2. The bridge spans over a deep wadi and has three simply supported spans of 16 m and a skew angle of 30°. The deck has a road width of 8·0 m for two traffic lanes. The bridge consists of a reinforced concrete slab cast monolithically over four main girders. The girders are connected transversely by three cross diaphragms in each span. A plan view and a cross-section of the bridge are given in Figs 1 and 2 respectively.

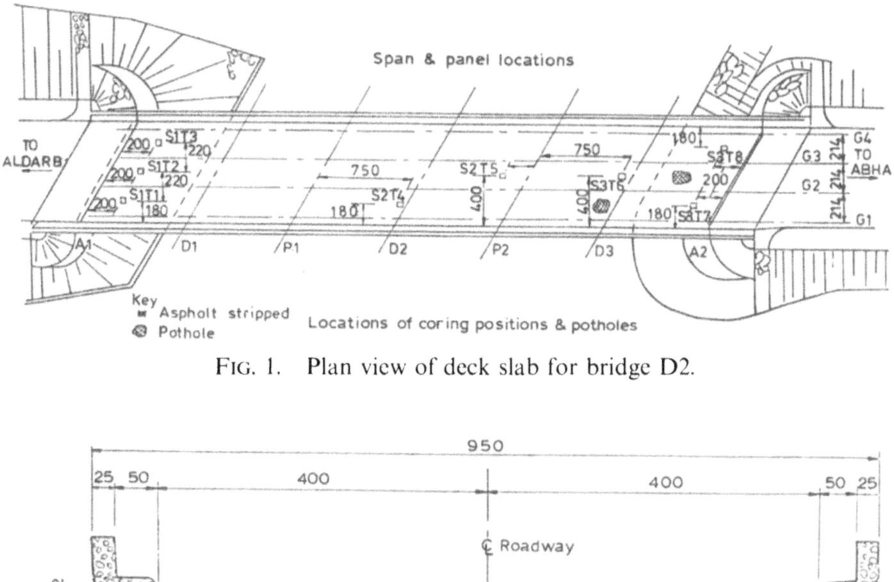

FIG. 1. Plan view of deck slab for bridge D2.

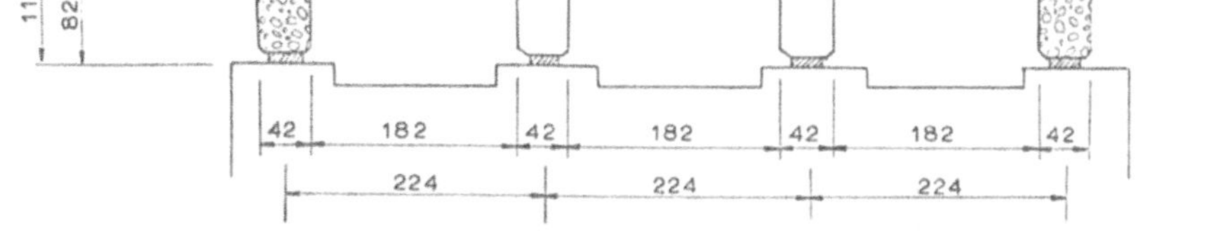

FIG. 2. Cross-section of deck slab for bridge D2.

FIG. 3. Plate showing localized failures (potholes) in the deck slab of D2.

At the time of our inspection there were two existing potholes in the deck slab where concrete has completely fractured and disintegrated, leaving the rebars exposed (Fig. 3). The position of the potholes is shown in Fig. 1. Numerous grid pattern cracks are noted on the soffit of the deck. Crack widths in the slabs ranged from 0·4 to 1·1 mm. This bridge had been abandoned due to the localized failures in the deck slab, and was bypassed by a road across the wadi.

Assessment of Concrete Quality and Strength

Twelve cores were selected from various locations on the deck slab (as shown in Fig. 1) for determination of concrete compressive strength. These core strengths ranged from 11·5 to 20·6 MPa with a mean of 17·4 MPa and a standard deviation of the mean of 2·7 MPa. Four extra cores were taken for crack-depth determination and all of these cores exhibited full-depth cracking.

The design specification required a minimum cover to reinforcement of 40 mm for top reinforcement and 30 mm for bottom reinforcement. The cover to the top reinforcement averaged 46 mm and the cover to the bottom reinforcement averaged 30 mm. Cover to the steel was clearly adequate.

The coarse aggregate gradation results showed a relatively oversanded mix. The ratio of coarse to fine aggregate (CA/FA) was 1·5:1.

Structural Analysis

A finite element analysis using ICES package program 'STRUDL' was carried out for bridge D2 to determine the service load moments, and to assess the flexural stresses in the deck slab as caused by truck loadings. Three

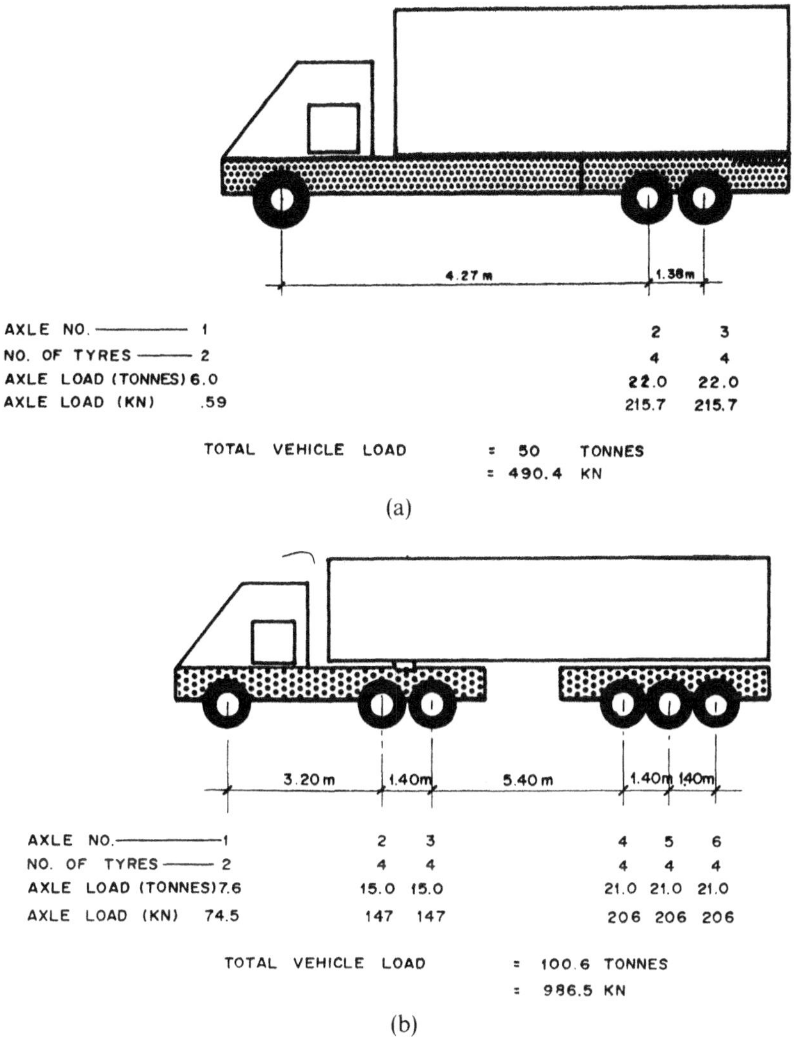

FIG. 4. (a) Bridge design vehicle DPK No. 1; (b) bridge design vehicle DPK No. 2.

loadings were considered: (i) AASHTO HS20-44 + 10% truck loading, which was the prescribed design loading; (ii) DPK vehicle No. 1; and (iii) DPK vehicle No. 2. The last two truck loadings (Fig. 4) represent some of the high vehicular loadings as observed from the axle load measurement study.[4,5]

Table 1 presents the results of the final analysis of the deck for the three different loadings. The maximum moments are listed for the deck slab and girders. Only the positive slab moment in both directions (x and y) are listed in order to check the stresses in the bottom reinforcement. The reinforcements in the deck slab are running orthogonal to each other and parallel to the x and y axes, as shown in Fig. 5.

TABLE 1
Live load analysis for bridge D2

Type of loading	Slab maximum positive moment (kN m/m)		Maximum girder moment (kN m)
	M_x	M_y	
AASHTO + 10% H20	17·67	12·18	454
DPK No. 1	31·15	16·65	823
DPK No. 2	33·09	17·65	1 120

The stresses in the bottom reinforcement of the deck slab were checked in both directions due to the maximum positive moment. In the transverse direction, the maximum positive moment is due to DPK No. 2 vehicle, $M_x = M_{DL} + (M_{LL} + I) = 48{\cdot}74$ kN m/m, where $M_{DL} = 5{\cdot}72$ kN m/m. Taking a slab width of 1 m and effective depth = 180 − 38 = 142 mm, the depth of the neutral axis from the top of the slab = 46 mm. This gives a steel stress (f_s) of 220·4 N/mm^2 (i.e. $0{\cdot}64 f_y$ if $f_y = 345$ N/mm^2), which is not critically high. Thus the transverse moment would produce a number of fine longitudinal tension cracks since the stress level is moderate.

In the longitudinal direction, the maximum positive moment is also due to DPK No. 2 vehicle, $M_y = M_{DL} + (M_{LL} + I) = 31{\cdot}06$ kN m/m, where $M_{DL} = 8{\cdot}11$ kN m/m. Taking a slab width of 1 m and effective depth = 180 − 53 = 127 mm, the depth of the neutral axis from the top of the slab = 49 mm. This

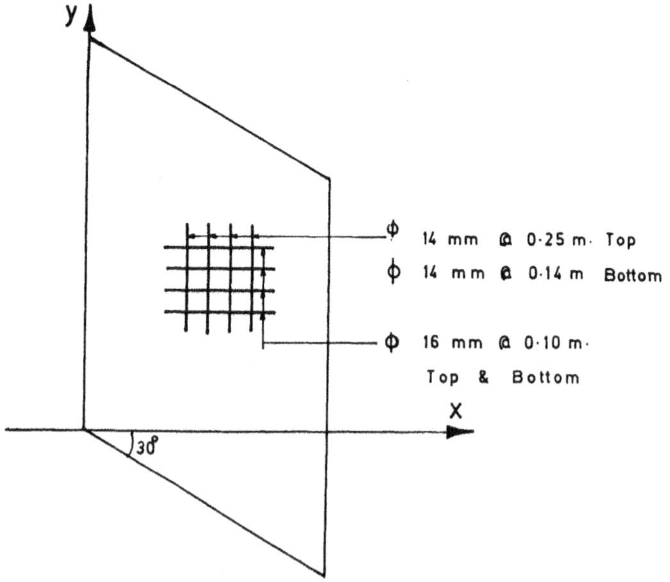

FIG. 5. Steel arrangements in the deck slab of bridge D2.

gives a steel stress of 252·0 N/mm^2 (i.e. 0·73f_y), which will cause severe transverse tension cracks.

Discussion of D2 Case Study

The deck slab of bridge D2 is characterized by the presence of two potholes, with the soffit exhibiting signs of extensive structural cracking, often in grid form. The crack widths in the slabs ranged from 0·4 to 1·0 mm.

The assessment of the strength of the cores does reveal that the concrete was indeed below specified strength. The core strengths ranged from 14 to 20 MPa with a mean of 17·2 MPa, whilst the design specification called for 24·6 MPa. It may be argued that core strengths will be lower than the in-situ concrete strengths as a result of the coring process; nevertheless, the concrete strength definitely falls short of the design specified strength, not only on an average strength basis but also because over 50% of the cores had strengths which were less than 75% of the design specified strength.

The cement content of the concrete estimated from chemical analysis of the cores was 255 kg/m^3, which is some 32% below the current MOC specification requirement of 375 kg/m^3 for 21 MPa strength concrete.

The ratio of coarse to fine aggregate was 1·5:1, which is acceptable but does indicate slight oversanding. The larger aggregates were extremely elongated, some exceeding 60 mm in linear dimension, and thus were far from being ideal materials for concrete.

The detailed structural analysis of the deck system for D2 did not provide us with conclusive evidence as to the causes of the formation of the two potholes. It does, however, vividly illustrate that deck overloading is the main cause for the shear and flexural cracks in the deck slab. Such high steel stresses under service loads ($f_s = 0{\cdot}73 f_y$) would most certainly lead to extensive flexural cracking.

However, at such high stress levels the linear elastic FE model as used herein becomes merely an indicator of the actual situation rather than a true index of stress distribution. A more appropriate criterion for structural behaviour in this seriously cracked deck would be in-situ monitoring of strain distribution under a vehicle of known load. Results of such a test would indicate the nature of redistribution of stresses in the deck system, and a more realistic evaluation of the actual situation could be made.

Hypothesized Mechanism for Pothole Formation

Turning our attention towards the assessment of the nature and mechanism of the spalling and the pothole formation, it is noted that most potholes are formed in the severely cracked regions as a consequence of a punching shear type of failure, where pieces of concrete are pushed through the slab reinforcing bars. A quick check on the deck's punching shear capacity reveals that this deck entertains a high margin of safety against punching shear type of failure on the basis of the ACI punching shear formula.[6] Therefore a punching shear design deficiency can be discounted.

However, it has been shown in a relevant study[7,8] that the punching capacity can be impaired by the existence of a flaw within the slab intricately developed by the active process of crack growth and crack nucleation. A deck slab will normally be subjected to two types of cracking: (i) non-structural cracking related to environmental factors such as plastic settlement, shrinkage and thermal effects, and (ii) structural cracking caused by the tensile stresses from the vehicular load action. Superimposed crack prints of these two represent the current state of cracking, which is continually being altered by the progressive crack growth due to dynamic overloading effects. This may lead to the formation of a concrete zone which is separated from the surrounding body along an enclosed perimeter of the nucleated crack surface. This separation along a closed perimeter constitutes, in effect, a flaw. Existence of random cracking at the slab soffit is conducive to the formation of such a flawed zone, which may be dangerous from the punching viewpoint.

MATERIALLY DAMAGED BRIDGE DECKS: EP2 BRIDGE CASE STUDY

General Description

This bridge is located in the Dammam district of the eastern province along the Abu–Hadriyah–Dammam road. The MOC reference number is 023-0093-0234 and is code named EP2. The bridge has three spans of continuous slab type with a central span of 13·45 m and two end spans of 8 m. The bridge deck is a voided slab system with a slab thickness of 110 cm. The voids are circular, 20 in number across the section, with each void being 70 cm in diameter. Shown in Fig. 6 is a typical cross-sectional view of the bridge deck.

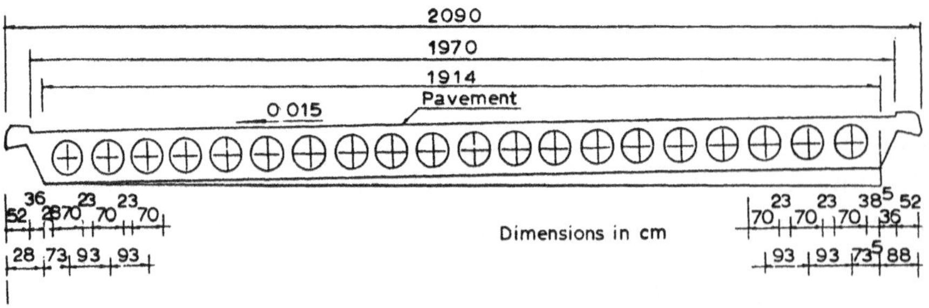

FIG. 6. Cross-section of deck slab for bridge EP2.

FIG. 7. View of asphalt cracking in EP2.

Attention was drawn to distress in the deck slab by evidence of severe cracking in places in the asphalt road surface (Fig. 7). The asphalt cover was stripped in such cracked regions in order to expose the concrete deck slab. The concrete slab itself showed few signs of cracking, but sounding a hammer on the concrete surface indicated delamination of the slab. Figure 8 shows the extent of the delamination in certain zones.

Since all indications pointed to corrosion-induced damage in the bridge deck, field measurements centred around corrosion-related parameters. Initially concrete was chiselled away in the delaminated zones and steel exposed at these sites. The reinforcement showed definite signs of corrosion and in some places the reinforcement had disintegrated entirely.

FIG. 8. Close-up view of delaminated concrete on deck slab of EP2.

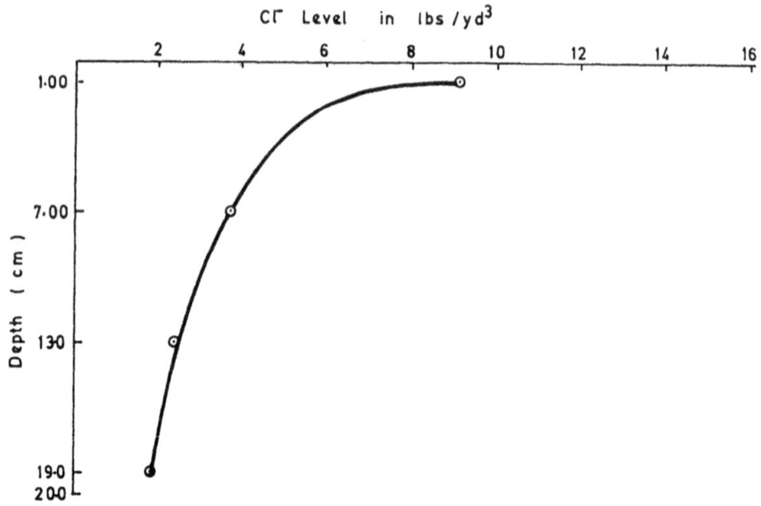

FIG. 9. Chloride profile in top 20 cm of EP2 deck slab.

Calomel half-cell potential measurements were made and all readings indicated a state of active corrosion. In addition, cores 5 cm in diameter and 20 cm in depth were taken from the bridge deck for chloride profile determination. It was not felt advisable to take full-depth cores due to the presence of the voided pipes. Figure 9 shows the chloride profiles for the top concrete depth of 20 cm. Bridge EP2 shows an extremely high gradient of Cl^- level, with maximum measured values of 9·12 lb/yd³ (5·42 kg/m³).

Discussion of EP2 Case Study

The threshold level of Cl^- to activate corrosion depends on the cement content, but for the normal range of cement content (from 335 to 490 kg/m³) the threshold level of Cl^- varies from 1·13 lb/yd³ (0·67 kg/m³) to 1·65 lb/yd³ (0·98 kg/m³), with lower threshold values for lower cement content. Thus the Cl^- level far exceeds the threshold level throughout the top 20 cm region in EP2, which indicates that the top reinforcement is susceptible to corrosion.

What is of great interest is the fact that there is a definite indication of chloride ingress into the deck from the top surface. If the chloride contamination were only to be from aggregate contamination and/or mix water, the gradient would not have been as pronounced. One can only hypothesize the source of this ingress, which includes the following:

(i) Chloride-contaminated curing water.
(ii) Airborne-contaminated pollutants deposited on the deck surface periodically, which are then washed into the permeable deck with condensation of moisture.

(iii) Periodic passage of transport vehicles loaded with saline water dropping on to the bridge surface.

Presence of contaminated aggregates and mix water would only serve to aggravate the problem further.

Inasmuch as up to 36 bridges in this sector are involved and several are showing similar symptoms it appears that items (i) and/or (ii) mentioned above may be plausible explanations. Item (ii) is considered a possibility because the entire region is identified as a Sabkha region and the likelihood of chloride-contaminated curing water (item (i)) should never be discounted in a dry and arid region.

It may be that the mechanism of thermal incompatibility of concrete components (TICC), as identified in the laboratory research component of this project,[9] is playing a role in increasing the permeability of the concrete decks over a period of time under repeated thermal cycling. This effect is known to exist in concretes made with limestone aggregates (as in the eastern province) due to the difference in the coefficient of thermal expansion of the aggregate and the cement paste. Diurnal and seasonal temperature variations induce microcracking in the concrete structure over a period of time, making it easy for chloride to ingress into the deck slabs.

Conclusions of EP2 Case Study

At this stage, based on analysis conducted thus far, it appears that there is a very good probability that top reinforcement in the EP2 bridge is in an active state of corrosion. Nothing can be said about the bottom reinforcement in the absence of full-depth cores for the chloride profile. In view of the fact that the bridge is continuous, corrosion of top steel in negative moment regions can have very serious implications. In terms of what needs to be done, one cannot say for certain that removal of delaminated sections of the deck with subsequent corrosion-associated repair would solve the problem. The reason for this is that there has been a substantial build-up of Cl^- level (at least in the top 20 cm of the slabs), to a level above the threshold deemed necessary for activating galvanic corrosion. In addition, there is a definite possibility that chemical corrosion may also be present in view of the dangerously high levels of sulphates.

In retrospect, it may be stated that there is an urgent need for detailed investigations of all affected bridges in this region. This should include full-depth chloride and sulphate profiles, and half-cell potential contour plots. Based on these results, the mode of repair can be decided, which could include conventional corrosion repair in decks where the Cl^- level has not built up excessively with chlorides and sulphates. Performance of cathodic protection in SO_4^{2-}-related chemical corrosion is not well known.

CONCLUDING REMARKS

During the course of the national project,[5] 21 bridge deck systems have been subjected to detailed investigations. These bridges cover a wide spectrum of varying parameters, as they not only differ in their geographic locations within the kingdom but their deck types, span lengths and deck widths also vary. The common denominator amongst these bridges is that they were undergoing a certain degree of distress to their decks at the time of investigation. Each of these bridges was treated as a unique case study and was diagnosed independently of other bridges in its region or of its own deck type.

The objective of this section is to seek inferences from the case studies so as to help identify certain global parameters common to the observed damage phenomena in an effort to minimize the probability of such occurrences in the future.

It may be concluded, in retrospect, that the shape and form of damage most likely to occur in a given situation is a function of bridge geometry, proportioning and environment. Shown in Table 2 is a damage likelihood chart based on the evidence collected as a result of the detailed study of the 21 bridges in the kingdom.

In order to minimize the likelihood of the occurrence of the most common forms of damage as identified during the tenure of this project, certain precautionary steps need to be exercised in order to avoid damage

TABLE 2
Damage likelihood chart for bridge decks in the kingdom of Saudi Arabia

Type of bridge deck	Environment	Possible mode of damage	Causal factors
(i) Girder/slab (ii) Box girder	Non-aggressive	(i) Pothole in slab (ii) Rectangular-grid cracking on slab soffit (iii) Shear/flexural cracking in girder	(i) Overloading (ii) Low-strength concrete (iii) Under-design
(iii) Slab	Non-aggressive	Rectangular-grid cracking on slab soffit	(i) Overloading (ii) Low-strength concrete (iii) Under-design (iv) Thermal effects
All types of concrete bridges	Aggressive	Reinforcement corrosion	(i) Chloride-contaminated materials and/or mix, curing water (ii) Improper mix design

occurrence in the future. These steps must cover all phases of the deck's life, namely its design, construction and operational maintenance.

ACKNOWLEDGEMENTS

The authors wish to thank Y. N. Ziraba and G. S. Al-Sourti, research associates of the national project, who have assisted in various activities of the project, including their involvement in the case studies. Thanks also are due to MOC, KFUPM and KACST for their technical and financial support to the project.

REFERENCES

1. Al-Tayyib, A. J., Rasheeduzzafar and Al-Mana, A. I., Deterioration of concrete structures in the Gulf States. First International Conference on the Deterioration and Repair of R/C in the Arabian Gulf, Bahrain Society of Engineers, Bahrain, October 1985, pp. 27–47.
2. Rasheeduzzafar, Dakil, F. H. and Al-Gahtani, A. S., Deterioration of concrete structures in the environment of the Middle East. *Journal of the American Concrete Institute*, **81**(1) (January 1984) 13–20.
3. *The CIRIA Guide to Concrete Construction in the Gulf Region.* CIRIA Special Publ. No. 31, London.
4. Pearson-Kirk, D., Al-Mandil, M. Y., Azad, A. K., Baluch, M. H., Sahir, A. M. and Ziraba, Y. N., Truck loadings for design of concrete bridges in Saudi Arabia. *Proceedings of a Symposium on Concrete and Concrete Structures in the Middle East*, King Saud University, March 1987. Riyadh, Saudi Arabia, 1987.
5. Al-Mandil, M. Y., Azad, A. K., Baluch, M. H., Sharif, A. M., Pearson-Kirk, D. and Al-Dhalaan, M. A., A study of cracking of concrete bridge decks in Saudi Arabia. National Project Final Report, KFUPM, Dhahran, Saudi Arabia, April 1989.
6. ACI Standard, Building Code Requirements for Reinforced Concrete (1983). American Concrete Institute, ACI 318-83, Detroit, Michigan, 1983.
7. Kareem, K., Load-induced cracking and failure of concrete deck slabs in girder-slab type bridges. MSc thesis, Department of Civil Engineering, KFUPM, Dhahran, Saudi Arabia, March 1989.
8. Al-Mandil, M. Y., Azad, A. K., Baluch, M. H., Pearson-Kirk, D. and Sharif, A. M., Punching shear failure of concrete girder-slab type bridge decks in Saudi Arabia. *Proceedings of the Fourth International Conference on Structural Faults and Repair*, University of London, June 1989.
9. Baluch, M. H., Al-Nour, L. A., Azad, A. K., Al-Mandil, M. Y., Sharif, A. M. and Pearson-Kirk, D., Concrete degradation due to thermal incompatibility of its components. *Journal of Materials, ASCE*, **115** (August 1989).

49

Traffic Load Simulation Programme

DIETRICH LEBEK
Kollwitzweg 11, D-5000 Köln 91, FRG

ABSTRACT

Most existing bridge codes are almost exclusively orientated towards bridge design = to-be-built bridges. The reliability degree is not the same for decisions on to-be-built bridges and existing bridges. A 'service concept' must be put at equal rank side by side with the 'design concept'. Both concepts complement each other. Neither can replace the other one. A traffic load simulation programme is an integral part of the 'service concept', allowing the evaluation of the individual bridge for its actual everyday traffic load. The summarised general considerations and a state-of-the-art review in brief underline the need for a 'service concept'. The aim of a simulation programme is defined and the basic requirements for its development are described. A proposal is presented for a complete traffic simulation programme and guidance given for its usage.

INTRODUCTION

Maintenance of road bridges in a condition to provide safe and uninterrupted traffic flow is the primary aim of every transportation agency. A growing task facing bridge engineers in the near future is the rational assessment of existing bridges and optimum allocation of resources among decisions regarding unqualified acceptance, restricting traffic, rehabilitation or closing existing bridges. Primary among the considerations is the safety aspect. The need is for flexible evaluation options which recognise the site-specific behaviour and account for the frequency and configuration of everyday traffic. The everyday safe load-carrying capacity serves as a basis for many keynote decisions about the bridge. A powerful tool for achieving this goal is a flexible traffic load simulation programme which is easily adjustable to the specific site conditions of the individual bridge.

SERVICE CONCEPT

General Considerations

Bridges are designed as structures with a potential for long service life. Bridge design is based upon codes and standards valid at the time of design. However, codes and standards must be reviewed, changed or adapted to changing requirements from time to time, i.e. the duration of their validity is limited as compared with the expected service life of a bridge. At present we experience a rather rapid change of service conditions and a principal change of our basic safety concept from a predominantly deterministic to a predominantly probabilistic safety concept.

The level of safety of a bridge and its components is determined by the ratio of the sum of acting service loads and the resistance. The resistance is not necessarily a time-independent value. Frequently there exists an interrelationship between the resistance and the kind and duration of service loads and service conditions (e.g. fatigue problems).

In addition, experience has shown that most technical products suffer from decreasing 'reliability' (R_t) with time. There are strong indications that bridges are no exception to this general rule. Such a decrease could be conveniently expressed as an exponential function in the general form $e^{-\lambda_i t}$, where t = number of years of the bridge in service and λ_i = function of failure mode and maintenance intensity.

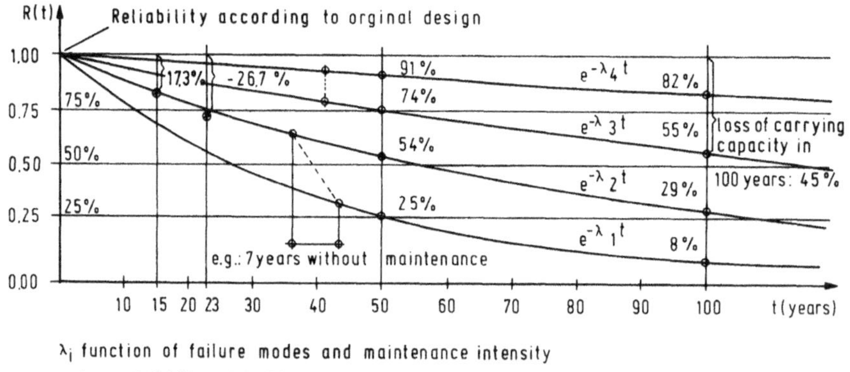

FIG. 1. Decrease in bridge 'reliability' over the years with regard to load-carrying capacity. ——, Decrease of the reliability due to wear and tear and other factors; – – –, erratic decrease due to overloading; — — —, due to complete omission of maintenance.

The failure mode will be influenced by parameters such as live load intensity and frequency, structural defects, fatigue, damage, deterioration, change of assumed structural system, etc. Figure 1 illustrates possible trends by a few examples of bridges in service. The starting point of curves, $R(t) = 1$, corresponds to the original reliability according to design assumptions and a basically sound structure. The four cases, randomly chosen, may illustrate the likelihood of the general validity of a function of failure mode and maintenance intensity for bridges in service and typify what may be called the 'creeping' or the 'erratic' change of the reliability of a bridge affecting its load-carrying capacity. Figure 1 also provides an indication of the 'individuality' of bridges in service.

State of the Art

In the federal road system (i.e. motorways and federal roads), the Federal Republic of Germany has a relatively new and modern bridge stock. The first 'official' bridge census was made in 1965. Tables 1–3 illustrate the development of the bridge stock over more than two decades.

Table 1 shows that about half of the present bridge stock has a service life of 25 years or more. Prestressed concrete bridges represent the youngest kind of bridge within the entire bridge stock. Only a marginal number of new steel and composite bridges have been built in the last 15 years. About 90% of all bridges presently in service are reinforced or prestressed concrete bridges. Table 1 also indicates that the mean service age of the total bridge stock will increase more rapidly in the future since new construction has reduced. Table 2 indicates the tendency within the last two decades to adapt bridge

TABLE 1

Development of bridge stock (federal road system only)
(percentage based on total number of bridges in service in 1987)

Year	Kind of bridge				Total bridge stock
	Steel	Composite	Concrete	Prestressed concrete	
1965	76	55	68	15	49
1970	95	79	80	32	63
1975	93	109[a]	87	54	76
1980	103[a]	103[a]	94	75	88
1987	100	100	100	100	100

[a] Between 1975 and 1987 some steel and composite bridges have been replaced by other kinds of bridge structures, foremost by prestressed concrete bridges.

TABLE 2
Development of the statistical mean bridge dimensions in metres (federal road system only)

Year	Kind of bridge								Total bridge stock	
	Steel		Composite		Concrete		Prestressed concrete			
	Span	Width	Span	Width	Span	Width	Span	Width	Span	Width
1965	51·9	11·3	83·2	18·4	14·5	13·1	64·1	15·1	25·1	13·7
1970	49·4	13·5	88·9	18·7	15·0	13·1	69·1	15·4	29·4	14·5
1975	51·4	14·8	82·2	18·9	14·7	14·7	74·9	18·0	34·6	16·7
1980	47·6	15·7	94·3	18·8	14·5	16·3	76·2	18·0	37·5	17·4
1987	46·6	17·5	101·1	18·7	14·2	17·2	75·6	17·8	39·6	17·7

dimensions to modern traffic requirements. One criterion is the development of the bridge deck width. With the exception of composite bridges, the statistical width of the bridge deck increased from about 11 to 14 m in 1965 to about 17 and 18 m in 1987.

This fact has implications as regards the total amount of design live load moment. The design live load consists of two components: an axle load configuration and a uniformly distributed live load. In the German bridge design code, the axle load configuration is independent of the bridge deck width. The amount of uniformly distributed live load is directly proportional to the bridge deck width. Taking the total design live load moment for the mean bridge in 1965 as unity, i.e. equal to 1·0, the total design live load moment for the mean bridge in 1987 would be 1·09, or about 10% greater. However, both bridges have to carry the same daily traffic service loads since the actual number of truck lanes remained unchanged for both bridges.

Another interesting aspect is: the present-day valid code would require a factor of 1·35 for the same bridge, or a 35% greater total design live load moment. This underlines the statement made above about the limited duration as compared with the expected service life.

Table 3 indicates that about three-quarters of the existing bridge stock belongs to the range of bridge lengths between 5 and 100 m, i.e. the range of small to medium span bridges. This is the range mainly affected by modern truck traffic.

Intensive measurements have revealed interesting facts. As regards directional traffic on motorway bridges, 75% have to cope with a yearly transport volume of 10 million t or more, 50% with 20 million t or more, 15% with 30 million t or more and 1% with 40 million t or more. This again

TABLE 3
Range of bridge lengths (federal road system only)
(percentage based on total number of bridges in service in 1988)

Length range between	Kind of bridge				Total bridge stock
	Steel	Composite	Concrete	Prestressed concrete	
5 and 30 m	30	26	61	26	45
30 and 100 m	12	45	9	61	29
5 and 100 m	42	71	70	87	74

underlines the 'individuality' of bridges in service. Between 1978 and 1984 the mean gross vehicle weight increased about 10% for a representative truck traffic configuration. The damage equivalent value, used for fatigue loading, increased by about the same amount. This corresponds to an average increase rate of more than 1·5% pa at present. Predictions made public recently indicate an increase of 30% for truck traffic intensity in the European Community in the near future.

SIMULATION PROGRAMME

The Aim of a Traffic Simulation Programme

The primary aim of a traffic simulation programme is the evaluation of the safety (or reliability) of road bridges and their components under the action of 'everyday service loads', taking into consideration the actual condition of the bridge and its components.

The expression 'everyday service loads' includes the following service conditions:

—'Normal' traffic conditions, i.e. free-flowing truck traffic in the usual configuration at the bridge site.
—'Specific' traffic conditions, i.e. truck traffic flow influenced by specific measures or conditions such as specific road alignment conditions, access to industrial or harbour areas, etc.
—Conditions caused by specific infrastructure conditions or requirements, either limited to a certain duration or unlimited in time.
—Conditions caused by construction measures or structural requirements.

The evaluation of the safety (or reliability) should include static as well as dynamic aspects and, if required, fatigue aspects for the entire structure, giving due consideration to redundancy aspects.

Another essential objective is the usage of the traffic simulation programme as an everyday tool by the practising engineer (foremost in the fields of public service and consultant engineering). Expertly used, this tool can provide the much needed data serving as a basis for many keynote decisions about the bridge.

Principles of a Simulation Programme and a Design Load Concept

A traffic simulation programme and a design load concept are based upon quite different principles. It is essential that this fact is clearly understood for the development of a simulation programme as well as for its proper usage. This fact is completely independent of the underlying basic philosophy on which the design load concept is based, e.g. on working stress design, on load factor design and on a basically deterministic or basically probabilistic concept.

—The principle of a simulation programme is the individuality.
—The principle of a design load concept is the general validity.

In consequence of this, the following statements can be made:

—The design of a bridge orientated towards a long service life should be based upon the principle of the general validity.
—An effective management for bridges should be based upon the principle of individuality.
—Design codes or standards and a traffic simulation programme should be viewed as sensible complementary principles rather than competitive or, even worse, contradictory principles.
—Design codes or standards can be based upon the regulation principle whereas a simulation programme must not be based upon the regulation principle.

Basis and Requirements for the Development of a Traffic Simulation Programme

The development of a traffic simulation programme is dependent upon certain requirements. The following list of general requirements may not be complete in every detail. However, the list contains the more important items. The requirements concern such main items as load characteristics, traffic composition, traffic flow characteristics, calibration of basic data, accuracy of the models, and hardware and software requirements.

—Distinction between types of truck (see also Fig. 2). Twenty-one types of truck form the basis for the simulation programme.

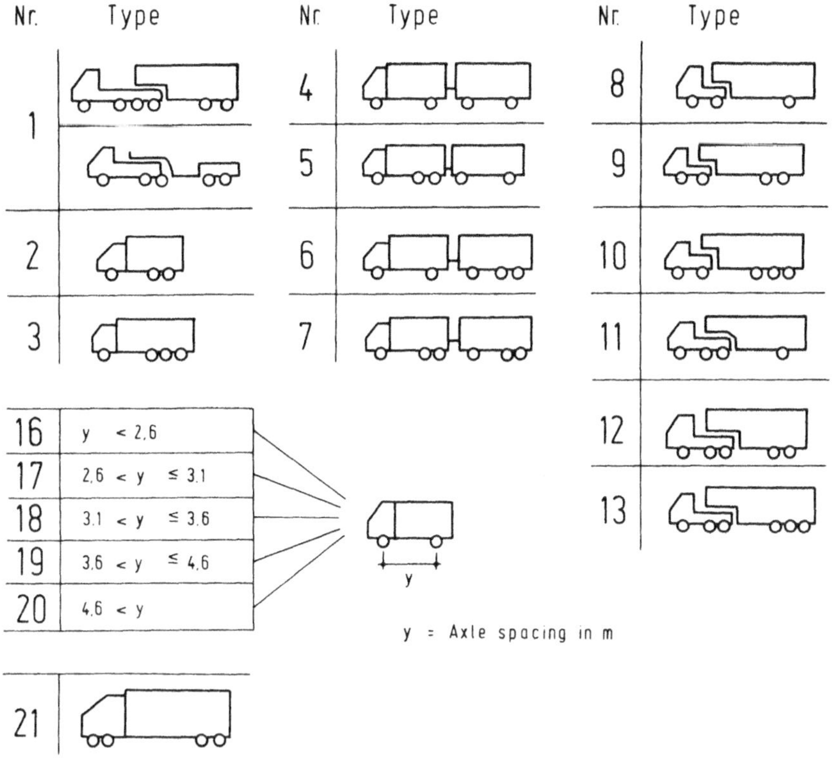

FIG. 2. Classification of truck types.

—Acquisition of accurate and sufficient data—preferably at sites which are representative of normal traffic flow—for the following items:

- vehicle gross weight
- axle load
- axle spacing
- vehicle speed

(separate for each truck type)

- distance between vehicles (separate for each lane)

—The load functions can be described for each type of truck by parameters that are independent of the location.

—The distribution functions of the load function parameters can be transformed into mathematical functions with sufficient accuracy.

—Verification and recalibration of the above data are required at suitable intervals (preferably measurements should be made at intervals not exceeding 10 years).

—Traffic composition and traffic frequency must be assessed separately at each bridge site and be verified at suitable intervals.

—Specific traffic conditions must be assessed at each bridge site and transformed into suitable mathematical models.

—Structural models are required that reflect the actual condition of the bridge at the time of the investigation. It is essential that the accuracy of the structure's model(s) and the accuracy of the load models are compatible.
—Each module of the simulation programme should be developed with regard to the practical capacity of a suitable type of personal computer. Consideration should be given to the possibilities offered by modern hardware to guide the user by including a suitable form of diagram or 'pictures' in the programme for the various modules.
—The resulting strains and stresses should be given in the form of distribution functions independent of the kind or the form desired, e.g. stresses, moments, forces, etc. Only distribution functions can make it possible to evaluate extreme values and their likely frequency of occurrence, and thus assist in a reliable assessment of the bridge and its components.

LAYOUT OF A SIMULATION PROGRAMME

The need for flexibility has been stressed repeatedly above. Flexibility is an essential element for every possible form of a suitable layout. The number of modules to be included in the programme is not a decisive item as long as the programme is capable of achieving its desired objective.

Two general ways appear to be possible for a layout:

—A clear definition of a desired objective resulting in a programme that suits this objective only. As a consequence each objective would require a specific programme.
—A complete layout for a simulation programme that is built up step by step whereby the necessary number and sequence of modules can be chosen to suit the desired objective(s).

In the following, the second method will be illustrated. However, the first may offer advantages for certain conditions. The proposed layout considers the step-by-step method. The modules 1–3 and 6 are devoted to modelling traffic loads and traffic conditions. Module 1 is the 'foundation' for all traffic loads and conditions. Module 6 is of similar importance if fatigue is considered as an essential item. Module 2 covers the traffic stream for normal conditions. Module 3 is intended to cover specific traffic conditions. Module 4 should be interpreted as a summary of the model(s) needed to describe the structural characteristics. This module will only suit certain conditions, i.e. in general each bridge will require its own specific built-up module. Nevertheless, specific elements could be programmed in this

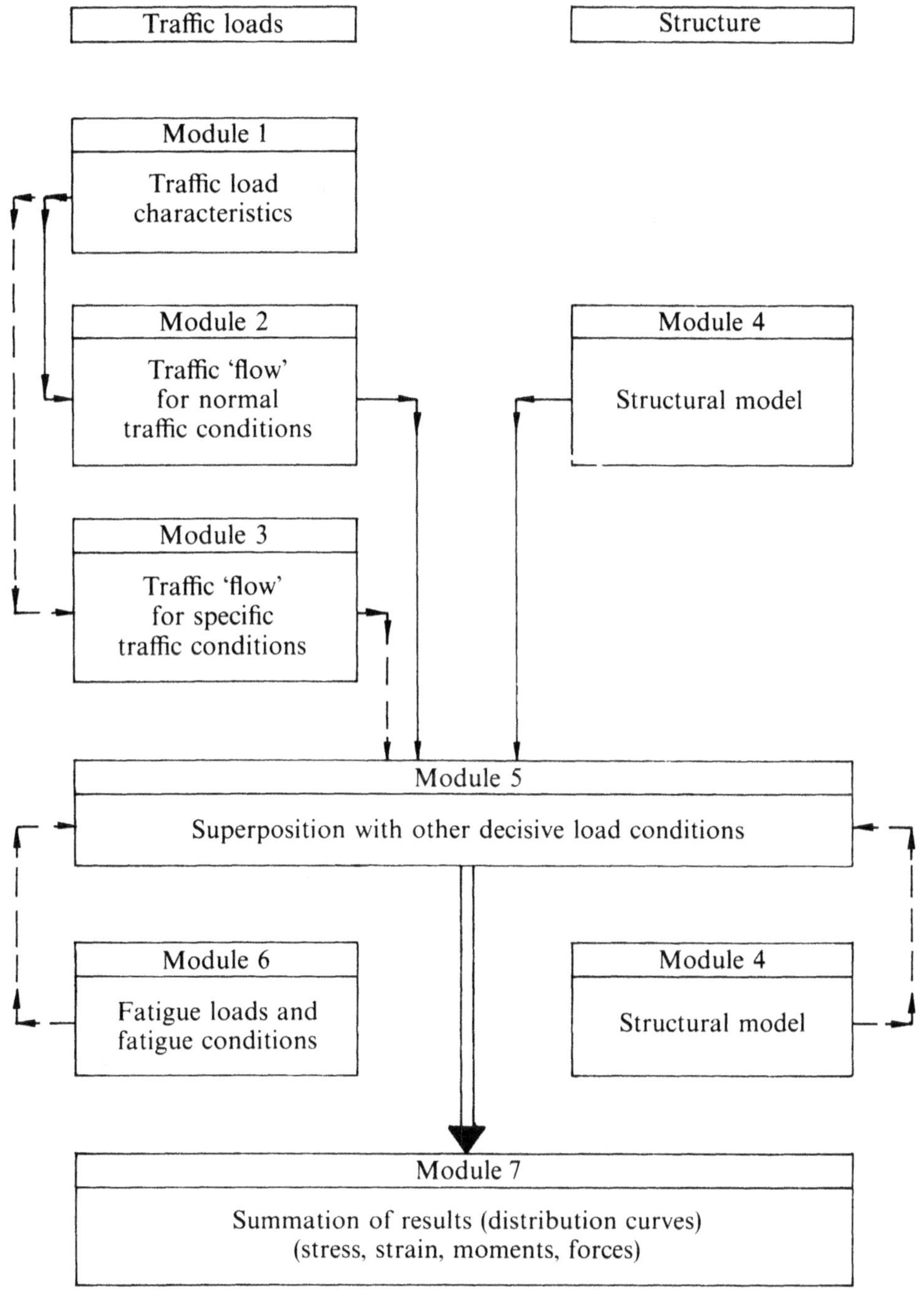

FIG. 3. Proposed layout for a complete traffic simulation programme.

module in such a way as to be applicable for more general use. Module 5 is reserved for the superposition of traffic loads with other defined loads or load combinations. This module will be needed for more in-depth investigations. Defined loads or load combinations could be, for example, prestressing, creep, shrinkage, temperature, stress redistribution, uneven settlement, etc. Finally, module 7 summarises all the modules needed to give final results for a given objective in the form of distribution curves of whatever is required, e.g. stress, strain, moments, forces, etc. The proposed

TABLE 4
Description of the modules forming the simulation programme

Module	Descriptive remarks	Dependent on, independent of
1	Load characteristics of each type of truck: distribution function of gross vehicle weight, axle load and geometry	Dependent on development in course of time, independent of site
2	Traffic flow (Q_i) in each lane for normal traffic conditions, Q_1 in truck lane and Q_2 in passing lane(s); Q_1 reflects the local traffic composition and its linear distribution functions, i.e. constant speed and distance between vehicles according to distribution functions	Traffic composition dependent on local conditions, speed and distance, independent of local conditions
3	Traffic flow for specific conditions not included in module 2, e.g. specific infrastructure, gradients and repair	Dependent on location, on specific situation; variable or constant with time
4	Structural model, e.g. influence lines for moments, shear, torsion or their combined actions; FE models, grid models, etc.	Dependent on structural characteristic, material, condition, etc.
5	Superposition of loads; load combinations, including temperature, creep, shrinkage and stress redistribution, could be of importance	Dependent on structural characteristic, material property, time and condition
6	Fatigue loads and/or fatigue conditions for fatigue-prone components	Dependent on traffic composition and density, kind of material and detail
7	Summary of resulting stresses, strains, moments or forces in the form of distribution functions	Same as for modules 5 and 6

layout in Fig. 3 shows a possible arrangement of the seven modules briefly described above. Table 4 sums up the modules in a descriptive manner.

USAGE OF A SIMULATION PROGRAMME

The scope of usage is largely dependent upon the imagination and flexibility of the user. An extension of the number of modules is just as possible, as is the enlargement of an individual module by introducing submodules. In any case further development of the simulation programme is most desirable. An interrelationship with management systems should be planned as a further step of development.

The following list of possible usage is meant as guidance to some practical applications, which may range from the individual bridge to a set of bridges up to the network level.

Individual bridge

—Safety of the structure and its components
—Redundancy and critical sensitivity
—Suitability and/or sensitivity to heavy transport
—Urgency and scope of special investigations
—Scope of repair, rehabilitation or strengthening
—Necessary load reductions, but also possible upgrading in allowable loads
—Possible change in inspection intervals

Route level or set of bridges

—Kind, degree and scope of critical 'weak points' in a route
—Degree of load-carrying capacity of a route
—Suitability for heavy transport (route criterion)
—Kind, scope and justification for measures to be taken (rehabilitation, strengthening, upgrading or downgrading)
—Sensitivity to possible changes of the route infrastructure

Network level

—Number and location of the 'safe', 'critical' and 'unsafe' bridges
—Kind and degree of changes with time in the level of safety in certain types of structure and in certain structures designed on the same principle
—Strong and weak points of comparable types of structure

—Implications of changes in the legally allowed loads
—Implications of changes in the traffic infrastructure

ACKNOWLEDGEMENTS

The author gratefully acknowledges the highly competent advice given and the helpful support received by Dr rer. nat. H. Steinhilber and Dr ing. Lehrke from the Fraunhofer-Institut für Betriebsfestigkeit, Darmstadt.

50

Canada's Advanced National Standard on Bridge Evaluation

PETER G. BUCKLAND

Buckland and Taylor Ltd,

1591 Bowser Avenue, North Vancouver, British Columbia, Canada V7P 2Y4

ABSTRACT

With time bridges deteriorate but traffic loads increase. There is insufficient money available to replace all deficient bridges, so in the last decade efforts have been under way in Canada to identify with more accuracy than before which bridges are deficient, which parts are critical and by how much they fall short.

In a simple step-by-step procedure several 'firsts' are accomplished for a bridge evaluation code. The safety index β *is selected as a function of the expected behaviour of a member and the consequences of its failure. Load factors reflect the confidence with which the loads can be predicted. A bridge on an ore haul route in the Yukon Territory serves as an example.*

INTRODUCTION

There are two reasons for evaluating a bridge, either the loads are increasing or the bridge is deteriorating—sometimes both.

The only loads over which mankind can exercise control are loads caused by traffic. But unfortunately for the bridge engineer there are all sorts of pressures to keep increasing the allowable traffic loads.

Because strengthening of bridges is a very costly business and there is never enough money anyway, it was decided in Canada to develop a more accurate method of deciding when a bridge really needs to be upgraded and when, even though it does not meet the provisions of the design code, the bridge is in fact adequate.

The result is known as clause 12, 'Existing Bridge Evaluation', of Canadian Standard CAN/CSA-S6-88: 'Design of Highway Bridges'.

In order to determine the reasonableness of answers given by the theory, it was calibrated against a series of bridges which had previously been evaluated by other means.

PROCEDURE

Clause 12 is in typical limit states format. That is to say,

$$\alpha_R R \geq \alpha_L L(1+I) + \alpha_{D_1} D_1 + \alpha_{D_2} D_2 \tag{1}$$

where

R = resistance
L = live (traffic) load effects
I = dynamic load allowance
D_1 = effects of all dead loads except unmeasured asphalt
D_2 = effect of unmeasured asphalt
α = the appropriate factor
α_R = $U\phi$, where ϕ is the resistance factor given in the code and U is an adjustment factor to 'fine tune' the values of ϕ to increase their accuracy.

It is common to consider only dead and live loads when using clause 12, but if it is thought necessary other loads, such as wind or earthquake, may be included.

The difference between clause 12 and other codes is that the load and resistance factors are selected within clause 12, and are not the same as those in the main body of the standard. The reason for this is that because of the high cost of upgrading a bridge it pays to be more exact in the calculation of factors. The basic equation (1) is often written in the form

$$\mathrm{LLRF} = \frac{U\phi R - (\alpha_{D_1} D_1 + \alpha_{D_2} D_2)}{\alpha_L L(1+I)} \tag{2}$$

where LLRF, the live load rating factor, is equal to 1·0 if the element of bridge being considered can exactly carry the load required and more than 1·0 if there is spare capacity. If it is less than one, the element being considered is substandard.

It can be seen that the LLRF is sensitive to the difference in two large numbers, the factored resistance and the factored dead loads. This explains the value of fine tuning ϕ with the modifier U.

LIVE LOAD FACTORS

Live load factors depend on three things:

(1) How well the live load is known. Factors are large when loads are uncertain and smaller when loads are well known. This is in keeping with limit states philosophy.
(2) The amount of warning likely as collapse is approached; a sudden failure in a non-redundant structure commands a larger factor than a ductile failure in a multiple-path system, for example.
(3) The consequences of failure. The failure of a deck element is a nuisance but not catastrophic; the failure of the main compression chord of a truss is far more serious and a greater safety margin is appropriate.

TRAFFIC LOADS

Clause 12 defines four kinds of traffic:

—NP (non-permit), normal 'legal' traffic; maximum loads are not known with confidence.
—PS (permit, single-trip), an overload for which a special permit is required, normally a large indivisible load with axle loads exceeding legal limits; loads often not well known.
—PM (permit, multiple-trip), a series of overloaded vehicles on a special permit, usually a bulk haul from a mine, for example; individual legal axle loads are not exceeded; loads are usually well controlled.
—PC (permit, controlled), an unusually heavy load, escorted; no other vehicles on the bridge; load often weighed and therefore accurately known.

The significance of these traffic groupings can be understood conceptually by reference to Fig. 1. Figure 1(a) shows a nominal resistance and nominal load. They are separated by a 'safety factor'. In fact the actual resistances and the actual loads can be described by curves representing probability density functions: most, but not all, resistances are greater than the nominal; most, but not all, loads are less than the nominal.

The shaded area where curves overlap is the area where the resistance can be exceeded by the load, which would lead to failure. The probability of this occurring must be kept sufficiently small as to be an acceptable risk (which will never be zero so long as the curves have no limits). Seen graphically, the 'safety factor' in Fig. 1(a) must be sufficiently large that the shaded area is acceptably small.

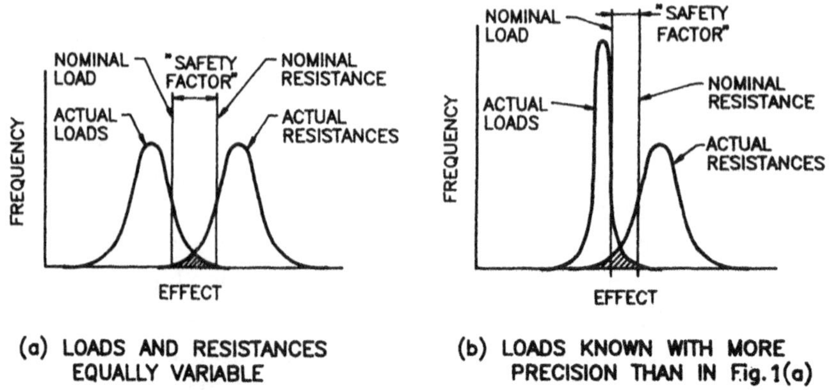

FIG. 1. The relationship between 'safety factor' and variability of loads and resistances with the real safety (i.e. probability of failure) kept constant.

The probability of failure, demonstrated by the shaded area and usually represented by β, the safety index, is chosen in clause 12 as a function of structural behaviour (e.g. redundant or not) and the importance of the consequences of failure. This is done by simple reference to tables and it requires some intelligent engineering judgement.

Figure 1(b) represents the case where the traffic load is well known—for example an escorted weighed vehicle at crawl speed with no other traffic on the bridge. In this case the safety factor can be reduced without changing the shaded area or the probability of failure. In other words, despite the reduction of the 'safety factor', the risk is unchanged and the bridge's safety is not altered from the condition in Fig. 1(a).

EXAMPLE—THE YUKON TERRITORY

One of the eight bridges evaluated in the Yukon Territory will serve as an example. The Yukon has an area more than double that of the United Kingdom and a population of 28 000. The two main industries are tourism and mining. The few roads are vital to the economy.

Lead zinc concentrate from the territory's largest mine is trucked 300 miles (500 km) to tidewater along the only north/south road through the mountainous wilderness. In order to stay competitive the mine wished to increase the amount it could carry on each truck. A truck leaves the mine every 20 min on average.

The trucks are defined by clause 12 as category PM (permit, multiple-trip) and the loads are well regulated. Figure 2 shows a PM truck from Curragh Mine at Faro passing over the Pelly River Bridge.

FIG. 2. A bulk haul truck with overload permit (PM) on the Pelly River Bridge at Faro, Yukon Territory.

Figure 3 compares the global distribution of the heaviest 10% of non-permit trucks with the distribution of the PM trucks from the mine. It can clearly be seen that to cover the range of vehicles a larger load factor is appropriate for the NP (typical) traffic than for the PM traffic.

Clause 12 is in fact built around the distributions of 'annual maxima', as shown in Fig. 4. Since one would expect the nominal loads to be exceeded at least once per year, it is not surprising that the entire distributions are greater than the nominal. But here again it can be seen that NP traffic should be given a greater load factor than the better regulated PM traffic.

Selecting the Safety Index and Load Factors

Table 1, which is a reproduction of Table 47A of CAN/CSA-S6, demonstrates how the safety index is chosen. INSP1, INSP2 and INSP3

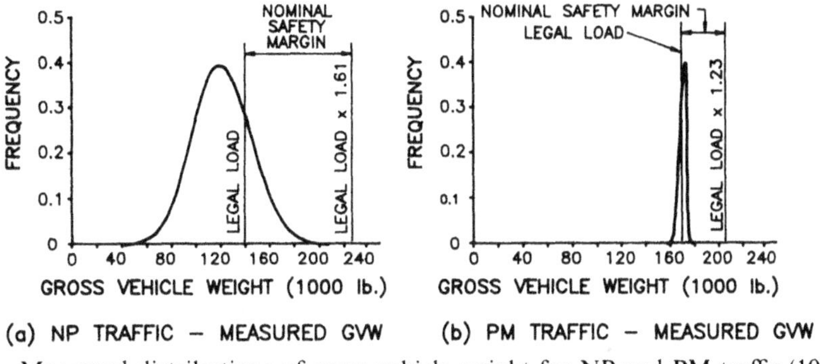

FIG. 3. Measured distributions of gross vehicle weight for NP and PM traffic (1000 lb = 4·45 kN).

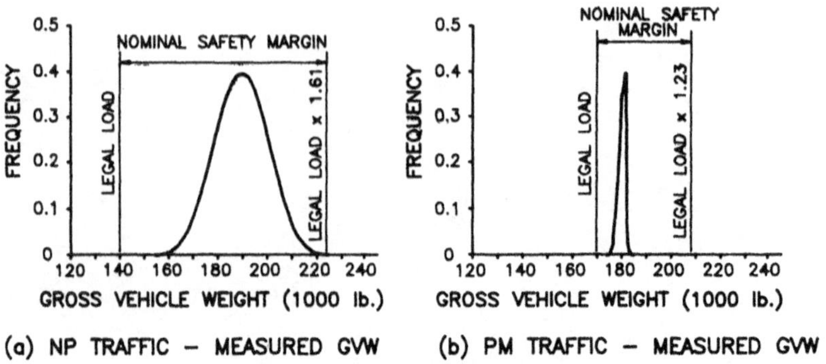

FIG. 4. Distribution of annual maxima of gross vehicle weights for NP and PM traffic (1000 lb = 4·45 kN).

denote the levels of inspection to which the bridge has been subjected. It is assumed that all bridges have routine inspection. INSP1 refers to elements that are not inspectable, INSP2 is routine inspection and INSP3 is for critical or substandard elements that have been inspected by the evaluator (who may notice clues about structural performance).

S1, S2 and S3 relate to system behaviour. S1 is where failure of one element can lead to total collapse, S2 is where one element failure will probably not lead to total collapse (e.g. multiple load paths) and S3 is where element failure leads to local failure only.

E1, E2 and E3 refer to behaviour of the element being considered as it fails. An E1 element is subject to sudden failure with little or no warning. An E2 element also fails suddenly but will retain post-failure capacity. An E3 element is subject to gradual failure with warning of failure probable.

TABLE 1
Target reliability index β for NP, PM and PS traffic

System behaviour	Element behaviour	Inspection level		
		INSP1	INSP2	INSP3
S1	E1	3·75	3·50	3·50
	E2	3·50	3·25	3·00
	E3	3·25	3·00	2·75
S2	E1	3·50	3·25	3·25
	E2	3·25	3·00	2·75
	E3	3·00	2·75	2·50
S3	E1	3·25	3·00	3·00
	E2	3·00	2·75	2·50
	E3	2·75	2·50	2·25

TABLE 2
Live and dead load factors

Load	Load factors	Target reliability index β						
		2·25	2·5	2·75	3·00	3·25	3·50	3·75
NP	α_L	1·31	1·37	1·43	1·49	1·56	1·64	1·71
PM	α_L	1·01	1·05	1·09	1·14	1·19	1·25	1·31
PS	α_L	1·13	1·18	1·23	1·29	1·35	1·41	1·47
PC	α_L	1·05	1·09	1·14	1·18	1·23		
D1	α_{D_1}	1·05	1·07	1·08	1·09	1·10	1·12	1·13
D2	α_{D_2}	2·32	2·51	2·70	2·90	3·09	3·28	3·47

Once β has been selected, the load factors are selected from Table 2. There is in fact an intermediate step which depends on the analysis method used, omitted here for simplicity.

Evaluating the Pelly River Bridge at Faro, Yukon Territory

The Pelly River Bridge at Faro, Yukon Territory (Fig. 2), provides an example of the use of clause 12. The bridge was evaluated for the heaviest allowed non-permit (NP) 'legal' vehicle, a truck with a gross vehicle weight (GVW) of 140 000 lb (63 500 kg) on eight axles, and for bulk haul trucks operating on a multiple-use overload permit (PM) and a GVW of 170 000 lb (77 000 kg) on eight axles.

Interestingly, it was found that in virtually all cases the lighter NP vehicle governed because of the large load factors it commanded. The figures following are for a top chord member in compression and for an interior floorbeam in bending:

	Chord	*Floorbeam*
Inspection level	INSP2	INSP2
Element behaviour	E1 (failure sudden)	E3 (failure ductile)
System behaviour	S1 (single load path)	S3 (multiple load paths)
Safety index β (from Table 1)	3·5	2·5
Appropriate factors (from Table 2)		
α_L—NP	1·61	1·37
α_L—PM	1·23	1·05
α_{D_1}	1·12	1·07
α_{D_2}	3·28	2·51
U	1·00	1·06

When these numbers were inserted into the formula (2) for live load rating factor, they yielded

Chord	LLRF = 0·56–1·29	for NP traffic
	= 0·62–1·35	for PM traffic
Floorbeam	LLRF = 1·08	for NP traffic
	= 1·27	for PM traffic

From this it can be seen that:

(1) NP traffic has a lower live load rating factor than PM traffic even though the nominal applied load is less.

(2) Some chords (those with LLRF less than 1·0) must be strengthened or, in some cases, braced to reduce the slenderness ratio. Both are fairly simple operations.

(3) The floorbeams do not need strengthening, which avoids an expensive procedure. However—and here is the important point—if conventional load and resistance factors had been used the LLRFs would have been 0·92 and 1·08 for NP and PM traffic respectively, and strengthening would have been required.

(4) For chords with an LLRF of 1·0 or greater the annual probability of failure is 0·000 23, corresponding to $\beta = 3{\cdot}5$. For the floorbeams the annual probability of failure, without modification, is about 0·0014. In other words, by accepting a lower value of β for the floorbeams the cost of strengthening now is avoided, but a 1 in 750 chance per year is accepted that the floorbeams may yield in bending, an event that will cause only inconvenience. The philosophy of clause 12 is that the preferable choice for the taxpayers' money is to not strengthen the floorbeams.

SUMMARY

(1) Clause 12 of CAN/CSA-S6-88 provides a simple to use method of evaluating bridges which takes account of the type of traffic and the expected warning and consequences of a failure.

(2) Using the Pelly River Bridge in the Yukon Territory as an example, it is shown that because multiple-trip bulk haul vehicles are better controlled than random 'legal' traffic higher nominal vehicle weights may be permitted without reducing the real safety levels.

(3) By accepting a higher probability of failure for members which give warning of failure and have multiple load paths which, if they should fail, will not cause failure of the bridge, expensive strengthening can be avoided.

For those having difficulty with this concept the question to be answered, at least qualitatively, is: would you prefer to have a floorbeam yield in bending or a main chord fail by compression buckling? If the former, by how much more would you prefer it?

ACKNOWLEDGEMENTS

The author wishes to express his appreciation to Mr Eric Gibson, manager of technical services, Yukon Transportation and Community Services, for permission to reproduce information about the Pelly River Bridge and for the opportunity to evaluate the eight bridges in the Yukon Territory, and to the Canadian Standards Association and Mr Lorne J. Hamblin, chairman of the Technical Committee of CAN/CSA-S6, for permission to reproduce parts of the standard. Clause 12 is a supplement to CAN/CSA-S6-88 and is available with its commentary from Canadian Standards Association, 178 Rexdale Boulevard, Rexdale, Ontario, Canada M9W 1R3.

The author is greatly indebted to F. Michael Bartlett, formerly of Buckland and Taylor Ltd, for his assistance in developing clause 12 and in the preparation of this paper.

APPENDIX: CHOICE OF TARGET RELIABILITY INDEX β

The target reliability index β is a function of the notional target annual probability of failure P. Note that clause 12 uses an annual probability, not a lifetime probability. P is based on life safety criteria and is defined as

$$P = AK/(W\sqrt{n})$$

where

A = activity factor, a measure of the risk associated with the activity (i.e. driving a car)
 = 3·0 for NP, PM and PS traffic
 = 10·0 for PC traffic
K = a calibration factor = 10^{-4}
W = a warning factor
 = 1·0 for no warning of failure expected
n = the number of people at risk
 = 10 for NP, PM and PS traffic on spans up to 100 m
 = 1 for PC traffic (other traffic kept off the bridge).

For elements supporting larger or more important bridges, n can be increased but β is insensitive to n and a value of 10 was chosen as a typical value for most bridges.

These values lead to:

For NP, PM and PS traffic	$P \approx 9{\cdot}5 \times 10^{-5}$	$\beta \approx 3{\cdot}75$
For PC traffic	$P \approx 1{\cdot}0 \times 10^{-3}$	$\beta \approx 3{\cdot}25$

These values of β are then reduced in a systematic way to account for improved warning of failure (which comes from inspection and from ductile behaviour) and for consequences of failure that are other than catastrophic.

51

Serviceability Assessment of Masonry Arch Bridges Using Vibration Tests

A. J. PRETLOVE and J. C. A. ELLICK
Department of Engineering, University of Reading, Reading, UK

ABSTRACT

Over the last 5 years an extensive programme of research at the University of Reading has explored the use of vibration methods in the assessment of the serviceability state of masonry arch bridges. Bridges in Derbyshire, West Sussex, Hampshire and Berkshire have been tested in the studies. Not only has a variety of methods of excitation been used but also a range of measurement and analysis techniques. The results from this work have been compared with the standard MEXE method of assessment and with building vibration damage criteria such as those contained in the German Standard DIN 4150. An account of the techniques which have been developed and used is given here. Some of them have been shown to be of no significant value whilst others are seen to be more promising. The better methods can, at the least, give additional information which can be considered alongside the present standard method. Recommendations are made as to their use. These methods require some development and verification before they can become suitable for routine assessments.

INTRODUCTION

There is considerable concern that some masonry arch bridges in Britain may be deteriorating rapidly as a result of increasingly heavy traffic and axle loads. An accurate assessment of the structural condition of these bridges is known to be extraordinarily difficult because of the wide range of configurations and materials that have been used in their construction. There is also a wide range of ages and, indeed, some of these bridges are classified as ancient monuments. Accurate methods of structural condition assessment are required if timely and cost-effective maintenance is to be achieved. If this cannot be done with sufficient precision then, in some cases, these structures may deteriorate increasingly rapidly as a result of developing structural faults, not only in the pavements but also in the arch structure itself. However, this is not to say that they will immediately fall down. The static strength of masonry arches is generally very much higher than the applied traffic loads, as has been shown by some recent tests to destruction.[1] The real problem is the progressive wear and tear of the structure. This can take many forms and some frequently observed ones are:

(a) mortar decay and erosion leading to arch barrel deformation or cracking;
(b) separation of spandrel walls, together with the outer parts of the barrel, from the rest of the arch barrel;
(c) fill deformation leading to an uneven pavement with consequent increases in dynamic wheel loads and possible damage to services, such as gas pipes; and
(d) foundation settlement.

The problem is to determine accurately which structures have reached a point at which they require maintenance or major repair. If the methods used in this determination are not accurate then, on the one hand, money may be needlessly spent on structures which are in adequate condition whilst, on the other, structures on the point of collapse may not be detected.

A simple empirical method for assessing the capacity of masonry arches to carry traffic was developed by the Military Engineering Experimental Establishment (MEXE) in the 1940s. It was based on Pippard's theory[2] and various experimental studies. Details of the method, as currently used, are set out in Ref. 3. In the method a provisional axle load is calculated from the span, ring thickness and depth of fill using a nomogram. This value is then modified by a series of factors which takes into account the actual shape of the arch, the type and condition of the materials, joints and workmanship, and the presence and position of cracks. This leads to a permissible axle load. Generally the method is reckoned to be conservative and it only assesses the load-carrying capacity of the arch barrel.

In this programme of work a variety of vibration techniques have been studied with the aim of supplementing the information given by a MEXE assessment and hence providing a more accurate indicator of state of serviceability. The principle behind the use of vibration techniques is simply that the vibration properties may change if the structural condition deteriorates. Thus the following properties have been studied:

(a) natural frequencies,
(b) damping,
(c) comparisons of time traces,
(d) non-linear behaviour, and
(e) response values to given inputs.

The earliest work made use of vibration excitation by normal traffic because this had the obvious advantages of zero traffic disruption. However, the wide variability of this excitation, depending as it does on axle loads, speeds and lines of action, precluded its further use. This was followed by work using standard vehicles to provide excitation but this was also unsatisfactory because response measurements made on the bridge included the dynamic properties of the vehicle as well as those of the bridge. A fresh approach was required: a test was needed that would give a repeatable response that was characteristic of that bridge and of no other. For this purpose an impact device was developed which dropped a mass through a fixed height on to the bridge surface. This has the disadvantage of requiring more equipment on the bridge. However, not only could spectral techniques be applied to the data but also the response traces could be subjected to various forms of analysis. A series of tests on a number of arches showed that each one produced a distinctly characteristic response.

In the latest work the impact test technique has been exclusively used with a standard drop weight which is designed to apply a known and fixed impulse to the bridge structure. This is described in more detail in the next section.

A previous article on this work[4] has described the full range of work on the various vibration response quantities listed above. The most promising indicators of condition were found to be:

(a) Peak-to-peak velocity in response to a standard impulse. This ties in with the German Standard DIN 4150,[5] which gives criteria for building damage due to vibration in terms of velocity.
(b) Fundamental natural frequency. The fundamental mode of vibration has in these tests always consisted of vertical motion coherent across the arch width and is thus termed 'bending'.

This paper concentrates on the analysis of these two features.

A wide range of tests has been made on arch bridges in England. Most of these arches are in a reasonably sound structural state but one or two are known to be reaching the point where maintenance or repair is needed. In this work the aim is to show that the two vibration methods listed above can pick out these less sound structures.

DESIGN OF THE IMPACT DEVICE

In the latest version of the impact device a mass, which can be varied from 12·5 to 75 kg in six steps, is lifted by pulley and held by a bomb release on a tripod system. It is dropped through 1 m on to a thin bed of damp sand that has been placed on the bridge pavement. The purpose of the sand is to prevent bounce and hence provide a known and repeatable impulse. Some care is needed with the dampness and thickness of the sand if repeatable results are to be achieved. This impact device is not as complicated as the force instrumented type developed recently by Bruel and Kjaer, and for this purpose does not need to be. The impulse time has been made short (10 ms) compared with the fundamental period for this type of structure, usually about 50 ms. The movement resulting from the impact is measured using a seismometer. In most tests the impact is applied at the arch crown adjacent to one parapet and the measurement is made adjacent to the opposite parapet. This offers not only minimal disturbance of the traffic flow but also simultaneous excitation of bending and torsion modes of the arch ring. The seismometer output is recorded on tape, and is taken back to the laboratory for digitisation and computer analysis.

The 75 kg drop test has been shown in earlier tests to give a bridge response roughly equivalent to that obtained from the passage of a 10-t lorry. It has therefore been used in these tests as it will give realistic values for response, particularly necessary in the analysis of vibration velocities.

TEST MEASUREMENTS AND ANALYSIS

A typical response time trace is shown in Fig. 1 complete with its calibration marker. The results of a Fourier analysis of this trace can be seen in Fig. 2, giving information on the natural frequencies of the structure. In this case the fundamental (bending) frequency is at 12 Hz. The torsion frequency can also be detected though it is beyond the edge of the measurement range of the instrumentation used, which is from 1 to 50 Hz.

The basic results from this procedure are shown for bridges numbered 1–11 in Table 1. For the purpose of comparison the table also shows the modified axle load (MAL) calculated using the MEXE procedure. This gives some indication of the state of serviceability of the bridge because it includes

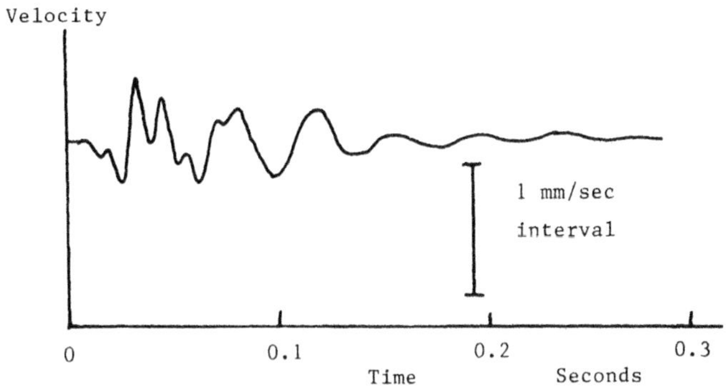

FIG. 1. Velocity–time trace for bridge 1.

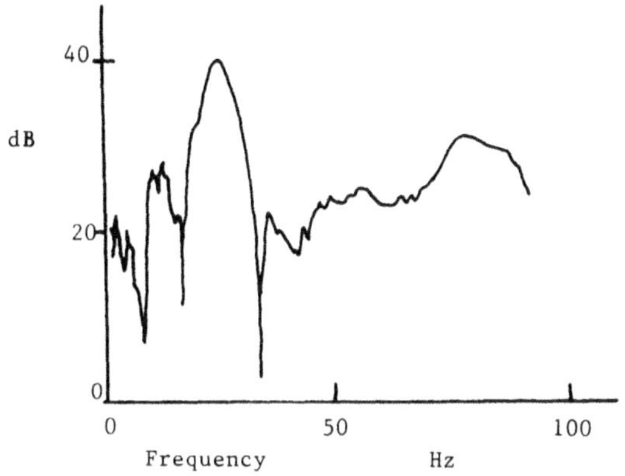

FIG. 2. Frequency spectrum for bridge 1, derived from Fig. 1.

TABLE 1
Summary of the major dimensions and results obtained for the bridges included in this study

Bridge	*d* (mm)	*c* (mm)	Span *S* (m)	CF	MAL (t)	P-P6 velocity (mm/s)	Natural frequency (Hz)
1	405	862	8·38	0·9	18	1[a]	12
2	355	1 064	5·26	0·9	38	3·33	28·9
3	457	737	9·40	0·9	14	4·92	11·8
4	470	800	8·28	1·0	12	3·60	—
5	405	1 230	6·70	0·65	12	2·27	14·6
6	335	830	6·03	0·7	12	1·96	17·4
7	565	785	8·62	0·7	7·5	5·51	15·3
8	340	760	6·57	0·8	12	3·41	15
9	353	648	6·55	0·8	11	4·69	15
10	340	740	6·55	0·6	8·5	1·34	15
11	432	686	6·71	[b]	[b]	5[a]	22

For a definition of the dimensions see Fig. 3.
[a] This figure has been estimated.
[b] Figures not available; bridge demolished 1987.

a condition factor (CF) which is obtained by visual inspection of the structure.

VIBRATION VELOCITY MEASUREMENTS

The peak-to-peak velocity values for a 75 kg drop test are shown in Table 1. It can be seen that there is not, in general, a clear relationship between peak-to-peak velocities and the modified axle load (MAL) obtained by the MEXE assessment method. However, there are some useful remarks which can be made. Bridge 7 has not only the highest measured velocity (5·5 mm/s) but also the lowest MAL (and there are strong grounds for weight restriction here). Bridges 3 and 9 have the next highest velocity values and both have relatively low MAL, particularly so in the case of bridge 9.

The peak-to-peak values can be directly compared with the DIN 4150 guideline limit of 3–8 mm/s. Several bridges are seen to have responses in this band, but not beyond it. These bridges are therefore close to the boundary of vulnerability. Values quoted here have come from drop tests only. It would be useful to verify these results using real traffic excitation. This would require the measurement of peak-to-peak vibration levels during busy times of the day for a large number of arch bridges. The data so obtained could be used not only as an indicator of the state of the structure but also as a measure of the current rate of damage (and hence the prospective serviceability). Studies of this kind would confirm the relevance of drop-weight tests to serviceability assessment.

FUNDAMENTAL NATURAL FREQUENCIES

The values given in the last two columns of Table 1 may be referred to as 'raw' data. They take no account of the different dimensions of the various arch bridges. In the case of natural frequencies this has to be taken into account if valid comparisons are to be made. This can be done by determining the mass and stiffness of an arch in relation to its dimensions. Figure 3 shows the symbols used for the relevant dimensions of an arch bridge.

In the fundamental (bending) vibration of the structure the major movement is confined to a volume near the crown. It is therefore the mass in this area which is important in the vibration. It is estimated that the generalised mass M is proportional to bcS. The shallowness, or otherwise, of the arch will only play a secondary part. The stiffness at the crown to a vertical force can be derived from Pippard's theory for the elastic action in

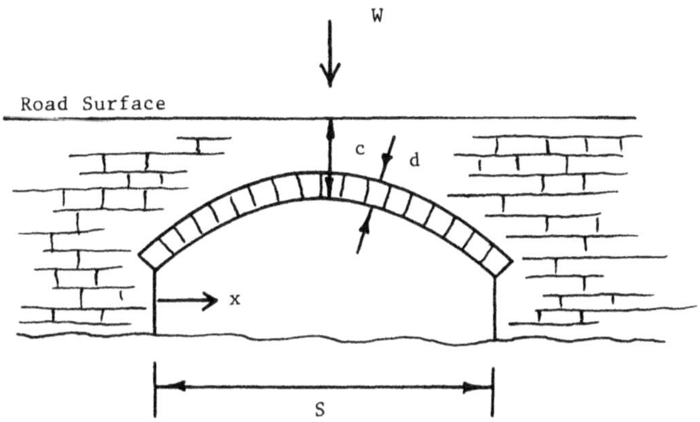

FIG. 3. Arch bridge nomenclature: b (not shown) = arch/roadway width; c = crown depth including the arch ring; d = arch ring thickness, assumed constant; and S = span.

the arch ring as given in Ref. 1. In summary, the deflection at the central point of load is given by Castigliano's theorem:

$$\delta = \partial U/\partial W$$

With strain energy

$$U = 2\int_0^{S/2} \frac{M_x^2}{2EI_0}\,\mathrm{d}x$$

For the arch ring

$$I_0 = bd^3/12$$

The arch ring bending moment

$$M_x = \frac{Wx}{128}\left[100\frac{x}{S} - 36\right]$$

Evaluation of the elastic deflection δ in this way shows that the stiffness K is proportional to bd^3/S^3.

The fundamental natural frequency is

$$\omega_1 = \sqrt{K/M}$$

and this will therefore be approximately proportional to $d^{3/2}/S^2c^{1/2}$. We have called this quantity the bending frequency factor (BFF).

Table 2 shows, for the bridges considered here, the calculated BFF, the test natural frequency from Table 1 and R, the ratio of the two. One would expect this ratio to be constant. If it is less than the general norm then it suggests that the test natural frequency is low and this may be an indicator of poor structural condition.

For most of the bridges the ratio R is remarkably consistent at a value of about 3 (in the mixed units used), indicating that the theory above is

TABLE 2
Test frequencies and calculated factors for bending

Bridge	Test bending frequency (TBF) (Hz)	BFF[a]	$R = \text{TBF}/\text{BFF}$
1	12	3·95	3·0
2	28·9	7·41	3·2
3	11·8	4·07	2·9
4	—	5·25	—
5	14·6	5·18	2·8
6	17·4	5·85	3·0
7	15·3	6·45	2·4
8	15	5·27	2·8
9	15	6·07	2·5
10	15	5·37	2·8
11	22	7·61	2·9

[a] BFF is defined in the text.

appropriate and reasonably accurate. Bridges 7 and 9 fall significantly below the mark. These bridges also were shown to have high velocity values and, in the case of bridge 7, the lowest modified axle load. There is, therefore, a consistency about all of these results which supports the view that these tests provide valuable indicators of serviceability state.

CONCLUSIONS

This work has concentrated on the use of impulsive tests on masonry arch bridges and analysis of the consequent vibration to assess the serviceability states of these structures. Two features have proved to be consistent and promising indicators of structural condition.

First, the measured peak-to-peak velocities (see Table 1) have shown that the highest value (bridge 7) corresponds to the least permitted axle load from the MEXE assessments. This value (5·51 mm/s) is not above the guideline band recommended in DIN 4150 for structures in this class (3–8 mm/s).

Second, the measured fundamental (bending) frequency of vibration has been found in all cases to be close to a prediction based on

$$f_1 = K(d^{3/2}/c^{1/2}S^2)\ \text{Hz}$$

where the dimensions and their units are defined in Fig. 3 and Table 1. The constant K is consistently close to a value of 3. The bridge which falls most short of the prediction given by this formula is again bridge 7, with a constant of 2·4. This indicates a possible lack of stiffness in this structure.

Two tests are therefore recommended for common use which may supplement the information given by a MEXE assessment.

(1) To measure the peak-to-peak vibration velocity at centre span in response to a drop-weight test using 75 kg dropped from a height of 1 m on to a 50 mm sand bed on the bridge at its centre span. The mass is to be dropped near one parapet and the measurement to be made near the other parapet. Any result in excess of 8 mm/s will give rise to concern about the serviceability state of the structure.

(2) The fundamental natural frequency of the structure f_1 (Hz) is to be measured by any suitable and available means. A drop-weight test on a 50 mm sand bed on the bridge at its centre span, as described here, is recommended. The constant K in the above formula is then to be calculated using measured values of c (mm), d (mm) and S (m). A value of K less than 2 will give rise to concern about the serviceability state of the structure.

It is an advantage to local authorities that these proposals are for relatively simple one-off tests. However, there is a need to develop standard procedures and software for the analysis of results. These developments should be pursued in further work.

In some respects the value of these recommended procedures requires some supporting verification. In the case of procedure 1 it would be useful to compare drop-weight velocity values with those caused by normal traffic. The case for procedure 2 would be strengthened if a comprehensively accurate method can be developed for distinct definition of the bending natural frequency. It is proposed that this could be achieved by means of simultaneous measurement using two seismometers, one at each parapet, and dropping the weight on the bridge centreline. However, this will unfortunately involve stopping the traffic in order to make the test.

ACKNOWLEDGEMENTS

The authors are grateful for help and advice from the County Surveyors Departments of Berkshire, Derbyshire, Hampshire and West Sussex County Councils. Various colleagues at Reading and Dr D. W. Cullington and Mr R. Eyre at TRRL have contributed to the work with discussions and physical assistance.

The work was carried out under contract to TRRL and is published by permission of the Director, TRRL, Crown Copyright 1990. Any views expressed are not necessarily those of the Department of Transport.

REFERENCES

1. PAGE, J., Load tests to collapse on two arch bridges at Strathmashie and Barlae. TRRL Research Report RR201, 1989.
2. HEYMAN, J., *The Masonry Arch.* Ellis Horwood, London, 1982.
3. DEPARTMENT OF TRANSPORT, The Assessment of Highway Bridges and Structures: Advice Note BA 16/84 and Departmental Standard BD 21/84, 1984.
4. PRETLOVE, A. J. and ELLICK, J. C. A., Vibration techniques to assess the structural condition of masonry arch bridges. *Proc. Inst. Acoust.*, **10**(2) (1988) 501–8.
5. German Standard DIN 4150, Part 3: Structural Vibrations in Buildings; Effects on Structures, 1986.

52

Assessing the Dynamic Properties of Existing Bridge Structures by Hammer Testing

J. R. Maguire

Lloyds Register (Industrial Division), Lloyds Register House, 29 Wellesley Road, Croydon, Surrey, UK.

ABSTRACT

This paper describes the use of hammer testing to assess the dynamic properties of existing bridge structures. After a brief reference to the background theory, data acquisition and processing considerations are examined. One case history is then presented relating to bridge beams at Basingstoke. The natural frequencies, mode shapes and damping values of the beams are summarised. It is concluded that hammer testing provides a quick and accurate method of assessing as-built structural dynamic properties, and it is envisaged that this technique could be successfully used on many existing bridge structures.

NOTATION

c	Damping coefficient (% critical)
$c(n)$	nth damping coefficient (% critical)
E	Young's modulus
f	Frequency (Hz)
$f(t)$	Applied force
$F(\omega)$	Fourier transform of $f(t)$
$h(t)$	Linear differential operator
$H(\omega)$	Frequency response function
I	Second moment of area
$x(t)$	Displacement
$X(\omega)$	Fourier transform of $x(t)$

INTRODUCTION

Developments in the use of computers for structural design and analysis have indicated[1] an urgent need for the knowledge of material properties and the actual behaviour of as-built structures, which can only be provided by tests carried out on those structures. Such tests which make use of applied static loads are often difficult and expensive to carry out, whereas the application of dynamic loads, whether steady-state or transient, is relatively simple; although it has to be said that the collection and processing of the response data require skill and experience if pitfalls are to be avoided. Dynamic testing also usually operates at low input force levels, producing low amplitude vibrations, usually in the elastic range of behaviour. The dynamic force applied to the structure under test can be basically of two types: first, steady-state forces such as are produced by rotating eccentric mass exciters or hydraulic actuators; second, transient dynamic forces which are produced by wind, explosions or direct impact. Steady-state studies have been previously reported.[2] This paper describes the use of transient forces, input to the structure by an instrumented hammer, to assess the dynamic properties of existing bridge structures, and it presents one case history of the use of this technique.

The vibration of a linear elastic structure may be described by the combination of different modes of vibration, each of which has the modal characteristics of frequency, mode shape and damping. Once these parameters are known the dynamic behaviour of the structure subjected to a known input may be predicted. The method of modal analysis[3–5] can be used to derive these modal parameters from a hammer test, as has been successfully demonstrated in the aerospace and automobile industries.[6,7] This method is extended here to the bridge engineering field.

BACKGROUND THEORY

One of the main objectives of modal analysis is to break down a signal measured from a vibrating structure into its components at various frequencies. This is usually carried out using Fast Fourier Transform (FFT)

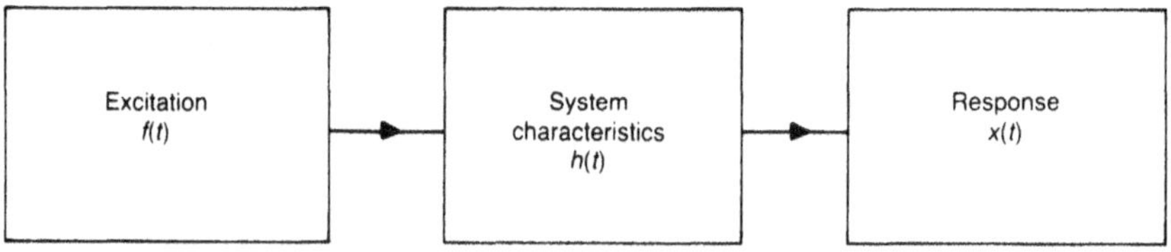

FIG. 1. Time domain representation of a linear SDOF system.

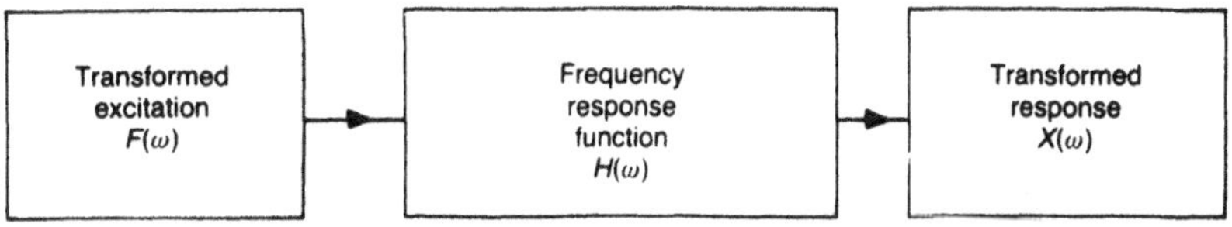

FIG. 2. Frequency domain representation of a linear SDOF system.

techniques,[8] based on the theory that any periodic signal may be looked on as a combination of a number of pure sinusoidal curves with harmonically related frequencies.[9] It may be demonstrated that any system (and signal) in the time domain (Fig. 1) has an equivalent representation in the frequency domain (Fig. 2). The theory of dynamic systems is well known and described elsewhere.[10–13]

DATA ACQUISITION AND PROCESSING CONSIDERATIONS

A full discussion of modal analysis and digital signal processing is outside the scope of this paper but is covered comprehensively by specialised textbooks.[11,13] However, a few of the more important practical considerations will now be briefly mentioned.

Digital Spectral Analysis

Accurate practical evaluation of spectral functions involves considerable computation and manipulation and is, therefore, ideally suited to digital computer analysis. It should be noted, however, that the digitisation procedure possesses a number of fundamental constraints (such as aliasing, leakage and noise) which can severely limit the quality of results and lead to wrong interpretation.

Resolution

The single most important factor affecting the accuracy of calculated modal parameters is the accuracy of the frequency response evaluation. It is not possible to extract the correct values of the modal parameters when there is inadequate and/or insufficient information to process. Adequate selection of frequency resolution, sampling interval and record length is therefore of prime importance.

Aliasing

A decision on the sampling rate for digital data acquisition is dependent on the analysis of the structure and is usually performed at equally spaced time intervals. One task is to determine this interval; too short an interval will lead to more data than can be economically processed, whereas too long an interval will lead to confusion between low and high frequency components in the original data. This latter problem is known as aliasing, and the preferred method of solving it is to filter the data before digitisation, using low pass filters which cut out unwanted high frequency components.

Leakage and Windowing

To convert from the time domain to the frequency domain, the FFT is used. One of the greatest sources of error in any digitally computed spectrum results from the fact that the measured signal is probably not periodic in the measurement period chosen and, therefore, violates a prime requirement of the FFT. Spectral estimates may consequently be modified by power leaking from other frequency components. A number of undesirable properties are exhibited, namely a broadening of peaks and the appearance of side lobes. In a spectrum containing a number of closely packed frequency components, leakage may smear together peaks and mask important detail. Leakage can be significantly reduced by taper-windowing the sampled time-domain record—that is, shaping it to become periodic in the measurement period chosen.

Noise and Averaging

One of the major characteristics of any modal testing system is that extraneous noise from a variety of sources is always measured along with the desired excitation and response signals. By taking a number of power spectrum averages, it may be shown[4] that the measured frequency response function estimates more accurately the true frequency response function, assuming the noise has a zero mean value and is uncorrelated to the measured input signal.

The Coherence Function

To determine the quality of the frequency response function, it is not sufficient to know only the relationship between input and output;[3] a major question is the degree to which the system output is caused by the system input. Noise and/or non-linear effects can cause large outputs at various frequencies, thus introducing errors in estimating the frequency response functions. The influence of noise and/or non-linearities, and thus the degree of noise contamination in the frequency response function, is measured by calculating the coherence function denoted by γ^2, where

$$\gamma^2 = \frac{\text{Response power caused by input}}{\text{Measured response power}}$$

The coherence function is easily calculable on a digital Fourier analyser (spectrum analyser) when frequency response functions are being evaluated. As the coherence function indicates the degree of noise in a frequency response function, it has two very important uses: first, it can be used qualitatively to determine how much averaging is required to reduce measurement noise; second, it can serve as a monitor on the quality of the frequency response function measurements.

Illustrative Example

A typical hammer testing setup is shown in Fig. 3. The hammer has a force transducer attached and is used to impact the structure. Excitation occurs with a nearly constant force over a limited frequency range. The effect of different types of hammer head is to alter this frequency range, as shown in Fig. 4. A very soft head will concentrate a high density of energy in a narrow

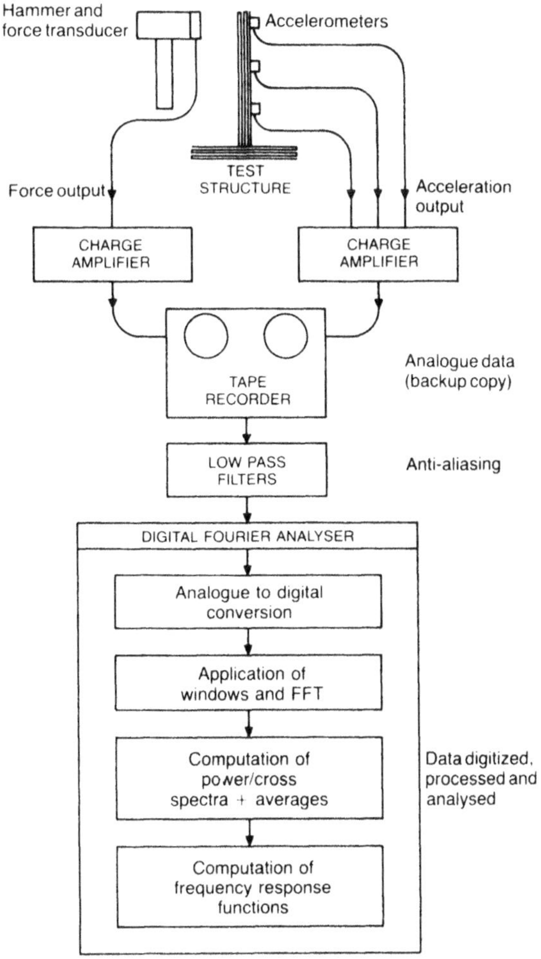

FIG. 3. Typical hammer testing setup.

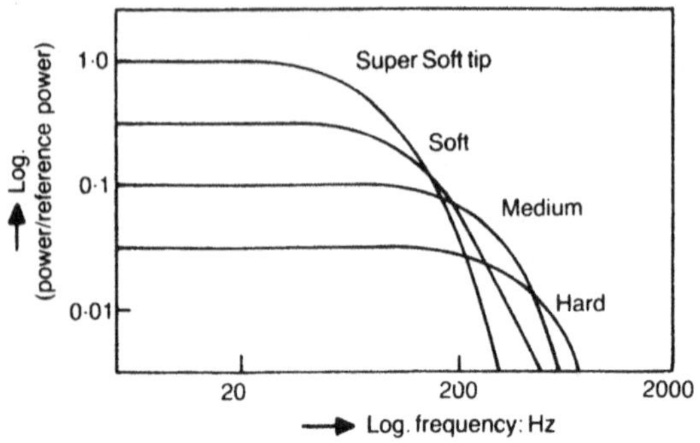

FIG. 4. Hammer autopower spectrum.

frequency range, whereas a very hard head will spread energy evenly across a wide frequency range. Typically, a large sledgehammer with a rubber tip will excite uniformly over a range of 0–100 Hz, whereas the same sledgehammer with a hard plastic tip will excite uniformly over 0–400 Hz. Details of the sledgehammer used during the prototype testing are given in the Appendix. If there is no mains supply near to the test structure, power required for the setup shown in Fig. 3 may be provided by a small portable generator.

TESTS ON BRIDGE BEAMS AT BASINGSTOKE

Hammer tests were conducted[15] during November 1983 on four simply supported precast post-tensioned concrete bridge beams at Basingstoke (shown in Fig. 5). Each beam weighed approximately 40 t and effectively spanned 27·6 m over timber (sleeper) supports. Prestressing was by three tendons, two straight and one draped, designed for a total force of 573 t. Each tendon (consisting of twelve 15·2 mm strands) was soundly grouted into its own corrugated duct. These beams were left over at the end of the Basingstoke ringway construction and were to be demolished as they were excess to requirements. The first three modes of vibration for the undamaged beams were determined, and their frequencies and damping values are given in Table 1. A limited number of mode shape measurements (at mid- and quarter-span points) showed that these three modes corresponded to the expected classical mode shapes for a simply supported beam.[20] Also given in Table 1 are theoretical frequencies based on the concrete cross-section, including the tendons (modular ratio = 15), which show close agreement with the measured values. The dynamic Young's modulus for the concrete[15] was taken as 43 kN/mm^2.

Hammer tests were also carried out during tendon exposure and cutting, and results for beam 3 are presented in Fig. 6. The effect of damage may be

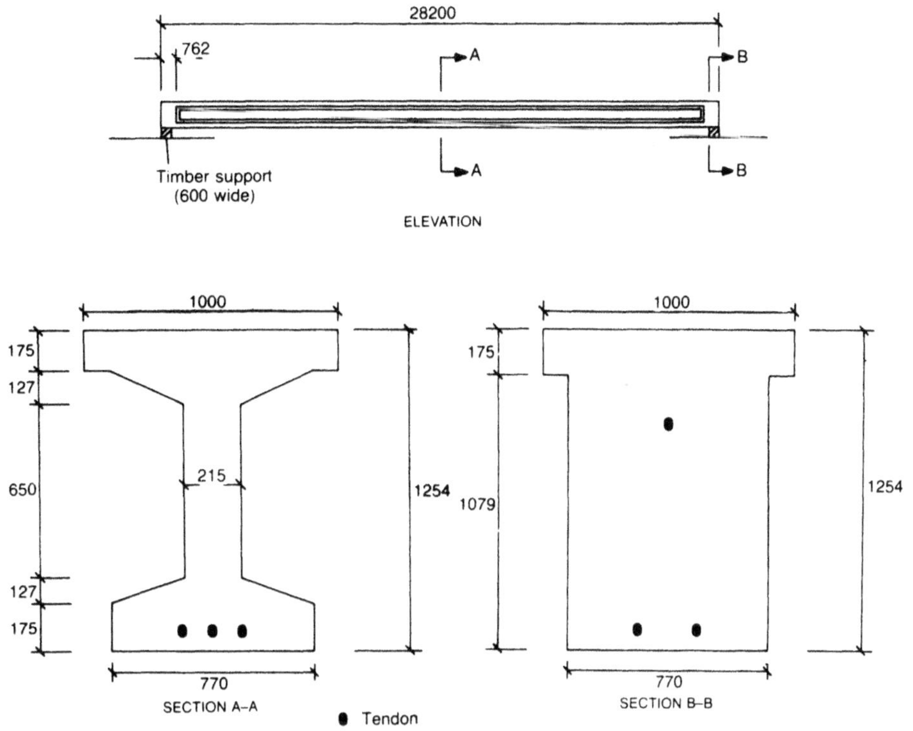

FIG. 5. Basingstoke beams (all dimensions in mm).

TABLE 1
Basingstoke beam tests (undamaged state)

(a) Measured and theoretical frequencies (Hz)			
	$f(1)$	$f(2)$	$f(3)$
Beam 1 (measured)	4·32	14·88	27·48
Beam 2 (measured)	4·40	15·20	28·10
Beam 3 (measured)	4·28	15·12	28·52
Beam 4 (measured)	4·40	15·56	29·64
Beams 1–4 (theoretical)	4·06	16·26	36·60
(b) Measured damping (% critical)			
	$c(1)$	$c(2)$	$c(3)$
Beam 1	1·85	1·21	1·97
Beam 2	1·99	1·48	1·95
Beam 3	1·87	1·06	1·68
Beam 4	1·59	1·41	1·69

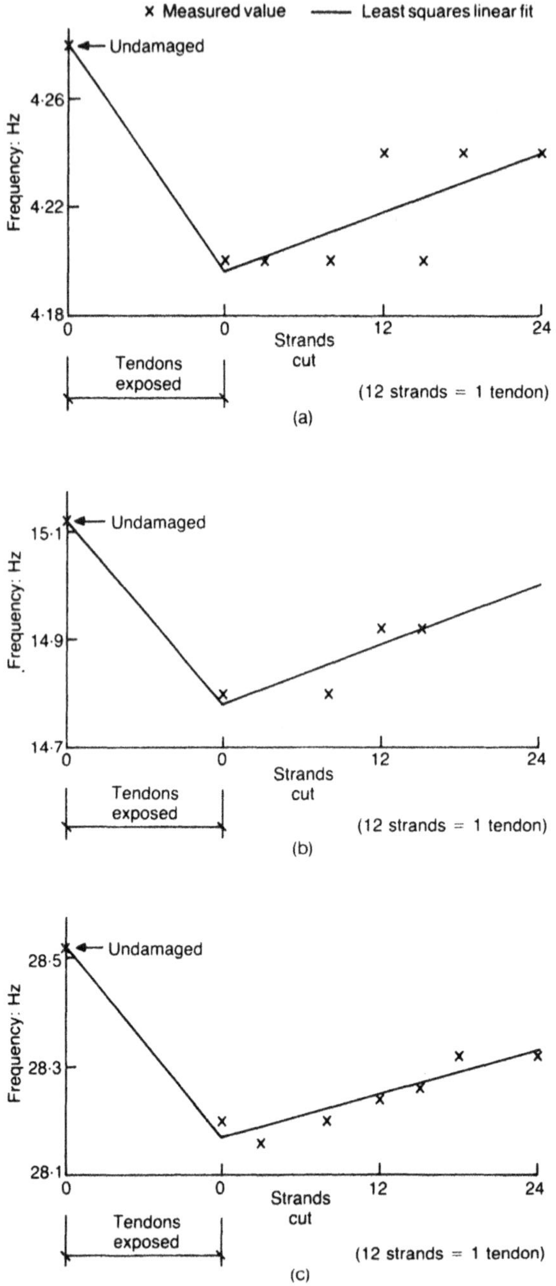

FIG. 6. Change in frequency during tendon exposure and cutting: (a) mode 1; (b) mode 2; (c) mode 3.

seen to affect the frequencies in two different ways: first, the effect of exposing the tendons (removing concrete and thereby removing stiffness) is to decrease the frequency; second, the effect of cutting the tendons (reducing axial load and thereby increasing stiffness) is to increase the frequency. It may be seen from the above that integrity monitoring is possible using hammer testing.

CONCLUSIONS

This paper has described the use of hammer testing to assess the dynamic properties of existing bridge structures and has presented one case history. It has been found that, given sensitive instrumentation, hammer testing is able to determine the structural dynamic characteristics of bridge beams. It is felt that hammer testing provides a quick and accurate method if assessing as-built structural dynamic properties, and it is envisaged that this technique could be successfully used on many existing bridge structures.

REFERENCES

1. Science and Engineering Research Council, *Long-term R and D in Civil Engineering*. London, June 1982, Memo P:BG:123.
2. Ellis, B. R. *et al.*, Forced vibration tests and theoretical studies on dams. *Proc. Instn Civ. Engrs, Part 2*, **69** (Sept. 1980) 605–34; **71** (June 1981) 575–95.
3. Ramsay, K. A., Effective measurements for structural dynamics testing. *Sound and Vibration* (Nov. 1975) 24–35.
4. Richardson, M. and Potter, R., Identification of the modal properties of an elastic structure from measured transfer function data. *20th Int. Symp. on Instrumentation*, Albuquerque, New Mexico, May 1974, pp. 239–46.
5. Walgrave, S. C. and Ehlbeck, J. M., Understanding modal analysis. American Society of Automobile Engineering, West Coast Meeting, August 1978, Technical Paper Series 780695.
6. Knauer, C. D. *et al.*, Space vehicle experimental modal definition using transfer function techniques. American Society of Automobile Engineering, National Aerospace Engineering and Manufacturing Meeting, Culver City, LA, November 1975.
7. Richardson, M. and Kniskern, J., Identifying modes of large structures from multiple input and response measurements. American Society of Automobile Engineering, National Aerospace Engineering and Manufacturing Meeting, San Diego, November–December 1976, Paper 760875.
8. Cooley, J. W. and Tukey, J. W., An algorithm for the machine calculation of complex Fourier series. *Math. Comput.*, **19**(90) (1965) 297–301.
9. Fourier, J. B. J., *The Analytical Theory of Heat*. Didot, Paris, 1822 (in French).
10. Bendat, J. S. and Piersol, A. G., *Random Data Analysis and Measurement Procedures*. John Wiley, New York, 1971.

11. CLOUGH, R. W. and PENZIEN, J., *Dynamics of Structures.* McGraw-Hill, New York, 1975.
12. RANDALL, R. B., *Application of B and K Equipment to Frequency Analysis.* Bruel and Kjaer, Hounslow, 1977.
13. MEIROVITCH, L., *Elements of Vibration Analysis.* McGraw-Hill, New York, 1975.
14. BEAUCHAMP, K. and YEN, C., *Digital Methods for Signal Analysis.* Allen and Unwin, London, 1979.
15. MAGUIRE, J. R., The dynamic characteristics of elevated piled tanks and other selected prototype structures. PhD thesis, University of Bristol, May 1984.
16. MAGUIRE, J. R. *et al.*, Assessing the dynamic properties and integrity of structures by the use of transient data. *Proc. 8th World Conf. on Earthquake Engineering*, San Francisco, 21–28 July 1984.
17. BATHE, K. *et al.*, SAPIV—a structural analysis program for static and dynamic response of linear systems. Earthquake Engineering Research Centre, University of California, June 1973 (revised April 1974). Report to the National Science Foundation, No. EERC-73-11.
18. RICHART, F. E. *et al.*, *Vibrations of Soils and Foundations.* Prentice-Hall, Englewood Cliffs, New Jersey, 1970.
19. HOUSNER, G. W., The dynamical behaviour of water tanks. *Bull. Seism. Soc. Amer.*, **53**(2) (Feb. 1963) 381–7.
20. BROCH, J. T., *Application of B and K Equipment to Mechanical Vibration and Shock Measurement.* Bruel and Kjaer, Hounslow, 1972.

APPENDIX: SLEDGEHAMMER USED DURING BRIDGE BEAM TESTING

The sledgehammer used during the bridge beam testing was a PCB type (GK291B50) and is shown in Fig. A1. The sledgehammer mass was 5·4 kg, providing a maximum impact force of 22 kN (typically 17 kN during testing). The shape of the force pulse generated by the hammer was approximately that of a half sine wave, of 5 ms duration when hitting a concrete structure through a soft plastic tip. The sledgehammer required no power supply, although the signal from the quartz force transducer was amplified by a battery-operated charge amplifier supplied with the sledgehammer.

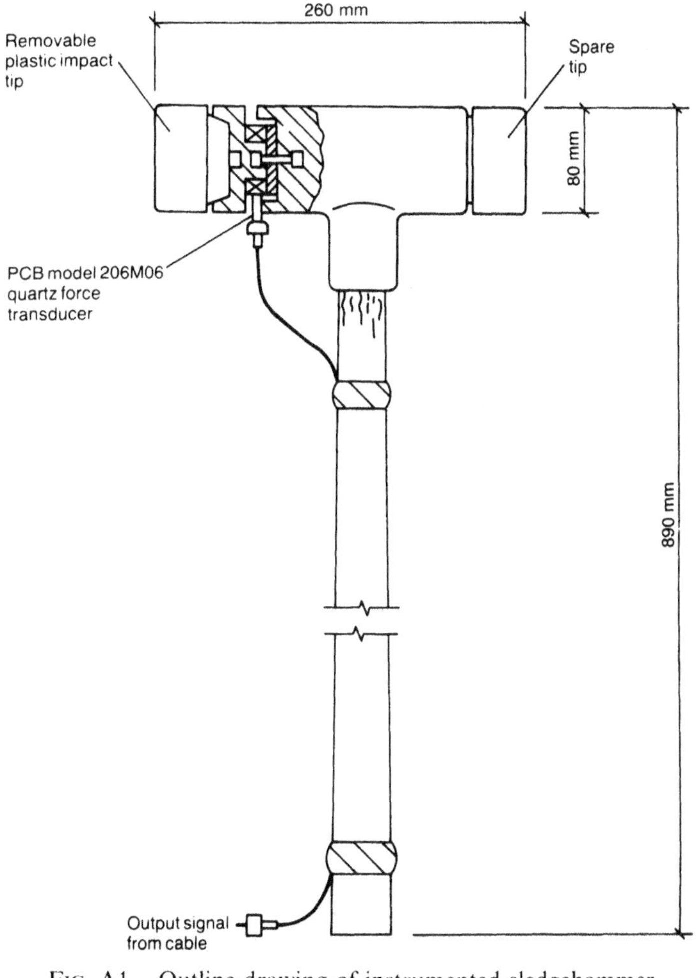

FIG. A1. Outline drawing of instrumented sledgehammer.

53

Serviceability Performance of a Steel Highway Bridge

I. Rosenthal and M. Itzkovitch

Faculty of Civil Engineering, Technion—Israel Institute of Technology, Haifa 32000, Israel

ABSTRACT

The response behaviour of a steel single span highway bridge (45·60 m long) was examined because of complaints of heavy vibration. The study consisted of field measurements under conditions of regular and controlled motor traffic as well as under forced vibration induced by a vibration generator. The data yielded the dynamic characteristics of the bridge. The peak acceleration (0·25 g), and vibration velocity (150 mm/s), obtained were compared with various code serviceability requirements.

INTRODUCTION

The steel bridge over the Kishon River (leading to Haifa airfield) was observed to vibrate heavily under motor traffic. In view of the almost total absence of pedestrians, the problem was not one of discomfort but rather whether or not this vibration could endanger the structure itself. Accordingly, the local authority sponsored an investigation on the serviceability performance of the bridge, including various field measurements.

THE BRIDGE AND TESTING PROGRAMME

The 15-year-old bridge is a 7 m wide (two-lane), 45·60 m long single span composite structure comprising of six 2·50 m deep welded steel girders and a

0·20 m r.c. deck; because of the small volume of pedestrian traffic a 1·50 m wide single sidewalk was included in the design (Fig. 1).

The test programme included field measurements using the instrumentation described in Table 1 so as to ensure coverage of the entire possible (and still unknown) range of vibration.

Three separate series were run:

Series A. Vibration measurements under regular traffic over 4 days, mainly during the morning and afternoon rush hours.

Series B. Vibration measurements under a controlled regime, for which purpose the bridge was closed to regular traffic for 7 h at night. The regime consisted of running convoys comprising two 32-t trucks and a 62-t semitrailer in various sequences (the 126-t total being equivalent to the design load) across the bridge at four specified speeds, 10, 30, 50 and 70 km/h.

FIG. 1. The bridge viewed from its northeastern corner.

FIG. 2. The vibration generator mounted on the bridge deck.

TABLE 1
Measuring instruments (installed on third girder) and their locations ($l = 45{\cdot}60$ m, length of bridge)

Measuring point	Distance from southern support	Instruments	Range	Recorder channel
1	6·90 m (0·15l)	Accelerometer	2g	9
		Vertical geophone	±1 mm	1
		LVDT (southern riverbank)	±50 mm	7
2	22·80 m (0·50l)	Accelerometer	2g	10
		Vertical geophone	±3 mm	2
		Horizontal geophone	±3 mm	3
		Strain gauge	±2 000 μm	6
3	30·30 m (0·66l)	Vertical geophone	±3 mm	4
4	34·80 m (0·76l)	Accelerometer	2g	11
		Vertical geophone	±1 mm	5
		LVDT (northern riverbank)	±50 mm	8

Series C. Forced vibration (sweep sine) applied to the bridge immediately following Series B, by means of a vibration generator attached to the deck at measuring point 4, so as to yield both the first and second vibration modes (Fig. 2).

RESULTS

All readings were recorded on magnetic tape and by a multichannel recorder, as illustrated for Series A in Fig. 3.

Under large vibration all vertical geophones went off the scale but operated again after passage of the vehicle in the vibration-free stage. This yielded the fundamental frequency of the bridge (2·6 Hz) as well as its critical damping ratio (1·1%). The vertical velocities calculated from the accelerations were throughout about ten times higher than those in the horizontal direction.

The tensile strain and displacement plots over the river banks (channels 6–8) represented the static component due to the load, on which the vibration induced by the inertia forces was superimposed. Their sum yielded the maximum stress while the ratio of the dynamic component to the static one yielded the impact of the moving vehicle or the dynamic increment. It can be seen that although high levels of acceleration were measured the tensile stresses in the girders were low. Table 2 lists some of the peak results of Series A.

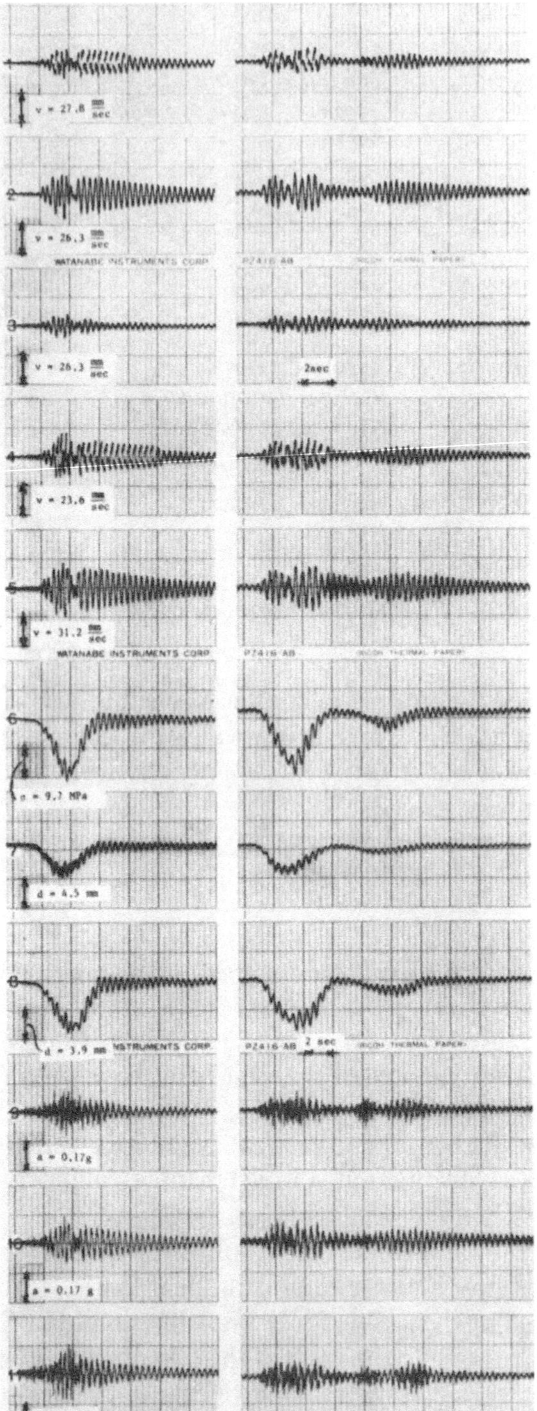

FIG. 3. Analogue recording of traffic vibration in Series A: heavy truck (left peak), semi-trailer tanker (middle peak) and bus (right peak).

TABLE 2
Peak midspan vibration results obtained in Series A

Vehicle	Speed[a] (km/h)	Girder tensile stresses (MPa)			Impact ratio	Acceleration (g)	Velocities (mm/s)		Static deflection[c] (mm)
		Static	Dynamic	Total			Vertical[b]	Horizontal	
Heavy truck	84	17·4	2·9	20·3	0·17	0·14	84	11	14
Semi-trailer, tanker	72	16·5	3·9	20·4	0·24	0·10	60	8	13
Bus	84	3·8	1·9	5·7	0·50	0·08	48	5	3

[a] Calculated according to traverse time from points 1 to 4.
[b] Calculated from acc./$2\pi f$, with $f = 2{\cdot}6$ Hz.
[c] Average value calculated from static displacement components measured at both river banks.

TABLE 3
Main midspan vibration results obtained in Series B

Sequence of vehicles	Moving load (t)	Speed[a] (km/h)	Girder tensile stresses (MPa)			Impact ratio	Acceleration (g)	Velocities (mm/s)		Static deflection[c] (mm)
			Static	Dynamic	Total			Vertical[b]	Horizontal	
T + SM + T	126	10	18·5	1·0	19·5	0·05	0·06	36	3	21
T + SM + T	126	30	19·4	1·9	21·3	0·10	0·09	54	4	20
T	32	50	9·7	1·9	11·6	0·20	0·15	90	7	6
SM + T	94	50	13·6	3·9	17·5	0·29	0·17	102	7	13
T	32	70	9·7	1·9	11·6	0·20	0·20	120	11	7
SM + T	94	70	11·6	3·9	15·5	0·34	0·25	150	12	13

Notes: See Table 2 above.
T = truck; SM = semi-trailer.

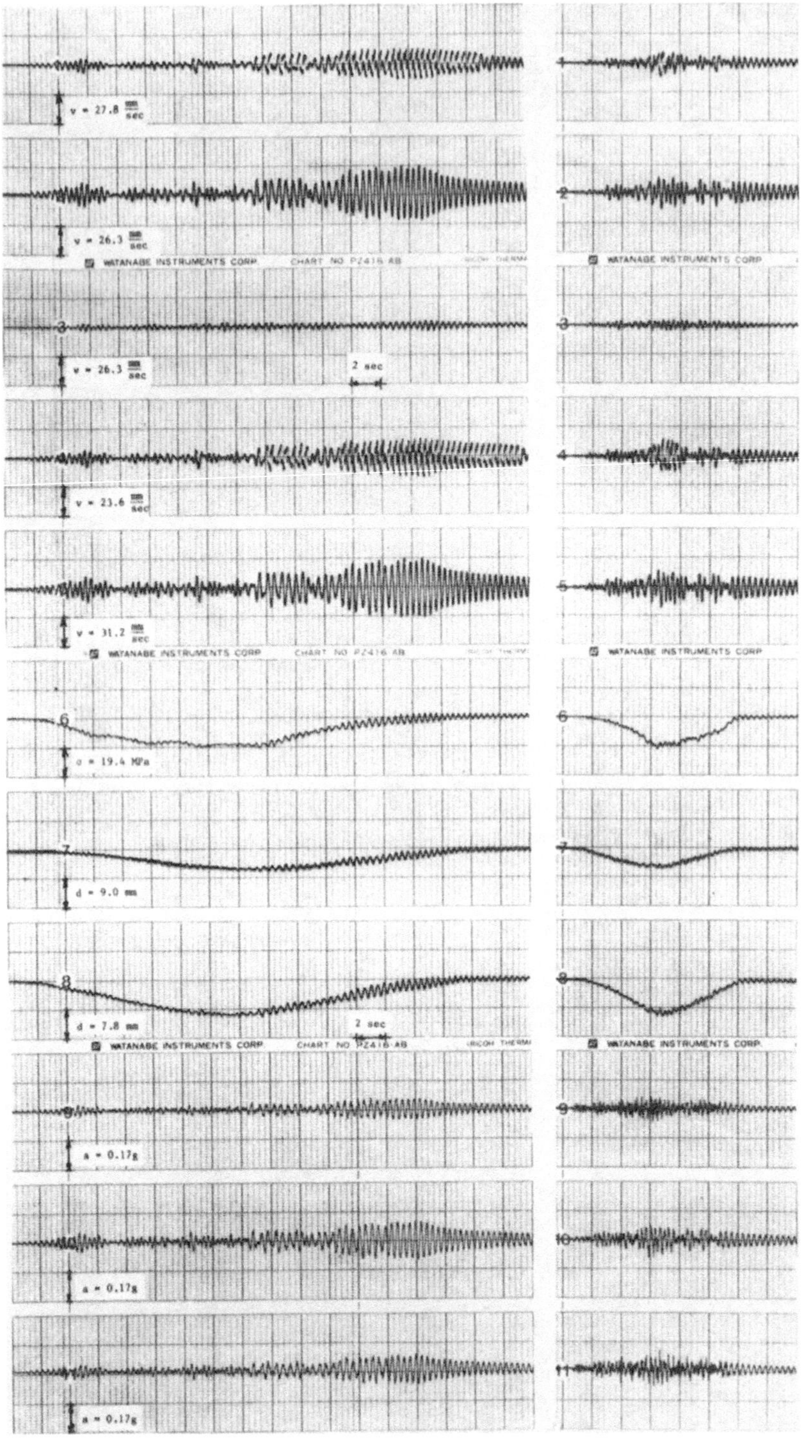

FIG. 4. Analogue recording of vibration of convoy in Series B running at 10 km/h (left) and 30 km/h (right).

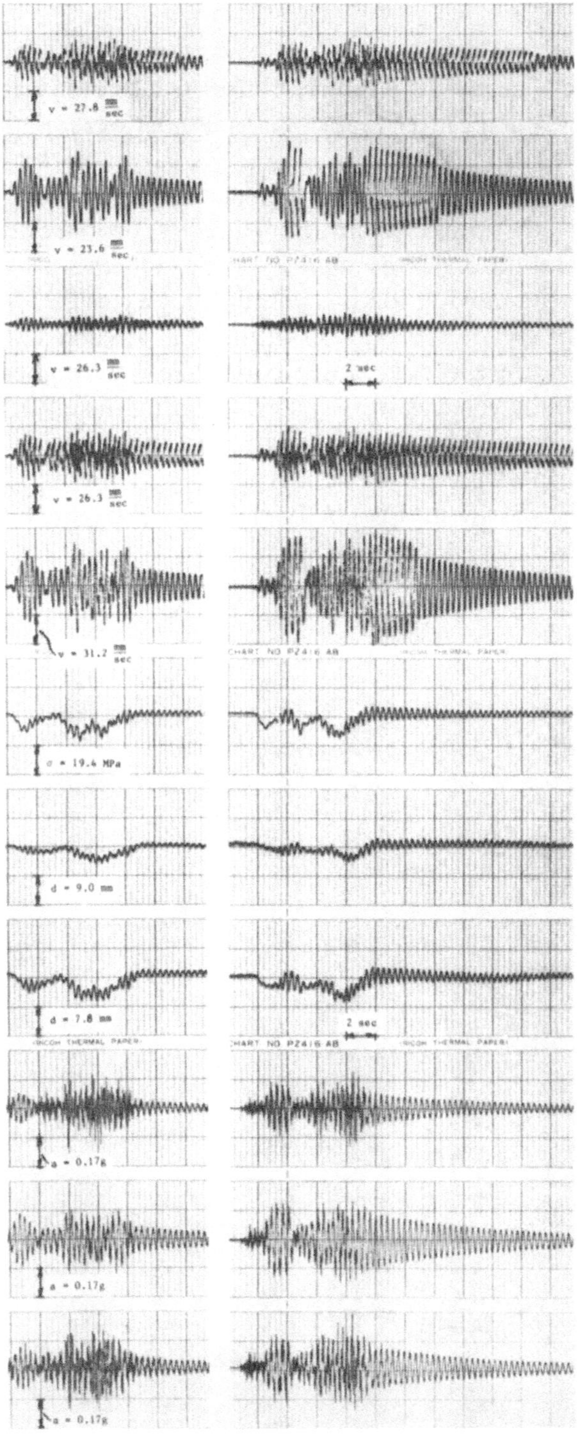

FIG. 5. Analogue recording of vibration of convoy in Series **B** running at 50 km/h (left) and 70 km/h (right).

The traffic volume during rush hours was about 1500 vehicles per hour. The percentage of heavy-duty vehicles (buses, trucks, etc.) was 15–25% in the mornings and 8% in the afternoons; the remainder were small vehicles (vans, private cars, etc.).

The main results of Series B are given in Table 3 and Figs 4 and 5, indicating that the traffic vibration—directly proportional to speed—reached acceleration levels as high as 0·25*g*. On the other hand, girder stresses due to the moving loads again did not exceed 21 MPa.

At the lower speeds (10 and 30 km/h, Fig. 4) each convoy ran close together, creating the effect of a single load, while at the higher speeds (50 and 70 km/h, Fig. 5) the vehicles were spaced, creating the effect of two distinct loads. The results in Table 3 are therefore arranged accordingly.

The Fourier spectrum of the accelerograms for points 1, 2 and 4 in the 70 km/h test is shown in Fig. 6. It clearly yields the first and second

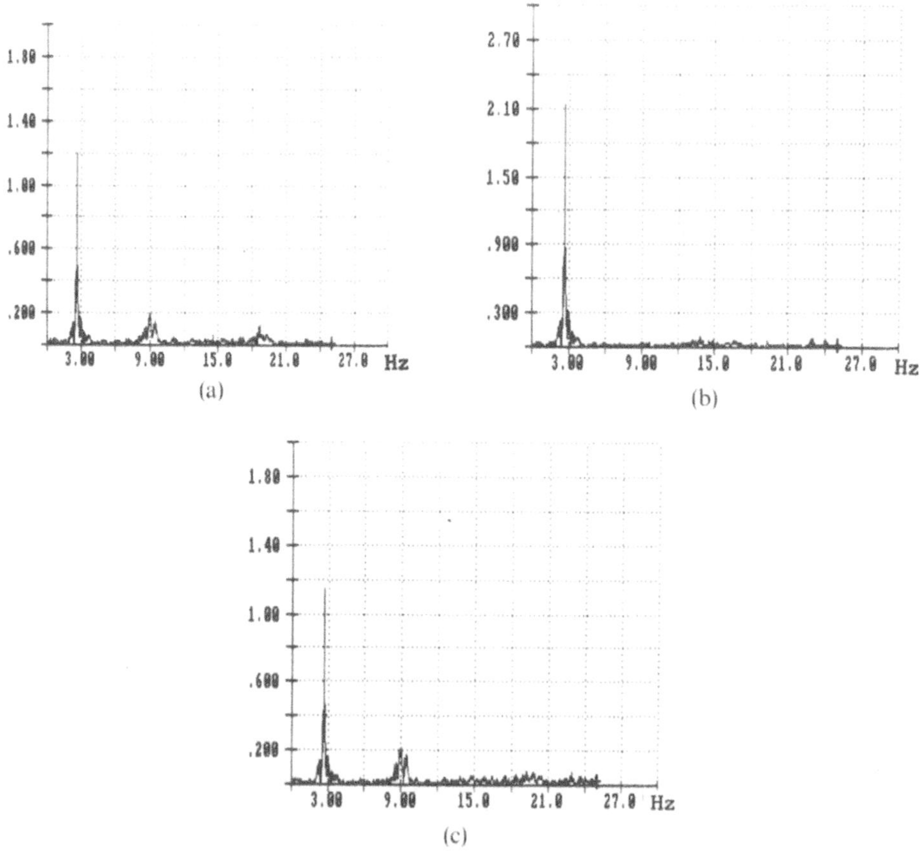

FIG. 6. Fourier amplitude spectra of accelerations at points (a) 1, (b) 2 and (c) 4 from 70 km/h test (Series B).

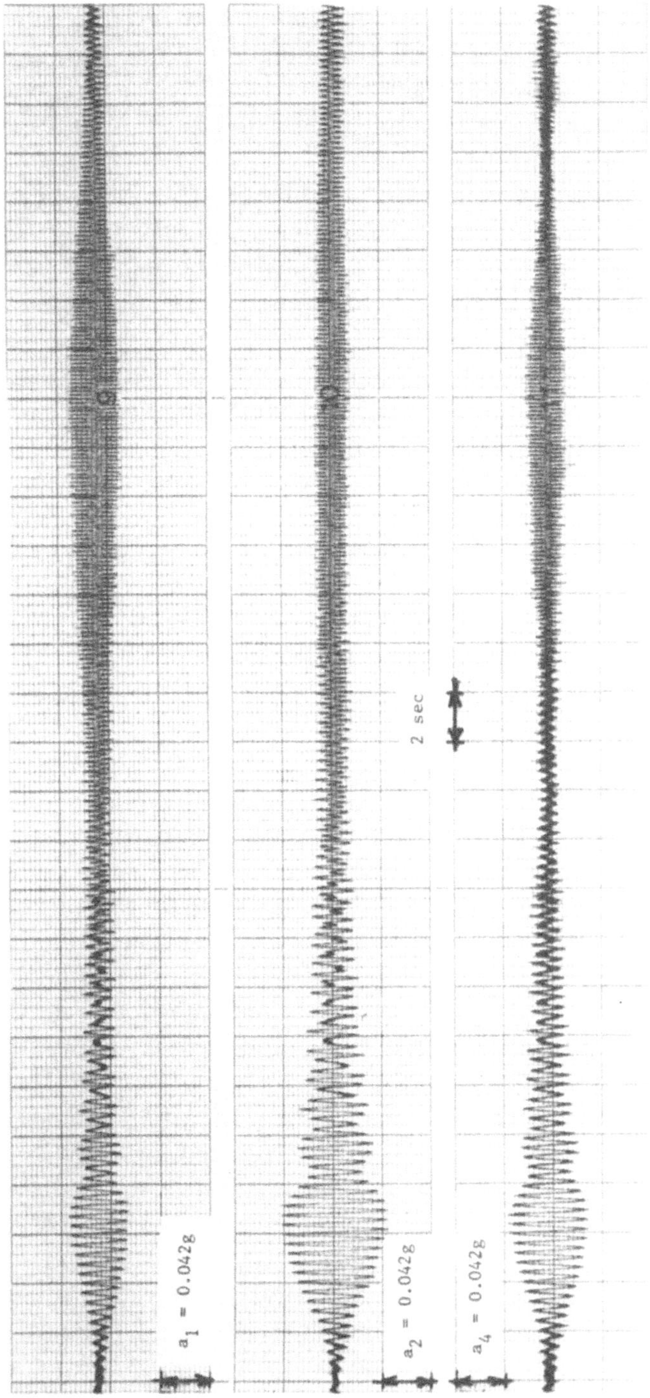

FIG. 7. Accelerograms recorded at points 1 (top), 2 (middle) and 4 (bottom) during sweep sine forced vibration test (Series C).

frequencies of the bridge, 2·6 and 9·0 Hz respectively, and indicates the third frequency at around 19 Hz. The amplitudes obtained at 2·6 Hz, lower at the extremities of the bridge (points 1 and 4) and highest at midspan (point 2), describe the first mode of vibration shaped as a half sine wave. The second mode, at 9·0 Hz, has the form of a complete sine wave as amplitudes exist at points 1 and 4 but none at point 2.

In Series C the bridge went into first resonance at 2·6 Hz, which is its fundamental frequency according to the accelerograms at points 1, 2 and 4 (Fig. 7). The second resonance state, at 9 Hz, could not be reached because of the mechanical limitations of the generator (maximum rotational speed 380 rpm or 6·3 Hz).

Significantly, the three amplitudes at points 1, 2 and 4 at 2·6 Hz were all of the same phase, with the peak at point 2 (midspan), indicating the half sine wave of the first mode, while those at points 1 and 4 at 6·3 Hz had opposite phases with an almost zero amplitude at point 2, indicating the complete sine wave of the second mode.

With the first frequency established beyond doubt the stiffness of the bridge was found from $K=(2\pi f)^2 W/g = 76{\cdot}6$ kN/mm, with $f = 2{\cdot}6$ Hz and $W = 282$ t, the total dead load of the bridge.

DISCUSSION

The vibration components, measured in the present study under regular and controlled traffic, exceeded all known relevant standards. Insofar as dynamic behaviour is referred to, if at all, in highway bridge codes, no detailed treatment is given, but only to footbridges, where the approach is physiological.[1–3] In some codes emphasis is on prevention of damage in structures, by limiting the velocity component to 20 or 30 mm/s in the frequency range up to 30 Hz,[4–6] while the US Bureau of Mines criterion for structural safety against damage from blasting limits velocity to 50 mm/s up to 3 Hz, and acceleration to 0·10*g* in the range between 3 and 100 Hz.[7] By contrast, the Ontario Bridge Code[8] allows for the serviceability limit state by limiting the maximum static deflection due to factored highway live load (including the dynamic load allowance) as a function of the first flexural frequency and the anticipated degree of pedestrian use. For the bridge in question, with $f = 2{\cdot}6$ Hz and hardly any pedestrian traffic, this code would limit the static deflection to about 23 mm. However, as the latter refers to a standard loading vehicle, there is no common basis for comparison with the actual 21 mm deflection of Series B, obtained with a convoy of vehicles.

In spite of the heavy vibration observed in the bridge (acceleration 0·25*g* or velocity 150 mm/s), the problem is not yet one of structural safety, due to the fact that the girder stresses are low. The traffic causes 21 MPa including the dynamic component and, together with 65 MPa from the dead load, yield a maximum stress of 86 MPa, namely 60% only of the allowable level for steel. As a result this case may be considered a *fatigue* problem, particularly regarding the connections, which are mainly welds.[9]

If the present situation continues—with vehicles running at 70 km/h at 2-s intervals during most of the day (and with the already large traffic volume likely to increase further in the future)—cumulative damage is bound to endanger the structure in the long run.

CONCLUSIONS

At first an attempt was made to reduce vibration by imposing a speed limit of 40 km/h, but this proved unenforceable and the attempt was abandoned after a trial period of 6 months. As an alternative design measures were recommended for moderating the response of the bridge, mainly with the aid of vibration dampers,[10] so as to improve the presently very low damping ratio (1·1%).

REFERENCES

1. The Standards Institution of Israel, Loads on Bridges: Highway Bridges: IS 1227, Part 1, 1985.
2. British Standards Institution, Steel, Concrete and Composite Bridges: BS 5400, 1978.
3. National Standard of Canada: CAN 3-S6-M78: Design of Highway Bridges, 1978.
4. German Institute for Standards, DIN 4150: Vibration in Civil Engineering—Part 3: Effects on Structures, May 1986.
5. Institution of Swiss Highway Engineers (VSS), Swiss Standard SN 640312: Vibration Effects on Structures, 1978.
6. GDR Chamber of Technology, Directive KDT 046/72: Effects of Blasting Operations on Buildings, East Berlin, 1972.
7. Eshelman, R. L., Vibration standards. In *Shock and Vibration Handbook*, 2nd edn, ed. Harris and Crede, Chapter 19 (Fig. 19.8, p. 19-12). McGraw-Hill International, New York, 1976.
8. Ontario Ministry of Transportation and Communications, Ontario Highway Bridge Design Code, OHBDC, 1983 (updated 1985).
9. Tilly, G. P. (ed.), Dynamic behaviour of concrete structures—report of the RILEM 65 MDB Committee, Chapter 5.1: Bridges. Elsevier, Amsterdam, 1986, p. 225.
10. Jones, R. T. and Pretlove, A. J., Vibration absorbers and bridges. *The Highway Engineer*, **26**(1) (January 1979) 2–9.

54

Monitoring of Traffic Induced Strain in the Steel Reinforcement of a Concrete Bridge Deck

JOHN CAIRNS

Department of Civil Engineering, Heriot-Watt University, Edinburgh, UK

ABSTRACT

This report describes measurements undertaken on a steel/concrete composite bridge to determine the range of stress to which reinforcement in the deck slab of the bridge is subject under vehicle loading. The study was prompted by the proposal that a fatigue classification for unwelded reinforcement be included in the Code of Practice for Fatigue. Measurements were undertaken with a vehicle of known weight prior to opening to general traffic and under service loading after 10 months of use. Measured strains were well below the level at which fatigue damage might occur, and were also below values that would be calculated for design purposes. The influence of surfacing and of oscillations of the bridge deck are discussed.

INTRODUCTION

In 1980 the Code of Practice for Fatigue, BS 5400: Part 10,[1] a part of the Code of Practice for Bridges, was published. The bulk of its requirements affected steel construction, for which cumulative damage calculations deemed to satisfy equivalent rules were specified. Although a classification for butt-welded reinforcement was included in the detailed rules, normal reinforcement was covered by a limiting stress range only. Subsequently the intention to establish a full classification for unwelded reinforcement was declared.[2] This paper describes measurements taken on a highway bridge structure to examine the need for such a rule. Measurements were considered necessary because of imprecise assumptions made in analysis in respect of load dispersion, impact factors, material properties, etc.

An earlier study in which an M-beam type bridge was instrumented found measured strains in deck slab reinforcement to be considerably lower than anticipated.[3]

DESCRIPTION OF BRIDGE DECK

Milton Bridge No. 1 is a composite steel and concrete bridge on the M74 at Lesmahagow, approximately 15 miles to the south of Glasgow. Two independent bridges carry northbound and southbound carriageways over the River Netham in three continuous spans of 47, 54 and 31 m. Each carriageway comprised hard shoulder, two traffic lanes and a 1 m safety strip. A section through the deck is shown in Fig. 1.

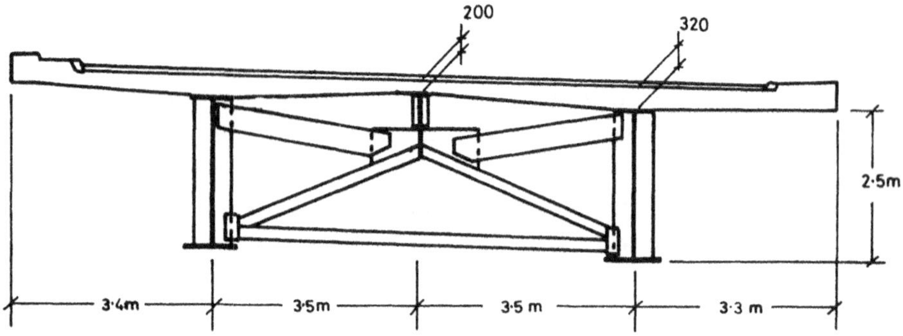

FIG. 1. Milton Bridge—section through deck.

The reinforced concrete deck slab of the bridge is supported on and acts compositely with twin longitudinal plate girders and a central stringer member, itself supported off the plate girders by cross-bracing spaced at intervals of between 6 and 7 m along the length of the bridge. Depth of the slab varies from 320 mm over the plate girders to 200 mm over the stringer. The deck slab was cast in five sections during August/September 1986 and opened to traffic during October that year.

Transverse reinforcement is 20 mm bars at 150 mm centres in the bottom and 32 mm diameter bars at 125 mm centres in the top of the slab. Longitudinal reinforcement is 16 mm diameter bars at 150 mm centres top and bottom within the spans, increasing to bottom reinforcement of 25 mm diameter bars at 150 mm centres in hogging regions near piers. Total depth of surfacing to the deck including waterproofing membrane was 120 mm. Grade 37·5 structural concrete with 40 mm minimum cover to reinforcement was specified for deck slab concrete.

INSTRUMENTATION OF BRIDGE DECK

Instrumentation to monitor reinforcement strains was installed during construction of the deck slab. Two sets of gauges were installed within the 47 m end span of the bridge. One set, denoted R1, was positioned 16·5 m from the southern abutment in a region where the deck is subject to sagging moments and the concrete deck slab will be maintained in longitudinal compression. Gauge set R2 was positioned 2·7 m from the centre of pier B, where loading will tend to place the deck slab in longitudinal tension. In both cases gauges were approximately midway between cross-bracings, as the greater flexibility of support to the slab from the stringer away from the cross-bracing was expected to give greater strains in reinforcement. Gauges were centred 2·0 m from the centreline of the plate girder and 1·5 m from the stringer, under the nearside lane. Preliminary calculations showed the maximum stress range to occur with the nearside wheel line around 2·0 m from the plate girder. BS 5400 rules for fatigue assessment require the nearside wheel line to be between 1·4 and 2·0 m from the line of the plate girder.

There were three gauge units within each of the two sets of reinforcement gauges R1 and R2. At both locations two gauge units measured strains of longitudinal and transverse reinforcement in the bottom mat. The third gauge unit at R1 gave a repeat measurement of bottom transverse reinforcement strains.

At location R2 the third gauge unit measured strains in the top layer of transverse reinforcement directly over the transverse gauge unit on the lower mat. The position of the neutral axis of the slab for transverse bending could be found from the results of the top and bottom transverse pair to determine whether compression membrane action was occurring and if surfacing had a significant stiffening effect on deck behaviour.

Gauge units were made from a section of welded wire fabric to which electrical resistance gauges were fixed (Fig. 2). The technique therefore did not require any treatment or modification to the main reinforcement, nor interfere with the contractors' programme. Gauge units were fully calibrated in the laboratory prior to installation. Two full strain gauge bridges were fixed to each unit, one set providing a back-up in case of damage. Values measured on first and back-up gauges were in good agreement. Crack inducers 40 mm high × 300 mm long were also installed directly under and perpendicular to each gauge unit to persuade the slab to crack at gauge locations. Strains measured on a gauge unit cast in a slab constructed and tested in the laboratory showed satisfactory agreement with calculated values and with strains measured on instrumented reinforcement.

After suitable conditioning, signals from the various gauges were

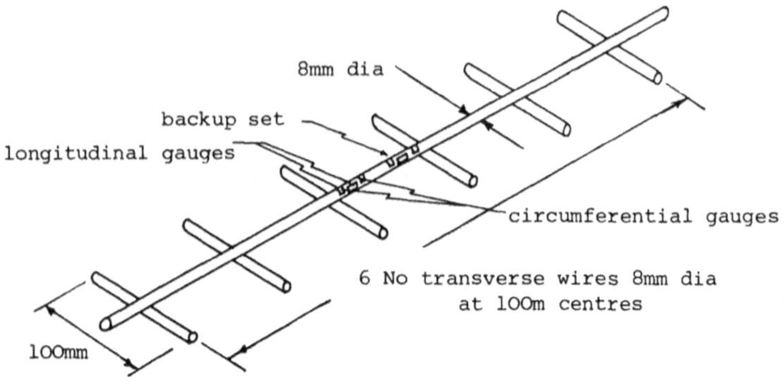

FIG. 2. Detail of reinforcement gauge unit.

recorded to tape on a seven-channel portable instrumentation recorder for analysis in the laboratory. Six channels recorded data from the six reinforcement gauges. The seventh track was used for an audio recording under the expansion joint at the southern abutment, and could be used to trigger a logging sequence. Noise level recorded was equivalent to a strain of 8×10^{-6}.

THEORETICAL CALCULATION OF DECK STRAINS

Analysis of Deck

For design purposes, the behaviour of the deck slab of beam and slab bridge decks under concentrated loads is usually split into local and global components. Local moments arise from behaviour of the slab under individual wheel loads between beams, global components arise from overall longitudinal and transverse flexing of the whole deck under vehicle loading.

A grillage model consisting of two longitudinal members continuous over three spans and neglecting the central stringer was used to calculate global moments. To this were added the results of a local analysis using a finite element model consisting of plate elements for the slab and offset beam elements for the stringer. Cross-bracing was represented by beam or spring elements, the stiffness of which was determined from a substructure analysis.

Maximum moment in a deck slab under a concentrated load depends on the area over which the concentrated load is applied. BS 5400 specifies wheel loads and contact pressures for a standard fatigue vehicle, and the manner in which the load may be assumed to disperse through surfacing and concrete to the neutral axis of the deck slab. The finite element model could not model load dispersion without refining the mesh to a much higher degree than otherwise necessary. It was therefore necessary to correct values calculated by the finite element model for the deck by deducting an allowance for loads applied to an area of finite size. The deduction was calculated as the difference between moments calculated using Westergaard's charts[4] for a point load on a thin slab and that for a loaded area determined by the BS 5400 rules and the appropriate span/depth ratio.

Only vehicle live loads were considered in the analysis, as it is only stress range that is of importance for fatigue.

Reinforcement Strains

Strains in reinforcement were calculated for slab bending moments determined from the structural analysis above. Analysis was carried out on the assumption that concrete had no tensile strength, representing concrete cracked in flexure, and on the assumption that the concrete was uncracked. Poisson's ratio was taken as 0·2 for uncracked concrete and 0·0 for cracked concrete. Strains were calculated at the centroid of the appropriate reinforcement. Where concrete was to be considered uncracked, reinforcement and concrete strains were assumed to be the same. A linear variation of strain through the depth of the section was assumed in both cases.

Vibrations of Bridge Deck

A further analysis of bridge deck behaviour was carried out to estimate the natural frequency of the deck. The deck was modelled as a three-span beam continuous over two internal supports. Two analyses were carried out, with section properties based on the full composite section throughout the three spans in one case and using values for steel only over supports (to allow for cracking of the deck slab in these areas) in the other. Lowest frequencies were determined as 2·20 and 1·74 Hz, respectively, bounding the measured value of 2·0 Hz.

TEST PROGRAMME

Measurements

Four sets of measurements were undertaken: sets 1–3 used a measured vehicle load and were conducted prior to opening of the road to general traffic, while set 4 readings were taken with the bridge in service. Measurements taken were:

1. Static measurements prior to surfacing of the deck (R1 only), taken 20 days after completion of deck concreting.
2. Quasi-static measurements of surfaced deck, 42 days after completion of deck concreting.

3. Dynamic measurements with vehicle travelling at approximately 30 km/h (20 mph) and 60 km/h (40 mph).

   The load vehicle used in 1–3 was a Leyland Atlantean double-deck bus. Details are:

   Wheelbase: 4·95 m
   Front axle loading: 3·28 t, single tyres, 2·05 m c/c
   Rear axle loading: 6·37 t, double tyres, 1·85 m c/c of pair
   Tyre pressures: 0·7 N/mm^2 (100 psi) all round; tyre size 1000 × 20.
4. Continuous monitoring of strains with bridge in service.

Analysis

Analysis of reinforcement strains was carried out using the procedure described earlier for both the load vehicle used in sections 1–3 and the standard fatigue vehicle described in BS 5400: Part 10.[1]

RESULTS

Load Tests 1–3: Measured Vehicle Loads

The maximum value of strain recorded in tension reinforcement as the load vehicle moved at walking pace across the deck along successive transverse lines is shown in Fig. 3. Maximum values occurred with the rear axle in line with respective gauges. Strains calculated analytically are also shown in Fig. 3 for the load vehicle at corresponding locations. The values are for the slab in the uncracked condition.

Measured strains in longitudinal reinforcement at R2 near pier B were around double the longitudinal values measured at R1 near midspan. It appears likely that the concrete section near the pier was at least partially cracked due to tensile stresses arising from deck surfacing loading and restrained shrinkage of the deck slab. Loads from deck surfacing at midspan gauges would tend to place the slab in compression and offset shrinkage stresses.

Transverse strains measured at the two sets of gauges are in closer agreement although differences may be noted, particularly with the compressive strains recorded at R2 with wheel loads on the far side of the stringer from the gauges. The difference may be attributable to increased stiffness of support from the stringer to the deck slab from a slightly heavier cross-bracing over the pier and from resistance to lateral movement of bearings under the plate girder.

A comparison of strains in top and bottom transverse reinforcement at R2 showed that the neutral axis of the slab lay around 120 mm from the top

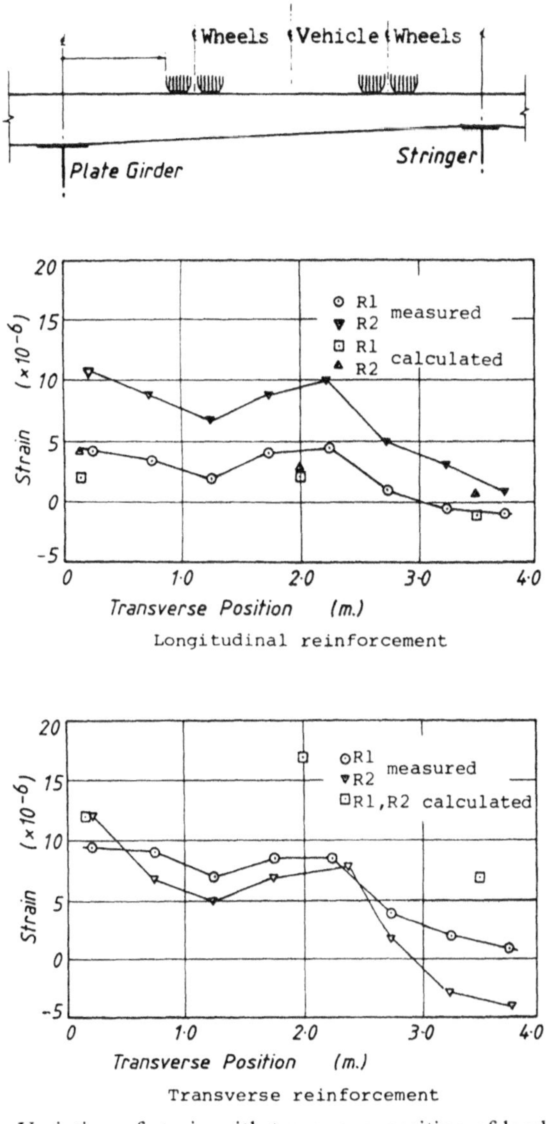

FIG. 3. Variation of strain with transverse position of load vehicle.

surface, an x/h ratio of 0·45. An x/h value of just under 0·5 would be expected for an uncracked slab. The measured value is in close enough agreement for it to be concluded that the slab behaved as if uncracked transversely.

It is considered unwise to attempt detailed comparison of measured and theoretically calculated strains in view of the very low levels of strain. A strain of 12×10^{-6}, the largest measured value, is equivalent to a reinforcement stress of only 2·5 N/mm^2.

It was anticipated that surfacing of the deck might reduce strains on two counts. Greater dispersion of local wheel loads through the thickness of the surfacing will increase the effective contact area of the wheel and reduce its intensity, thus reducing local moments; and composite action of concrete slab and asphalt surfacing may create a slab of increased section modulus, thus reducing measured strains under a given loading.

Lower strains were recorded in the surfaced slab, but the position of the neutral axis was not altered. It follows that the surfacing did not alter the flexural stiffness of the slab. Strains on transverse reinforcement were approximately 35% less on the surfaced deck. Longitudinal strains, however, showed a reduction of only 10%. Using Westergaard's method for calculation of moments under concentrated loads and dispersion allowances in BS 5400, the difference between surfaced and unsurfaced conditions would not be expected to exceed 10%. Bearing in mind the possible development of transverse cracking and its influence on longitudinal strains, and the increase in modulus of elasticity of concrete between the two sets of readings, it is difficult to draw significant conclusions from these results.

Studies on fatigue of steel bridge decks have shown that the dispersion characteristics of the surfacing are temperature and rate of loading dependent. No significant difference was found between quasi-static values measured in series 2 and corresponding values measured with the vehicle at speed in series 3.

Service Loadings

Results of measurements during service are summarised in the histograms of frequency of occurrence of stress range in Fig. 4. Number and intensity of the stress range in longitudinal and in transverse reinforcement was similar at both locations. It is apparent from Fig. 4 that measured strains are low.

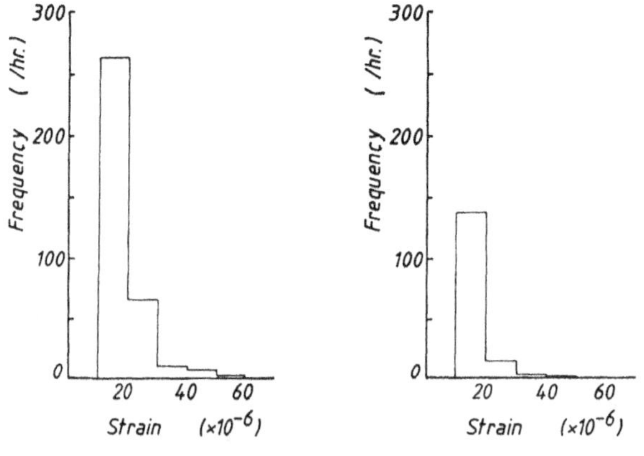

FIG. 4. Histograms of measured strain range.

The largest strain range recorded was less than 60×10^{-6}, equivalent to a stress of 12·5 N/mm^2. BS 5400: Part 10 requires that the range of strain to which high yield reinforcement is subject should not exceed 325 N/mm^2. Measured strains therefore did not exceed 4% of this value.

Longitudinal reinforcement showed greater numbers of greater stress ranges at both midspan and near the pier than did the corresponding transverse gauges. A detailed section of the complete record is presented in Fig. 5, showing variations in strains as two articulated vehicles with three-axle trailers crossed the bridge at around 1·5 s apart. Only the variation in strain is of significance, as it was not possible to determine true datum values.

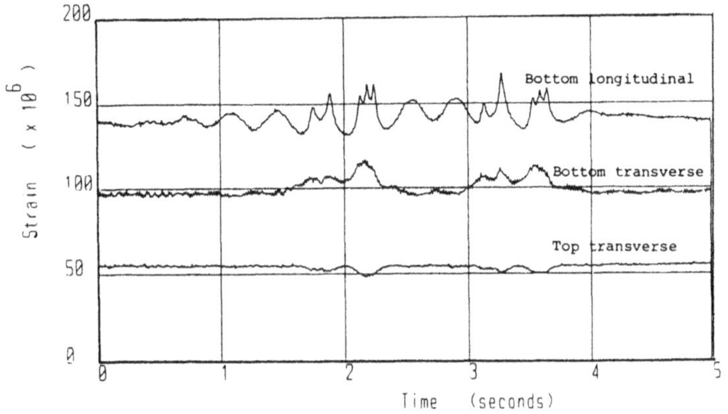

FIG. 5. Typical strains recorded by reinforcement gauges R2 near pier B for the passage of three-axle trailer articulated vehicles.

Traces for the longitudinal gauges clearly show strains arising from longitudinal oscillations of the deck at a frequency of approximately 2 Hz. Subjectively, it had been noticed that the deck was 'bouncy' enough to have an unsteadying effect while walking. The vibrations accounted for the much larger number of events recorded for longitudinal gauges (Fig. 4), and constructive reinforcement of the pattern of strains arising directly from wheel loads and from oscillation may also account for the higher stress ranges recorded overall on longitudinal gauges.

Readings from the pair of transverse reinforcement gauges at R2 were compared for several of the larger events recorded. The measurements indicated a neutral axis depth of 85 mm, against 120 mm determined from measurements 9 months earlier. Using an elastic section analysis for a cracked slab assuming zero concrete tensile strength, the neutral axis was estimated at 65 mm from the top surface of the concrete. The measured value of 85 mm would therefore be reasonable if either the concrete were assumed to have a small tensile strength or if some compression membrane behaviour were occurring.

TABLE 1
Calculated strain ranges for standard fatigue vehicle

	Calculated strains ($\times 10^{-6}$)			
	R1		R2	
	Transverse	Longitudinal	Transverse	Longitudinal
Cracked section	250	130	250	130
Uncracked section	53	11	53	12

Strain range calculated for the BS 5400 standard fatigue vehicle by the methods outlined in section 4 are listed in Table 1. Vehicles using the bridge could (legally) have been up to 25% heavier in both overall and individual axle loadings.

A comparison of Table 1 values with maximum values in Fig. 4 shows transverse reinforcement strains were smaller than calculated, even if the concrete slab were assumed to be uncracked. Measurements at R2 made it appear that the section was cracked. Measured values of longitudinal strain lie between those calculated for the slab in the cracked and uncracked condition.

SUMMARY OF FINDINGS

1. Measured stress ranges were less than those calculated analytically.
2. Oscillations of the bridge deck served to increase the number and intensity of measured stress range on longitudinal reinforcement.
3. Surfacing does not influence the stiffness of the deck but does assist in dispersion of the contact load over a wider area.
4. Fatigue of deck slab reinforcement is unlikely to be a problem unless accompanied by other deterioration.

ACKNOWLEDGEMENTS

Financial support from the Transport and Road Research Laboratory and the Science and Engineering Research Council and the assistance of the Director of Roads, Strathclyde Regional Council, and his staff is gratefully acknowledged.

REFERENCES

1. BS 5400, Steel, Concrete and Composite Bridges: Part 10. Code of Practice for Fatigue, British Standards Institution, London, 1980.
2. Scottish Development Department, Technical Memorandum (Bridges), SB 12/83, Edinburgh, 1983.
3. Cairns, J., Fatigue of bridge decks: in-service measurements in a reinforced concrete bridge deck. *Proc. 2nd Int. Conf. on Short and Medium Span Bridges*, Ottawa, 1986. Canadian Society for Civil Engineering, Montreal, 1986.
4. Westergaard, H. M., Computation of stresses in bridge slabs due to wheel loads. *Public Roads*, **VII**(1) (1930) 1–23.

REPAIR AND REHABILITATION

55

Cracks in Steel Orthotropic Decks

PIERRE MEHUE

Service d'Etudes Techniques des Routes et Autoroutes, Bagneux, France

ABSTRACT

Many cracks have been discovered in steel orthotropic decks during the past 12 years. They occurred either at the rib-to-deck plate junction or at the rib-to-floor beam junction, depending on local facts. Frequent inspections on a large number of decks allowed observation of their appearance and progress in order to set up a plan of action and repair. In the light of finding this damage certain measures can be recommended to ensure good structural behaviour of orthotropic decks, such as using thicker plates and making edge preparation compulsory.

INTRODUCTION

Orthotropic decks were introduced in France towards the end of the 1960s on long-span highway bridges and on movable bridges because of the dead-weight savings they made possible in the designs. It is for the same reason they were used on span units for the temporary flyover viaducts which were erected in the early 1970s as a provisional solution to traffic problems at many urban crossroads all over the country, with a total area of approximately 120 000 m^2 from 1970 to 1976.

Several of these latter units, which carried heavy traffic, were found to have cracked a few years after they were put in service, showing a repetitive character in damage. Lately similar cracks have also been discovered in the deck of a motorway bridge built in 1966. In both cases cracks occurred (1) at the rib-to-deck plate junction and (2) at the rib-to-floor beam junction, just in the zones where the wheels of trucks pass, which made people wonder about the reliability of some construction details of such structures.

MAIN CHARACTERISTICS OF BRIDGES

Temporary Flyover Viaducts

The span units for demountable viaducts, which range in length from 6 to 30 m for a single width of 3·50 m and a depth of 1 m, consist of two main plate girders with an orthotropic deck employing closed trapezoidal longitudinal ribs and transverse floor beams. They are joined end to end according to the crossing conditions, and can be used separately for single-lane viaducts or assembled side by side in order to form two- or three-lane viaducts with one- or two-way traffic (Fig. 1).

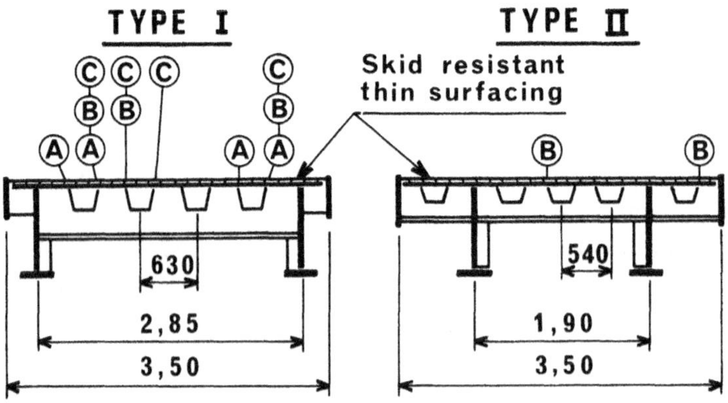

FIG. 1. Cross-sections of span units and location of cracks.

The span units are generally one of two types, which only differ basically in the distance between girders, the number and size of ribs, and the depth of floor beams. Type I structures are used most frequently. Both types have a floor beam spacing of 3 m and use a 10 mm thick deck plate with a skid-resistant thin surfacing, and are entirely shop manufactured so as to be as light as possible and easy to lift and handle. The ribs, which are made of 630 × 6 mm folded plates for type I and 570 × 7 mm for type II, are generally continuous through the floor beams, except for some curve on plan span units in which they are cut at the floor beams to be welded to the webs in order to accommodate the curvature of the road. The rib-to-deck plate junction is made by one-sided fillet welds. Steel grade A 52 S was used for all the components, only the deck plate of the span units under 20 m long being made of steel grade A 42 S.

Richemont Bridge

This is a three-span continuous 52·65–58·00–52·65 m bridge which was built over the Moselle River in 1966 to carry a one-way two-lane roadway of

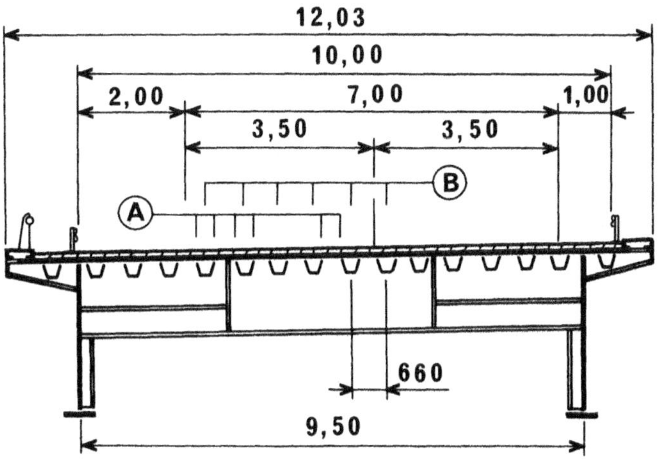

FIG. 2. Cross-section of Richemont Bridge and location of cracks.

the A31 motorway with an overall clear width of 10 m. It consists of twin plate girders 9·50 m apart and approximately 2·80 m deep, and comprises an orthotropic deck employing longitudinal closed ribs (Fig. 2).

The 12 mm thick deck plate is supported by trapezoidal ribs spaced 0·66 m and floor beams spaced 3·625 m in the centre span and 3·51 in the two end spans, with a 50 mm thick wearing surface. The ribs, which are made of 550 × 6 mm folded plates, are discontinuous and welded to the floor beam webs by fillet welds, the connection with the deck plate being also made by one-sided fillet welds. The structure is fully welded, using steel grade A 52 S except for the deck plate, which is fabricated from steel grade A 42 S.

CRACKS AT THE RIB-TO-DECK PLATE JUNCTION

The cracks were found either in the deck plate or in the fillet welds at the top of the trapezoidal ribs, depending on the bridges and the structure concerned.

Cracks in the Deck Plate

These cracks have been discovered only on span units for temporary viaducts which were subjected to heavy traffic. They are exactly located above the weld lines of ribs and affect sometimes one side sometimes both sides, with a relatively symmetrical disposition on straight single-lane unit spans (mark A). On two-way viaducts they are slightly shifted towards the axis of the roadway (mark B), and the shift is more evident towards the left on one-way two- or three-lane viaducts, and towards the inner edge of the deck on curved on plan units (mark C).

Over 500 cracks were checked within the past 12 years, both on short and long spans, between floor beams or right above them, in steel grade A 42 S and in steel grade A 52 S, with no regularity in occurrence. Investigations carried out on a sample of deck taken from a damaged unit showed that these cracks resulted from fatigue phenomena initiated by local poor welding and lack of penetration, due to no edge preparation and excessive gap between rib and deck plate, which led to a notch effect in the root of the fillet weld. So the cracks start at the underface of the deck plate inside the rib, where they cannot be seen, coming out under the road surfacing which is quickly damaged. This obviously means that water can infiltrate into the cracks and soon fill the ribs (Fig. 3). Under such circumstances these cracks

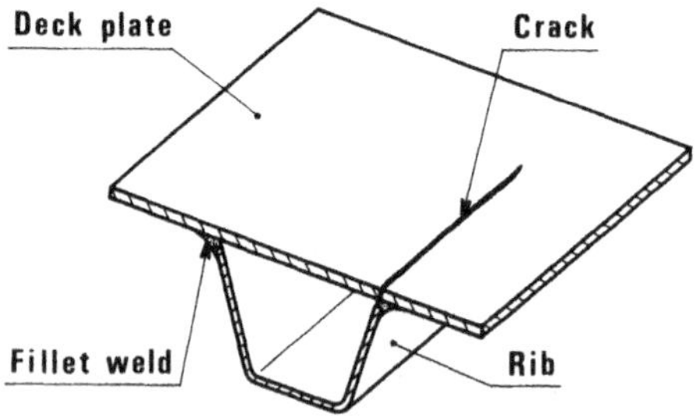

FIG. 3. Crack in deck plate.

are relatively easy to find, positioned at the bottom of a scar or a crevice in the road surfacing whenever they are over 40 mm long. Most of the time the water drained into the rib maintains a permanent humidity between the lips of the cracks and, even in summer, oozes at its lower end, accompanied in very hot weather by a constant bubbling. The first cracks were found in 1977 on two viaducts erected in 1971 which supported a very heavy traffic load (800–1500 trucks per lane per day), but as the cracks were already long (420–560 mm) it is likely that they were then at least 1 year old. Many other cracks have been discovered later in the deck plate of many viaducts 8, 10 or 12 years after their construction, depending on the intensity of loading imposed on the structures. The way in which cracks progress from the moment they have been detected is difficult to estimate because any accurate investigation requires the traffic to be stopped, which is politically difficult, particularly on single-lane viaducts. However, it was possible on a few occasions to undertake some measurements (as shown in Table 1) which gave three examples of crack growth with quite different rates of extension.

The width of the crack is generally about 0·5 mm but it may reach 1 or 2 mm at the central part of long or old cracks, and mostly with roughly bevelled edges. At the same time a slight difference of level (1–4 mm) between the cracks lips can be observed, due to deck plate deformation under heavy

TABLE 1
Evolution of cracks in deck plate

Viaduct	Cracks	Length (mm) and date of measurement		
		17/2/81	29/4/81	1/6/81
1	1	310	350	440
	2	350	350	400
	3	480	520	600
	4	480	750	870
		22/6/82	6/10/83	24/5/84
2	1	130	240	240
	2	190	270	270
	3	200	320	410
	4	240	330	370
		16/11/84	20/6/86	10/10/86
3	1	120	200	240
	2	120	410	460
	3	270	430	550
	4	270	500	675

wheel loads (Fig. 4). Finding the first cracks created some concern with the viaduct owners, but due to frequent inspections it was evident there was no danger to users as long as the cracks were few in number, rather short and fine, and grew at a slow gradual rate. Consequently it was generally possible to wait for fair conditions to undertake repairs, after careful examination of risks and possibilities. As a matter of fact the situation only became worrying when separated successive cracks were growing towards each other to form a single crack, longer and wider, or when several parallel cracks were rapidly progressing in line and slit the deck plate into strips, about 300 mm wide and over 1 m long, which were likely to sag considerably under heavy loads or bow up with thermal effects. In those cases, fortunately infrequent, trucks were obviously no longer allowed on the viaducts.

Lastly, attention must be drawn to the fact that, in spite of severe inspections when the wearing surface was removed in order to renew the surfacing, no cracks have been found in the 12 mm thick deck plates of four long-span bridges 14 and 17 years after construction.

Cracks at Welds

In this case the cracks are located at the lower toe of the fillet weld at the

FIG. 4. Crack in a 10 mm thick deck plate.

junction of the rib and the deck plate, or very near the toe in the weld (Fig. 5). These cracks are generally difficult to find because of the dim light under the bridges, the dark colour of paint coatings and the poor state of finish, and also because of difficult access to the bridge soffit. However, about two dozen cracks were found on temporary overpass bridges, 8 years after erection for the first ones, ranging in length from 40 to 350 mm and with an opening of less than 1 mm. Although only a few growth measurements were made, it seems that cracks are able to extend rapidly, reaching 1·50 m over a 2-year period. For example, the crack in the photograph shown in Fig. 6 was 300 mm long on 8 July 1988 and 500 mm on next 12 September. In most cases cracks progress by running along the weld toe, but it may happen that they propagate into the rib wall following a curved line. As before, these cracks are likely due to the thinness of the deck plate, which causes high transverse flexural stresses, and to poor weld penetration, which makes the junction rather unsymmetrical.

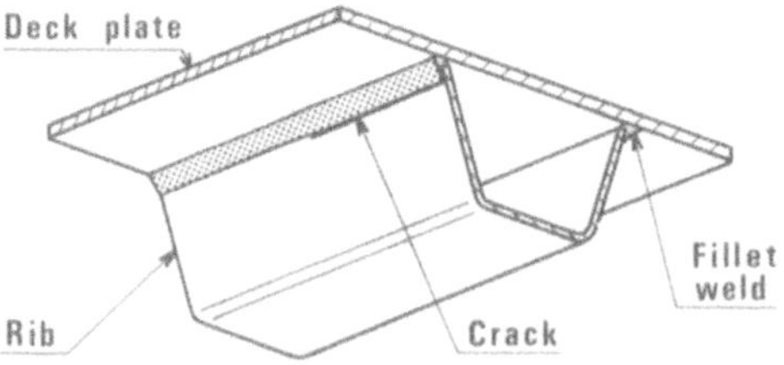

FIG. 5. Crack in fillet weld.

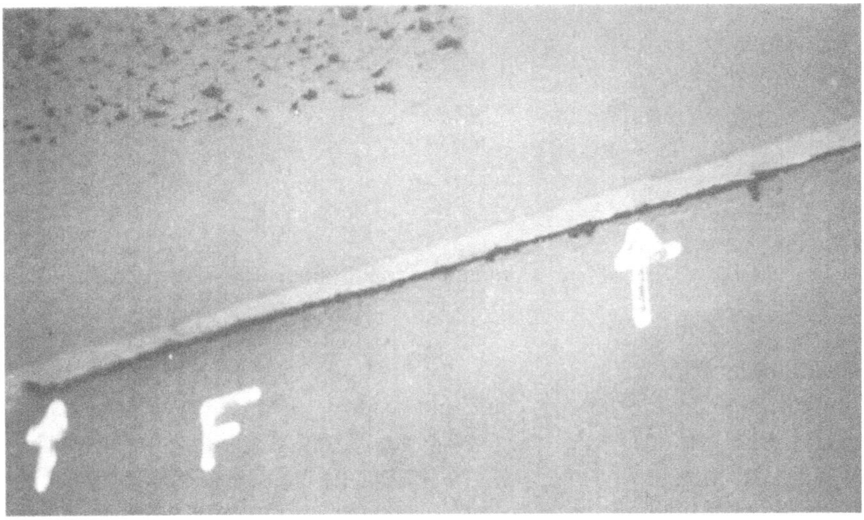

FIG. 6. Crack at the toe of a fillet weld.

In 1984 a dozen similar cracks were observed on Richemont Bridge (shown as mark A in Fig. 2). In 1987 fifteen cracks of varying lengths from 570 to 2240 mm had occurred in the bridge. Because a previous inspection in 1981 had revealed no damage, it would seem to be evident that crack growth progressed very quickly. The early investigations carried out to try to determine the reason for these failures showed that the fillet welds were thin and poorly made at many locations, and also that the ribs were very economically designed.

CRACKS AT THE RIB-TO-FLOOR BEAM JUNCTION

The cracks were found with both continuous and discontinuous ribs, being more widespread in the latter disposition.

Continuous Ribs

About a dozen cracks have been discovered on span units in temporary viaducts within the past 8 years. These cracks were located at the toe of the junction weld to the rib wall close to the floor beam web. The cracks propagated at the lower end of the weld, on both sides, and were 30–60 mm long when they were discovered. The cracks generally progressed slowly towards the deck plate and in only two cases were found to affect the whole length of the weld (Fig. 7).

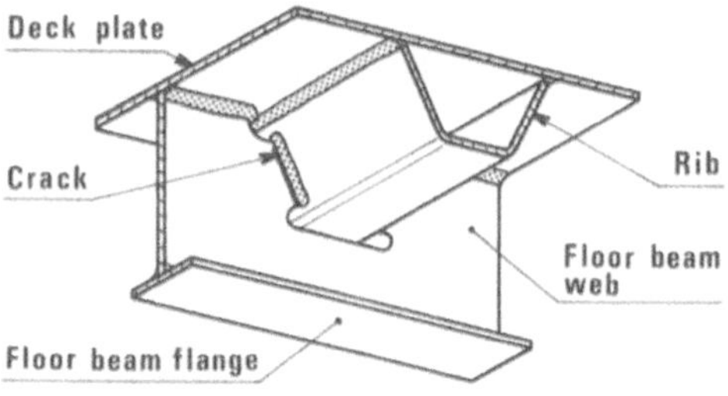

FIG. 7. Crack at the continuous rib-to-floor beam junction.

Discontinuous Ribs

The cracks developed in the fillet welds joining the ribs positioned between the webs of the floor cross-beams. They started at the lower end of the round outline, at or very near the toe of the weld, close to the floor beam web (Figs 8 and 9), then they gradually propagated into the rib-to-deck plate and floor beam-to-deck plate welds. Most of the time they occurred on one side of the rib and the floor beam, but in some cases they were discovered on each face of the web, or on both sides of the ribs. In these latter cases a crack rapidly appeared at the bottom weld, which made the end of the rib quite free. Finally, it may happen that the cracks in the fillet weld escape from the round outline, into the web of the floor beam, or into the bottom flange of the rib.

About a hundred of these cracks have been found on span units of temporary bridges, the first ones in 1978 with a viaduct erected in 1971 (Fig. 10). In 1987 ninety cracks were repaired on Richemont Bridge (shown as mark B in Fig. 2) and twenty new cracks have appeared subsequently, growing at a rate of 5–10 mm per month (Fig. 11). The causes of this damage may be due to both weld shrinkage and fatigue effect, together with the rib discontinuity generating high residual stresses due to the welding arrangement and severe stresses due to live loads, being transmitted through poorly made fillet welds.

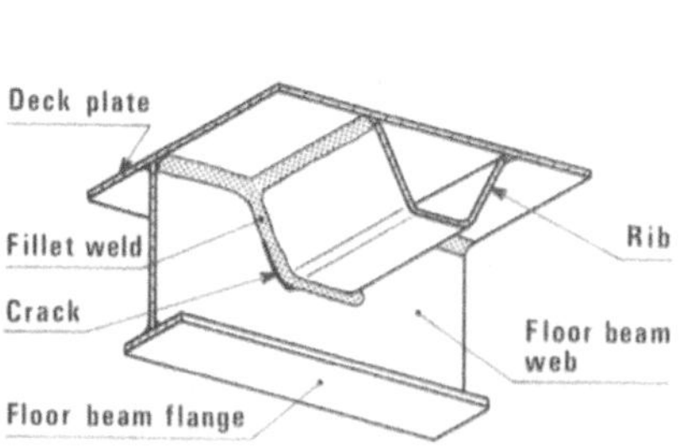

FIG. 8. Crack in the weld of discontinuous ribs.

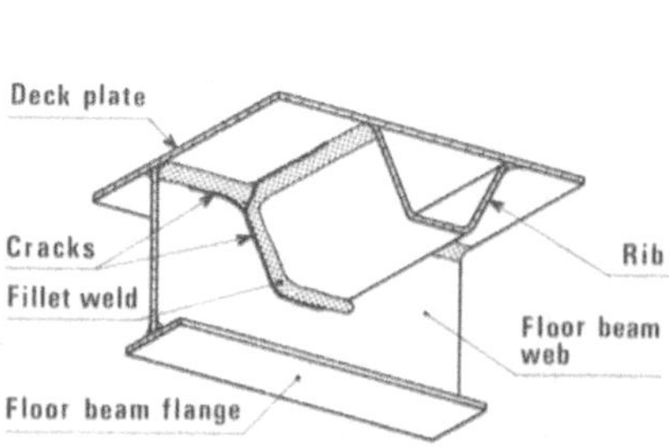

FIG. 9. Propagation of cracks in the welds of discontinuous ribs.

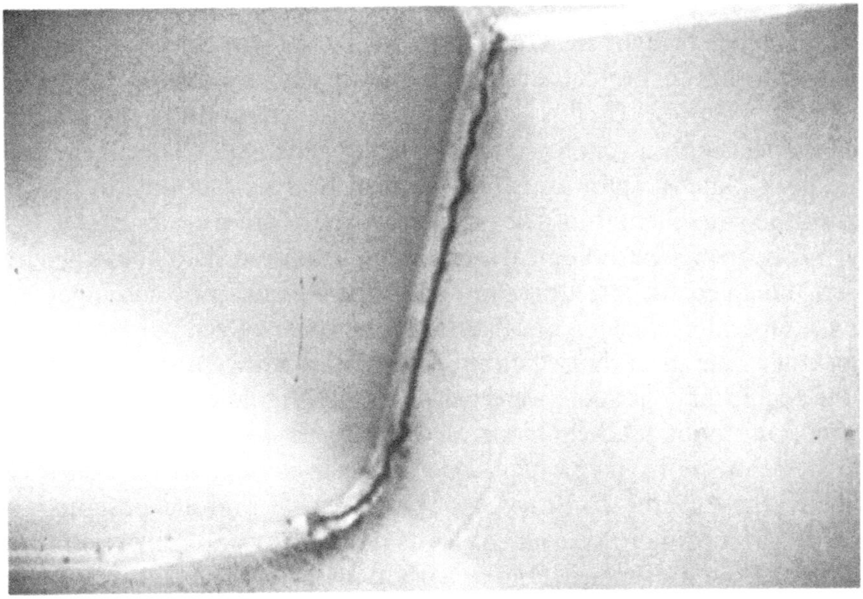

FIG. 10. Crack in the weld of a discontinuous rib of a temporary viaduct.

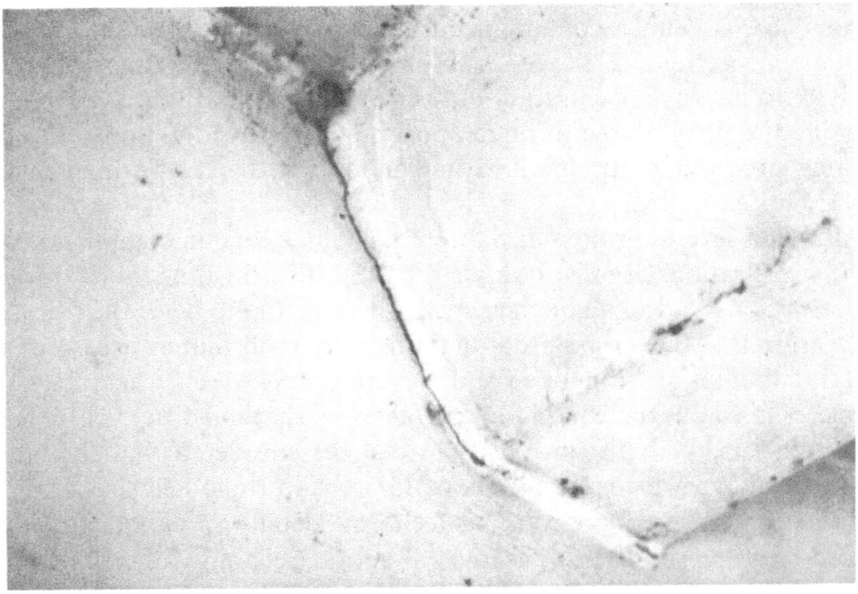

FIG. 11. Crack in the weld of a discontinuous rib of Richemont Bridge.

REFLECTIONS AND COMMENTS

The damaged bridge structures previously described could give the impression that orthotropic steel bridge decks are unwise for bridge designers to consider. So it is important to draw attention to the point that damage affects only a very low proportion of welds—for example the cracks in the deck plate of span units for overpass bridges account for less than 0·2% of the whole length of the rib fillet welds. Moreover, as interesting as they may be from a pathological angle, the cracks described above generally neither damaged the structural integrity of the decks nor endangered the user's safety, and most of the time they can be easily repaired. It is also important to mention that when the Richemont Bridge and the span units for the temporary viaducts were designed, in 1965 and 1969 respectively, information about the behaviour of orthotropic decks was limited. With regard to the span units, which were to be used successively on several viaducts, they were, contrary to widespread opinion, designed and fabricated according to the same rules as permanent bridges, except for the thickness of the deck plate. Due to early damage in the wearing surface it soon became apparent that the deck plate was at fault. At this time the deck plate of orthotropic decks was always considered by the steel designers and fabricators to be redundant.

Several lessons were drawn from this experience in the early 1980s, and it was strongly advised that designers ensured that (1) the deck plate is not thinner than 12 mm, a minimum thickness of 14 mm is recommended to minimise deflections; (2) the edges of the plates forming the ribs are bevelled in order to improve penetration of fillet welds; (3) the ribs are continuous, crossing the floor beams through appropriately shaped cutouts; and that welding procedures are specified and monitored to ensure high quality performance.

Although several bridges have been built since 1980 in compliance with these specifications, it must be stated that most of the time fabricators are reluctant to use a deck plate thicker than 12 mm and to bevel the rib edges. They argue that there is no proof that this work is absolutely necessary and that, in addition, it will make the deck more expensive.

In conclusion, as there has been no damage of the kind observed, on many orthotropic decks built since 1971, it can be considered that the cracks described above, which affect the two oldest orthotropic steel decks built in France, are due to inappropriate technical details. Consequently there should be no reason to be concerned about the reliability of well designed and properly detailed orthotropic bridge decks.

56

An Analysis of the Behaviour of Reinforced Concrete Beams Following Deterioration and Repair

JOHN CAIRNS

Department of Civil Engineering, Heriot-Watt University, Edinburgh EH14 4AS, UK

ABSTRACT

Results of a study of the redistribution of stress in reinforced concrete beams following repair are presented. A non-linear analytical model which permits removal and replacement of parts of a reinforced concrete section has been set up, and is used to compare the performance of repaired and equivalent 'as new' beams. The effects of relieving dead loads during a repair are considered.

INTRODUCTION

Corrosion of reinforcement is the main cause of deterioration of structural concrete. In some instances loss of section of the reinforcement requires provision of additional or replacement bars. At present little guidance is available to the engineer on the redistribution of stresses within a member following repair, on the effectiveness of replacement reinforcement, and on the structural integrity of repaired members.

The aim of this paper is to report findings of an analytical study of this problem, in which relative deformations and stresses at sections of 'as new' and 'repaired' reinforced concrete beams were examined.

Where deterioration of a member is of a severity to require that additional reinforcement be provided, current practice is generally to prop the member during the repair.[1]

These temporary props may perform either or both of two functions:

(a) to support a member in order that it can continue to carry the loads imposed despite loss of section and hence of strength, and
(b) to unstress a member while a repair is carried out.

The support role (a) may not be required if the member has adequate reserves of strength, or if the loads the member is required to support can be reduced for the repair period. Provided props are relatively stiff, it will be adequate if they are initially placed in contact only with the member, and take up load as the member deflects.

The object of unstressing a member (b) is to ensure that replacement or supplementary reinforcement provided as part of the repair will assist in carrying dead load. Unless dead loads are relieved by propping the member during the repair and while the replacement concrete gains strength, new reinforcement and concrete cast in during the repair will be unstressed when the member carries dead load. For props to be effective in this role they must actively impose a load on the member to counteract dead loads. Hydraulic jacks with pressure gauges to monitor applied prop forces will generally be required.

Even under carefully controlled propping it is possible that the load may not be fully removed from a member. The nature of the structure may be such that the prop force is shared between the member under repair and other parts of the structure. It will then be difficult to ensure that a member is unstressed without resort to stress relief strain measurement techniques.

Cracks in reinforced concrete members may become filled with fine detritus or crystalline growth which wedges a crack open and prevents closure on unloading.[2] Reinforcement crossing such wedged cracks will then be unable to shed load when prop forces are applied.

Concrete is not a linear elastic material. Loading and unloading relationships are different, as is apparent from the short-term stress–strain plot in Fig. 1. Creep will accentuate the difference between the loading and unloading portions. Shrinkage of concrete also occurs over a period of time. It is evident that even if the member can be unstressed by propping it cannot be unstrained.

It becomes clear that even in the best controlled work similarity between 'as new' and 'repaired' members will be imperfect, and that any assumption of structural integrity on this basis is questionable.

Is propping to counteract dead loads necessary for a repair? An analogy

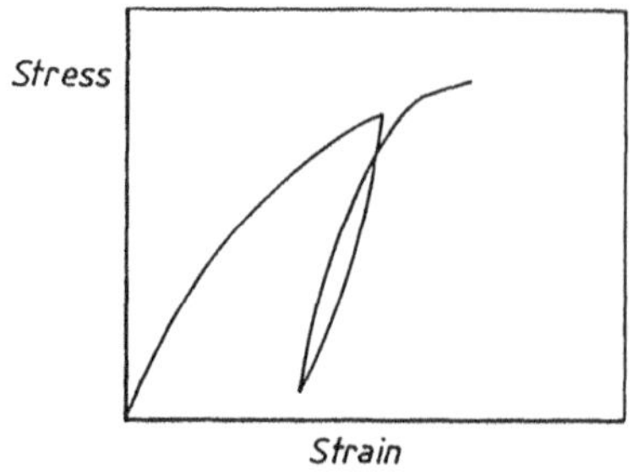

FIG. 1. Short-term stress–strain relationships for concrete loading and unloading.

with composite construction would suggest that although stresses and deformations at service loads would differ, ultimate strength would not. The analogy is imperfect, however, and there are differences between composite construction and structural repairs that must be borne in mind. For example, a precast concrete beam supporting an in-situ composite slab will be relatively young when the slab is cast, and differential shrinkage will be less than that between the original concrete of a member under repair and the replacement concrete, unless shrinkage-compensated repair materials are to be used.

This paper describes the first stage of a study to examine behaviour of structurally repaired members. The term structurally repaired is intended to denote repairs to members that have been significantly weakened by deterioration. The repair area will generally be larger than patch size, defined as $0{\cdot}25\,m^2$.

A mathematical model has been constructed to analyse stresses at a section of a reinforced concrete beam in a constant moment region at various stages of its history, through long-term deformations, loss of section, cutting out of concrete and casting in of supplementary reinforcement, etc. The analyses reported here assume the beam to be free of external restraints. Results presented in the report relate only to cases of deterioration where compression reinforcement is affected. Results of an analysis relating to deterioration of tension reinforcement have previously been reported.[3] The model is being used to design test specimens and procedures for physical tests in the laboratory, results of which will be reported in due course.

GENERAL DESCRIPTION OF MATHEMATICAL MODEL

The mathematical model is largely based on methods of analysis of reinforced concrete sections outlined in BS 8110.[4] Two variations of the model are used, one for serviceability and one for ultimate behaviour. Both consider only bending behaviour and are based on the assumption that plane sections remain plane. It is assumed that an adequate anchorage length is provided to supplementary reinforcement to realise the assumption that plane sections remain plane. Non-linear stress–strain relationships are used for concrete and reinforcement. The models are based on a finite element type approach. The model allows parts of the section of a beam or column member to be removed, to represent spalling of concrete, loss of reinforcement section due to corrosion or cutting out of contaminated concrete, and subsequently to be reinstated, to represent replacement or supplementary reinforcement and replacement concrete.

The ultimate strength analysis uses the rectangular parabolic stress–strain

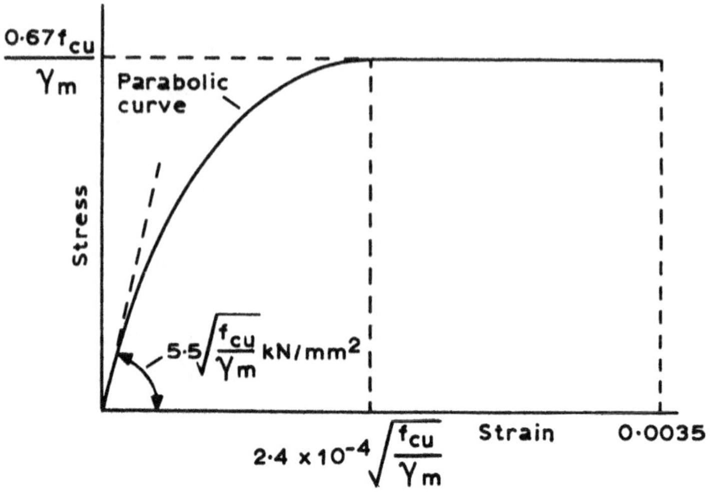

FIG. 2. Stress–strain curve for concrete.[4]

curve for concrete of BS 8110 (reproduced in Fig. 2) and the bilinear stress–strain curve for reinforcement (reproduced in Fig. 3). The tensile strength of concrete is ignored. This model is used in calculations for reinforcement stress at all loads and for all calculations at reinforcement stresses at or above the characteristic yield strength.

Creep and shrinkage were taken into account by the use of an effective modulus method and an equivalent force method respectively. Short-term unloading/reloading follows the slope of the initial tangent modulus up to the short-term curve. Coefficients of thermal expansion are fairly similar for steel and many concretes, and changes in temperature will produce negligible changes in stress in members that are not externally restrained. Thermal strains have therefore been neglected in the analysis. Cracks in reinforced concrete may become filled with fine detritus or crystalline growth.[2] On unloading of the member, the cracks attempt to close but

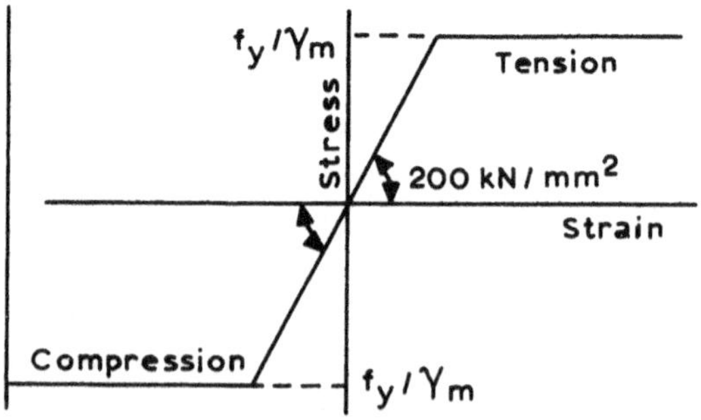

FIG. 3. Stress–strain curve for reinforcement.[4]

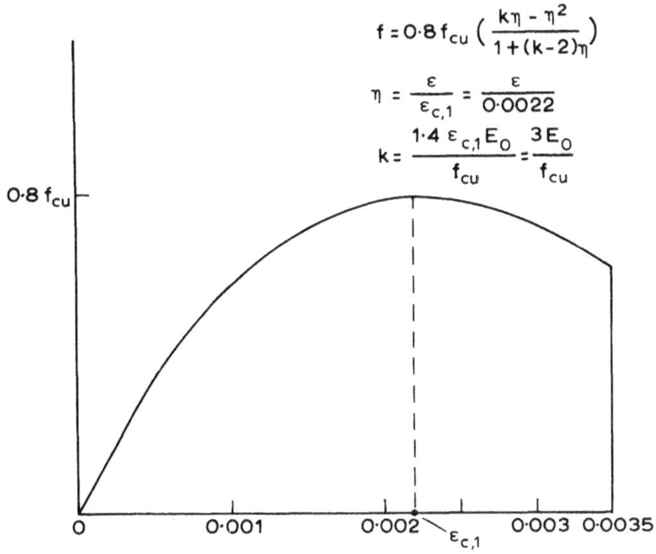

FIG. 4. Alternative stress–strain curve for concrete.[4]

remain wedged open. The analysis can allow for this by effectively reducing the width of an open crack.

The model for serviceability behaviour differed from that for ultimate behaviour in that a different stress–strain relationship was used, and allowance was made for tension stiffening of the concrete. The stress–strain relationship was again taken from **BS 8110** and is reproduced in Fig. 4. Where stress in the concrete at the level of tension reinforcement would exceed a value of 1·0 N/mm^2 in the short term or 0·55 N/mm^2 under long-term loading, stresses in the tension zone were calculated on the assumption of a triangular stress distribution varying from zero at the neutral axis to 1·0 or 0·55 N/mm^2 at the centroid of tension reinforcement in short and long term respectively.

Chloride-contaminated concrete must be removed from around the reinforcement before replacement concrete can be cast. A compression bar not confined by surrounding concrete will be a slender compression member, and its contribution to member strength will be limited by elastic instability considerations. For simplicity the model assumes that compression reinforcement carries no stress unless it is confined by concrete.

ASSESSMENT PROCEDURE

The performance of a 'repaired' beam with supplementary reinforcement is assessed by comparison with the performance of an identical beam in an 'as new' condition subjected to similar long-term loading. In making the comparison it has been assumed that both original and replacement concrete and reinforcement in the repaired beam have the same properties as those in the original. It is also assumed that the repaired member will be rebuilt to the same section profile as the old. These assumptions will frequently not hold in practical situations, as a repaired member may need to be of greater dimensions than the original where cover was inadequate, and replacement concrete will differ from original, often with reduced shrinkage properties and greater strengths. These differences would improve the performance of a repaired beam.

Service loads were calculated by first determining the ultimate bending strength of the section under consideration, with all partial safety factors on material strengths included in calculations. Partial safety factors of 1·5 on concrete strength and 1·15 on steel strength were used. This ultimate bending strength was divided by a load factor of 1·5 to obtain the service bending moment, on the assumption that dead and imposed loads are of similar magnitude. Permanent loads used in the calculation of long-term deformations were taken to be full dead load plus a proportion of imposed load, usually taken as 25% in results presented here.

Partial safety factors were taken as 1·0 in all further calculations.

The procedure followed in analyses of 'as new' sections was:

(a) Apply permanent load, long term.
(b) Apply increment to full service load, short term.
(c) Increase load to 'first yield' of reinforcement.
(d) Increase load to ultimate, taken to occur when the maximum compressive strain in concrete reaches a value of 0·0035.

The procedure for analysis of 'repaired' sections was:

(e) Apply permanent load, long term.
(f) Allow loss of section from corrosion of reinforcement. A 25% general loss of reinforcement section along a bar has been assumed throughout.
(g) Allow for any crack wedging.
(h) Apply increment to full service load, short term.
(j) Reduce load to level at which repairs are conducted, short term. This load was taken as full dead load, equivalent to 50% of service load, for unpropped repairs and 10% of service load for propped repairs.
(k) Cut out concrete around bars.
(l) Install supplementary reinforcement and replacement concrete.
(m) Apply load increment to permanent load, long term.
(n) Apply load increment to full service load, short term.
(o) Increase load to first yield of reinforcement.
(p) Increase load to ultimate.

The assessment of the performance of beams by this analysis is based on a limit-state approach. Bending strength is compared at the ultimate limit state, stages (p) and (d) above for 'repaired' and 'as new' beams respectively. Crack width and deflection are compared at service load, determined as described above. Section curvature calculated in the analysis is used as the basis for deflection, and mean strain in the extreme tension fibre of the concrete is used for crack width comparisons.

In addition, the stiffness of sections as the load is increased from permanent load to full service load are compared, based on the secant to the moment/curvature relationship between (m) and (n) for the 'repaired' section and between (a) and (b) for the 'as new' beam. This comparison is introduced as a better measure of the deformability of a member during service than total deflection.

Figure 5 shows a schematic stress–strain history for concrete at a point in the compression zone of a repaired beam. The notation of Fig. 5 follows the stages outlined above for a 'repaired' beam. Initial elastic deformation is followed by creep and shrinkage, with little change in stress under sustained application of load. As load is briefly increased to full service load, the stress–strain plot follows a line parallel to the initial tangent modulus until it reaches the short-term stress–strain curve, at which point it turns to follow that curve. Unloading as imposed load is removed and props are inserted follows down a route parallel to the initial tangent modulus. Stresses may, under certain circumstances, go negative, although strains will remain positive. On reloading/removal of props, the line rises parallel to the initial

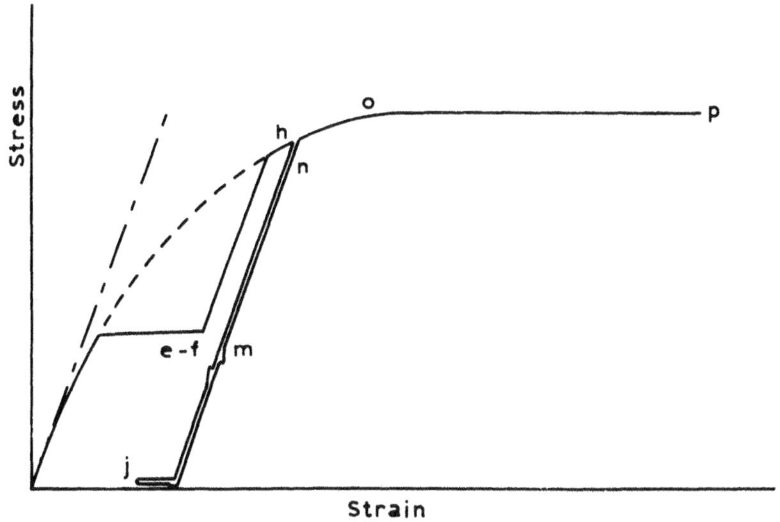

FIG. 5. Stress–strain history for element of concrete in compression zone.

tangent modulus until the short-term curve is reached, when that curve is followed until ultimate load is reached.

An analysis was carried out for a rectangular beam section under different environments and loading conditions. Creep and shrinkage strains are influenced by the relative humidity of the environment to which the beam is subject. While dead loads act permanently, the proportion of imposed load which should be considered as permanent will depend on the nature of the use of the structure. Each analysis was run for a member in 'as new' condition and for a member repaired under propped or unpropped conditions. The section used had 2·4% tension reinforcement and 0·9% compression reinforcement when 'new'. Ratio of effective to overall depth of section was 0·92. Corrosion was taken to reduce the area of compression reinforcement by 25%. This loss was balanced by supplementary reinforcement installed at repair. All reinforcement was taken to have a characteristic strength of 460 N/mm^2. Concrete was taken to have a characteristic cube strength of 30 N/mm^2. When analysed with partial safety coefficients taken as unity, the neutral axis of the section at ultimate load was at 42% of the effective depth. Concrete was taken to be removed to a depth of 0·10 times overall section depth, equivalent to removal of 75 mm depth of concrete from the compression zone of a beam of 750 mm overall depth.

Results are presented in Tables 1–3 as the ratio of values for 'repaired' beams, either propped to relieve dead load or unpropped during repair, to corresponding values for beams subject to similar long-term loading, creep and shrinkage, but in otherwise 'as new' condition. Unpropped conditions assume 50% of service loads carried during repair; propped conditions assume 10% of service load at repair.

TABLE 1
Ratio of calculated deformations and loads for 'repaired' versus 'as new' sections: influence of permanent load

Permanent load	Propped/ unpropped	Ratio for 'repaired'/'as new'					
		Service load				First yield	Ultimate
		Curvature	Crack width	Reinforcement stress	Stiffness		
DL + 25% IL	P	1·13	1·04	1·04	1·01	0·98	1·00
	UP	1·35	1·10	1·09	1·04	0·96	1·00
DL + 75% IL	P	1·12	1·04	1·05	1·02	0·98	1·00
	UP	1·40	1·12	1·11	1·06	0·95	1·00

Environment: temperate.
No crack wedging.

Table 1 shows results for beams subject to varying proportions of permanent load. Dead load, taken to be 50% of service load throughout these analyses, is always in place, but the proportion of imposed load that may be considered to be permanent will depend on the nature and use of the structure. The analysis used a creep factor of 1·0 and a shrinkage strain of -110×10^{-6}, values appropriate to outdoor conditions in a temperate climate. Cracks were assumed free to close on unloading.

At the serviceability limit state, Table 1 shows that the curvature, and hence the deflection, will be greater following repair, particularly if the beam was not propped. However, it may be seen that repaired beams are slightly stiffer as the load is increased from permanent to full dead + imposed load. If the load is subsequently cycled between these limits, repaired and new beams will have the same stiffness. The greater curvature of the repaired beam is therefore attributable to deformations of the member in the deteriorated state and not to a reduction in stiffness following repair. If sag is likely to be of concern visually, steps could generally be taken to repair the member to a flatter profile.

Crack width is closely related to reinforcement stress, and both are greater in the repaired beams. In the past, crack widths have been regarded as of importance for durability, but this is now questioned.[5] Given the random nature of cracking, the calculated reduction of around 5% gained from propping during repair would not be noticeable. The increase in reinforcement stress at service load in the repaired beams is brought about by movement of the centre of compression away from the compression face of the member, and consequent reduction of the lever arm between tension reinforcement and the centre of compression. The movement is greater in unpropped repairs, where new concrete is less highly stressed. The movement also accounts for the reduction in the load at which tension reinforcement starts to yield. Load at first yield still comfortably exceeded service load, however.

As the load on the beam increases towards the ultimate and stresses reach a plateau, new concrete attains the same stress as the old, although it is still less highly strained. Table 1 shows ultimate bending strength to be unaffected by the repair.

A similar comparison is made in Table 2 for beams subjected to different climates. A creep factor of 2·0 and shrinkage of -350×10^{-6} have been used in the calculations for members in a 'hot and dry' climate. These values are also reasonable for indoor conditions in the UK. The performance of repaired beams relative to 'as new' beams is slightly worse in the less humid environment. The difference is principally due to shrinkage of concrete used for reinstatement. As mentioned earlier, ordinary concrete was assumed for repair although shrinkage-compensated mixes would often be specified.

TABLE 2
Ratio of calculated deformations and loads for 'repaired' versus 'as new' sections: influence of creep and shrinkage

Climate	Propped/ unpropped	Ratio for 'repaired'/'as new'					
		Service load				First yield	Ultimate
		Curvature	Crack width	Reinforcement stress	Stiffness		
T	P	1·13	1·03	1·04	1·01	0·98	1·00
	UP	1·35	1·11	1·09	1·04	0·96	1·00
H and D	P	1·18	1·03	1·06	1·05	0·96	1·00
	UP	1·53	1·14	1·13	1·07	0·94	1·00

Permanent load: DL + 25% IL.
No crack wedging.
T, temperate; H and D, hot and dry.

Table 3 shows the effect of cracks becoming blocked and prevented from full closure. The crack wedging value in the table represents the proportion of the width of a crack at stage (g) that has filled with other material. A value of 100% denotes a crack that has completely filled; a value of 0% denotes perfect closure of the crack. Tension reinforcement is unable to release some of the stress carried as prop forces are applied if cracks are wedged open.

TABLE 3
Ratio of calculated deformations and loads for 'repaired' versus 'as new' sections: influence of crack wedging

Crack wedging (%)	Propped/ unpropped	Ratio for 'repaired'/'as new'					
		Service load				First yield	Ultimate
		Curvature	Crack width	Reinforcement stress	Stiffness		
0	P	1·13	1·03	1·04	1·01	0·98	1·00
25	P	1·13	1·03	1·04	1·01	0·98	1·00
50	P	1·19	1·05	1·04	1·03	0·98	1·00
75	P	1·21	1·06	1·05	1·05	0·98	1·00
100	P	1·22	1·07	1·05	1·07	0·98	1·00
0	UP	1·35	1·11	1·09	1·04	0·96	1·00

Permanent load: DL + 25% IL.
Environment: temperate.

Stress in tension reinforcement at repair was calculated to increase from 25 N/mm^2 where 'perfect' crack closure would occur to 112 N/mm^2 where the crack was fully wedged open. Ratios for the beam repaired unpropped are shown in the table for comparison. Crack wedging had no effect on 'unpropped' repairs at values of crack wedging below 75%, under the loading regime used. The effect of crack wedging is to reduce the effectiveness of propping as a means of destressing a member for repair.

CONCLUSIONS

An analysis of the behaviour of repaired members suggests that the ultimate bending capacity of a beam will not be reduced by failure to relieve dead load on the beam. Total deflection will increase if dead load is not relieved although the stiffness of the repaired beam does not reduce. The bending moment at onset of yield of reinforcement will reduce and crack widths will increase.

Experimental work is being carried out to verify the analytical results reported here.

REFERENCES

1. Lewis, D. A. and Boam, K. J., Cathodic protection of reinforced concrete. *Proc. Conf. on Deterioration and Repair of Reinforced Concrete in the Arabian Gulf*, October 1987, Vol. 1, pp. 79–98 and discussion; Vol. 2, 1988, pp. 38–9.
2. Hodgkiess, T. and Arthur, P. D., Fatigue and corrosion effects in reinforced concrete beams partially submerged in seawater and subjected to reverse bending. Offshore Technology Report, OTH 87242, HMSO, London.
3. Cairns, J., Analysis of structurally repaired reinforced concrete beams. *Proc. Conf. on Deterioration and Repair of Reinforced Concrete in the Arabian Gulf*, October 1989.
4. BS 8110, The Structural Use of Concrete. British Standards Institution, London, 1985.
5. Beeby, A. W., Corrosion of reinforcing steel in concrete and its relation to cracking. *The Structural Engineer*, **56A**(3) (March 1978) 77–81.

57

An Investigation into the Effectiveness of Silane for Reducing Corrosion Activity in a Chloride-Contaminated Reinforced Concrete Bridge Structure

GUY P. HAMMERSLEY, MICHAEL J. DILL
Laing Technology Group Limited,
Page Street, Mill Hill, London NW7 2ER, UK

and

JOHN J. DARBY
Oxfordshire County Council,
Speedwell House, Speedwell Street, Oxford OX1 1NE, UK

ABSTRACT

An investigation was undertaken on the A34 Wolvercote Viaduct into the effectiveness of silane for reducing chloride-induced corrosion activity. Concrete patch repairs were undertaken in areas of cracking and spalling on two crossbeams, but visibly intact chloride-contaminated concrete was left in place. The upper surface of the beams was waterproofed and the sides and soffits treated with silane. Permanent Ag/AgCl reference electrodes and resistivity electrodes were installed to monitor the performance of the remedial measures. Preliminary results after 1 year show a general small increase in resistivity and reduction in electrode potentials, indicating that some drying out of the concrete has occurred.

INTRODUCTION

In concrete bridge structures suffering from reinforcement corrosion induced by the penetration of deicing salt, the removal and replacement of all severely chloride-contaminated concrete is often difficult, structurally or technically undesirable and economically unviable. Of the alternatives, one possible approach is to repair only cracked or spalled concrete and leave visibly undamaged but chloride-contaminated concrete in place, relying on a surface treatment to maintain the concrete at a sufficiently low moisture content to minimise corrosion activity. The treatment used should not only prevent the ingress of further water and chlorides but should permit water vapour to pass out, hence allowing the concrete to dry out. Water-repellant surface impregnations such as silane have been promoted as having the desired qualities, but there is a lack of published data on their use in this particular remedial application.

A trial was undertaken[1] with the objective of investigating the effectiveness of this remedial treatment for reducing corrosion activity in a typical chloride-contaminated highway bridge structure. The trial comprised three stages:

1. Survey.
2. Remedial work and installation of instruments.
3. Monitoring.

DESCRIPTION OF STRUCTURE

The structure chosen was the Wolvercote Viaduct, which carries the A34 Oxford Ring Road over the A40, the Oxford Canal, a British Rail main line and a stream. The viaduct, completed in 1962, consists of two separate structures carrying the north- and southbound carriageways, each

FIG. 1. General view of the Wolvercote Viaduct.

comprising 12 spans consisting of six precast post-tensioned beams supporting a cast in-situ reinforced concrete deck slab. The spans are supported by trestle piers (Fig. 1). The trial was carried out on the crossbeams of piers 7 and 8 beneath the northbound carriageway.

SURVEY

The survey was undertaken during May 1988 and consisted of the following:

1. Visual inspection and hammer sounding.
2. Cover survey.
3. Half-cell potential mapping using a silver/silver chloride (SSCE) reference electrode.
4. Sampling by drilling and analysis of the dust for chloride content.
5. Carbonation testing.

The survey revealed the local deep penetration of chloride ions in areas of the beams subjected to ponding or run-down of water leaking down the deck joint above. The original design detail consisted of a small dam around the top of the beams with a number of short drainage spouts. Inadequate falls and blocked spouts resulted in considerable ponding of salt-contaminated

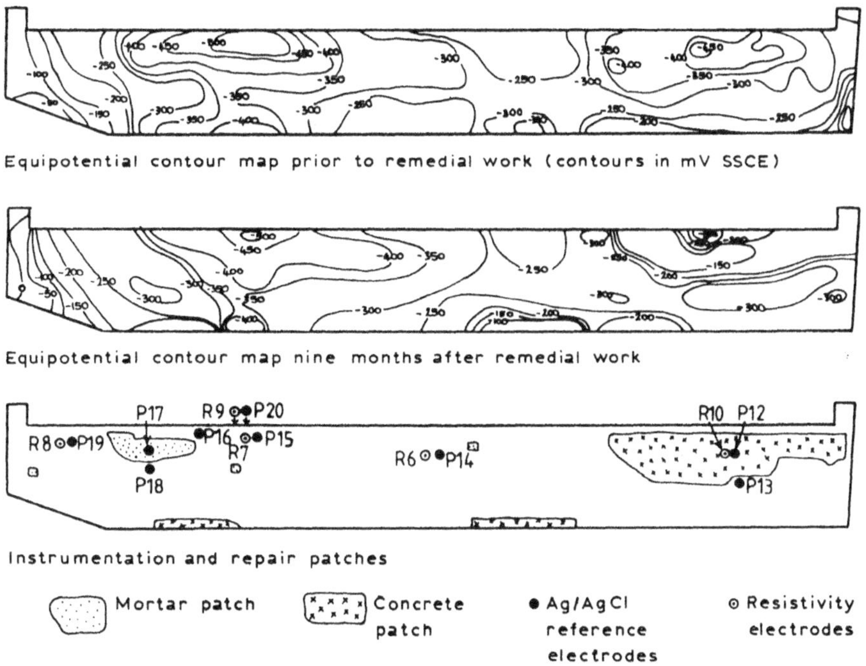

FIG. 2. South face of pier 8 crossbeam.

FIG. 3. Large breakout on the south face of pier 8 crossbeam showing localised pitting corrosion of the reinforcement.

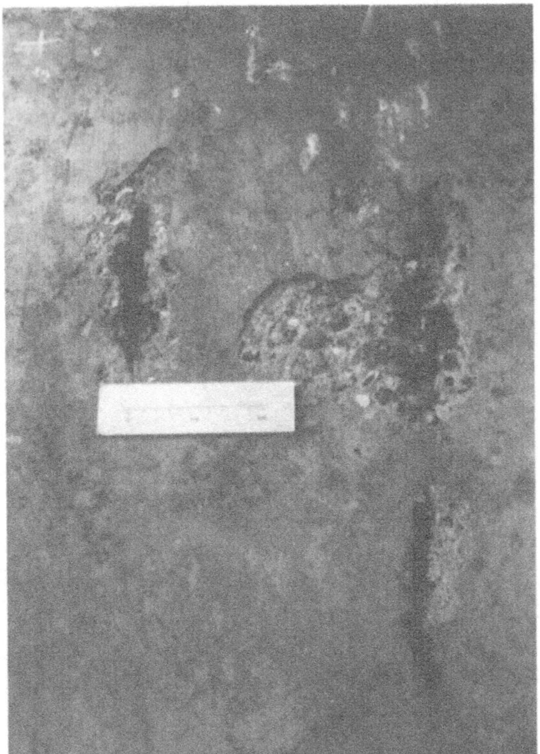

FIG. 4. Small spalls over corroding reinforcing bars with low cover. Note the location of the drainage spout above the spalls.

water on top of the beams. Water emitted from spouts ran down the sides of the beams. In affected areas chloride contents in the medium to high category of BRE Digest 264[2] were frequently present at the depth of the reinforcement (30–60 mm). The chloride contents reduced with depth into the concrete, confirming the external source.

Half-cell potential mapping (Fig. 2) showed a wide variation in electrode potential around the beams with a number of localised 'high spots' characterised by steep potential gradients and electrode potentials reaching −540 mV (SSCE). These 'high spots' could be correlated with areas of medium to high chloride contents and subsequent breakouts revealed pitting corrosion (Fig. 3). In areas of lower cover (less than about 20 mm) corrosion inevitably resulted in spalling (Fig. 4). However, where the cover was higher the pitting corrosion did not always result in surface damage, and was only revealed by breakouts.

The original quality of the concrete was judged to be reasonable, as evidenced by generally low carbonation depths of less than about 5 mm.

REMEDIAL WORK

The remedial work was undertaken between June and August 1988, and consisted of the following principal operations:

1. Cutting out cracked and spalled concrete, and grit blasting reinforcement.
2. Repairing the above areas with proprietary cementitious repair mortar or concrete appropriate to the size of the breakout.
3. Cutting away of the dam around the top of the beams and removal of the existing drainage spouts.
4. Steam cleaning of concrete surfaces.
5. Application of silane to the sides and soffits of the beams.
6. Application of an acrylic waterproof coating to the upper surface of the beams.
7. Fixing flashings, gutters and downpipes around the top of the beams.

All the materials used complied with Department of Transport Standard BD 27/86[3] or the Department of Transport Specification for Highway Works,[4] as appropriate. The materials are detailed in the Appendix.

INSTRUMENTATION

In order to monitor the effectiveness of the remedial measures, the instruments were installed in the two beams during the course of the

TABLE 1
Instruments installed in the beams

	Host material						Total
	Original concrete		Repair concrete		Repair mortar		
Pier	7	8	7	8	7	8	
Silver/silver chloride reference electrodes	5	10	–	1	2	2	20
Resistivity electrodes	3	5	–	1	1	–	10

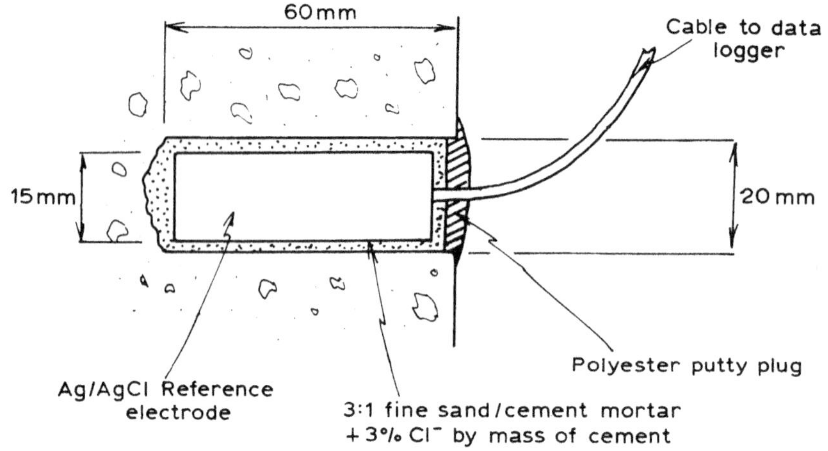

FIG. 5. Detail of half-cell potential reference electrode.

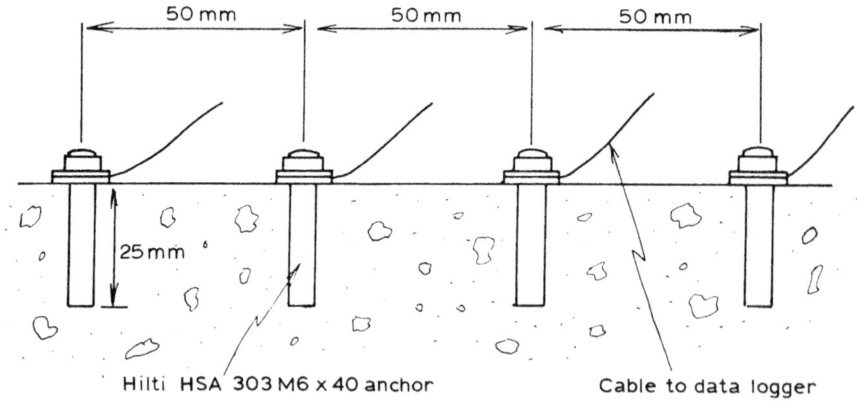

FIG. 6. Detail of resistivity electrodes.

FIG. 7. Part of the south face of pier 8 crossbeam showing the large concrete repair patch, new drainage gutter and permanent instrumentation consisting of 2Ag/AgCl reference electrodes and one set of resistivity electrodes (covered by the plastic trunking).

remedial work (Table 1). In addition, the air temperature and relative humidity were monitored in the vicinity of the beams.

The distribution of the instruments on the south face of pier 8 is shown in Fig. 2 and details of their installation are illustrated in Figs 5 and 6. Figure 7 shows part of the south face of pier 8 on completion of the remedial works and instrumentation.

The instruments were monitored by two battery-powered data loggers, one being dedicated to each beam. The loggers were located in traffic counter boxes on top of the viaduct on the narrow pavement between the parapet fence and Armco barrier.

HALF-CELL POTENTIAL SURFACE MAPPING

In addition to the fixed instruments, it was decided to take available opportunities to carry out a half-cell potential mapping exercise on the south face of pier 8. As it was uncertain whether half-cell potential measurements could be taken satisfactorily through a silane impregnated surface, measurements were made on a fixed grid shortly before and after the silane treatment. This revealed that the silane itself did not significantly affect the measurement of half-cell potentials.

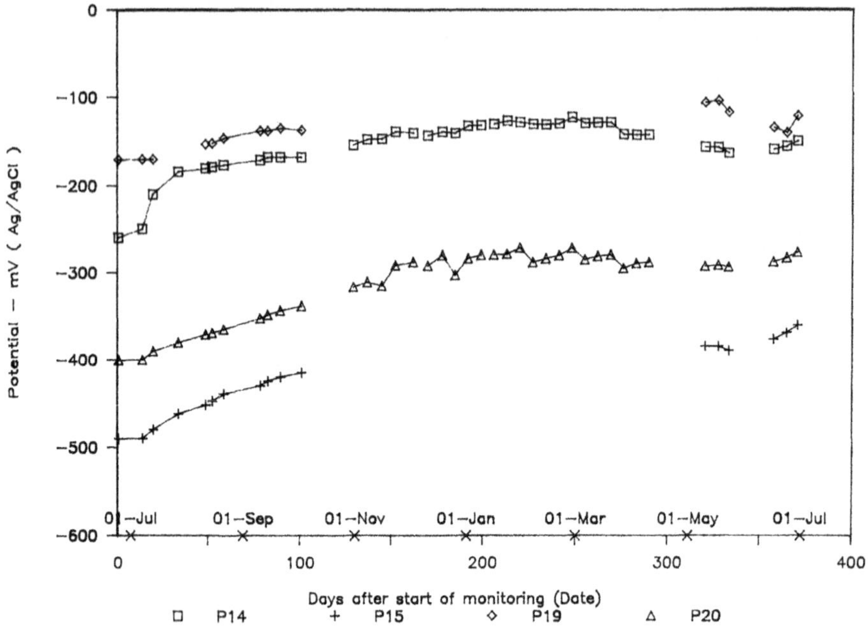

FIG. 8. Average weekly electrochemical potential measurements for selected electrodes (pier 8).

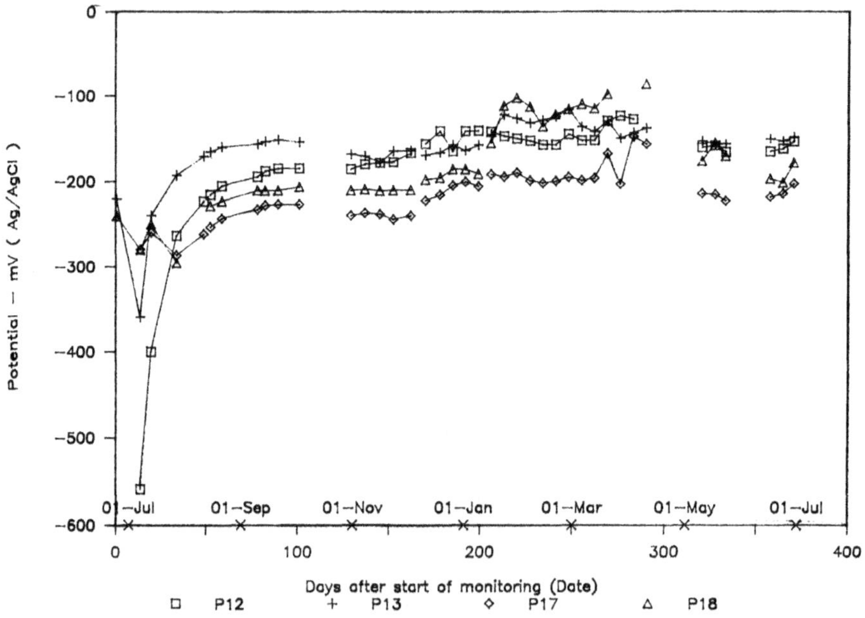

FIG. 9. Average weekly electrochemical potential measurements for selected electrodes (pier 8).

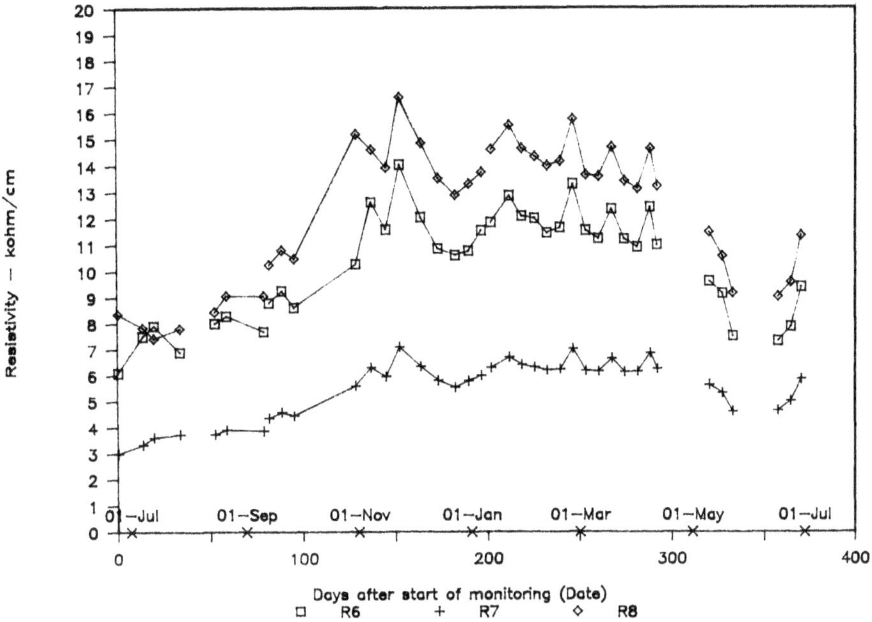

FIG. 10. Average weekly resistivity measurements for selected electrodes (pier 8).

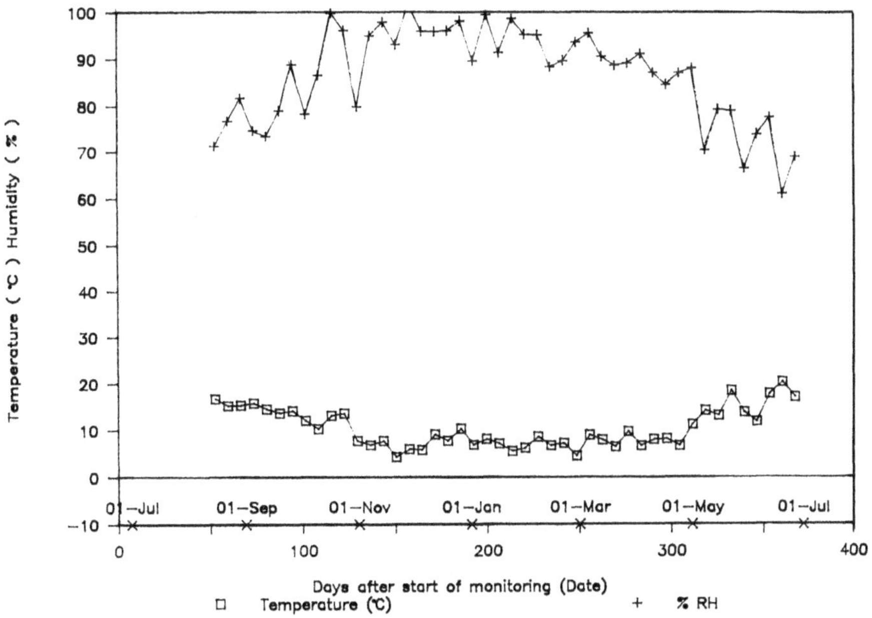

FIG. 11. Average weekly temperature and relative humidity.

PRELIMINARY EVALUATION OF RESULTS

Equipotential Contour Mapping

In those areas where patch repairs were undertaken the electrode potential in the new patch materials, as might be expected, were appreciably lower than in the original concrete replaced. This can be clearly seen in Fig. 2 by comparing the equipotential contour maps before and after the remedial work. Electrode potentials have generally reduced slightly (become less negative) since completion of the remedial work. Some areas immediately adjacent to repair patches are now local potential 'high spots' and consequently may become future anodic corrosion sites, as is often observed adjacent to patch repairs in chloride-contaminated concrete.

Permanent Instrumentation

A preliminary evaluation of the permanent instruments after approximately 1 year from completion of the remedial work has revealed a reduction in electrode potentials for all but one of the reference electrodes of between 50 and 150 mV. Electrode P12, installed in a large concrete repair patch on pier 8, has reduced from −560 to −150 mV. Plots for individual instruments from the south face of pier 8 are shown in Figs 8 and 9.

Two of the ten resistivity electrode sets have failed. The remaining instruments have revealed a general increase in resistivity of about 2–4 kohm/cm (Fig. 10). The data recovered from the resistivity electrodes have generally exhibited greater fluctuations between readings than those obtained from the silver/silver chloride reference electrodes.

Temperature and relative humidity data are presented in Fig. 11.

DISCUSSION

The observed increase in resistivity is indicative of drying out of the concrete and the general reduction in electrode potentials is probably a consequence of this. Figures 8–11 show that the above effects appear to be independent of the surrounding air temperature and relative humidity with a similar trend during winter as well as summer months.

Although the observed changes in resistivity and electrode potential have not been large and locally there is still a high risk of active corrosion, the preliminary results do at least indicate a trend in the right direction. It was anticipated that any significant drying out of the concrete would only be observed in the long term (years rather than months) and it is too soon to draw any conclusions on the success of the remedial measures in reducing corrosion risk. However, it can be concluded that some drying out of the concrete has already occurred. As a spin-off, useful experience has also been gained in monitoring the effects of full-scale concrete repairs.

ACKNOWLEDGEMENT

The Wolvercote Viaduct is owned by the Department of Transport, who are acknowledged for funding the trial.

REFERENCES

1. LAING TECHNOLOGY GROUP LIMITED, A34 Wolvercote Viaduct, Oxford. Investigation into the effectiveness of silane for the treatment of chloride-contaminated concrete—report on survey, remedial work and instrumentation. Report ref. RD88/74/LS, October 1988.
2. BUILDING RESEARCH ESTABLISHMENT, The durability of steel in concrete. Diagnosis and assessment of corrosion-cracked concrete. BRE Digest 264, 1982.
3. DEPARTMENT OF TRANSPORT, Materials for the repair of concrete highway structures. Departmental Standard BD 27/86, 1986.
4. DEPARTMENT OF TRANSPORT, Specification for highway works, 1986.

APPENDIX

Contract Details

Agent: Oxfordshire County Council
  Remedial design, supervision and monitoring
Consultant: Laing Technology Group Limited
  Remedial design, survey instrumentation and monitoring
Contractor: Hertfordshire Concrete Repairs Limited
  Remedial works

Materials

Replacement concrete (for large repairs)
  SBD Five Star Repair Concrete
Repair mortar (for small repairs)
  SBD Mulsifix Acrylic Repair Mortar DG351
Water repellant (for sides and soffit)
  SBD Silane
Waterproofing (for upper surface)
  Stirling Lloyd Eliminator

Instruments

Half-cell reference electrodes

Ag/AgCl general-purpose reference electrode manufactured by Silvion

Resistivity electrodes

Wenner electrode arrangement using 6 mm diameter Hilti HSA 303M6 × 40 stainless steel anchors

Environmental monitoring

SKH 103 temperature and humidity probe supplied by Campbell Scientific and located in a Gill radiation shield

Data logging

Campbell Scientific CR10 data loggers

58

Bridge Strengthening Using Load Relieving Techniques

BRIAN PRITCHARD

W. S. Atkins Consulting Limited, Transportation Engineering Division, Woodcote Grove, Ashley Road, Epsom, Surrey KT18 5BW, UK

ABSTRACT

The paper describes several new techniques for strengthening existing bridges to withstand increased loading by imposing dead load relief or load sharing. The techniques cover external prestressing, the installation of extra shear connectors and the use of shock transmission units. They all benefit from requiring minimum, if any, traffic disruption.

INTRODUCTION

Strengthening of the world's bridge stock is a growth industry. This is inevitable as the years pass because existing bridges are expected to carry traffic of increasing loading and intensity for which they were not originally designed. The same passage of time also means that existing bridges are increasingly subjected to weakening environmental hazards, ranging from winter de-icing salt to polluted atmosphere carbonation.

Strengthening of an existing bridge may become necessary because of increasingly apparent overloading or because major repairs are required and the opportunity is taken to strengthen the bridge to higher standards while traffic restrictions are in operation. The traffic restriction aspect is usually dominant and often precludes straight bridge replacement. It also strongly influences the method of strengthening and those methods which involve little or no traffic restriction are strongly favoured.

Three recent techniques for strengthening existing bridges by load relief or load sharing with minimum, if any, traffic disruption are described.

DECK BENDING RELIEF BY EXTERNAL PRESTRESSING

General

Conventional prestressing of a bridge deck imposes a permanent direct compression together with a bending moment which counters, or relieves, the applied dead load moments. The two effects can be most beneficial to tension-weak concrete decks and together they allow the prestressed concrete bridge deck to carry further superimposed dead and live load moments without exceeding the permissible bending stresses or load factors.

The bending moment reduction effect of added prestressing can also be used to advantage in relieving dead load bending in existing overloaded decks of reinforced concrete, steel or composite concrete deck/steel girder structures. This dead load bending relief can be sufficient to reduce the deck bending under full dead and live loading to permissible limits. Alternatively, a bridge deck can be upgraded to carry increased superimposed dead and/or live loading.

In general, the direct compression effect of the added prestressing is not helpful. Reinforced concrete allowable compressive stresses are usually lower than with prestressed concrete and extra compression in steel structures can lead to plate stability problems. It is therefore beneficial to mobilise as much of the prestressing bending moment reduction as possible and there is every advantage in locating the prestressing tendons at the beam extremities, or even beyond.

External Prestressing Applied to an Existing Composite Deck

Rakewood Viaduct carries the M62 motorway between Lancashire and Yorkshire across a 36 m deep valley (Fig. 1). The 256 m long six-span

FIG. 1. Rakewood Viaduct.

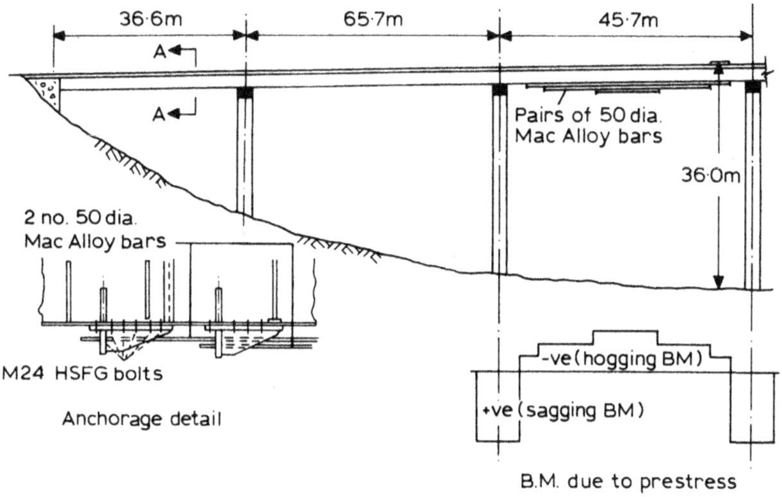

FIG. 2. External prestressing.

continuous deck, completed in 1969, consists of ten 3 m deep steel plate girders carrying and composite with an in-situ reinforced concrete deck slab. DTp (NWRO) required upgrading to cater for a proposed increase in traffic lanes carried and the more onerous requirements of the newly introduced BS 5400 bridge code. The main shortfall was identified as an approximate 40% overloading in the steel girder compression flanges over the piers. Upgrading by 'unloading', using external prestressing, was found to provide an economical strengthening procedure with minimal disruptions on this heavily trafficked motorway. Figures 2 and 3 indicate the strengthening

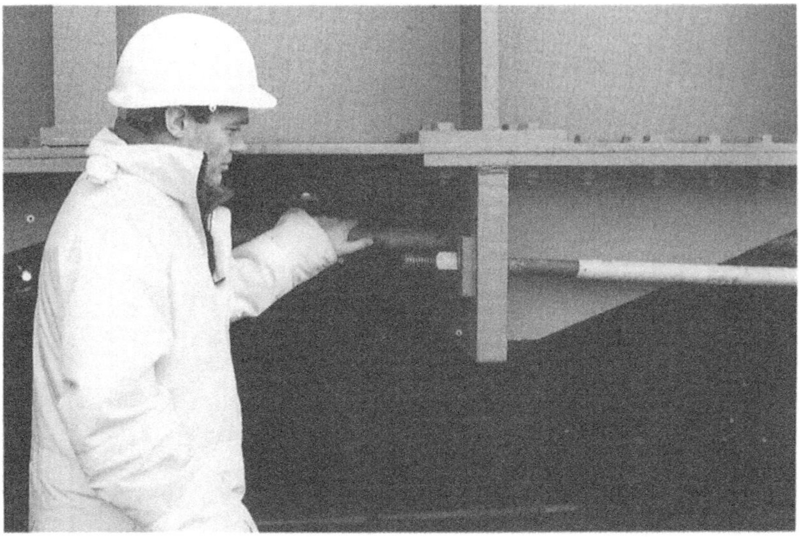

FIG. 3. Prestressing anchorages.

procedure, which first requires the attachment of fabricated steel anchors to the locally stiffened underside of each steel beam bottom flange by HSFG bolting. Three pairs of 50 or 36 mm diameter Macalloy prestressing bars of overlapping lengths are then attached under each flange between piers. Upon stressing, hogging bending is set up in the midspan regions of the beam. However, it is the parasitic sagging moment over the piers, caused by deck continuity, which performs the required 'unloading' to acceptable stress limits in the bottom girder flanges over the piers.

The prestressing anchorages apply high local loading to the in-service deck girders. They are therefore attached at non-critical global bending and shear locations where the additional local stresses can be readily accommodated. The dispersion of these high anchorage loads into the girder flanges and webs and the associated local design had been examined using three-dimensional finite element techniques. Special consideration has also been given to the provision of anti-corrosion protection and intermediate supports to prevent wind vibration of the stressing bars. The completed deck strengthening is shown in Fig. 4.

It so happens that a similar deck unloading procedure is being applied to an understrength three-span composite girder viaduct in Iowa State, USA, this year. Prior experimental work on large-scale models has already been undertaken and covered in several recent papers by Professor F. W. Klaiber and his colleagues at Iowa State University. It has been proposed that both parties undertake and compare monitoring of prestressing bar loads during and after construction.

FIG. 4. Strengthened Rakewood Viaduct.

FIG. 5. External prestressing of reinforced concrete bridge beam.

External Prestressing Applied to a Reinforced Concrete Deck

Figure 5 shows how a similar external prestressing technique was used to 'unload' the rectangular beams of an understrength two-span continuous reinforced concrete deck in South Wales. In this case prestressing was by cables located on the sides of the beams and anchored and deflected by steel assemblies attached by epoxy grouted bolts passing through the beams.

SUBSTRUCTURE TRACTION AND BRAKING LOAD RELIEF USING SHOCK TRANSMISSION UNITS (STUs)

General

A large number of our existing stock of viaducts feature long sequences of simply supported deck spans, often carried on a series of high and substantial piers. This is particularly evident in major river crossings where high navigation clearances require long approach viaducts (Fig. 6). The piers under each simply supported span inevitably carry fixed bearings for one span alongside free bearings for the adjacent span. This means that the design longitudinal traction and braking must be individually applied to each deck span throughout the viaduct. Main resistance is offered by the pier carrying the fixed bearings of that particular span.

Current integrity assessments of a number of these viaducts often indicate that the piers are understrength due to increases in the deck longitudinal

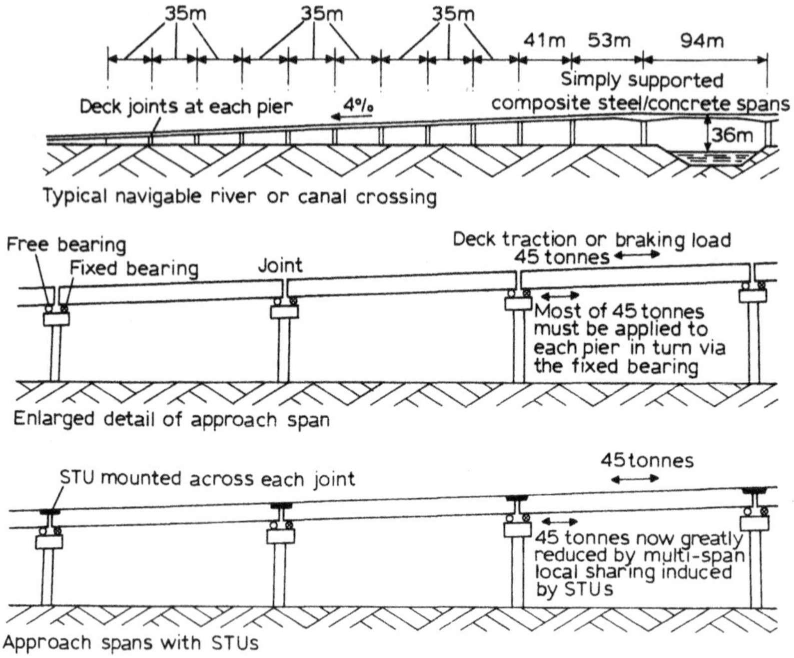

FIG. 6. Substructure strengthening using STUs.

loading since the original design, sometimes accompanied by damage generated by road salt, carbonation or ASR. A substructure of this type with, say, ten equal height piers has a total resistance capacity of approximately ten times the original deck design traction and braking longitudinal loads. This total resistance capacity can be mobilised by providing load transfer STUs across the deck joints.

Shock Transmission Units (STUs)

STUs are mechanisms which are connected across movement joints between structural elements. They transmit slow acting joint movements like temperature and shrinkage with negligible resistance and, when required, transmit momentary impact forces like traction, braking and earthquake with negligible movement.

A simple, economical and minimum maintenance bridge STU was developed in the UK some years ago. Instead of oil the STU utilises the peculiar properties of 'bouncing putty', a silicone compound which will readily deform under slow pressure but becomes rigid under impact. The unit consists of a steel cylinder containing a loose-fitting piston fixed to a transmission rod, the void round the piston being filled with the silicone putty. Under slow movement this putty is squeezed around the piston and displaced from one end of the cylinder to the other (Fig. 8).

FIG. 7. Docklands Light Railway strengthening and Canary Wharf building in progress.

Load Relief Using STUs for Viaduct Piers of the London Docklands Light Railway

The newly completed viaducts carrying London's Docklands Light Railway (Fig. 7) were designed for a train service which, due to a breath-taking increase in adjacent development at Canary Wharf, will now require considerable expansion before 1990. This will mean heavier and

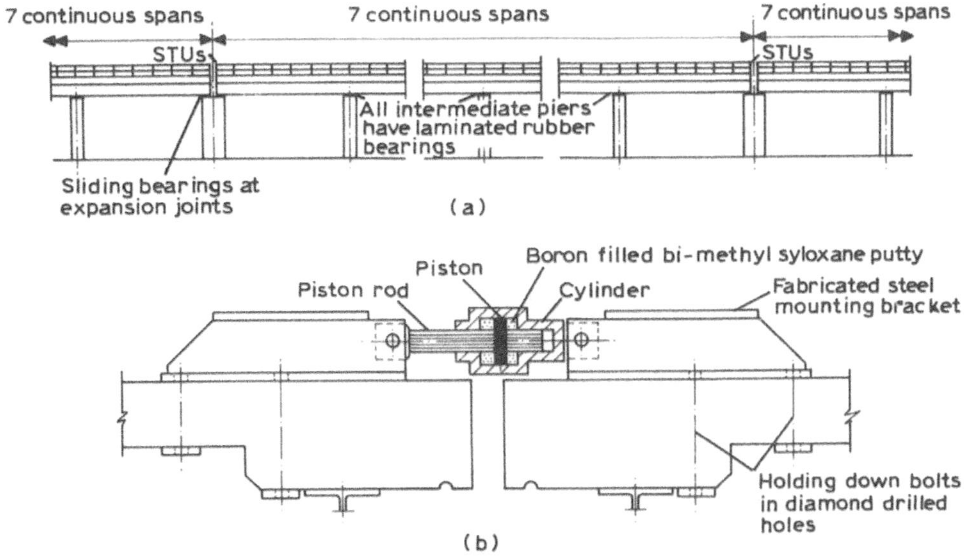

FIG. 8. Detail of STU and mounting.

more frequent trains, which will add braking and traction effects in excess of those originally catered for.

Figure 8 shows a typical as-built seven-span deck unit continuous between expansion joints. Train traction and braking loads are currently shared among the slender piers, which generally support the deck via rubber bearings. STUs are being installed at rail level between joints such that, when the new increased longitudinal traction and braking loading is applied to one particular seven-span unit, load is beneficially transmitted and shared with adjacent seven-span decks sufficient to require no pier and foundation strengthening in any substructure. This simple procedure represents a tremendous saving in cost and interference with the existing train service.

COMPOSITE DECK FATIGUE RELIEF USING ADDITIONAL SHEAR CONNECTORS

General

Existing composite bridge decks, like the two described earlier, often require strengthening or fatigue life enhancement of the shear connection between the concrete deck slab and steel girders. This can be undertaken by installing additional new shear connectors, ideally from the underside of the slab/girder interface to minimise traffic interference.

Fatigue Life Enhancement of Viaduct Decks of the London Docklands Light Railway

The existing new viaduct decks, completed in 1987, are generally of continuous composite construction with an in-situ reinforced concrete deck slab supported by and composite with twin steel universal or plate girders.

The original design of the decks to BS 5400 established that fatigue considerations were a critical factor, particularly in the deck shear connectors. As a result of the increase in weight and frequency of trains after 1990, mentioned earlier, the fatigue life would suffer considerable reduction. Strengthening measures to restore the fatigue life back to the originally designed 120 years were required. Additional shear connectors installed between the original 19 mm welded stud connectors would relieve the loads on these connectors sufficiently to accomplish this.

Technically, the easiest solution was to drill out holes through the deck slabs from above and install new stud connectors, in clusters of three or four, to optimise the hole size and economies of installation. Practically, this would be extremely disruptive and costly to undertake with a live service in operation.

The provision of new shear connectors by drilling-in from under the top

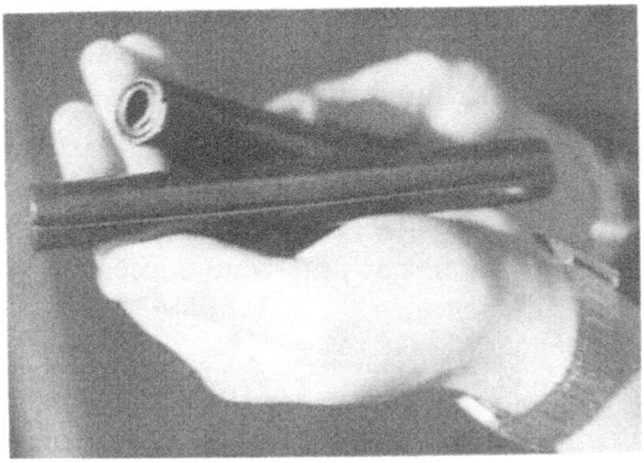

FIG. 9. 20 mm spring steel pins.

flange of the steel girders was also examined. Several types of connectors were considered, including 20 mm diameter spring steel pin fasteners. These offered the advantage of a readily achieved force fit into the hole drilled through the steel flange and lower section of the concrete deck slab with no requirement for grouting, glueing or welding.

The benefit of causing no interference to the train service had to be balanced against the unknown shear and fatigue parameters of such a previously untried system. Strength and fatigue testing were carried out on push-out samples by the Welding Institute at Cambridge. Two types of 20 mm spring pin were shown to have superior strength and fatigue properties to the 19 mm studs. These are shown in Fig. 9.

FIG. 10. Jacking-in 'Spirol' spring pins.

Both pins obtain their force fit by jacking the lead-in chamfer into drilled holes with slightly smaller diameters (Fig. 10). In one, the pin slot closes as the pin diameter reduces during driving. In the other, the spring mechanism is generated by the compression of a $2\frac{1}{2}$ turn spirally coiled strip of steel. Good interface shear connection is established with a degree of pull-out resistance afforded by the spring-loaded friction between the pin and the hole face. In the event, the spiral pins were successfully installed with no interruption to the train services.

59

The Integrated Construction and Conversion of Single and Multiple Span Bridges

MARTIN P. BURKE JR

Burgess & Niple Limited, 5085 Reed Road, Columbus, Ohio 43220, USA

ABSTRACT

In the United States and Canada, integrated bridge construction is becoming one of the bridge engineer's primary responses to joint related bridge damage caused by the use of deicing chemicals and the restrained growth of rigid pavements. The relative success that has been experienced with integral bridges—bridges without deck joints—is now being reflected, not only in the increasing numbers of longer integral bridges but also in the integral conversion of existing jointed bridges. As revealed in a recent survey made during the development of the National Cooperative Highway Research Program's Synthesis 'Bridge Deck Joints', the use of integral conversion techniques for existing jointed bridges is generating widespread interest but as yet only a moderate number of tentative applications. It appears that the initial success of such techniques will see an accelerated use of integrated conversion as an effective alternative to bridge joint rehabilitation.

INTRODUCTION

Integral bridge construction may be defined as the practice of constructing bridges without deck joints. When using such construction to eliminate intermediate joints in multiple span bridges, it is accepted that the continuity achieved by such construction will subject superstructures to secondary stresses. These stresses are due to the response of continuous superstructures to thermal and moisture changes, and gradients, settlement of substructures, post-tensioning, etc. When such construction is used to eliminate deck joints at abutments, it is likewise accepted that such structures will, in addition, be subjected to secondary stresses due to restraint provided by abutment foundations and backfill against the cyclic movement of bridge superstructures. The justification for such construction is based on the recognition that for small and medium span bridges of moderate lengths significantly more damage and distress has been caused by the use of deck joints than by the secondary stresses they were intended to prevent. In addition, elimination of costly joints and bearings, and the details and procedures necessary to permit their use, generally result in more economical bridges. Consequently, more bridge engineers are now willing to relinquish some of their control of secondary stresses, primarily to achieve simpler and less expensive bridges with greater overall integrity and durability.

CONTINUOUS SUPERSTRUCTURES

Current design trends received their primary impetus and direction almost six decades ago. In May 1930, a brief ten-page paper published in the *Proceedings of the American Society of Civil Engineers* generated considerable discussion in academia. It also created a minor revolution in the design and construction of small and medium span bridges. The paper, 'Analysis of continuous frames by distributing fixed-end moments' by Hardy Cross,[1] presented a simple and quick method for the analysis of integral type structures such as continuous beams and frames. The method was quickly adopted by bridge engineers and the bridge practice of many transportation departments began to change. Prior to Hardy Cross' moment distribution, multiple span bridges were generally constructed as a series of simple spans. Following the introduction of moment distribution, bridge engineers began eliminating troublesome deck joints at piers by providing continuous superstructures.

Line A in Fig. 1 shows the beginning of the routine use of continuous construction in the United States and Canada, and the per decade increase in the number of transportation departments that have adopted the use of continuous construction. As shown, 26 of 30 departments responding to a recent mail survey,[2] or 87% of responding departments, now routinely use continuous construction for short and medium span bridges.

Currently the state of Tennessee appears to be leading the way in constructing long continuous bridges. For example, the Long Island Bridge at Kingsport, Tennessee, was constructed in 1980 with 29 continuous spans without a single intermediate transverse joint. The total length of this bridge is about 2700 ft center to center of abutment bearings. Deck joints and movable bearings have been furnished, but only at the two abutments. It has been aptly named 'The Champ'.

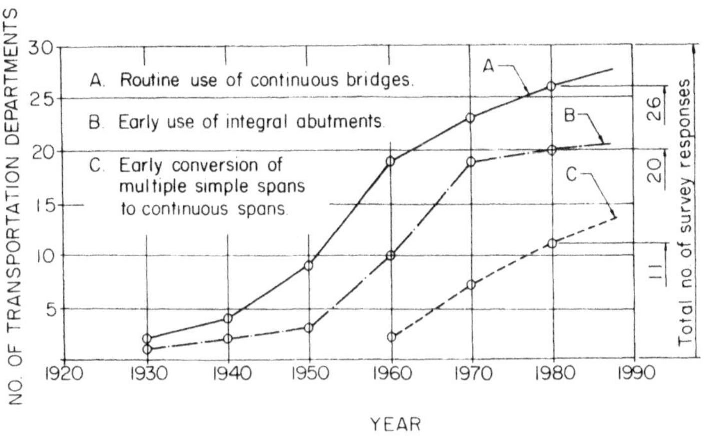

FIG. 1. Design trends for continuous bridges.

Approximately 50% of transportation departments in the United States and Canada limit maximum bridge length without intermediate transverse deck joints to 600 ft or less. However, almost half of the departments responding to a recent mail survey[2] allow the unjointed bridge length to exceed 1000 ft.

INTEGRAL BRIDGES

During the past 2–3 decades, many bridge engineers have become acutely aware of the relative performance of bridges built with deck joints at abutments and those built without them. In most respects, bridges without joints, integral bridges, have performed more effectively since they remain in service for longer periods of time with only moderate maintenance and occasional repairs. Some of this experience was forced upon bridge engineers by circumstances beyond their control.

Due to the growth and pressure generated by jointed rigid pavements, many bridges built with deck joints at abutments have been and are being severely damaged. Many abutment backwalls have been fractured. Other abutments have been split from top to bottom. In longer bridges with intermediate deck joints, piers have been cracked and fractured as well.

In geographical areas with low seasonal temperatures and an abundance of snow and freezing rain, the use of de-icing chemicals to maintain dry pavements throughout the winter season has also had a significantly adverse effect on the durability and integrity of bridges built with deck joints. Open joints and sliding plate joints of shorter bridges and open finger joints of longer bridges have allowed deck drainage, contaminated with deicing chemicals, to penetrate below deck surfaces and wash over supporting beams, bearings and bridge seats. The resulting corrosion and deterioration have been so serious that some bridges have collapsed while others have had to be closed to traffic to prevent their collapse. Many bridges have required extensive repair; and most of the bridges that have remained in service have required almost continuous maintenance to counteract the adverse effects of these chemicals. To help minimize or eliminate these corrective efforts, a whole new industry was created.

Beginning in the early 1960s, the first elastomeric compression seals were installed in bridges in the United States to seal deck joints. Since these first installations, numerous types of elastomeric joint seals have been developed and improved in an attempt to achieve a joint seal design that would be both effective and durable. Most designs have been disappointing. Many leaked. Some required more maintenance than the original bridge built without seals. By and large, the many disappointments associated with various types of seals have caused bridge engineers to consider other options.

Costs of various types of bridges showed marked differences. For two bridges built essentially the same, except with one provided with separate abutments and deck joints and the other provided with integral abutments, the jointed bridge was usually more expensive. In addition, bridges with integral abutments suffered only minor damage from pavement pressure, were essentially unaffected by de-icing chemicals, and functioned for extended periods without appreciable maintenance or repair. Consequently, more bridge engineers began to appreciate the merits of integral bridges for short to moderate bridge lengths. Gradually design changes were made and longer integral bridges were built and evaluated. In 1946, Ohio's initial length limitation for its standard continuous concrete slab bridges was 175 ft. In a 1973 study of integral construction,[3] four states responded that they were using steel bridges and 15 states were using concrete bridges in the 201–300-ft range. In a 1982 study of integral abutment bridges,[4] even longer bridges are reported:

> Continuous steel bridges with integral abutments have performed successfully for years in the 300-ft range in such states as North Dakota, South Dakota, and Tennessee. Continuous concrete structures 500 to 600 feet long with integral abutments have been constructed in Kansas, California, Colorado, and Tennessee.

Currently 11 states are building continuous bridges with integral abutments with lengths in the 300-ft range. Missouri and Tennessee report even longer lengths. Missouri reports steel and concrete bridges in lengths of 500 and 600 ft, respectively, while Tennessee reports lengths of 400 and 800 ft for similar bridges. Finally, line B of Fig. 1 shows that 20 of 30 transportation departments, or 67% of survey responding departments, are now using integral construction for continuous bridges.

The attributes of integral bridges have not been achieved without cost. Parts of these bridges operate at very high stresses, stresses that cannot easily be quantified. These stresses are significantly above those permitted by current design specifications. In this respect, bridge engineers have become rather pragmatic. They would rather build cheaper integral bridges and tolerate these higher stresses than build the more expensive jointed bridges with their vulnerability to destructive pavement pressures and deicing chemical deterioration. In a 1985 magazine article,[5] Clellon Loveall, then engineering director for the Tennessee Department of Transportation, reflects this attitude when he writes:

> In Tennessee DOT, a structural engineer can measure his ability by seeing how long a bridge he can design without inserting an expansion joint.... Nearly all our newer (last 20 years) highway bridges up to several hundred feet have been designed with no joints, even at the abutments. If the structure is exceptionally long, we include joints at the abutment but only there.... Joints and bearings are costly to buy and install. Eventually they are likely to allow water and salt to leak down onto the superstructure and pier caps below. Many of our most costly maintenance problems originated with leaky joints. So we go to great lengths to minimize them.

Even though bridge engineers have conditioned themselves to tolerate higher stress levels in integral bridges, occasionally their design control is not sufficient to prevent these high stresses from resulting in structural distress and structural fracture.

STRUCTURAL DISTRESS

Responses to an early survey about the construction of continuous bridges with integral abutments indicated a rather widespread concern by bridge engineers for the potentially high stresses that would be present in longer bridges.[3] This concern, more than any other, appeared to be responsible for the early lack of enthusiasm for using integral abutments for longer continuous bridges (line A, Fig. 1). Although the great majority of bridges with integral abutments perform adequately, many of them operate at high stress levels. For instance, an abutment supported on a single row of piles is considered flexible enough to accommodate longitudinal thermal cycling of the superstructure and dynamic end rotations induced by the movement of vehicular traffic. Nevertheless, the steel piles of such an abutment are routinely subjected to axial and flexural stresses approaching, equaling or exceeding yield stresses.[4,6] Occasionally a combination of circumstances results in visible distress.

Responding to a 1973 survey, a number of bridge engineers said that some wingwalls of their integral abutments had minor cracks.[3] This problem was corrected with the use of more generous wingwall reinforcing steel. Other engineers reported pile cap cracking which appears to have been eliminated by rotating steel H-piles to place the weak axis normal to the direction of bridge movement. In a number of other instances, primarily in cast-in-place construction, columns, stringers and bridge seals have cracked and fractured when integral construction was used without proper allowance for the restraint produced by rigid supports.

Currently precast concrete or prefabricated steel superstructures are generally replacing small cast-in-place bridges in many states and provinces. Consequently, problems associated with initial shrinkage are gradually being eliminated. However, where cast-in-place construction continues to be used, flexibility of substructures remains a critical part of bridge design. For example, in Clellon Loveall's recent Civil Engineering article,[5] he says:

> Structural analysis of our no joint bridges indicates that we should have encountered problems, but we almost never have. Once we tied the stub abutment of a bridge into rock, and the structure cracked near its end, but we were able to repair the bridge and install [a] joint while the bridge was under traffic. The public never knew about it. That was one of few problems.

Development of new forms of construction will be accompanied by instances of structural distress, and this has certainly been true for continuous bridges with integral abutments. However, as is shown by line A of Fig. 1, the increased use of integral abutments suggests that 60% of transportation departments are satisfied with the performance of integral construction and are using such construction in one form or another for longer and longer bridges. With continued care and consideration, the trend shown by line A no doubt will continue.

INTEGRAL BRIDGE DETAILS

Illustrated in Figs 2 and 3 are integral abutment details for six transportation departments. It is probably not accidental that a fair amount of similarity is evident in these designs since structural details from early successful designs are adapted by other bridge engineers for use by their departments. Even though there are similarities, there are also differences which reflect the types of bridges being built and the care and concern being given to the choice and development of specific details. It should also be realized that these sketches are 'bare bones' presentations. They do not reflect other important design aspects such as skew, construction procedures, etc., that are considered in the application of these details for specific bridges. These aspects cannot be illustrated and properly described in a paper as brief as this one. Nevertheless, since these aspects can have a considerable effect on the performance, integrity and durability of integral designs, it is appropriate to mention some of them for those engineers considering such designs for the first time. Passive pressure, pile stresses and cycle control joints are three topics under which these aspects may be considered.

(A) Passive Pressure

To minimize passive pressure developed in abutment backfill by an expanding integral bridge, design engineers have used a number of controls, devices and procedures. Including but not limited to the following, they have: limited bridge length, structure skew and the vertical penetration of abutments into embankments; used select granular backfill and uncompacted backfill; provided approach slabs to prevent vehicular compaction of backfill or to permit the use of backfill voids behind abutments; used embankment benches to shorten wingwalls and used suspended turn-back wingwalls; and they have used semi-integral abutment designs to eliminate passive pressure below bridge seats.

(B) Pile Stresses

Knowing that longitudinal forces in superstructures are somewhat directly related to the resistance of abutment pile foundations to longitudinal movement, design engineers have: limited the foundation of integral bridges to a single row of slender vertical piles; limited the pile types; oriented the weak axis of H-piles normal to the direction of movement; used prebored holes filled with fine granular material for piles; provided an abutment hinge to control pile flexure; limited structure skew; and used semi-integral abutment designs for longer bridges to minimize foundation restraint to longitudinal movement.

(C) Cycle Control Joints

Figure 4 illustrates design details being used by four states to provide for the cyclic movement of integral bridges, and for the cyclic movement of approach slabs where the slabs are attached to integral bridges. A number of other states are using similar details, but most states do not appear to have developed special approach pavement designs to accommodate the cyclic movement of integral bridges.

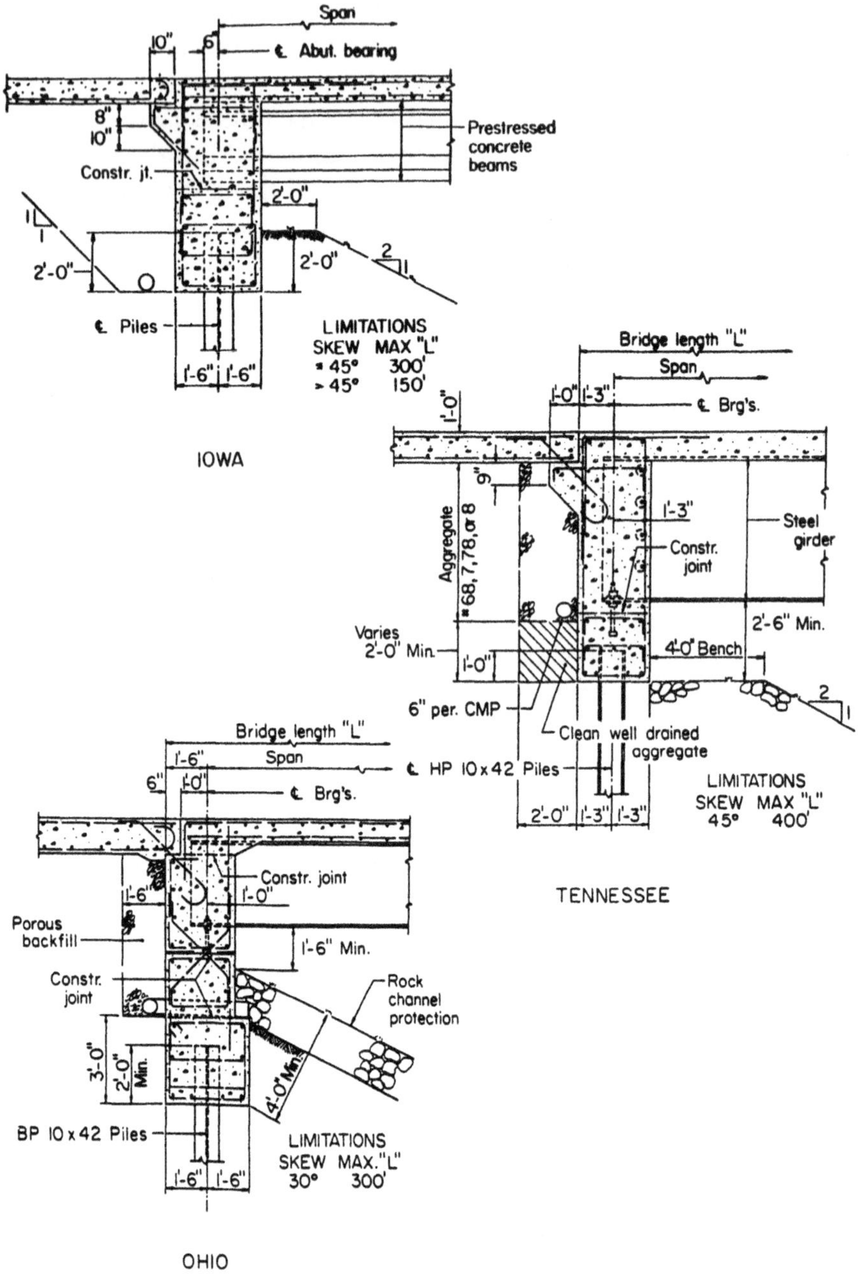

FIG. 2. Integral abutments.

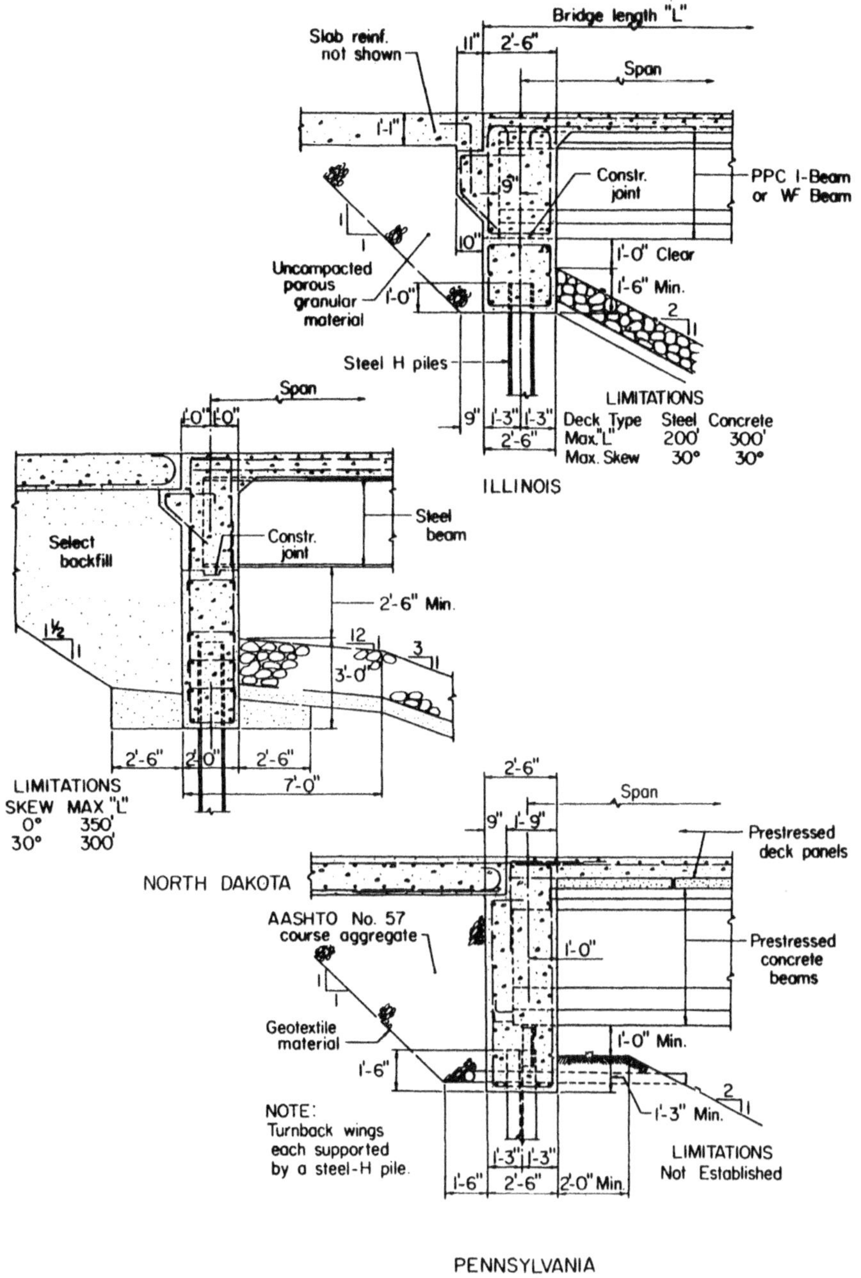

FIG. 3. Integral abutments.

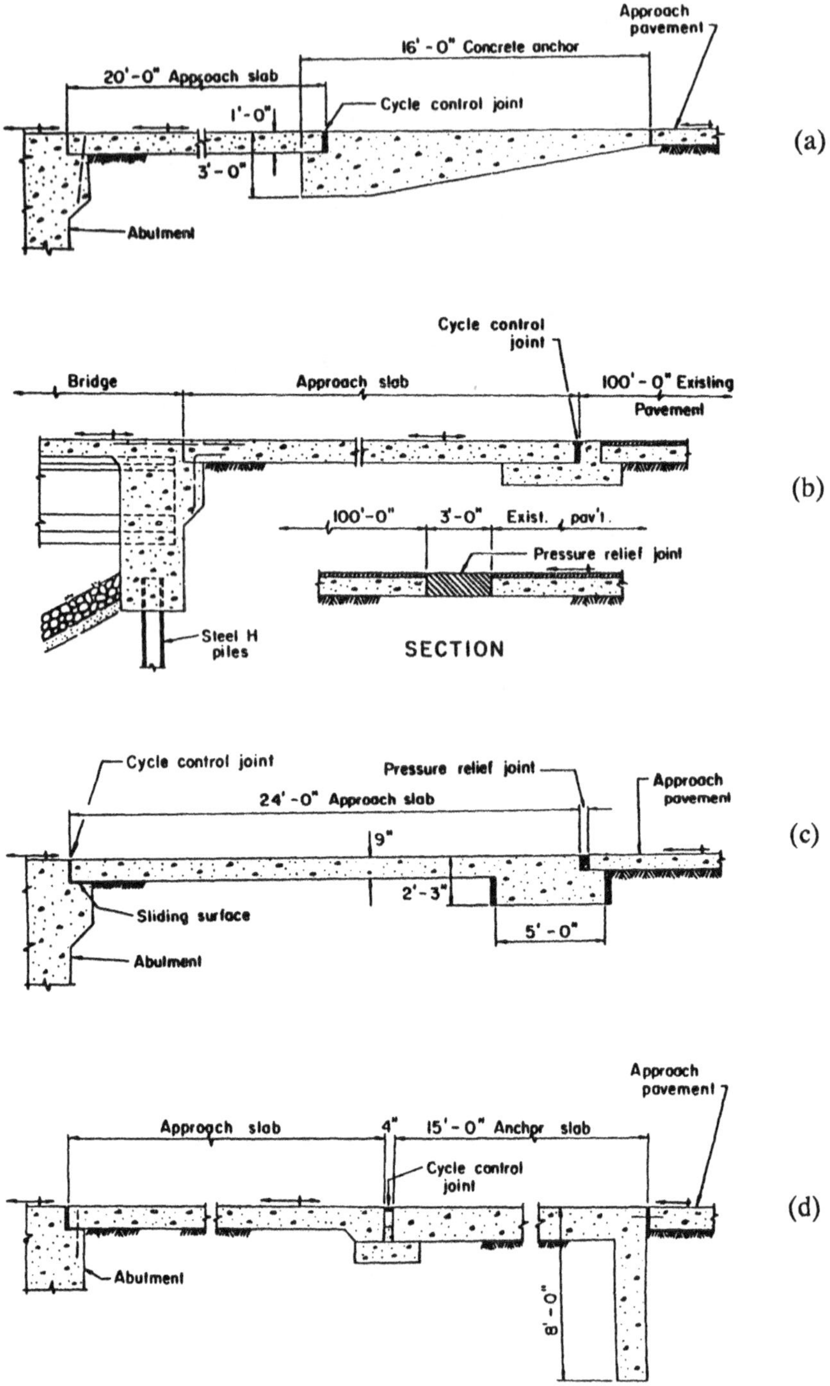

FIG. 4. Integral bridge approaches.

The purposes of all of the designs shown in Fig. 4 are the same although specific details of each design are substantially different. Each recognizes cyclic bridge movement and, where the approach slab is attached to the bridge, cyclic approach slab movement as well. Each contains a reinforced concrete approach slab for spanning the abutment backfill, and each provides for the growth and pressure generated in rigid approach pavements. All but one provides a sealing system for the cycle control joints.

The initial appropriateness of these designs will be reflected in project costs, and in the stress levels developed in response to cycling of bridges and growth of approach pavements. Their actual effectiveness or success will ultimately be measured by their ability to function for long periods under traffic without periodic maintenance or major modifications. An evaluation of these designs and a background for that evaluation is given in Ref. 7. The importance of the bridge/approach pavement interface design is emphasized by a recent decision of the California Department of Transportation to reconsider its use of longer integral bridges. Apparently cycle control joint seal failures have been responsible for substantial erosion of abutment backfill.

A number of questionnaires about integral bridge practices have been circulated in recent years. The responses reflect the policies, attitudes and opinions of those engineers responsible for bridge design policies. They also show how some of those attitudes and opinions have changed during the last decade.[2–4,8,9] References 4, 8 and 9 also contain valuable bibliographies for those interested in a more in-depth study of current research on integral bridge behavior and abutment piling performance.

INTEGRAL CONVERSIONS (RETROFIT)

Following the trend toward the use of continuous construction and the use of integral abutments, as illustrated by lines A and B of Fig. 1, transportation departments are also beginning to convert existing multiple span bridges from simple to continuous spans. Line C shows that this effort began in the 1960s and has gathered strength in the past two decades. Presently 11 of 30 departments, or about 37% of the transportation departments, have converted one or more bridges from multiple simple spans to continuous spans.

Although line C of Fig. 1 suggests considerable activity, it actually shows only the relative number of departments that have made such conversions. It is not indicative of the number of bridges that have been converted. For example, when asked the question 'In recent years, have you converted any bridges from multiple simple spans to continuous spans to eliminate intermediate deck joints?', positive responses were received from only two departments. The Ontario Ministry of Transportation and Communications responded:

> We are modifying a few structures from simple spans to continuous spans, eliminating the intermediate deck joints in the process...

The Texas Department of Highways and Public Transportation responded:

> In recent years, we have eliminated numerous intermediate joints. Generally, this is done while replacing the slab. We simply place the slab continuous across the bents. On a few occasions, we have removed only the joint and surrounding deck area, added reinforcing, and replaced that portion of the deck, thus tying the adjacent spans together.

The Tennessee Department of Transportation also has been actively converting simple span bridges to continuous spans. To describe some of this work, Edward Wasserman, the engineering director of structures, presented a recent paper on 'Jointless bridges' in which he describes and illustrates a number of such conversions.[10]

To give this movement some direction, the Federal Highway Administration has issued a Technical Advisory on the subject.[11] That advisory in part recommends that a study of the bridge layout and existing joints be made '...to determine which joints can be eliminated and what modifications are necessary to revamp those that remain to provide an adequate functional system.... [Where] feasible, develop continuity in the deck slab. Remove concrete as necessary to eliminate existing armoring, and add negative moment steel at the level of existing top-deck steel sufficient to resist transverse cracking [Figure 5a]'.

The detail shown in Fig. 5(a) reflects the procedure described by Texas. Note that the detail shows only the slab portion of the deck is being made continuous. The simply supported beams remain simply supported. For such a construction it is important to ensure that one or both of the adjacent bearings supporting the beams at a joint are capable of allowing horizontal movement. Providing for such movement will prevent horizontal forces from being imposed on bearings due to rotation of the beams and continuity of the slab.

The state of Utah also has converted some simple span bridges to continuous ones by using a design similar to the one shown in Fig. 5(b). For deck slabs with a bituminous overlay, a waterproofing membrane can be used to waterproof the new slab section over the piers. With a design like this, it is understood that the deck slab would be exposed to longitudinal flexure due to rotation of the beam ends responding to the movement of vehicular traffic. However, for short and medium span bridges, the deck

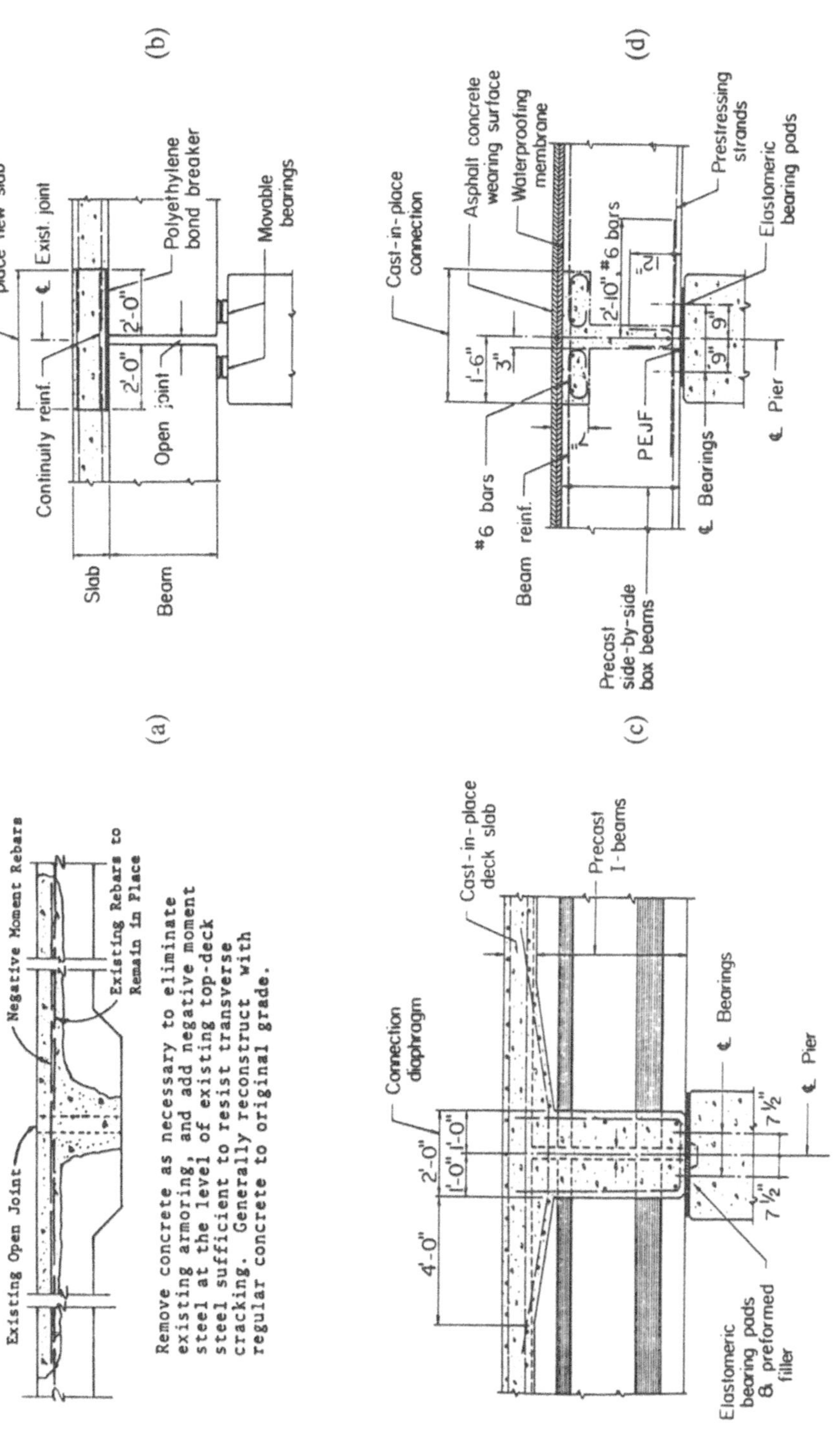

FIG. 5. Integral conversions at piers.

cracking associated with such behavior is preferred by some over the long-term adverse consequences associated with an open joint or a poorly executed sealed joint.

In a new construction, conversion of simple spans to continuous spans is rather commonplace. Figure 5(c) shows the design detail used by the state of Wisconsin for prestressed I-beam bridges. A substantial concrete diaphragm is placed at the piers between the ends of simply supported prestressed beams of adjacent spans. It extends transversely between parallel beam lines. Then a reinforced concrete deck slab is placed to integrate the beams and deck slab, thereby providing a fully composite continuous structure. This type of prestressed I-beam construction appears to be standard for many transportation departments.

Figure 5(d) shows the standard design detail used by the state of Ohio to achieve continuity for simply supported prestressed box beams. These box beams are placed side by side and then transversely bolted together. Finally, continuity reinforcement is placed and the concrete closure placement is made.

A 1969 paper by Clifford L. Freyermuth gives a rather complete description of the considerations necessary to achieve continuity in a bridge composed of a continuously reinforced concrete deck slab on simply supported precast prestressed beams.[12] Conversion of existing bridges either by a complete deck replacement or by replacing portions of the deck adjacent to deck joints over piers can be accomplished by following the procedures developed for new structures. Obviously, for existing bridges, creep effects will be negligible. Shrinkage effects for other than complete deck slab replacements should also be negligible. Such continuous conversion not only eliminates troublesome deck joints but the continuity achieved also results in a slightly higher bridge load capacity since positive moments due to live load are reduced by continuous rather than simple beam behavior.

The details and methods described above provide either partial or complete continuous behavior for live loads and superimposed dead load. If justified, continuity and composite behavior can be achieved for all loads by providing temporary intermediate supports which are then removed after all of the structural elements have been completed.

Although too recent to consider in terms of a design trend, conversion of nonintegral to integral or semi-integral abutments for both single and multiple span bridges has begun. Figure 6 illustrates design details used for two recent conversions by the Ohio Department of Transportation. Reconstruction of these abutments was made necessary by the substantial damage induced by pavement growth and pressure, by de-icing chemical deterioration, or by both. Instead of replacing backwalls and joints, and in some cases bearings and bridge seats as well, it was decided to pattern the

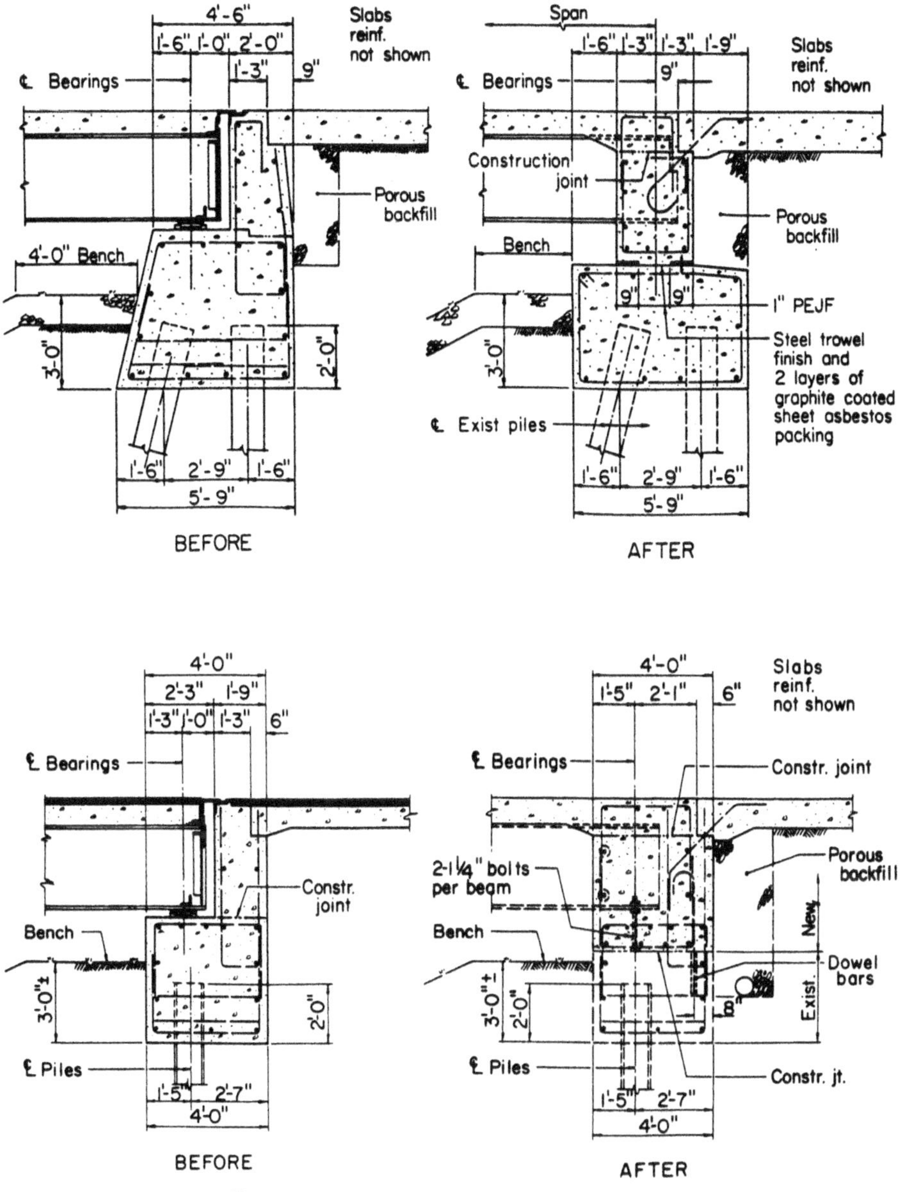

FIG. 6. Integral conversions at abutments.

reconstruction after the design details used by the department for its new integral bridges. In this way subsequent concern about the effects of pavement pressure and de-icing chemical deterioration has been minimized.

Lastly, a number of transportation departments have begun to retrofit multiple span steel beam or girder bridges constructed with intermediate hinges under unsealed deck joints. For one such example, end span hinges and deck joints (originally intended to accommodate embankment consolidation and abutment settlement) are being replaced with bolted splices and a continuous concrete deck. Since this particular structure is 20 years old, the embankments are essentially fully consolidated and the original justification for deck hinges no longer exists.

SUMMARY

As the trends shown in Fig. 1 continue, it appears that the use of continuous construction for multiple span bridges will become standard for all transportation departments in the very near future. It also appears that the use of integral abutments for single and multiple span bridges will increase when comprehensive and conservative guidelines for their use become more readily available, and when their long-term performance is more fully documented.

Since design and construction of fully continuous bridges has become routine, and continuous conversion of simple spans in new construction is becoming more commonplace, it is surprising that similar conversion techniques are not used more often to convert existing bridges to continuous bridges. Presumably the next decade or two will see a burgeoning in retrofitting simple multiple span bridges to continuous bridges (line C, Fig. 1) and from nonintegral to integral abutments. When more information on the operating stress levels of integral bridges is developed, and when more fully described design details and procedures for integral conversions become available, bridge engineers will be able to more fully justify their consideration of such construction. Until then much intuition and prudent judgement will continue to be used to ensure that integral construction and conversion techniques will provide the service life needed to justify their adoption and continued use.

REFERENCES

1. CROSS, H., Analysis of continuous frames by distributing fixed-end moments. *Proc. Am. Soc. Civ. Engrs* (May 1930).
2. BURKE, M. P. JR, *Bridge Deck Joints*. National Cooperative Highway Research Program Synthesis of Highway Practice, Transportation Research Board (publication forthcoming).
3. EMANUAL, J. H., HULSEY, J. L., BEST, J. L., SENNE, J. H. and THOMPSON, L. E., *Current Design Practice for Bridge Superstructures Connected to Flexible Substructures*. University of Missouri-Rolla, Rolla, Missouri, 1973.
4. WOLDE-TINSAE, A. M., GREIMANN, L. F. and YANG, P. S., *Nonlinear Pile Behavior in Integral Abutment Bridges*. Iowa State University, Ames, Iowa, 1982.

5. Loveall, C. L., Jointless bridge decks. In *Civil Engineering*. American Society of Civil Engineers, New York, November 1985.
6. Jorgenson, J. L., *Behavior of Abutment Piles in an Integral Abutment Bridge.* Transportation Research Record 903, Transportation Research Board, Washington, DC, 1983.
7. Burke, M. P. Jr, *Bridge Approach Pavements, Integral Bridges and Cycle Control Joints.* Transportation Research Record 1113, Transportation Research Board, Washington, DC, 1987.
8. Greimann, L. F., Wolde-Tinsae, A. M. and Yang, P. S., *Skewed Bridges with Integral Abutments.* Transportation Research Record 903, Transportation Research Board, Washington, DC, 1983.
9. Wolde-Tinsae, A. M. and Klinger, J. E., Integral abutment bridge design and construction. Report No. FHWA/MD-87/04, Maryland Department of Transportation, 1987.
10. Wasserman, E., Jointless bridges. *Engineering Journal*, **24**(3) (1987). American Institute of Steel Construction, Inc.
11. Federal Highway Administration Technical Advisory T5140.16, *Bridge Deck Joint Rehabilitation (Retrofit).* US Department of Transportation, 1980.
12. Freyermuth, C. L., Design of continuous highway bridges with precast prestressed concrete girders. *ACI Journal*, **14**(2) (1969). Prestressed Concrete Institute, Chicago, Illinois.

60

Inspection and Rehabilitation of Steel Trusses for Highway Bridges

A. G. LICHTENSTEIN

A. G. Lichtenstein and Associates, Inc., Consulting Engineers, Fair Lawn, New Jersey, USA

ABSTRACT

Steel trusses played an important role in the development of highway transportation in the United States between 1870 and 1940. In this period of time, numerous patented structural steel (and wrought iron) structures were designed and constructed by engineers/contractors on major state highways and rural county roads. Many of these structures have been replaced over the years but still a sufficient number remain today.

The subject of this paper is to describe the methods and procedures involved in extending the useful life and upgrading of these older truss bridges into structures that can safely support modern and heavier loadings.

The author has selected three types of trusses for a detailed presentation.

INTRODUCTION

Old bridges should not be allowed to die. They should be rehabilitated and put back into service. This is particularly true for trusses on the county roadway systems, where rural conditions still prevail and traffic has only increased moderately over the years. Three examples will be presented as follows:

1. A pin-connected Baltimore truss erected in 1899 by the Berlin Bridge Company of Berlin, Connecticut. This bridge is located at Stuyvesant Falls, Columbia County in the State of New York.
2. A pin-connected eye-bar deck truss bridge constructed in 1896 by the Canton Bridge Company of Canton, Ohio, and first strengthened in 1920. This bridge is located in Boonton, Morris County, New Jersey.

3. A Phoenix truss with patented compression members built in 1894 by the Stagg Company. It is located on Elm Street in Oradell, Bergen County, New Jersey.

STUYVESANT FALLS BRIDGE

Two steel trusses 34 ft (10·4 m) high, top to bottom chord, and 19·4 ft (5·9 m) C_LC_L carry a 17 ft (5·2 m) wide roadway over Kinderhook Creek in one simple span 202 ft (61·6 m) long. A 4·7 ft (1·4 m) sidewalk is cantilevered off the upstream truss. The trusses and floorbeams were of the original construction; however, longitudinal steel stringers and a grating deck had replaced the original timber beams and floor planks.

An in-depth inspection of the bridge was undertaken by our firm and identified the following structural deficiencies:

1. The bottom chord was so severely corroded at the south bearing that it became discontinuous and the truss member and shoe moved outward longitudinally, and the rocker tilted in a vertical direction.
2. The connections of three vertical members of the truss to the floorbeams, at the panel points, were extensively corroded and non-functioning.
3. Many roadway stringers were rusted and cracked in the last panels at the abutments, where salt and dirt accumulated around them.
4. All floorbeams were in good condition but only capable of safely carrying a 7-t (AASHTO H) live load.
5. The trusses computed well except for 16 subpanel verticals, whose rating was 9 t.
6. Other elements of rehabilitation were required, such as restoration of the twisted rocker, repairs to some sidewalk and railing areas, and cleaning and painting of the entire bridge.

The county superintendent of highways then decided to repair and upgrade the bridge with county forces. The county had an abundance of steel members stored in its yard and these could be tapped for use on this project.

The first order of business was to jack up the south end of the trusses approximately 1 ft (0·3 m) and relieve the pressure on the twisted shoe. This completed, the truss member and rocker were pushed horizontally by another jack to a calculated point and a chord mending device was installed to keep them there permanently.

The jacks and auxiliary beams were removed. Then the verticals and floorbeams were reconnected with permanent type details, and the subverticals of the two trusses reinforced with special 'horseshoe' collars (see Fig. 1).

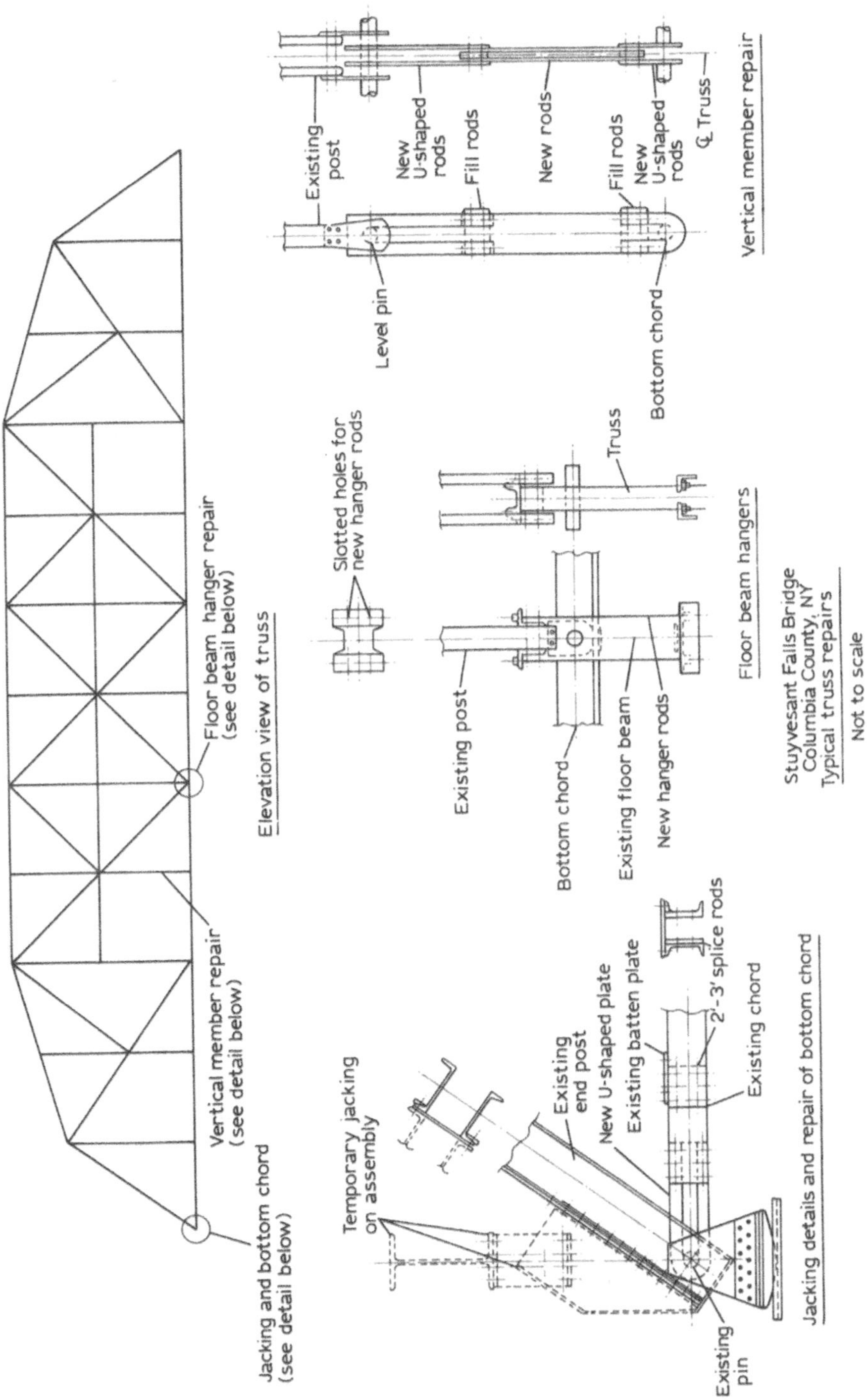
Jacking and bottom chord (see detail below)
Vertical member repair (see detail below)
Floor beam hanger repair (see detail below)
Elevation view of truss
Temporary jacking on assembly
Existing pin
Existing end post
New U-shaped plate
Existing batten plate
2'-3' splice rods
Existing chord
Jacking details and repair of bottom chord
Slotted holes for new hanger rods
Existing post
Bottom chord
Existing floor beam
New hanger rods
Truss
Floor beam hangers
Stuyvesant Falls Bridge
Columbia County, NY
Typical truss repairs
Not to scale
Level pin
Bottom chord
Existing post
New U-shaped rods
Fill rods
New rods
Fill rods
New U-shaped rods
℄ Truss
Vertical member repair

FIG. 1

Next in line was the strengthening of the floorbeams. Without removing the grating (fair condition) and stringers, a platform was erected under each floorbeam. Four angles were bolted onto the webs of the beam as close to the flanges as possible. The corroded stringers were left in place, but new beams of similar depth were set alongside the bad ones and shimmed into place. The railing and sidewalk were repaired in spots where needed. Cleaning and painting, however, was deferred for next season. The bridge was reopened to traffic with a capacity of 15 t (inventory), which is adequate to handle local traffic including buses, fire engines, ambulances and other community type vehicles. For the occasional runaway heavy truck, the operating rating of the bridge is over 20 t. The time necessary to complete this work was 4 months.

The county solved the Stuyvesant Falls Bridge problem by doing most of the work itself and gained a great deal of satisfaction at a reduced cost of construction and restoration of a historic structure. The superintendent and Columbia County received awards from state and national organizations commemorating their achievements on the Stuyvesant Falls Bridge.

THE BOONTON BRIDGE

The 90-year-old Boonton Bridge carries Washington Street over the Rockaway River and the Jersey City Reservoir. In 1920 the bridge was strengthened by the addition of intermediate chords and new deck, and other structural work.

The original bridge consisted of four pin-connected deck truss spans providing a total structure length of approximately 470 ft (143·3 m). The floor system was constructed of steel plate girder floorbeams and steel stringers, which supported an asphalt-filled metal pan deck. The deck provided a 23 ft 6 in (7·2 m) roadway and was flanked by a 5 ft (1·5 m) cantilever sidewalk. The substructure consisted of stone masonry abutments and three non-reinforced concrete piers (Fig. 2).

A. G. Lichtenstein and Associates, Inc., made an in-depth inspection, and prepared a rating and feasibility study for upgrading. An underwater investigation by AGLAS found the piers in good condition. The superstructure was determined to be substandard, however, and was posted for 3 t loading. The study recommended that the bridge can be economically rehabilitated to a 15 t capacity. The rehabilitation project included the following:

1. Replacement of the asphalt pan deck with a concrete-filled grating floor.
2. Upgrading the steel stringers and floorbeams.
3. Repair and strengthening of truss members, shoes and connections.

(a)

(b)

FIG. 2

Total construction cost for the complete bridge and approaches was approximately \$2 600 000, which compared favorably with an estimated cost for a new bridge of \$7 000 000. Because of its historic nature, as well as its location over a reservoir, the acquisition of permits from state and federal agencies would have taken approximately 5 years. The obtaining of permits and the preparation of bid documents for the rehabilitation project required a little more than 1 year. The actual construction was also completed in about 1 year. It is interesting to note that the Federal Government contributed 90% of the cost of construction, while the remaining portion was jointly provided by the State of New Jersey and Morris County. The Morris County engineer was Mr George E. Burke and the general contractor was Karl Koch Erecting Company of Carteret, New Jersey. Our firm prepared the bid documents and checked all the working and shop drawings, while the New Jersey Department of Transportation supervised the contractor's operations and provided a full-time resident engineer.

A modern bridge crossing over the Rockaway River for heavy interstate traffic was constructed within 2 miles of the bridge site. The Boonton Bridge services county type of traffic and will be providing the community with excellent service for many years to come.

ELM STREET BRIDGE

The Elm Street Bridge over the Hackensack River is a single span, simply supported, half-through (pony) truss. The truss consists of five panels, each 15 ft (4·6 m) long for a total length of 75 ft (22·9 m). The trusses are spaced 23 ft (7·0 m) between centerlines and are 10 ft (3·0 m) high. The roadway accommodates two lanes of traffic, one in each direction, and a 5 ft (1·5 m) sidewalk was cantilevered off the west truss.

This structure is known as a Phoenix bridge, wherein the end posts, verticals and upper chord members are composed of Phoenix columns of wrought iron. These patented columns were fabricated utilizing four channels which were turned by machine into quarter round sections, riveted together, to result in a circular member. The open grating deck was found to be in good condition with only light to moderate rust. The stringers supporting the deck were found in generally good condition. However, they rated 6 t. The floorbeams were found to have moderate rust but had a computed live load capacity of 6 t. The truss appeared in fair condition with some members showing evidence of impact from vehicles. However, the capacity of the truss to carry live load was only 2 t. The two abutments were composed of ashlar stone masonry and were found to be in good condition, except that many joints needed remortaring (Figs 3–5).

The assignment was to upgrade this 90-year-old bridge to a live load capacity of at least 12 t, which would permit the safe crossing of school vans and small buses. The deck and the stringers could economically be replaced. The floorbeams, on the other hand, were strengthened by the addition of four angles which were riveted just under the top and bottom flanges. The

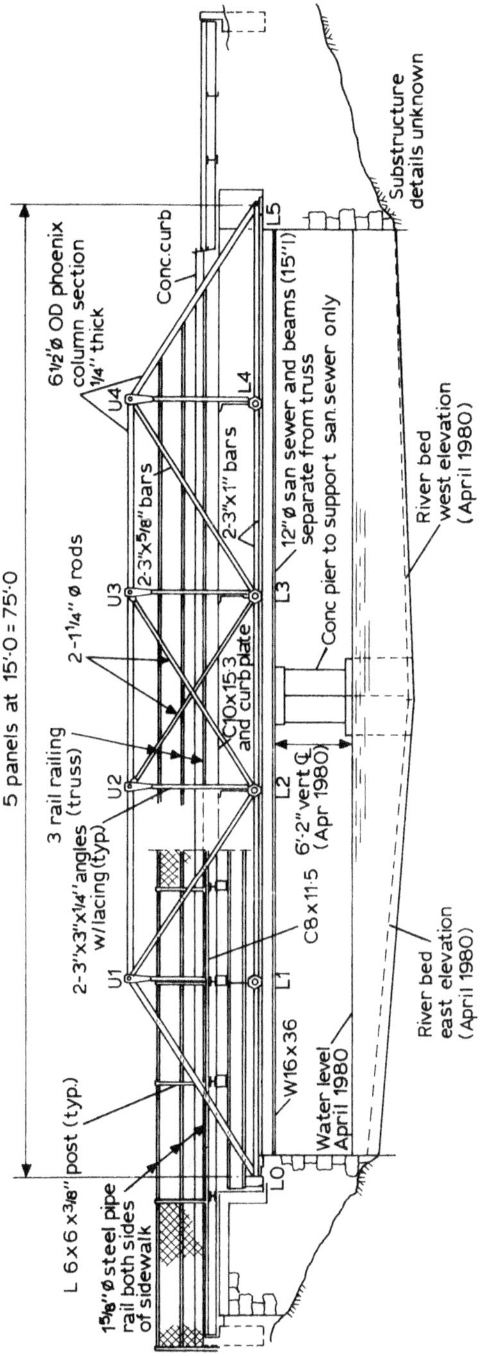

Elevation

FIG. 3

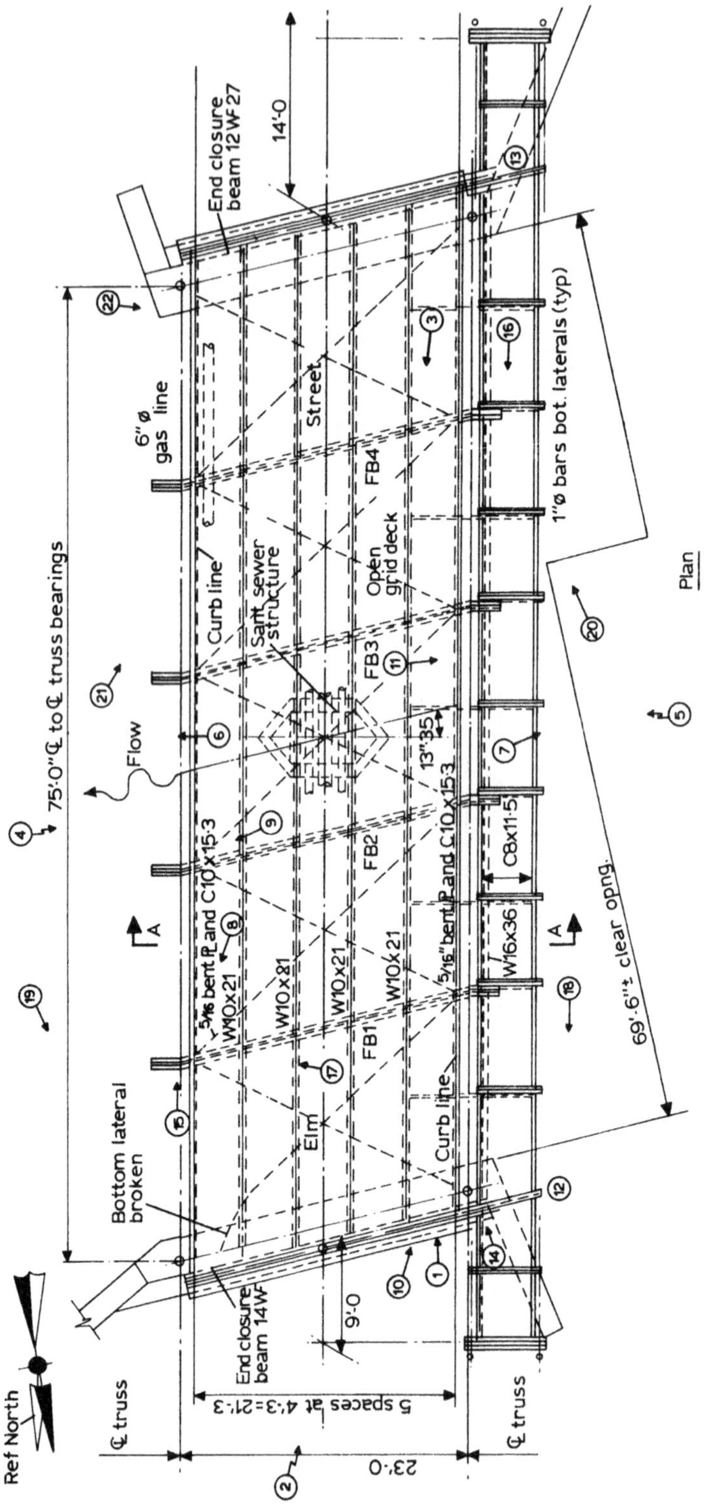
Ref North
℄ truss
℄ truss
23'-0
5 spaces at 4'-3=21'-3
9'-0
14'-0
75'-0" ℄ to ℄ truss bearings
69'-6"± clear opng.
End closure beam 14WF
End closure beam 12WF27
Bottom lateral broken
Flow
6"ø gas line
Curb line
Curb line
Street
Elm
Sani. sewer structure
Open grid deck
FB1
FB2
FB3
FB4
W10x21
W10x21
W10x21
W10x21
5/16 bent Pl and C10x15·3
5/16" bent Pl and C10x15·3
13"
35
C8x11·5
W16x36
1"ø bars bot. laterals (typ)
Plan
A
A

FIG. 4

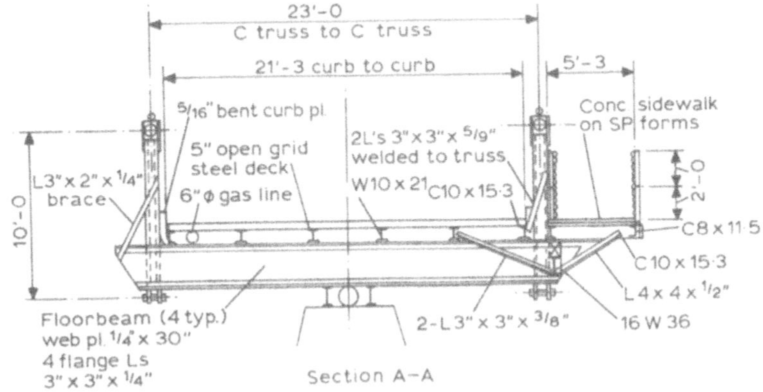

(a)

(b)

FIG. 5

main concern was how to reinforce the historic truss without dismantling it or distorting its architectural appearance. The solution proposed was to add two 36 in steel beams on the outside of the floorbeams under the trusses, which beams would span from abutment to abutment. Each floorbeam was then connected into these two auxiliary beams, creating a live load system wherein the old truss and the new beams were sharing the load (Fig. 6).

To replace this historic Phoenix bridge would have required many years of negotiations at a substantially increased construction cost for limited serviceability. In this way Bergen County has retained and restored a historic

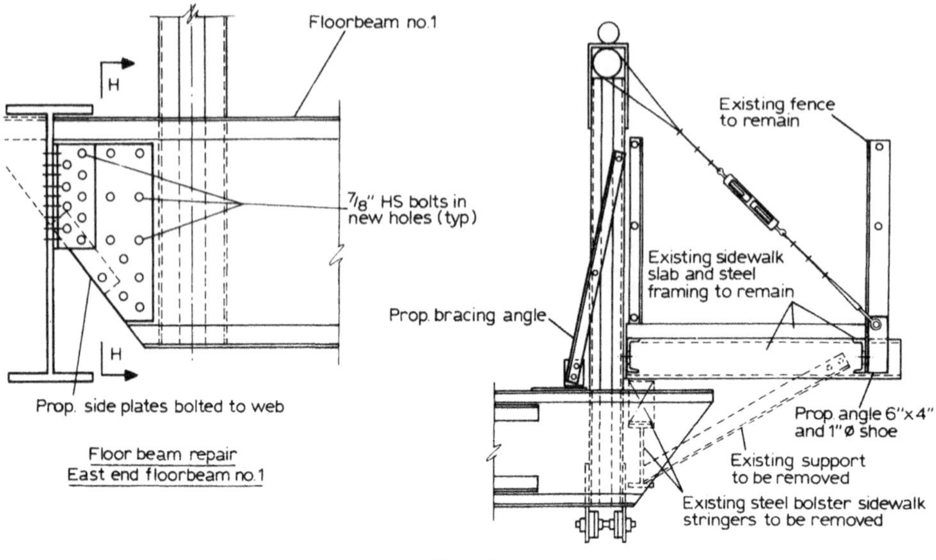

FIG. 6

property and, with normal maintenance and no overloading, the Elm Street Bridge will provide this community with many years of customized usage.

The three examples described above show that, under favorable circumstances, bridges can be retrofitted and placed into the transportation systems with much success. The additional benefit of preserving the American heritage of bridge construction is an unexpected bonus.

61

The Renovation of a Victorian Swing Bridge

BRIAN SIMPSON and MICHAEL F. BLYTH
Husband & Co., Consulting Engineers and Architects,
Alliance House, 12 Caxton Street, London SW1H 0QP, UK

ABSTRACT

Cross Keys Bridge is an iron and steel swing bridge carrying the A17 Newark to King's Lynn trunk road over the River Nene. Built in 1897 to carry a railway line and single lane road, it was converted to two lanes of highway in 1963 after the railway line was closed. In 1982 renovation of the old bridge for a limited future life was the most economic option for providing a route for the proposed Long Sutton and Sutton Bridge bypass over the River Nene. It also enabled the bridge, by then a class II listed structure, to be 'preserved' in working condition.*

Deterioration of some parts of the substructure and the mechanical wedging system was so advanced that an early contract was let to deal with these urgent works. The subsequent main renovation contract included

—replacing deck to improve loading capacity,
—repairs and repainting of existing metal work,
—increasing headroom for road users,
—separating pedestrians from road traffic,
—modernising the mechanical drives and control systems, and
—fendering to reduce risk of shipping collision.

The paper describes these works and the repairs which were devised after the full state of corrosion damage had been revealed. The repairs, including replacement of the bridge deck in stages, were undertaken without interruption to the navigation requirements to swing the bridge and, with the exception of a few weekend closures, single way working for road traffic was maintained throughout winter periods.

HISTORY OF THE BRIDGE

In 1827 part of the work of improving the Fens included a new man-made cut confining the River Nene at its mouth to provide a more direct and reliable navigation channel northward to the Wash. This work included the first Cross Keys Bridge.

In 1864 the expansion of railway ventures brought a rail track crossing this channel at Sutton Bridge utilising an existing movable bridge, the second such crossing at this location. The present structure was constructed by the railway company in 1897 to carry a single track of railway. A roadway and footpath were accommodated alongside the railway track and separated from it by the middle of three longitudinal girders. Road users were obliged to pay a toll and there is a commemorative plaque celebrating the abolition of this toll in 1903. In 1963, after closure of the railway (in 1959), the bridge was redecked to accommodate two lanes of road traffic separated by the central girder system. The present bridge was therefore constructed with railway use and railway clearances in mind and then converted to road use. The appearance of the control cabin for the swing span located over the carriageways reflects its original additional function as a signal box for the railway.

A feasibility study in 1963 for improving the A17 to bypass the villages of Long Sutton and Sutton Bridge included a proposal to replace the swing bridge with a new bascule bridge. This scheme did not proceed at the time and the question of crossing the river as part of Sutton Bridge bypass was re-examined in 1982. By this time there was increasing interest in industrial archaeology and the bridge and the associated building nearby housing the hydraulic accumulator were listed as class II* buildings. Under this category a public enquiry would be necessary to obtain approval for demolition.

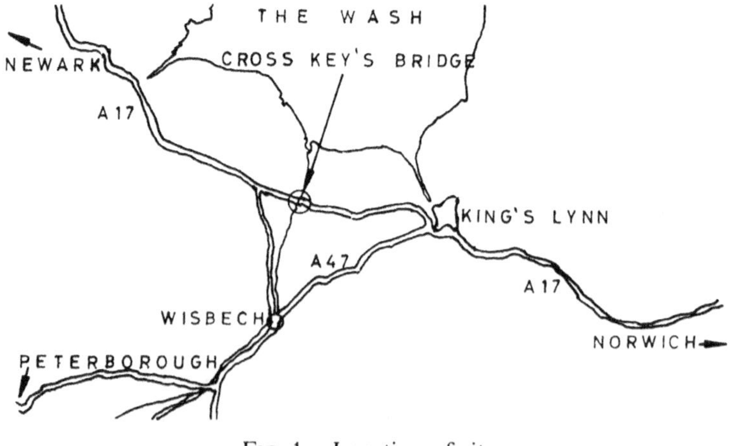

FIG. 1. Location of site.

OPTIONS FOR FUTURE ROAD TRAFFIC

Three options for the river crossing were considered.

1. A new single-leaf bascule bridge on the published line of the then preferred route for the bypass.
2. A new swing bridge to provide a skew crossing adjacent to the existing bridge.
3. Refurbish and strengthen the existing bridge, improving clearances and separating pedestrians from vehicular traffic.

The preferred route for the bypass produced a crossing about 300 m south of the existing bridge and the navigation authority required the demolition of the existing bridge as part of the scheme unless the new opening bridge gave virtually no obstruction to navigation. This arose from the navigation requirements peculiar to this site where, because of distances from the mouth of the river and the berths at Wisbech and the presence of a sand bar at the mouth of the river, ships' movements in both directions generally occur at high flood tide. An additional navigation hazard is a small bend in the channel at the bridge site.

If the existing bridge were demolished, pedestrians would have a considerably longer journey to cross the river and it may have been necessary to provide an additional pedestrian bridge close to the existing bridge site to meet this need.

A site for a new swing bridge immediately upstream of the old was found which could be accommodated within the restrictions imposed by the existing fendering system. With this option it would have been possible to

FIG. 2. Cross Keys Bridge before renovation.

retain the old bridge in either an operational condition or swung clear of the channel and kept for possible occasional use. With a new bridge close to the site of the existing crossing, special provisions for pedestrians would not be necessary.

The third option was to repair and improve the existing bridge to give it a further, if limited, useful life.

The three options were compared and estimates of cost prepared. It was decided to proceed with the refurbishment and strengthening of the existing structure and this was executed in two stages. Prior to this consideration the existing bridge had been inspected and shortfalls identified.

SHORTFALLS IDENTIFIED

The structure was inspected in 1982 and the following shortfalls identified:

1. Generally the structure was corroding and in need of structural repair and repainting.
2. The overhead girders supporting the control cabin and the overhead bracing trusses restricted vehicular headroom.
3. Through girder configuration of the swing and approach spans physically restricted the road to two narrow single lane carriageways with substandard side clearance.
4. Pedestrians were at risk from adjacent traffic lanes because of narrow footpaths and shallow kerbs.
5. The water hydraulic machinery was old, worn and costly to maintain.
6. Electrical wiring and electrical components were obsolete.
7. Access and lighting for operators and maintenance staff were unsatisfactory.

Following an assessment of the load capacity of the structure it was found that road decks were not capable of carrying 45 units of type HB abnormal loading required for this route.

A separate assessment of the risk of shipping collision found the existing fendering systems inadequate to protect the structure from accidental damage.

URGENT WORKS CONTRACT

The inspection identified the substructure subject to tidal action as in urgent need of repair and it was decided to deal with this aspect of the renovation in advance of the main strengthening and improvements. The wedging system operated by a low pressure water hydraulic system was replaced by a new high pressure hydraulic system. The rotational drive system remained on the water system until the main renovation contract. To house the new hydraulics and the associated electrical control system, a new pumphouse was installed below road deck level between the turntable and the tail end of the swing span. The main elements of the urgent works contract were

—repairs to the substructure,
—new pumphouse and access,
—new wedge gear,
—new control system, and
—cleaning and painting substructure.

The turntable on which the bridge rotates is supported on nine piles cased in cast iron and braced together. The floor of this turntable is comprised of arched iron plates between an arrangement of radial and circumferential box girders on which the roller track of the turntable carriage is supported. The arched plates had a concrete infill to give a reasonably level floor. At high tides the water level is higher than the underside of this turntable floor and the iron work had suffered considerable corrosion. The lower part of the main box girders had been inundated at high tides with similar consequences.

The contract included removing the concrete infill and all the arched steel floor plates to assess the extent of loss of material and decide on remedial measures. Although the floor plates were corroded beyond repair it was found that the supporting structural members were in reasonable condition so the floor was reinstated using new arched floor plates connected to the existing steel framework. The original floor plates had been riveted but the replacements were bolted using grade 8·8 bolts. A bituminous sealing compound was used to bed the plates and caulk the joints.

MAIN RENOVATION CONTRACT

The main renovation contract required the repair and repainting of the structure, provision of new steel trough stiffened decks surfaced with epoxy bauxite, a new footpath attached to the outside of the existing north truss and plate girders, and the provision of greater headroom beneath the cross-bracing and control cabin. Also included within the contract was abutment strengthening, the completion of the refurbishment of the mechanical and electrical drive system, and the provision of a new control cabin together with major extensions to the existing fendering system.

The structure is on the main A17 trunk road, and road traffic flow had to be maintained except for limited weekend closures. The bridge had to be operational at all times to allow the passage of waterborne traffic on the River Nene on their trips to and from Wisbech. The A17 is a busy route for holiday traffic to and from the East Anglian coast with peak traffic flows during summer, particularly at weekends. For this reason it was only possible to close one lane of traffic during the winter months from October to April. Because of the amount of work to be undertaken during such a partial road closure period, the contract was programmed to run over two winter periods with lane closures during consecutive winters. It was also necessary to let the contractor have complete possession of both carriageways for a few weekend winter periods in order to carry out operations that could not be done safely with traffic on the bridge. An example of this was raising the control cabin and its supporting framework to increase vertical headroom over the carriageways.

During the first major closure period in the winter of 1987/88 the contractor raised the control cabin and the overhead bracing, repaired and part painted the north truss and part of the centre truss and approach span plate girders along the north carriageway, and installed the new footpath. A limited amount of work was carried out to the centre truss and the north face of the centre plate girders. It was not possible to complete all the work planned for this period but it was possible to reopen the north carriageway to road traffic in time for the summer holiday period.

Details of repairs to the existing corroded structure could not be determined in advance of letting the contract except in general terms because the full extent of corrosion damage could not be revealed until the existing deck had been removed and the old metalwork blast cleaned. The provision of access for dismantling, cleaning and repair was also essential for inspection and assessment of damage. The areas of difficult access below deck level were those which had not had the benefit of routine painting and were the most corroded. The existing road deck protected the centre sections of the cross-girders and the intermediate longitudinal members but the ends of these girders and the lower gusset plate connections on the trusses were vulnerable to attack from road salted surface water and consequently were in extremely poor condition.

The sequence of operations was dictated by the nature of the redecking work. The old deck was taken up sequentially in panels starting from the approach spans at the east end. Cleaning, inspection and repairs to the structural members supporting the deck were then undertaken before the new deck panels were introduced, again starting from the east end. Once deck panels had been removed access to the area of activity was limited and careful organisation was needed to ensure progressive repair and reinstatement.

Work on the swing span required an additional degree of control to maintain the swing span in a state of balance whilst deck panels were removed and replaced. Furthermore, the addition of the footpath on the north side of the structure meant progressive counterbalancing transversely to keep the centre of gravity of the whole rotating structure reasonably close to the centre of rotation at all times. This was necessary to ensure safe operation and to maintain the loading pattern on the pile group under the roller path.

Two methods of counterweighting were tried. For the replacement of the first lane of road deck, water tanks were used but simple weights were adopted for the second lane because the water tanks were found to impede access.

To check the balance of the bridge, an attempt was made using strain gauges but these were frequently damaged and the method eventually employed entailed checking levels at the extreme ends of the swinging span, enabling quite small variations in the state of balance to be monitored.

In the winter of 1988/89 work on the south truss and approach span plate girders together with the centre truss and plate girders was completed. Work on the centre truss was achieved by providing a timber sheet barrier bolted to the north side of the truss, which allowed work to continue without interference to traffic flow. At the end of the second major lane closure period outstanding work in the first (north) lane closure was completed.

Work not affecting the free passage of vehicles was carried out throughout the contract period. Complete road closures were allowed to enable the ends

FIG. 3. View illustrating restricted traffic lanes.

of the swinging span to be repaired and painted, and to enable the mechanical and electrical equipment to be tested.

UNUSUAL ASPECTS OF REPAIR WORK

Some unusual aspects of the renovation work relate to the standards for health and safety introduced since the bridge was built. Reference has been made to box girders with access manholes. Many of these are now regarded as too small and exceptional measures had to be taken to deal with internal surfaces. It was first thought that because of these difficulties a sprayed grease paint, which could be used without a high standard of surface preparation, would be employed inside the boxes. However, access to the inside of some box members was better than anticipated because large areas of web and bottom flange had corroded completely away. Previous repairs in the turntable area had compensated for this loss of material with additional web and external flange plates.

Another difficulty related to the age of the structure was the uncertainty of the material used in its construction and its weldability. Samples of material were taken from a number of locations and identified by micro examination and chemical analysis. This investigation by Messers Sandberg revealed a mixture of wrought iron and low carbon steel with sulphur and phosphorus content higher than would be permitted to current standards for weldable structural steels. As welding was a technique that was being considered as an option for certain repairs, welding procedure trials were undertaken on the samples which established satisfactory procedures for field use.

Although the composition of some of the existing paint system was known, there was doubt about other paints. An investigation of paint samples did not initially reveal the presence of lead but when large-scale paint removal started it was found that the existing paint system contained lead and it was necessary to take the normal precautions to deal with this potential health hazard. To minimise airborne lead pollution most of the thicker coats of paint were removed by hand chipping prior to blasting to clean metal.

The lower chords of the lattice girders and the lower gusset plated connections had suffered quite badly from corrosion. As the deck was removed from each carriageway these areas were cleaned and inspected before a repair scheme for each connection was devised. These repairs were rendered more difficult because the bridge had to be kept operational and the loads in the members being connected were subject to variations due to bridge openings and live load on the adjacent carriageway as well as variations from thermal effects.

Wherever it was necessary to remove rivets these were replaced by high strength fitted bolts and anti-vibration nuts. When gusset plates were reduced in thickness, part of the plates were replaced and butt welded to the remaining sound material. These welds were executed before bolting up the repair.

These particular repairs were often a question of doing what was feasible rather than what would be done for a new construction and considerable ingenuity had to be used by all concerned to make an effective repair. For example, the general advice that welding should not be 'mixed' with riveting or bolting had to be tempered to produce a repair sequence in which the effects of welding (particularly welding shrinkage) could be accommodated before bolting. It was also found expedient to secure rivets which had virtually lost their heads from corrosion but which were otherwise sound and tight in their holes by placing a bead of weld around the perimeter of the exposed shank.

ACKNOWLEDGEMENTS

The work described in this paper was undertaken for the Director of Transport, East Midlands Regional Office of the Department of Transport, and the Director of Highways and Planning of Lincolnshire County Council. The authors gratefully acknowledge the permission of the Chief Highway Engineer of the Department of Transport to publish this paper.

62

Cost-Effective Strategies in Bridge Management

R. S. REEL and C. MURUGANANDAN
Bridge Management Section, Structural Office, Ministry of Transportation, Ontario, Canada M3M 1J8

ABSTRACT

This paper assesses rehabilitation projects in an economic framework. Two techniques are outlined. The first is the present value analysis which compares alternatives at the project level. The second is the incremental benefit/cost ratio analysis which prioritizes the cost-effective rehabilitation alternatives at the project and network levels. Various parameters required for the analysis are discussed. Computer programs utilizing Lotus 1–2–3, Version 2.01, required for the analysis are described. The application of the techniques is illustrated by examples of bridge rehabilitations.

INTRODUCTION

An economic evaluation is an important step in the decision-making process for work that involves major expenditures. At the project level, the costs for alternative levels of improvements to a bridge are compared to determine the most economical option for the bridge based on either the present value analysis or the incremental benefit/cost ratio analysis. At the network level, the costs for different levels of improvements for all the bridges in the network are compared to prioritize work based on the incremental cost/ benefit ratio analysis.

PRINCIPLES OF COST-EFFECTIVE STRATEGIES

Theory of Present Value Analysis

The present value analysis involves the calculation of the cost of alternative schemes in present-day monetary terms, i.e. the amount that is required in today's value to obtain goods and services at any future date. It allows for the comparison of alternative schemes on an equitable basis.

The present value of an expenditure C in year n at a discount rate r is

$$PV = \frac{C}{(1+r)^n}$$

The present value of a number of expenditures C_{1n}, where $n = 1, 2, \ldots, N_1$ for a period of N_1 years, is

$$PV_1 = \sum_{n=1}^{N_1} \frac{C_{1n}}{(1+r)^n}$$

Incremental Benefit/Cost Ratio Analysis[1]

The incremental benefit/cost ratio is the ratio of the additional benefits realized in moving from one improvement alternative to another divided by the corresponding difference in costs. This method not only optimizes the selection of alternatives efficiently but also ranks the projects beginning with the most net beneficial. It is used both at the project and network levels.

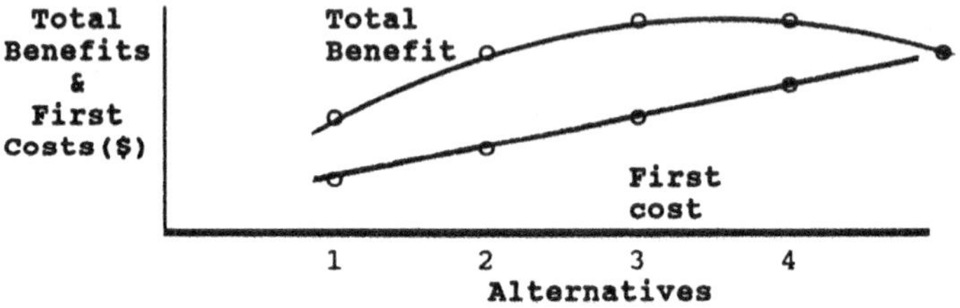

FIG. 1. Total benefits and first cost.

Figure 1 shows the total benefit and first cost curves plotted for the various alternatives for a bridge. Initially the increment of benefit is higher than the increment of cost. As costs increase the incremental benefits decline and are less than the incremental costs. The slopes of these benefits and first cost curves support the theory of diminishing returns. For a particular level of improvement there exist points on the benefit and cost curves where the slopes of the two curves are equal, i.e. IB = IC. At this level of improvement the net benefit is a maximum. This is evident from Fig. 2. Any option below this level where IB/IC > 1 is a desirable option.

The procedure is to list rehabilitation alternatives in the order of increasing costs and calculate the incremental benefit/cost ratios. Alternatives for which the incremental benefit/cost ratios fall below one are discarded. Usually, as the level of cost increases the incremental benefit/cost

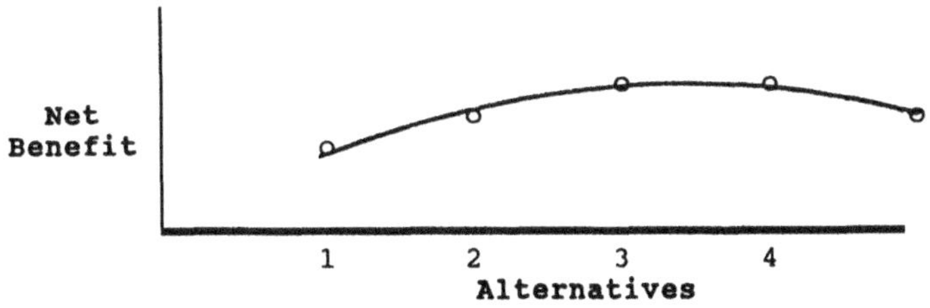

FIG. 2. Net benefits.

ratio decreases. However, if the ratio IB/IC increases with increase in cost an adjustment is made to that particular option. The incremental benefit/cost ratio is calculated by considering the previous option but one instead of the previous option. The options are sorted in descending order of IB/IC. For an unlimited budget the most net beneficial alternative is the one with the largest initial cost and whose incremental benefit/cost ratio > 1. For a limited budget the order of preference is the order from the highest to the lowest incremental benefit/cost ratio.

PARAMETERS REQUIRED FOR COST-EFFECTIVE ANALYSIS

Parameters Required for Present Value Analysis[2]

The following parameters are required to perform the present value analysis for each alternative:

—capital cost,
—maintenance cost,
—life cycle,
—residual life, and
—discount rate.

Capital Cost

The following should be estimated for each alternative in constant monetary terms:

—engineering design cost,
—construction cost, and
—miscellaneous costs such as demolition, right-of-way, approaches, utilities, stream diversion, detours, etc.

Maintenance Costs

Costs associated with maintenance are the routine maintenance costs. These would include minor repairs, maintenance, touch-up painting, etc., carried out on a regular basis.

Life Cycle

The life cycles for the treatments should be estimated. Usually it is the time between two successive replacements or rehabilitations.

Residual Life and Value

The various alternatives considered may have useful lives at the end of the time frame. This is termed as the residual life. There are no specific methods of assessing this. A thorough knowledge of the performance of past rehabilitations and experienced engineering judgement are probably the best way of assessing the useful residual life. From the residual life the residual value of the structure for the particular alternative can be determined. There are several methods available for determining the residual value. The method used here is the second cycle replacement method.

TABLE 1
Residual value

Option	Year of replacement (2nd cycle)	Replacement cost	Residual years	Value at year N_1	Differential value (option 1 as base)	Residual value at year 0
1	N_1	C	0	C_1	C_{D1}	C_{R1}
2	N_2	C	$N_2 - N_1$	C_2	C_{D2}	C_{R2}

r = discount rate
C = replacement cost

$C_1 = C$ $\qquad$ $C_2 = C_1/(1+r)^{N_2-N_1}$
$C_{D1} = C_1 - C = 0$ $\qquad$ $C_{D2} = C_2 - C$
$C_{R1} = 0$ $\qquad$ $C_{R2} = C_{D2}/(1+r)^{N_1}$

Table 1 shows the residual value calculations for options 1 and 2, whose second cycle replacement will be in years N_1 and N_2.

Discount Rate

The discount rate for government projects depends on several factors,[3] such as the magnitude of investment return, tax rates, capital market conditions, preferences for current and future consumption, methods used to finance projects, etc. A discount rate of 6% is recommended for projects owned by government agencies. These rates may be different for other agencies. Sensitivity analysis may be carried out by varying these rates.

Parameters Required for Incremental Benefit/Cost Analysis

For the incremental benefit/cost analysis, the following additional parameters are required:

—agency costs,
—user costs,
—agency benefits, and
—user benefits.

Agency costs

Agency costs are the same as for the present value analysis.

Agency benefits

Maintenance and various types of rehabilitations extend the useful life of the bridge. These expenditures would postpone major expenditures for replacement. The difference between the discounted future cost of a rehabilitation option and that of a replacement option is the agency net benefit. The agency net benefit plus the initial cost is the agency total benefit:

$$\text{Agency benefit} = PV_R - PV_1 + C_1$$

where C_1 is the initial cost of the rehabilitation option, PV_1 is the discounted present value of costs associated with a rehabilitation option and PV_R is the discounted present value of costs associated with the replacement option.

User costs

User costs are costs incurred by the user due to deficiencies or substandard conditions at the bridge. The following are the user costs.

Accident costs. Accident costs resulting from bridge width restrictions, poor approach alignment, etc.

Functional restriction costs. Functional restriction costs due to load restrictions and detours for certain classes of vehicles increase the travel time, and hence operating costs. These vary for different locations and countries.

User benefits

User benefits of a bridge rehabilitation option are the reduction in costs to the users due to the rehabilitation. In determining user benefits it is assumed that deficiencies will be eliminated when the bridge is replaced.

The reduction in the number of accidents due to a certain type of improvement is used as a measure of user benefit for that type of improvement. The dollar value placed on different types of accidents is crucial in estimating user benefits. These may vary for different countries:

Annual benefits = (change in accident rate) × (ADT) × (365) × (accident cost)

The change in accident rate is measured by the difference in the number of accidents per million vehicles. The accident cost depends on the severity of the accident. It is very difficult to place monetary values on accidents. Several attempts have been made to quantify these. The two common methods are (a) human capital approach and (b) willingness to pay approach.

The human capital approach takes into consideration the direct and indirect costs. This approach does not consider the intangibles offered to the society and the loss in the quality of life. The willingness to pay approach includes the value of life in the estimates. As such the willingness to pay is more conservative.

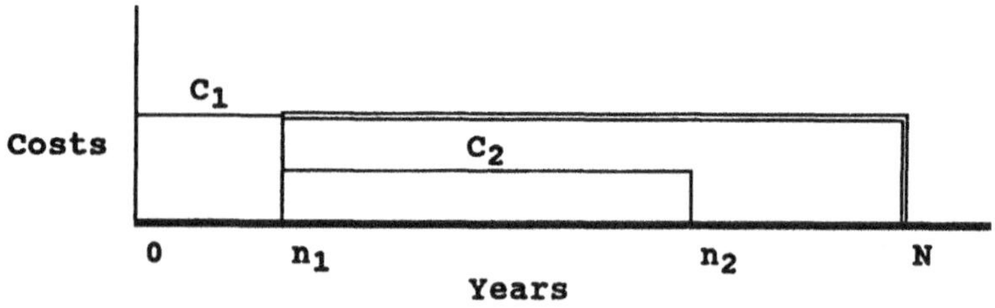

FIG. 3. User benefits.

Figure 3 illustrates the method used in determining user benefits.

The user benefit is the present value of the annual benefits of $(C_1 - C_2)$ for years n_1 to n_2 and C_1 from n_2 to N:

$$\text{User benefit} = \left\{\sum_{n=n_1}^{n_2} \frac{C_1 - C_2}{(1+r)^n}\right\} + \left\{\sum_{n=n_2}^{N} \frac{C_1}{(1+r)^n}\right\}$$

$$= (C_1 - C_2)\left\{\frac{1-(1+r)^{(n_2-n_1)}}{r}\right\}\frac{1}{(1+r)^{n_1}}$$

$$+ C_1\left\{\frac{1-(1+r)^{(N-n_2)}}{r}\right\}\frac{1}{(1+r)^{n_2}}$$

where N is the period considered for life cycle analysis, C_1 is the annual user costs associated with the bridge, n_1 is year of rehabilitation which would extend the life, C_2 is user costs after rehabilitation and n_2 is year of replacement.

Present Value Analysis Using PRVAL Program

PRVAL is a template overlay developed to perform financial analysis for bridge rehabilitation projects using Lotus 1–2–3, Version 2.01, on a worksheet format. There are four different options available to carry out the financial analysis at different levels of sophistication.

Incremental Benefit/Cost Ratio Analysis Using COSBEN Program

COSBEN is a program developed to perform incremental benefit–cost analysis for bridge rehabilitation projects using Lotus 1–2–3, Version 2.01, on a worksheet format. At the project level the analysis can be carried out with or without user costs. The output from the project level analysis is used to perform the network analysis.

EXAMPLES OF COST-EFFECTIVE ANALYSIS

The application of the techniques at the project and network levels are illustrated by the following examples (in $1000).

Example 1: Present Value Analysis at the Project Level

Select the most economical option from the three options given in Table 2 using the PRVAL program.

(a) Using the present value analysis technique the PRVAL04 program gives the output shown in Table 3.

The analysis in Table 3 shows option 3 is the preferred choice.

Example 2: Incremental Benefit–Cost Analysis at the Project Level

Select the most economical option from the three options given in Example 1(a) using the COSBEN01 program.

TABLE 2
Cost data

Years	Option 1	Option 2	Option 3
0	1 000	200	300
5			
15	100	1 000	
20			1 000
30	200	100	
35			100
45		200	

TABLE 3

Years	Option 1		Option 2		Option 3	
	Cost	PV	Cost	PV	Cost	PV
0	1 000	1 000	200	200	300	300
5						
15	100	42	1 000	417		
20					1 000	312
30	200	35	100	17		
35					100	13
45			200	15		
Total PV		1 077		649		625
Total PV*		1 109		642		612

* The present values adjusted for residual value and uncertainty in costs.

TABLE 4
Total benefits

Options	Net benefit	Cost	Benefit (total)
1	0	1 000	1 000
2	458	200	658
3	489	300	789

TABLE 5
Incremental benefit–cost ratio

Options	Net benefit	Cost	Benefit (total)	IC	IB	IB/IC
2	458	200	658	200	658	3·29
3	489	300	789	100	130	1·30
1	0	1 000	1 000	700	210	0·30

TABLE 6
Order of priority (with limited budget)

Priority	Options	Net benefit	Cost	IB/IC
1	2	458	200	3·29
2	3	489	300	1·30
3	1	0	1 000	0·30

TABLE 7
Input data

Bridge number	Option number	Net benefit	First cost	IB/IC
1	2	950	100	10·5
1	4	1 550	500	2·5
1	5	2 450	2 000	1·2
2	2	1 100	200	6·5
2	3	1 925	750	2·5
3	5	800	250	4·2
3	2	1 400	500	3·4
3	4	1 580	800	1·6
4	2	1 640	200	9·2
4	4	2 660	500	4·4
4	5	2 810	1 000	1·3
5	2	2 300	200	12·5
5	3	3 620	500	5·4
5	5	4 520	750	4·6
6	3	2 320	400	6·8
6	2	3 020	600	4·5
6	4	3 710	900	3·3
7	2	1 725	250	7·9
7	3	2 425	450	4·5
7	5	3 235	1 200	1·3
8	2	4 380	300	15·6
8	3	5 395	650	3·9
8	5	5 695	850	2·5
8	4	6 865	1 600	1·2

TABLE 8
Output

Priority	Bridge number	Option number	Net benefit	First cost	IB/IC
1	5	5	4 520	750	4·6
2	6	4	3 710	900	3·2
3	1	4	1 550	500	2·5
4	8	5	5 695	850	2·5
5	2	3	1 925	750	2·5
6	3	4	1 580	800	1·6
7	7	5	3 235	1 200	1·2
8	4	5	2 810	1 000	1·2

Budget spent = \$6 750 000.
Balance = \$250 000.

It is evident that options 2 and 3 are cost-effective options. If there are no limitations on the budget then option 3 would yield the maximum benefits. The advantage of this method is that it ranks the projects.

Example 3: Network Analysis

For a budget limitation of $7 000 000, prioritize the various options for the preceding eight bridges using the COSBEN03 program.

The bridges are individually analysed to obtain the cost-effective options as shown in Example 2 using either COSBEN01 or 02 programs. These results are used as input for COSBEN03. The program arranges them in the order of decreasing incremental benefit/cost ratio and allocates the funds to obtain the best cost-effective combination.

REFERENCES

1. FEDERAL HIGHWAY ADMINISTRATION, Bridge management systems. Draft Report No. FHWA-DP-71-01, Washington, DC, 1987.
2. MINISTRY OF TRANSPORTATION, *Structural Financial Analysis Manual.* Ontario, 1988.
3. HAVEMAN, R. and MARGOLIS, J., *Public Expenditure and Policy Analysis*, 3rd edn. Houghton Mifflin Company Ltd, Boston, Chapter 12, 1983.

63

Tension Drop in Cable-Band Bolts on Suspension Bridges

YUJI KAGAWA

Bridge Technical Affairs Division, Honshu-Shikoku Bridge Authority, 45th Mori-Bldg, 1–5 Toranomon 5-Chome, Minato-ku, Tokyo 105, Japan

and

AKIHIRO FUKUSHI

Bridge Engineering and Construction Department, Kobe Steel Ltd, 2–8 Iwayanakamachi 4-Chome, Nada-ku, Kobe 657, Japan

ABSTRACT

The tension (axial force) in cable-band bolts on suspension bridges decreases with time. Conventionally bolt tension was measured periodically and bolts were retightened when the measured tension was lower than that specified. The authors undertook a survey of the bolt tension on several suspension bridges and analysed the data. The results indicated that the bolt tension decrease has a certain tendency and, in particular, tension drop decreases markedly after retightening. This paper describes the method of investigation which the authors used to determine the tension in cable-band bolts together with the analysis of the data. In addition, proposals are given on how to determine the optimum bolt tightening time and an appropriate bolt tension value to use as criteria.

INTRODUCTION

Cable-bands play an important role in transmitting the tension of hanger ropes to the main cable, and the cable-band is held to the main cable with the tension of the cable-band bolts. The cable-band bolt tension generally decreases with the passage of time. Consequently bolt tension control becomes very important as it is necessary to retighten cable-band bolts which have lost their original tension. To date, however, bolt tensions have been measured several times at optional intervals and retightened as required.

The authors have carried out measurement, retightening and monitoring of cable-band bolt tension on five suspension bridges under the control of the Honshu-Shikoku Bridge Authority (Fig. 1 shows the twin bridges, which are the Minami Bisan-Seto Bridge and Kita Bisan-Seto Bridge). Figure 2 shows a general view of the Minami Bisan-Seto Bridge, showing the overall suspension bridge shape and the cable-band position and shape.

This paper reports the results of measuring the cable bolt tension, explains the tendency of bolt tension to decrease with time, and proposes a concept for determining the optimum bolt tension and a procedure for determining the proper retightening time.

FIG. 1. Minami Bisan-Seto Bridge and Kita Bisan-Seto Bridge.

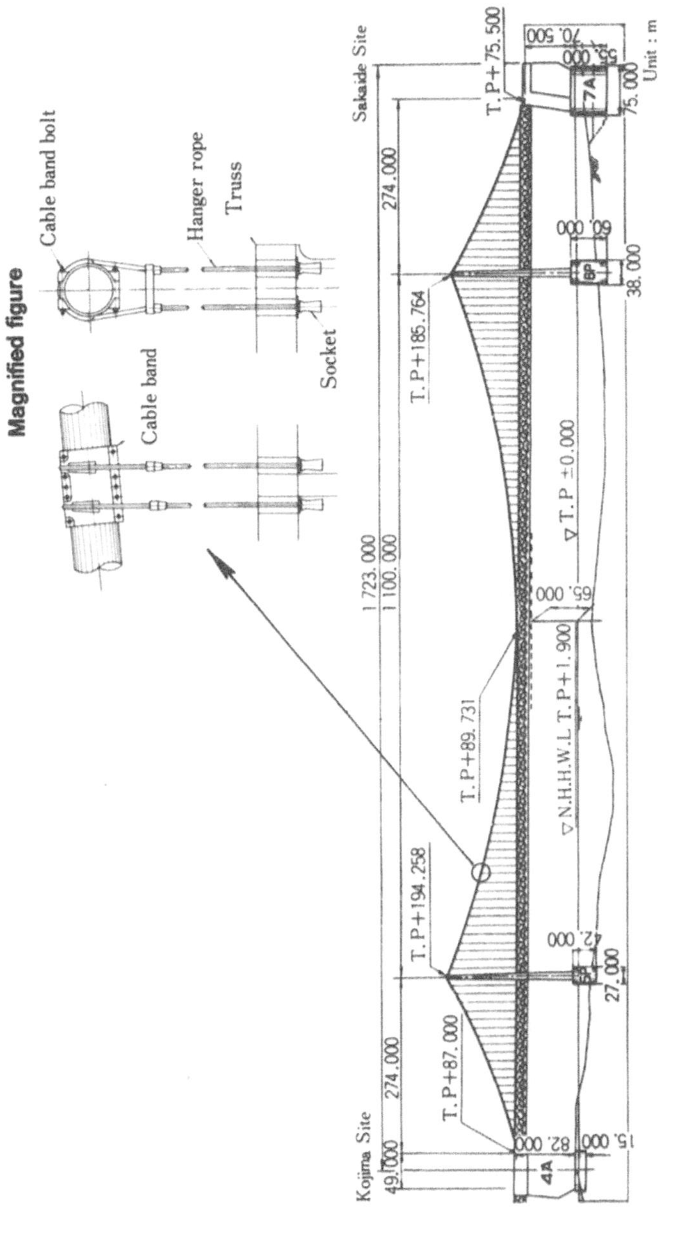

FIG. 2. General view of Minami Bisan-Seto Bridge.

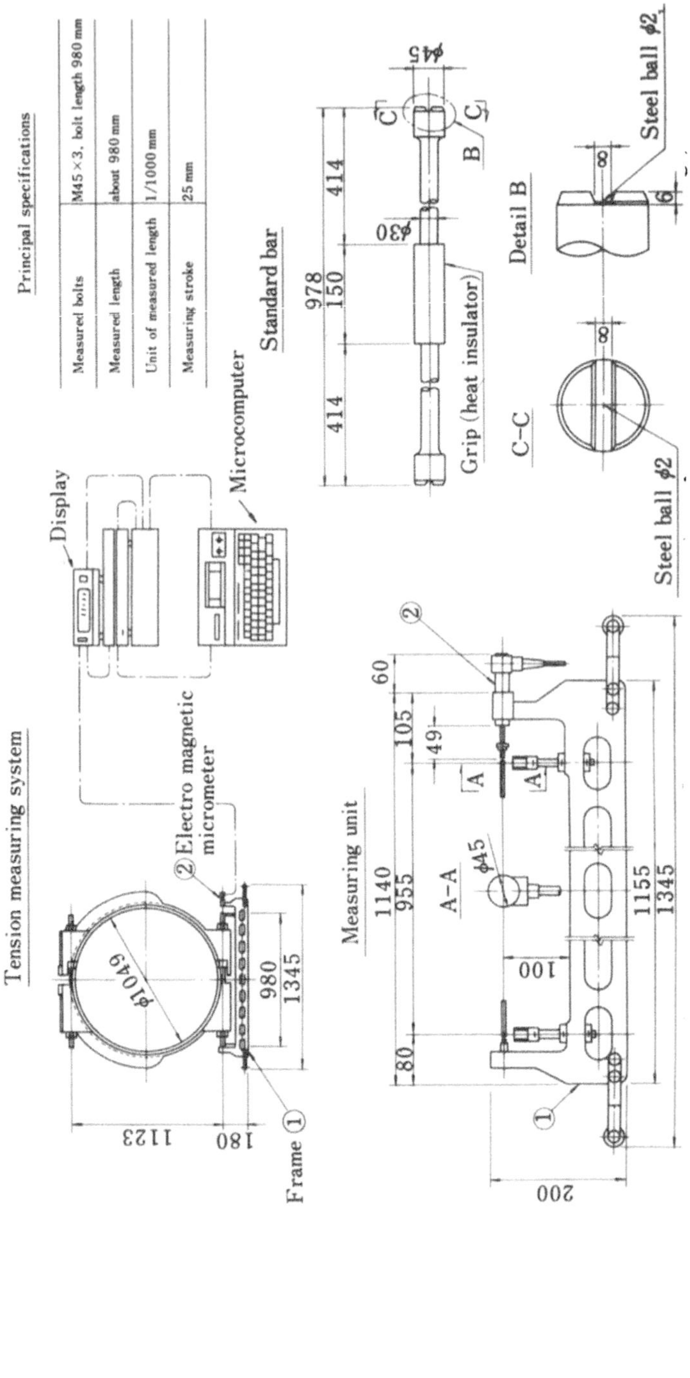
Tension measuring system
Principal specifications
Measured bolts M45×3, bolt length 980 mm
Measured length about 980 mm
Unit of measured length 1/1000 mm
Measuring stroke 25 mm
Display
Microcomputer
② Electro magnetic micrometer
Frame ①
φ1049
1123
180
980
1345
Standard bar
978
414
150
414
φ30
φ45
Grip (heat insulator)
B
C
Detail B
C–C
8
6
Steel ball φ2
Steel ball φ2
Measuring unit
A–A
φ45
1140
955
105
60
49
80
100
1155
1345
200
A
①
②

MEASURING METHODS AND EQUIPMENT

In the study cable bolt tension was obtained by measuring the elastic extension of the bolts and converting this measured value into a tension. To measure the elastic extension of the bolts, the electromagnetic digital micrometer shown in Fig. 3 was used. The micrometer was equipped with a portable microcomputer which can store data such as bolt number and initial length in memory on micro-cassette tapes. A length measuring flow chart, criteria for accepting or rejecting the data and calculating equations were input to the computer. If length is measured properly, bolt length is immediately converted to tension and output.

FIG. 4. Cable inspection vehicle.

As a means of transporting workers, the length measuring equipment and other materials to the cable-band positions, the cable inspection vehicle shown in Fig. 4 was used.[1] The cable inspection vehicle was developed by the Honshu-Shikoku Bridge Authority and is a type of movable scaffold that travels along the main cable using hand ropes as travelling rails. For the drive method, in place of a conventional wire rope winching take-up system, frictional force is generated by strongly squeezing the hand rope with two pairs of caterpillars incorporating special rubber shoes. Travelling force is then generated by rotating and driving the caterpillars, thereby ensuring excellent mobility and travel stability.

RESULTS OF MEASUREMENT

Tendency of Tension Drop

Table 1 lists the bridges for which the drop in cable-band bolt tension with time was investigated, together with the investigation time.

Figure 5 shows the relationship between the mean bolt tension and time for each of the six bridges investigated. From Fig. 5 it can be seen that:

(1) The higher the tightening torque, the higher the residual tension.
(2) A similar tendency of tension drop with time is observed in all bridges.
(3) The drop in cable-band bolt tension with time can apparently be plotted as a straight line on a semi-logarithmic coordinate paper with time as the abscissa.

Figure 6 illustrates the residual rate of bolt tension in one band of the Innoshima Bridge.[4] This bridge, selected as an example, is a very typical case, but virtually all bridges showed similar behaviour. The figure shows that in one cable-band a smaller tension drop occurs with the upper bolt than with the lower bolt, and with the edge bolt more than with the intermediate bolt. For the Kanmon and Innoshima bridges, cable-band bolts were retightened and later checked for bolt tension transition. Figure 7 shows the results and proves that the rate of bolt tension drop after retightening markedly decreases when compared with the rate before retightening.

Tightening Torque and Tension Drop Rate

Factors in the bolts' tension drop with time are thought to include:

(1) Decrease in the cable diameter caused by increase in the dead load of the bridge.

TABLE 1
Measured bridges and investigation time (elapsed time: days)

Name of bridge	After bridge completion				After retightening				Year
	1st	2nd	3rd	4th	1st	2nd	3rd	4th	
Kanmon[2,3]	500	1 200	—	—	10	800	—	—	1973
Innoshima[4]	1 000	1 400	1 800	2 100	90	240	400	700	1983
Ohnaruto	1 000	1 700	—	—	Not yet retightened				1985
Shimotsui-Seto	500	—	—	—	Not yet retightened				1988
Kita Bisan-Seto	600	—	—	—	Not yet retightened				1988
Minami Bisan-Seto	500	—	—	—	Not yet retightened				1988

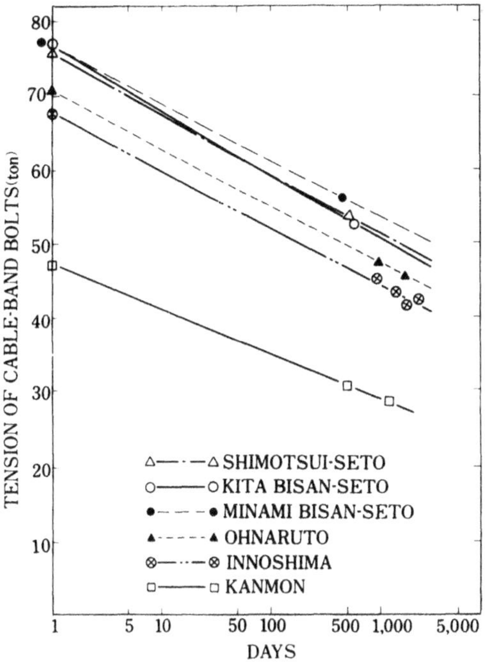

FIG. 5. Decrease in mean bolt tension of each bridge (before retightening).

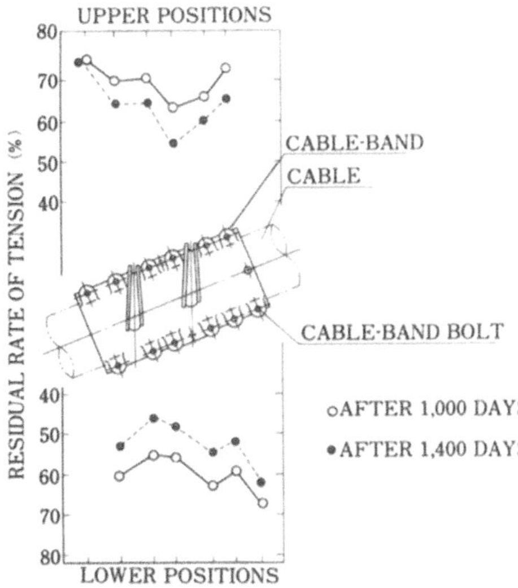

FIG. 6. Comparison of residual tension rate in bolts of one band (Innoshima Bridge).

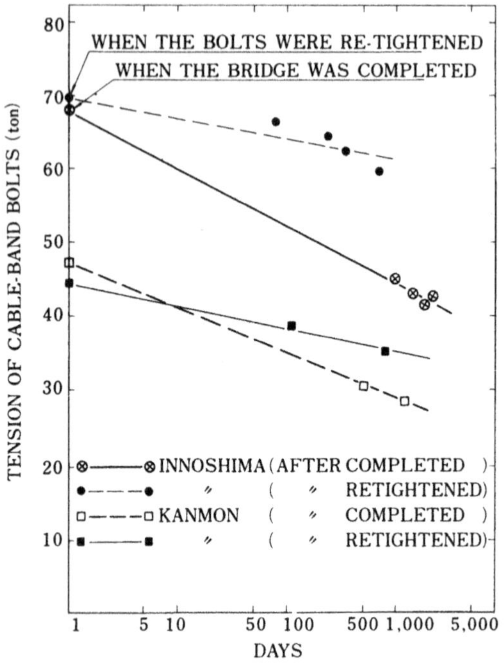

FIG. 7. Reduction of mean tension in bolts before and after retightening.

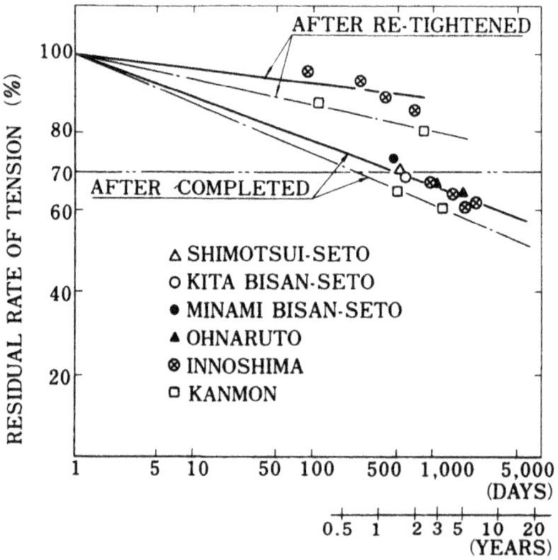

FIG. 8. Residual rate of tension in cable-band bolts.

(2) Rearrangement of the cable wires within the cable-band caused by movement of the bridge from fluctuations in the live load.
(3) Compressed creep of the zinc layers of galvanised cable wires.

Replotting Figs 5 and 7, choosing the tension residual rate as the ordinate, results in Fig. 8. This shows the following:

(1) Though some variations are found in each bridge, generally the bolt tension decreases after bridge completion by about 25–35% in 500 days and 30–40% in 1000 days.
(2) The tension drop after retightening is noticeably lower than before retightening, demonstrating that retightening has a great effect on suppressing the tendency for tension drop to increase with time.

DISCUSSION

Effects of High Tightening Torque

Figure 8 indicates that the conditions of increasing tension drop fall into one of two groups: firstly, those of the Kanmon Bridge and, secondly, all of the other five bridges. Table 2 shows the differences in design. Figure 8 and Table 2 show that tightening the bolts with high torque (stress) during both construction and retightening effectively prevents an increase in tension drop over a period of time.

Safety Factors and Retightening Time

The design standard of the Honshu-Shikoku Bridge Authority specifies that the cable-band slip safety factor v must ensure $v \geq 3$ even if the bolt tension reduces to 70% of that at the time of tightening.[5,6] That is,

$$v = \frac{0{\cdot}7\mathrm{m}\mu\eta T}{Th \sin\phi} \geq 3 \qquad \mathrm{m} = \frac{2}{\mu}(1 - \mathrm{e}^{-(\pi/2)\mu})$$

where μ = coefficient of friction,
η = number of bolts,
T = force of designed bolt tension,
Th = vertical force created by hanger rope, and
ϕ = inclination of cable at the cable-band installation point.

Consequently it is evident that the proper retightening time should be when $v = 3$. Judging from Fig. 8, the retightening time is about 1·5 years after bridge completion (except for low design bolt tension, which is represented by the Kanmon Bridge). However, a slip safety factor of less than 3 does not directly mean band slippage and, in reality, retightening is only required

TABLE 2
Difference in bolt design between Kanmon and other bridges

Name	Diameter of bolts (mm)	Designed tension of bolts (t)
Kanmon	50	40
Others	45	$65 \sim 74$

within the period from 1·5 to 3 years after completion. It can be seen from Fig. 8 that after retightening any further retightening would not be required for the time being.

CONCLUSION

The authors repeatedly measured the tension in cable-band bolts for several suspension bridges and subsequently retightened bolts on parts of the bridges depending on the results obtained. They also measured the bolt tension after retightening. Analysing this measurement data indicated the following:

(1) Retightening the bolts at the correct time definitely decreases the tension drop rate thereafter, eliminating fears of bolt tension drop indefinitely (for more than 10 years, or semi-permanently if prevention of cable-band slip is the only requirement).
(2) It was also seen that tension drop is smaller when the bolt is tightened to a high torque.
(3) With the foregoing description the authors suggest the following methods in designing cable-band bolts and controlling bolt tension in the long suspension bridges which will be constructed in the future:
   (i) Design for as high a bolt tightening tension as possible.
   (ii) Retighten bolts during the period from 1·5 to 3 years after construction. During this period it is not at all necessary to check bolt tension.
   (iii) After retightening, check the tension at intervals of about 10 years and decide the necessity of further retightening from the results.

REFERENCES

1. FUKUSHI, A., Cable inspection vehicle, *KOBELCO Technology Review*, No. 4, Kobe Steel Ltd, Kobe, August 1988, p. 41 (in Japanese).

2. On Kanmon Bridge Cable-Band Retightening Works, Japan Highway Public Corporation, Kobe Steel Ltd, March 1977 (in Japanese).
3. Investigation on cable-band tension of Kanmon Bridge, Japan Highway Public Corporation, Kobe Steel Ltd, November 1978 (in Japanese).
4. MATSUI, T., HIRANO, S. and KANEKO, M., Shape measurement and axial force research of cable-band bolts on Innoshima Bridge. Honshi Technical Report, Vol. II, No. 44, Honshu-Shikoku Bridge Authority, Tokyo, October 1987, pp. 26–34 (in Japanese).
5. Review on suspension bridge cable, Honshu-Shikoku Bridge Authority, March 1978, pp. 168–73 (in Japanese).
6. HIRAI, A., *Steel Bridges (III)*, Gihodo, 1967, pp. 719–22 (in Japanese).

64

The Design of a Flexible Surface Mix for Use at Bridge Expansion Joints

A. R. WOODSIDE and W. D. H. WOODWARD

Department of Civil Engineering and Transport, University of Ulster, Newtownabbey, County Antrim BT37 0QB, UK

ABSTRACT

Leakage at bridge joints causes corrosion and enhanced by rock-salts has caused major structural problems in the infrastructure. Development of a flexible surfacing material was considered to be a possible alternative to restrict the ingress of water, i.e. the material could accommodate the changes in movement experienced without cracking. This paper reports the work carried out in designing a flexible surfacing mix known as 'H-MAC'. It initially summarises the requirements of bridge joints and the use of conventional surfacing methods. Using the limitations of existing materials and the structural requirements of the joint, the development of the flexible mix is outlined. Comparative laboratory testing is discussed using various binder types and aggregate gradings. These resulted in a single sized 6·3/3·35 mm mix using 23% Icosit Membrane H as a binder. Bridge joint trials were carried out using this prototype mix and the results reported.

INTRODUCTION

In the UK roads network there are about 50 000 concrete bridges. These can be considered as being of two groups—those built before World War Two and those built after. The first are mostly 50–60 years old, reinforced with plain mild steel and have simple structural forms. The second and more recent group are constructed with high strength materials and many have more complex structural forms demanded by larger spans and more ambitious requirements.

For the 90 years or so since they were introduced concrete bridges have performed well and there were few problems until the 1970s. During the 1960s the use of rock-salt was introduced for deicing. It was not until some years later than the deleterious effect of salting became apparent through the development of corrosion of the steel reinforcement bars and carbonation of the concrete.

These problems tend to occur more often with modern bridges. Studies have shown that it is not due to modern concrete being less durable. Instead modern bridges on motorways and trunk roads are designed with more ambitious structural forms and so are plagued by leaking expansion joints, resulting in piers and crossheads exposed to chlorides in a manner not experienced in many of the older bridges.

The leakage of bridge joints causing corrosion and enhanced by deicing salts has created a major problem. Considerable research is now being carried out into early detection and preventive methods, and repair of affected bridges.

Development of a flexible surfacing material was seen as a possible alternative to restricting the ingress of water, i.e. the material could accommodate the changes in movement experienced without cracking. This paper reports the work carried out in developing such a surfacing and known as 'H-MAC'.

FAILURE OF SIMPLE EXPANSION JOINTS

Expansion joints are used in bridges to accommodate movements due to deformation of the bridge deck caused by changes in temperature, the passage of traffic, and in concrete structures, creep and shrinkage.

The type of bridge joint used depends on the structural materials as well as the span of the bridge. In Northern Ireland most of the bridges requiring expansion joints are made of reinforced concrete and have individual spans of less than 20 m. For this type of structure simple expansion joints can accommodate the full range of expansion and contraction under normal conditions.

A simple expansion joint consists of a vertical gap between the end of a section of a bridge deck and an abutment, or between two sections of deck. The end of the deck is therefore able to move horizontally and the gap will open and close accordingly. The expansion joint is usually covered by a layer of paving material to provide a continuous smooth surface for vehicles and pedestrians. It also prevents the ingress of water and dirt.

However, it has been found that normal bitumen-based surfacing materials deteriorate and must be replaced within a period of 1–3 years. This is caused by the constant movement across the joint and probably occurs

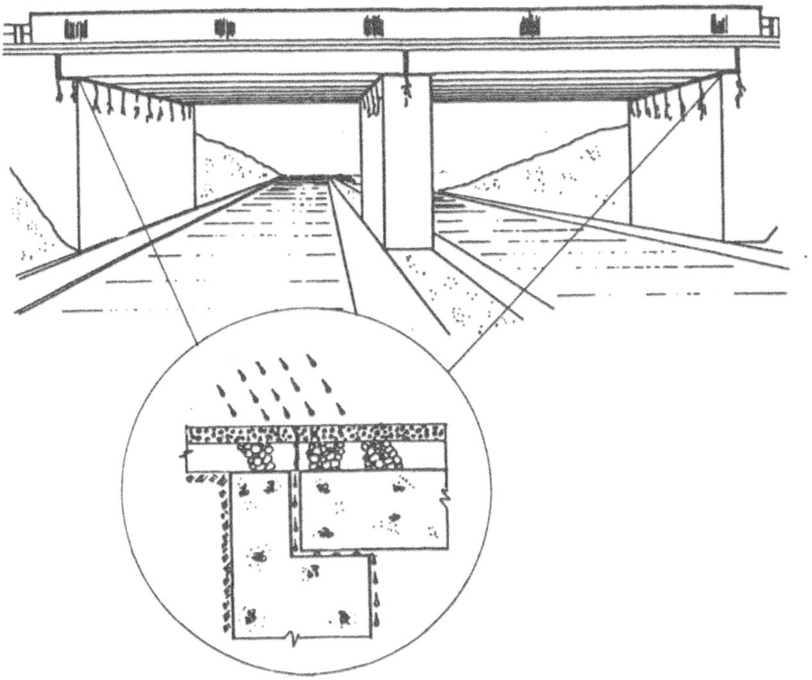

FIG. 1. Typical highway bridge. (Detail shows expansion joint where pavement has cracked.)

most rapidly at lower temperatures when the bitumen becomes stiff and brittle.

Alternative materials such as polymer modified bitumens have been used instead of bitumen as the binder but they are expensive and must be applied at carefully controlled temperatures.

Figure 1 illustrates a typical reinforced concrete bridge with expansion joints at either end. It can be seen how surface water, which may contain corrosive materials such as de-icing salts, is able to penetrate down into the structure so leading to unsightly staining and deterioration of the concrete structure.

MOVEMENTS EXPERIENCED AT SIMPLE EXPANSION JOINTS

Movement at expansion joints is mainly due to cyclic thermal expansion and contraction, and occurs throughout the life of the structure. Shrinkage and creep will also cause movement in concrete structures but 50% of the total movement will probably occur within the first month before the joint is covered. In addition, vertical and rotational deflections may be caused by vehicles crossing the bridge but these are generally small enough to be ignored.

Simple expansion joint systems may be designed for a 20-year return period as they would have to be replaced if the highway was resurfaced.

The movement and corresponding temperature changes for a bridge constructed over the M5 motorway in England was recorded. Movement occurred in daily cycles; expansion during the day and contraction at night. Daily movements of 2 mm were common and the surfacing had to accommodate movements of up to 4 mm over a week. Also much larger movements occurred over an annual cycle with 12·4 mm of movement recorded between February and June.

In developing a flexible surfacing, movements such as these would have to be accommodated if cracking is to be avoided.

DEVELOPMENT OF A FLEXIBLE SURFACING

In developing a flexible surfacing, the following criteria had to be met:

(i) Accommodate changes in expansion of ± 5 mm or greater.
(ii) Provide a durable impermeable surface.
(iii) Be relatively easy to manufacture and lay.
(iv) Be cost effective.
(v) Be mixed at ambient temperatures (5–25°C).

The investigation may be summarised as follows.

Materials

Aggregate—For all mixes Castlenavan Quarry gritstone was used.

Binder—(a) 200 binder supplied by Tennants Tar Distillers.
(b) Icosit Membrane H—a hand-applied two-pack solvent-free fast-curing polyurethane resin combination used for the production of elastic crack bridging coatings.

Types of Mix

Code	Binder (%)	Binder type	Mix type
B5.1/234	5·1	200 pen	14 mm dense macadam
H5.1/234	5·1	Membrane H	14 mm dense macadam
H7.1/234	7·1	Membrane H	14 mm dense macadam
H7.1/235	7·1	Membrane H	14 mm open macadam
H23/3.35	23·0	Membrane H	6·3/3·35 mm single sized

Manufacture of Test Mixes

For testing purposes slabs 40 × 90 × 300 mm were made. This size of mould is used for the wheel tracking test and was thought to be suitable to simulate in the laboratory a model bridge joint.

All moulds manufactured were allowed to 'cure' for 14 days before testing. An exception was the 6·3/3·35 mm single sized wheel tracking mould, which was tracked after 23 h in order to simulate early trafficking.

Types of Test

Model joint test

This is a non-standard test in which the prepared specimen is bonded to steel plates through which load is applied on an Instron machine and pulled apart at a constant rate. Test carried out at 20°C (Fig. 2).

Flexural test

This is a non-standard test in which the specimen is clamped at both ends and a vertical load is applied at the centre of the span by a Dartec until failure. Test carried out at approximately 20°C (Fig. 3).

Wheel tracking test

This is a simulation test in which a rubber wheel is repeatedly dragged

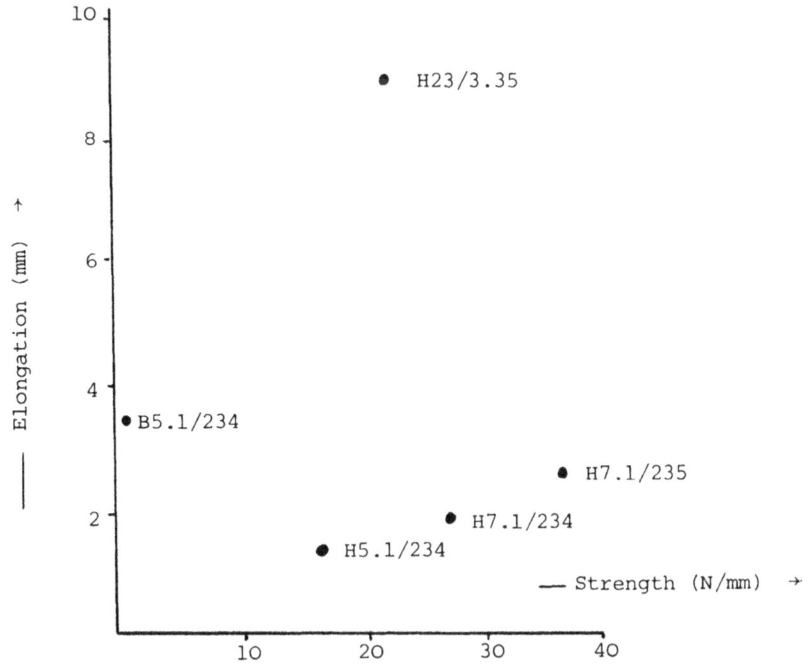

FIG. 2. Elongation plotted against strength for the model joint test.

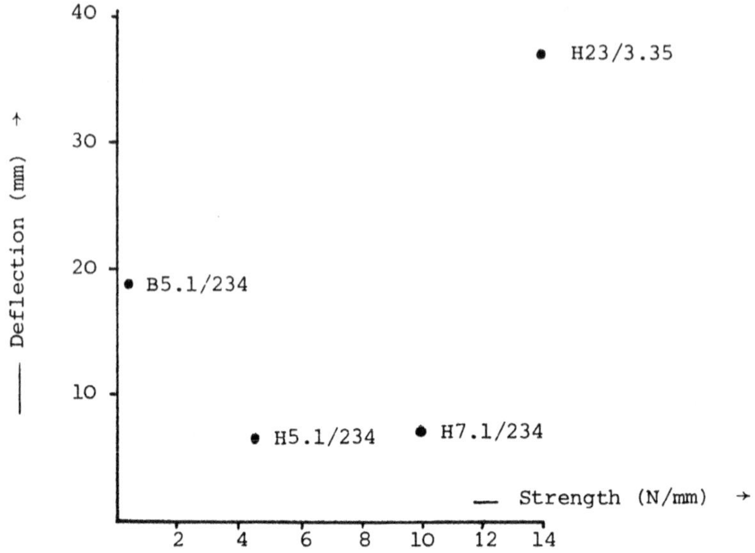

FIG. 3. Deflection plotted against strength for the flexural test.

across the surface of the specimen contained in a water bath at 40°C to assess susceptibility to concentrated trafficking or rutting.

Testing

The following tests were carried out:

Code	Model joint test	Flexural test	Wheel tracking test
B5.1/234	√	√	√
H5.1/234	√	√	—
H7.1/234	√	√	—
H7.1/235	√	—	—
H23/3.35	√	√	√

Note: √ denotes test carried out.

Results

Model joint test

The model joint test results are shown below:

Code	Strength (N/mm)	Elongation (ductility) (mm)	Stiffness (secant modulus) (N/mm^2)
B5.1/234	0·6 (1·0)	3·4 (1·0)	0·18 (1·0)
H5.1/234	15·9 (26·5)	1·3 (0·38)	12·2 (67·9)
H7.1/234	27·6 (46·0)	1·8 (0·53)	15·3 (85·2)
H7.1/235	37·0 (61·7)	2·5 (0·74)	14·8 (82·5)
H23/3.35	22·4 (37·3)	8·9 (2·62)	2·52 (14·0)

Notes: The elongation is that occurring between the two backing plates, which were 50 mm apart.
The figures in parentheses have been normalised to results for B5.1/234.

The following observations may be made:

(a) If the Membrane H-based macadam mixes are compared to the 200 pen/14 mm dense macadam mix they are 26–62 times stronger and 68–83 times stiffer at 20°C (testing temperature).
(b) For Membrane H, elongation and strength are increased by 38% and 74% respectively if the binder content is increased by 2% for the 14 mm dense macadam mixes.
(c) For Membrane H, elongation and strength are increased by 39% and 34% respectively when an open textured macadam (7·1% binder) is used instead of a dense macadam (7·1% binder).
(d) The single sized 6·3/3·35 mm mix with a binder content of 23% gave an elongation of 8·9 mm, which is 2·6 times that of the 200 pen macadam, a strength 37 times stronger and a stiffness 14 times greater.

Flexural test

The flexural test results are shown below:

Code	Strength (N/mm)	Deflection (mm)	Stiffness (secant modulus) (N/mm^2)
B5.1/234	0·74 (1)	17·9 (1)	0·041 (1)
H5.1/234	4·5 (6·08)	6·0 (0·34)	0·75 (18·3)
H7.1/234	9·82 (13·3)	6·6 (0·57)	1·49 (36·3)
H23/3.35	13·9 (18·8)	37·5 (2·09)	0·37 (9·02)

Note: The figures in parentheses have been normalised to results for B5.1/234.

The following observations may be made:

(a) If the Membrane H macadam mixes are compared to the 200 pen/14 mm dense macadam mixes they are at least 6–13 times stronger and 18–36 times stiffer.
(b) For Membrane H a 5·1% 14 mm dense macadam is 2·2 times stronger and gives 10% more deflection than a 7·1% binder mix.
(c) The 6·3/3·35 mm single sized mix with 23% binder gave a deflection of 37·5 mm before failure and was 18 times stronger than the 200 pen binder macadam.

Wheel tracking test

The 6·3/3·35 mm single sized 23% Membrane H mix was compared to the 200 pen/14 mm dense macadam. The comparison test was carried out in its most severe form, i.e. at 40°C with the tyre trailed across the surface. The results are as follows:

Code	Test assessment
B5.1/2.34	Rut 10–20 mm resulted
H23/3.35	Rut <1 mm—no noticeable deterioration

Note: The Membrane H mix was tested after 23 h curing.

DISCUSSION OF DEVELOPMENTAL WORK

When Membrane H was used as the binder substantial improvements in strength and stiffness occurred in comparison to the 200 pen mix. This would produce stronger bridge joints. However, a flexible bridge joint surfacing must accommodate expansive changes and in this respect the Membrane H macadams could not provide similar levels of elongation and deflection. In practice, cracking of a traditional surfacing is more probable at low temperatures due to embrittlement of the binder. As Membrane H is not affected by temperature change then, despite the lower levels of elongation and deflection, the stronger mix may give a longer lasting bridge joint.

In combining strength, stiffness, elongation and deflection the single sized 6·3/3·35 mm mix by far out-performed the macadam type mixes. An elongation of 8·9 mm occurred across the 50 mm distance between the test backing plates with an elongation of approximately 50–100 mm occurring before complete separation.

Other notes of interest with this type of mix were:

(i) Single sized chippings, i.e. no grading problems.
(ii) Required minimal mixing.
(iii) Required no compaction other than levelling by hand.
(iv) Provided a high level of texture depth.

ROAD TRIALS

Full-scale road trials were carried out during November 1988 as part of a routine maintenance programme on the Maghery to Portadown bridge crossing the M1 motorway in Northern Ireland.

Prior to laying the single sized 23% H-MAC material the bridge joint was sealed with ICOSIT Membrane H, a primer applied to improve the bond between the sealant and the H-MAC, and a debonding agent 25 mm wide added to increase the movement across the joint.

H-MAC was mixed in batches, hand laid and levelled. A layer of fine aggregate (3 mm gritstone) was added to improve skid resistance. After 2 days of curing the traffic was allowed over the joint.

Of the problems which arose the most apparent was that due to the material's low viscosity, consequently there were considerable handling problems associated with mixing and laying.

After 10 weeks the joint was inspected and it was obvious that the material had failed. In comparison with laboratory trials at 20°C the bridge joint trial material showed poor cohesion properties. This illustrated the influence of environmental factors in the 10-week period after the trial when the air temperature range was +6·9 to −7·0°C, with the day of the trial bitterly cold with a strong wind and high chill factor. It is thus recommended that the bridge be allowed to cure in temperatures exceeding +5°C.

It was obvious to the authors that the environmental conditions did not allow adequate curing to occur. This was also evident from wheel tracking tests which failed on material extracted from site and cured under similar conditions.

Since the trials patching of similar material has been carried out under more favourable environmental conditions, and this appears to have cured more effectively and consequently is performing perfectly on site.

CONCLUSIONS

In comparison with other binders (i.e. 200 pen in a 14 mm macadam) Membrane H performs more effectively than other conventional flexible surfacings. Developmental work has shown that this type of binder may be used to provide a superior flexible surfacing over bridge expansion joints. However, the traditional prohibitive factor of cost must be viewed with respect to the potential damage to the bridge structure as a conventional bituminous joint may lead to corrosion problems at a later date and thus shorten the effective life span of such a structure.

ACKNOWLEDGEMENTS

The authors wish to acknowledge the help and cooperation given to them by Sika Ltd, the Department of the Environment (NI) Roads Service, Mr E. T. Stewart and Mr P. Brown of Larsen Associates, and Mr Brian McDowell, without whose help this research could not have been carried out and this subsequent paper produced.

65

The Repair of a Composite Concrete–Steel Bridge

P. H. BESEM, M. WOUTERS and C. WARNON

Ministry of Public Works,
Bridge Office, rue Côte d'Or 253—4200 Liège, Belgium
Bridge Office, WTC 3, Blvd S. Bolivar 30—1210 Bruxelles, Belgium
Roads Direction, Av. Gouverneur Bovesse 37/31—5100 Jambes, Belgium

ABSTRACT

Bridge 23 is an overpass over the E42 motorway on the route from Dunkirk (France) to Wurzburg (Germany) and is situated near the interchange of Daussoulx in Belgium. The bridge, built in 1969, belongs to a standard series of composite concrete–steel structures designed to pass above a motorway of 40 m wide. In 1984 the superstructure was crashed into by a trailer of a lorry, which suddenly came loose and slightly rose from the carriageway, thus hitting the girders of the bridge crossing the motorway. The damage was considerable: the first two main girders were irreparably deformed. The Road Department, which had commissioned the building of the structure, decided on a complete repair of the bridge by replacing all damaged elements. This repair was in line with what could be claimed from the insurance company: a restoration of the bridge to its original state before the accident. Moreover, after replacing the damaged girders, the entire steel construction was protected against corrosion by using a method based on metallising and painting in situ.

DESCRIPTION OF THE DAMAGE (Fig. 1) AND CONCLUSIONS AFTER THE ACCIDENT

Concrete Bridge Deck

After inspection of the bridge deck, no visible damage to the concrete as a result of the accident was found. In the corbellings there were many old cracks through which water was seeping.

Main Girders

The first two main girders at the east side were considerably deformed

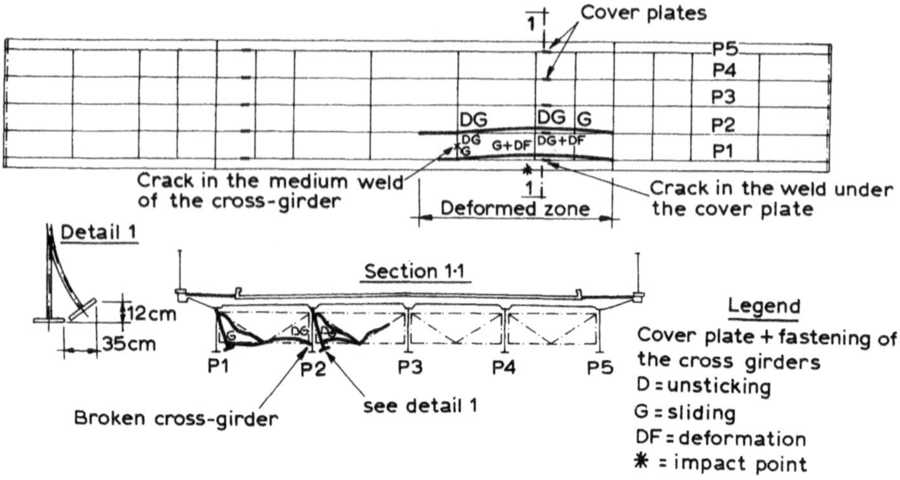

FIG. 1. Description of the damage.

from the pier onwards to the middle of the structure. The deformation was both axial and vertical. The cover plates near the point of impact had slid and loosened from the web of both girders, thus causing the cracking of the welding seams situated at the lower part of the cover plates.

Cross Girders

The six cross girders between girders 1 and 2 and 2 and 3 were heavily deformed. The second cross girder between girders 1 and 2 was broken near its contact with girder No. 2. There were cracks in the welding seams in the middle of the third cross girder between girders 1 and 2.

Fastening of the Cross Girders to the Main Girders

The cross girders were fastened to the main girders by means of T-irons bolted on these girders. Most of these T-irons had slid away from the web of the girders. Near the most deformed areas of the main girders these T-irons were also seriously deformed and they had loosened from the web of the girders.

Bearings on the Piers and on the Northern Abutment

The bearing of girder No. 1 on the northern pier had undergone a displacement to the centre of the bridge as well as to the west side. The bearing of girder No. 2 on the northern pier had undergone a horizontal rotation towards the west. The corresponding bearings on the northern abutment had undergone a differential settlement on their southern side but there were no definite indications that these failings were due to the accident.

CHARACTERISTICS OF THE STEEL STRUCTURE

Bridge 23 belongs to a series of composite concrete–steel bridges designed to pass above motorways 40 m wide. A study was made of these structures in order to standardise their elements as much as possible. Bridge 23 has five continuous main solid-web girders, placed at 2·71 m intervals, resting on four bearings, braced by lattice work and supporting a concrete slab of 18 cm thick contributing to the region of positive moments. The connection is made by means of pistol-welded dowels. The bearing distances are 22–41 and 22 m for a total length of 85 m.

The height of the main girders varies linearly over 7 m on each side of the piers. These main girders have two assembling joints situated in the centre span at 13·50 m on both sides of the axis of the structure. The bracing consists of 14 lattice cross girders, six in the centre span, two in each side span and one on each bearing. Temporary wind bracing was placed when mounting the structure. The building of the structure has taken place in six successive phases:

1. Building of the infrastructure.
2. Mounting of the three sections of the main girders on the piers, abutments and temporary support in the middle of the centre span, and fitting of two bolted joints by means of high-tensile bolts (HT bolts).
3. Control of the stresses in the steel girders by a 20 cm variation of height of the intermediate supports.
4. Casting of the concrete slab.
5. Removal of the central support and the assembling wind braces.
6. Placing of a waterproofing layer—surface dressing—cut stones—bridge rail.

STUDY OF THE REPAIR WORKS

When studying the repair works, two solutions were considered:

First solution: restoration of the structure to a state as close as possible to its original state by replacing the damaged parts (main girders and cross girders). To do this it was necessary to demolish and replace a considerable part of the slab and the equipment (cut stones, bridge rail).

Second solution: reinforcement of the structure by bracing the lower girder flange of the girders and by replacing the damaged cross girders only. The first solution required large-scale and expensive works, necessitating a temporary interruption of the motorway traffic during certain phases of the work (dismantling and reassembling of the steel elements, for instance). The second solution, on the other hand, would only have required traffic interruptions in one direction. This solution, however, would have amounted to reusing a structure containing damaged but repaired elements, and therefore one had to be absolutely sure that the remaining parts were still in good condition (particularly the slab, dowels, weldings, assemblies and bearings on the northern pier).

As the second solution involved more risks and because the consequences of the accident were covered by the insurance, the department decided on the first solution. This solution thus consisted in restoring the structure to a state as close as possible to its original state (before the accident).

Therefore it was necessary to demolish the part of the structure containing the slab at the east side over a width of about 5 m and to replace it by new or reused elements. To do this the same scheme had to be followed as for the original works. During the repairs certain stress regulations had to be carried out as well.

The study of the repair works resulted in a working scheme containing the following phases:

1. Building of a closed working platform above the carriageway with a clearance 9·50 m wide and 4·30 m high for each traffic lane.
2. Removal of the equipment (bridge rail and cut stones which were to be used again) over the whole length of the bridge and removal of the expansion joints.
3. Removal of the surface pavement and the waterproofing layer over the whole length of the bridge.
4. Unbolting and removal of the cross girders between girders P2 and P3.
5. Demolishing the slab over a width of 5·20 m at girders P1 and P2 over the whole length of the bridge, while taking care not to damage the dowels and the reinforcing bars coming out of the remaining part of the slab.
6. Placing a temporary support under girders P1 and P2 (which were to be replaced) in the longitudinal axis of the motorway and raising of the girders P1 and P2 with 20 cm.
7. Unbolting of the joints of the web and boom plates of the girders P1 and P2.
8. Removal of the central part of the P1–P2 unit and removal of the northern part (girders and cross girders) of the P1–P2 unit.
9. In the workshop fabrication of the new girder sections and the new cross girders. Control and restoration of the non-deformed cross girders.

10. Sub-assembly in the workshop of the central and side units of girders P1 and P2, and the cross girders.
11. Placing of the central unit and bolting to the remaining southern sections of P1 and P2.

    Placing of the northern side unit. Connection by bolts of the several elements. To assemble the different parts the girders were placed 20 cm higher than their final levels at the three piers (two definite piers and one temporary pier placed in the central reserve of the motorway) and afterwards those three supports were lowered by means of long-travel jacks installed on the temporary pier. A temporary wind bracing was placed between P2 and P3.
12. Casting of the slab, after shuttering and reinforcement, while providing a longitudinal joint in the middle of the slab between P2 and P3.
13. Removal of the central support after the concrete had hardened.
14. Mounting of the cross girders between P2 and P3.
15. Concreting of the longitudinal joint of the slab between P2 and P3.
16. Casting, placing and installation of the equipment: waterproofing layer, protective layer, surface dressing, cut stones, bridge rail, expansion joint, water diversion.
17. Removal of the working platform and putting the site back in order.

It is worth mentioning that the composition of the new girder elements is somewhat different from the original composition. This choice was based on certain technical requirements. However, this has resulted in only a few differences with regard to the stresses and in some minor adaptations with regard to the cambers.

During the works, as the new concrete had not yet stopped creeping or shrinking, a temporary overloading was placed on the new part before the longitudinal joint of the slab was cast, and before the cross girders were fixed between P2 and P3 in order to compensate these effects.

EXECUTION OF THE WORK

Traffic Problems on the Motorway

When organising the repair works, special attention was paid to the traffic situation:

—The E42 is a very busy motorway with an average daily traffic of 17 900 vehicles, of which 22% are lorries.

—The bridge is situated at a few hundred metres of the motorway interchange of Daussoulx between the E42 and the E411 Brussels–Luxemburg. Near the bridge the traffic flows of both motorways converge together.

—The bridge is situated north of the town of Namur. The possible relief routes guide the traffic either to rural roads with insufficient capacity or through the centre of the town where the roads are already saturated.

The department therefore decided to maintain the traffic on 2 × 2 lanes during the works, except during the dismounting of the damaged girders and the mounting of the new girders. To that end the contractor built a closed working platform preventing the traffic on the overpass and enabling the works to be carried out.

This working platform was built as a walkway resting on scaffolds placed in the central reserve and on the hard shoulder.

For the dismounting and mounting of the four sections of the main girders, the traffic on the motorway was interrupted for 30 min under control of the Gendarmerie (state police force). It would indeed have been too dangerous to transport girder elements of 8–10 t at the end of the jib of a crane above moving vehicles. The traffic was interrupted during off-peak hours.

Technological Conception of the New Steel Elements

Technological evolution

Nowadays the assembling methods which were applied when the initial structure was built are not used any more. As one-third of the main girder had stayed intact, the dimensioning of the new elements was influenced by the global dimensions and the thickness of the remaining parts.

The original main girder was made of a web plate in AE26 or AE36 steel, according to the different parts and assembled by butt-joints to half-sections in AE26 steel (upper flange plate) and AE36 steel (lower flange plate) which were cut out of rolled steel HE beams.

The current conception is rather based on a connection by fillet weldings of the web plate to the two flange plates. The modern quality of the fillet weldings made it possible to adopt this solution whereas butt-joints were preferred before because of the possibility of examining the composition by X-rays.

The plates also have a slightly different thickness. Because of the rather small quantities required for the repair works it was decided to standardise the products in order to limit the delivery period.

Figure 2 shows the concept of the connection between the remaining parts and the new ones. The differences in thickness in the joints made it necessary to use packing, sometimes at both sides of the joint, in order to maintain a minimum packing thickness of 4 mm.

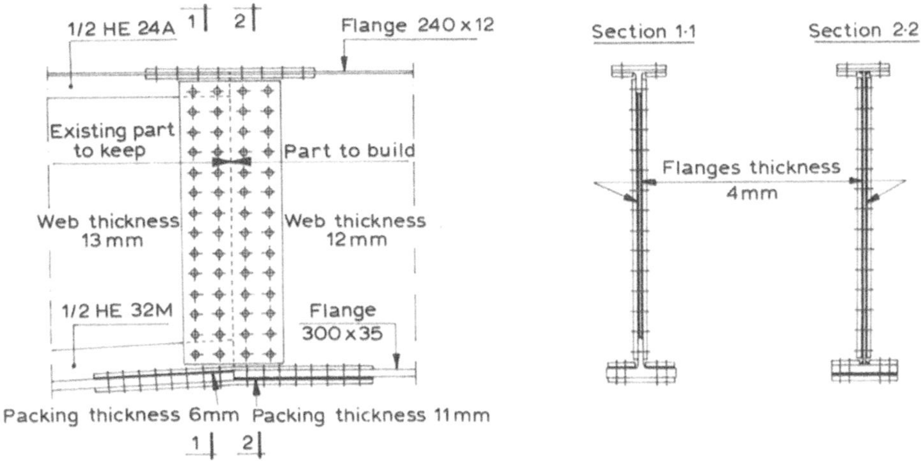

FIG. 2. Connection between remaining and new parts.

Choice of the materials and assemblage concepts

All the welded steel which was used belonged to the qualities AE235D or AE355D according to the Belgian NBN A21-101 standard.

All workshop assemblies were carried out by applying semi-automatic and automatic welding methods involving the use of basic products. As required by the regulations of the Ministry of Public Works, all welders and operators were chartered in advance for the entire work. To be chartered it was necessary to prepare standard test pieces while using the same steel types and welding products as were to be used in the actual repair works, thus allowing the department to check the exterior look of the joint and its mechanical characteristics.

With regard to the automatic welding method, these test pieces also served to check the characteristics of the actual joint. All steel types were to be dried or preheated before the welding when AE355 steel was used.

The basic products were kept in heated chambers until use. All the assemblages *in situ*, those providing the connection with the remaining parts as well as those between the new elements, were carried out according to the new NBN E27-071 to 073 standards by using HT bolts of the 10·9 quality. To secure the bolts, the couple method[1,2] was applied. It is worth mentioning that it was the first time in Belgium that HT bolts were used, of which the nuts were covered in the factory by a lubricant based on teflon.

This method has the advantage of reducing considerably the value and the dispersion of the friction coefficient k in the formula $Ma = kdPv$, enabling determination of the screw-holding power Ma (N m) according to the diameter of the bolt, d (mm), and the tension force which will have to be exerted, Pv (N). This force generally amounts to about 0·7 times the value of the (lower) yield point of the bolt steel. Thanks to the teflon, the value goes from $0{\cdot}18 \pm 0{\cdot}02$ to $0{\cdot}10 \pm 0{\cdot}005$.

Description of the Main Phases of the Repair Works

The works began with the removal of the equipment and the demolition of the surface dressing and the waterproofing layer. To do this the bridge was freed of all its dead loads and cut lengthwise into two separate decks. The cross girders were unbolted, but because of the welding seams it was impossible to remove them. Therefore the hindering seams were gouged.

The slab from the damaged part of the bridge was demolished by means of a concrete breaker. The parts of the slab which were to be demolished were cut out in rectangular pieces, the reinforcement bars were burnt through with a torch and the concrete blocks were removed. Near the upper boom plates the remaining concrete was removed by means of a pick hammer in order not to damage the dowels of the girder elements, which were to be used again.

The two deformed girders were jacked in the axis of the motorway, thus bringing both elements to rest. After unbolting the cover plates, the girders were lifted and placed on the remaining deck, then cut into transportable blocks and disposed of. After having removed the girders, the contact surfaces of the 20-year-old HT bolt assemblages were examined. It appeared that the contact surfaces had resisted well to corrosion. Little rust was discovered, especially in the support zones of the washers (Figs 3 and 4).

FIG. 3. HT bolt assemblage.

FIG. 4. Cross girder connection.

Moreover, the contact surfaces had been painted before the cross girders were connected. The behaviour of the paint appeared to have been excellent. At the same time the new girders were fabricated in the steel workshops. They were transported to the building site and placed on five supports before being assembled as continuous girders by means of cover plates. Because of the small weight of the structures, it only took 30 min to put the elements in their right places and to assemble them. Next the central jack was lowered by 20 cm, causing the supports on the pier to come down 20 cm as well. This prestressing by deformation of the girders caused stresses opposing the dead loads in the critical sections.

The orthotropic slab was cast between the two girders, on the verge corbels and by a corbelling between P2 and P3, in successive castings of about 15 m long and beginning at the edges.

Next the temporary central support was removed, the cross girders between P2 and P3 were replaced, and the longitudinal joint of the slab was concreted after partial loading of the bridge in order to adjust the levels of the two decks. The existing concrete deck was repaired by means of a modified hydraulic mortar reinforced with synthetic fibres. This repair product is characterised by a good adhesion power to the existing concrete and by an excellent compressive strength (50–60 MPa).

The whole bridge deck was cleaned under pressure with water in order to remove the non-adherent particles and the remains of the waterproofing layer. The deck was made waterproof by means of an impregnating varnish covered by a prefabricated asphalt membrane welded to the support by a torch flame. This membrane consists of a polyester reinforcement impregnated with filled elastomer asphalt covered with an anti-adherent film and with mineral aggregates. Protection of the waterproofing layer is by a bituminous mix covered by a wearing course of the same type.

The repair works also offered the possibility to replace the neoprene supports, of which the hoop reinforcements had deteriorated by corrosion, by new supports completely encased with rubber; to renew the anchorages of the bridge rail by galvanised all-thread rods sealed with modified hydraulic mortar; to repair the spallings of the concrete on the borders of the slab and to hide the repairs by means of an acrylic paint; and to repair the blue stone boards under the bridge rail by glueing the broken elements by means of epoxy glue.

Finally, a waterproof expansion joint was placed at the mobile support side and a 'Thorma' joint at the fixed support side of the bridge. All equipment of the structure were then put back in place.

Anti-corrosive Protection of the Structure

Introduction

Until a few years ago all steel structures of the Ministry of Public Works received, after an abrasive material blasting, an anti-corrosive protection consisting of four layers of so-called 'formula paints'.[4]

The composition of these paints was decided by the department, which consequently guaranteed they would last well. Although those products had been very satisfactory, more and more 'performance paints' are used nowadays. In this system the composition and the performance are guaranteed by the supplier.

The use of these paints is normally subject to an insurance policy covering the damages on the structure for a specified period.[5] By this policy the anti-corrosive behaviour, the adhesion, the non-chalking, the non-development of cryptogamic substances as well as the preservation of the final hue must be guaranteed.

Inspection of the existing structure

When the structure was built all steel surfaces were protected by a painting method based on red lead, basic lead carbonate (white lead) and flax seed oil.

After 18 years the existing paint had become completely porous, the structure was covered with rust pits and large steel surfaces were completely oxidised. This situation was discovered after an analysis by electronic microscope scanning (EMS) of old paint samples. Semi-quantitative analysis on a sample of existing paint indicates the nature of the discovered substances and gives a close estimation of their weight percentages.

As there are no references in that field, the results should be considered as an indication (Tables 1 and 2, Figs 5 and 6).

Study of the costs of a protection system

Before choosing between a 'formula' painting method and a 'perfor-

TABLE 1
Paint face

Substance	Electronic lay	Wt %
Na	K	2·507
Al	K	2·760
Si	K	9·597
P	K	0·902
Pb	M	72·986
K	K	0·233
Ca	K	0·748
Ti	K	7·954
Fe	K	2·313

TABLE 2
Steel face

Substance	Electronic lay	Wt %
Al	K	3·171
Si	K	9·733
Pb	M	22·137
K	K	0·417
Ca	K	0·520
Ti	K	0·802
Mn	K	0·313
Fe	K	31·537
Zn	K	0·463
Pb	L	30·908

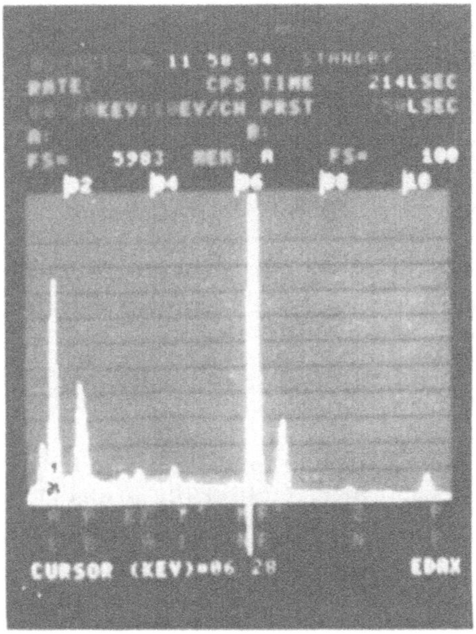

FIG. 5

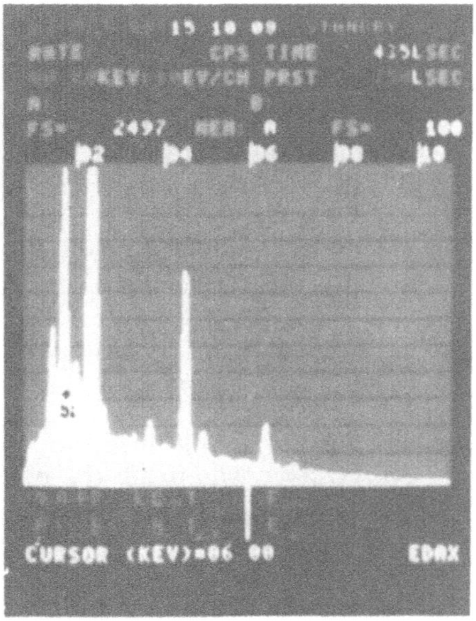

FIG. 6

mance' painting method it is, of course, necessary to compare the investment cost as well as the durability and the offered guarantees.

The investment cost of the considered performance method (total thickness 240 μm) amounts to about 150% of the cost of the method consisting of applying four layers of formula paints (total thickness 160 μm). The first method offers a guarantee of 20 years compared to a probable durability of 5–10 years before the first restoration for a traditional method.

Supposing that the traditional system requires (1) little maintenance (touching up of the upper layer) every 5 years and a complete renewal after 15 years, (2) a repainting of the upper layer after 20 years with the 'performance painting' method and (3) an annual interest rate of 8% for the invested sums, we conclude that after 10 years the method used in Daussoulx becomes more economical.

A saving of 20% is made after 20 years. Moreover, the lower frequency of the maintenance is a considerable advantage with regard to the comfort of the road users.

Description of the chosen protection method

The following method for the maintenance of the structure was chosen:

—abrasive material blasting;
—metallising with Zn–Al (the minimum thickness of the metal layer is 80 μm);

—application of a 20 μm layer of epoxy paint diluted to 30% and pigmented with micaceous iron oxide containing an amino hardener;
—application of a 60 μm layer of the same paint but non-diluted; and
—application of an 80 μm layer of polyurethane paint with two components. (An acrylic polyurethane resin interwoven with an aliphatic polyisocyanate resin used as a binder).

The choice of the method was based on the following arguments:

—taking into consideration the rusty state of the structure and the conclusions of the above-mentioned EMS analysis, complete stripping was necessary;
—the choice of the Zn–Al alloy was justified because of a better impact resistance; and
—the aim of the application, very shortly after metallising, of the first epoxy layer was to fill the porosities left after metallising in order to prevent humidity inclusion; it also guaranteed good adhesion of the first layer.

Considering these needs (filling and adhesion), the paint formula had to comply with the following characteristics:

—an appropriate solvent balance;
—a well defined pigmentation with high volume concentration; and
—a sufficient binder percentage.

The result was good wettability of the metallised surface and excellent filling power.

The first layer of 20 μm was covered with a second 60 μm layer of the same paint in order to form an intermediate layer of 80 μm (dried coat) of epoxy paint with a base of micaceous iron oxide, ensuring a good protective coating. Indeed pigmentation with a base of micaceous iron oxide forms a barrier against the permeation of humidity, which constitutes an essential quality of an anti-corrosive painting method.

The finishing coat had, in addition to good chemical resistance, to comply with aesthetic and mechanical standards, namely chalking resistance and colour stability. Moreover, the chosen polyurethane formula allowed good spray painting and brushing *in situ*, without blistering. It also guaranteed good coverage of the edges.

It may be useful to repeat the main principles of the metal spraying (metallising) method with pure zinc or Zn–Al alloy. This anti-corrosion protection method, also known under the name of schoopage, consists of melting the metal wires and spraying the melted metal by means of a flame spray gun; this method can be used in the workshop as well as on the building site, regardless of the type of elements or their dimensions, and can be applied on both new and old steel structures.

Metallising is always preceded by abrasive material blasting (sand, corundum, steel shot, cinders) of the steel surfaces, on the one hand, to eliminate all traces of rust and impurities in order to obtain a high degree of cleanliness and, on the other hand, to give them a degree of rugosity enhancing the adhesion of the coating.

As metal wires are used for spraying, the zinc or the zinc alloy (85-15) starts melting at the level of the spray gun nozzle under the influence of a gas torch, the fuel being propane or acetylene and the combustible being oxygen. Thus the melted metal is shot by compressed air from a distance of 15 cm onto the steel target and the droplets solidify when they hit the target. This way a structured and consistent coating is gradually formed by successive sweepings along the surface which has to be coated. The Zn or Zn–Al layer thus obtained after metallising is slightly porous and rough enough to allow good adhesion of the paint layers.

Technical requirements of the Ministry of Public Works

Basic products: The instructions of the Ministry of Public Works[5] dictate that the paint supplier must guarantee certain qualities for each layer as well as for the whole treatment. The method used is tested in advance for confirmation by the department. The tests are carried out by the Bridge Office of the Ministry of Public Works and they are confined to technological tests on small plates covered either with one layer or with all the layers of the protective treatment. Table 3 gives a survey of the results obtained on test plates and of the results announced by the paint factory.

Application: Both the metal spraying and the application of epoxy or polyurethane paints are subjected to very strict conditions with regard to the relative humidity, the ambient temperature when applying the paint and the maximum time between two successive phases.

Because the whole treatment was carried out on the building site and Belgium has a very unreliable climate, the following application conditions were imposed:

—degree of sand blasting, SA 3;
—time between sand blasting and metallising, maximum 2 h;
—time between the metallising and the first paint layer, 1 h maximum;
—ambient temperature, higher than +5°C; and
—steel temperature, 5°C higher than dewpoint.

Organisation of the building site for the painting

In order to observe the application requirements, the contractor had organised his planning very meticulously. He had built his scaffolds athwart of the structure; they were provided with three separate working platforms of confined width.

TABLE 3
Comparison between the announced data and the results of the tests

Tests	Unit	Epoxy		Polyurethane		System	
		GV	MV	GV	MV	GV	MV
Cupping test (ISO 1520)	mm	6	9·2	5	4	1·5	4
Adhesion test by cross-cut (ISO 2409)	0–5	1	0	1	0	0	0
Pull-off test for adhesion (ISO 4624)	N/mm^2	—	—	—	—	60	54
Pendulum damping test (ISO 1522)	s	115	165	230	105	—	—
Bend test on conical mandrel (ISO 6860)	mm	15	10–12	5	58–70	—	—
Rapid temperature variation resistance				Good	N	Good	N
Resistance to 20% HCl (ISO 2812)		—		—	—	Very good	N
Resistance to 20% NaOH (ISO 2812)		—		—	—	Very good	N

GV = guaranteed value.
MV = measured value.
N = no degradation such as blistering, cracking, flaking, etc.

A wooden platform 3 m wide and 13 m long (corresponding to the width of the bridge) had been fastened to a tubular scaffold travelling on rails. Special protection by means of canvasses had also been provided in order to isolate the workmen during the sand blasting, metallising and painting operations.

The necessary time to set up the building site (protection, moving of the scaffolds and collecting of the blasting material) took 249 working hours.

Two sandblasters and a part-time helper (70% of his time) worked for 400 h to treat the 1661 m^2 steel surfaces of the bridge, which constituted a blasting rate of 4 m^2/h. They used 50·5 kg/m^2 of abrasive material of the 'Vasilgrit' type, having a grain size between 0·5 and 2 mm and which was stored in silos of about 20 t. The material was blasted at a pressure of 7 atm (produced by a compressor) by means of a sand blast machine with a capacity of 240 litres, of which the blast-pipe was made of carbide and had a diameter of 8 mm (entry) and 12 mm (exit). In total, the compressor consumed 2825 litres of diesel oil for the sand blasting and the metal spraying operations.

A plan was worked out to observe the intervals between the sand blasting and metallising operations. Steel girder sections, each of 6 m long, were successively stripped by the sandblasters so that metallising could take place within 2 h after the sand blasting.

One metal sprayer and the helper (for the remaining 30% of his working time) sprayed the Zn–Al on the successive girder sections of 6 m long, as described in the plan.

In total, it took 184 working hours to spray a coating of minimum 80 μm thick over the whole surface of the steel parts of the bridge (1661 m^2). Therefore 1045 kg of Zinacor 850 wire with a diameter of 4·76 mm were necessary, which corresponded to a covering rate of 9 m^2/h. In the flame spray gun, the wire progressed at a speed of 2·64 m/min. This allowed the operator to obtain the required coat thickness by two crossed sweepings. The gas torch used acetylene as fuel and oxygen as combustible. The complete metallising operation required the use of 18 acetylene bottles and 26 oxygen bottles.

Two painters had to work 285 h to treat the total surface of the metallised structure. The first filling layer of 20 μm thick was applied with brushes whereas the intermediate layer of 60 μm thick and the finishing polyurethane layer (100 μm thick) were applied with a low-pressure spray gun (opening 1·5 mm). The following quantities were needed to paint the structure:

—epoxy, 707 litres;
—polyurethane, 365 litres;
—diluent, 350 litres; and
—cleanser, 230 litres.

During each working phase specific checks were carried out by the Bridge Office:

—The condition of the stripped areas and the rugosity of the surfaces were regularly checked. In cases of doubt visual control was completed by rugosity measure. The required degree of cleanliness was based on the Swedish Standard SIS 055900 B/SA3.
—The thickness of the metal layer was checked by means of an electronic thickness meter (magnetic effect) of the Elcometer type.

The measured thickness always varied between 75 and 140 μm. Indeed the required minimum thickness of 80 μm made it necessary for the metal sprayer to aim at a much higher average of about 120 μm.

The adhesion of the metal layer was checked by means of cross-cut adhesion tests carried out according to the Belgian Standard NBN T22-107:

—During the application of the intermediate paint layers, the thickness of the layers was checked several times to make sure that the painter was observing the required thickness limits.

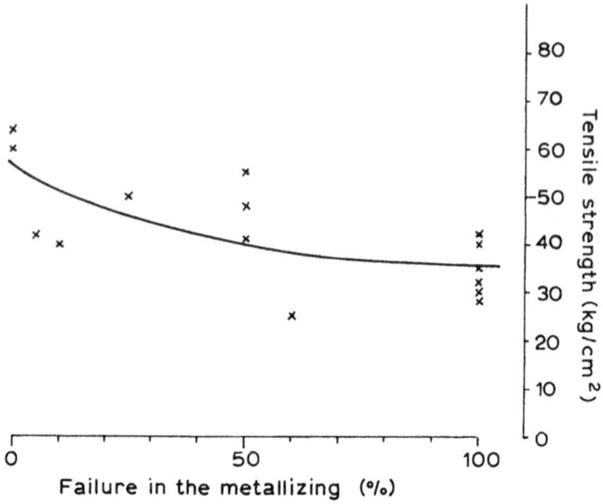

FIG. 7. Relation between the adhesion and the percentage of disbonding of the metal layer.

Taking into account the thickness variations of the metal layer, it was indeed very difficult to determine with accurate precision the thickness of each paint layer.

—A final check of the entire coat showed that its thickness varied between 200 μm and more than 400 μm. The theoretical value being 240 μm, some very restricted zones with a thickness between 200 and 240 μm were tolerated.

The adhesion of the coat was checked by means of tensile tests according to the International Standard ISO 4624/78. The values which were obtained varied between 25 and 64 kg/cm^2. The breaks occurred either in the metal layer or in the polyurethane layer; the lowest values were obtained in the case of debonding of the metal layer. From the results of those measurements it was concluded that the average lower limiting value of the adhesion amounted to 35 kg/cm^2 for the metal layer and 55 kg/cm^2 for the polyurethane layer (Fig. 7).

TESTING OF THE STRUCTURE

As in the case of a new structure, a loading test was carried out on the bridge by means of 25-t lorries. The test was in four parts:

—centre span;
—centre span and one side span;
—two side spans; and
—asymmetrical loading.

CONCLUSIONS

In spite of the considerable size of the elements which had to be replaced, a thorough previous study of the construction phases had made it possible to reduce the traffic interruption under the structure to a minimum.

The repair works have offered the opportunity to apply new techniques in Belgium: high-tensile (HT) bolts covered with teflon, metallising with Zn–Al of the steel structures on the building site, use of polyurethane paints, and waterproofing layer consisting of a prefabricated elastomer asphalt membrane.

PARTIES INVOLVED

Commissioning Authority

Ministry of Public Works Road Department— Direction of Namur
Avenue Gouverneur Bovesse 37, B-5100 Jambes

Inspection of the Works

Bridge Office—External Services, rue Côte d'Or 253, B-4200 Liège

Planning and Design

Bridge Office—Central Services, WTC—Tour 3, Boulevard Simon Bolivar 30, B-1210 Brussels

Controlling Engineering Office

Bureau SECO, rue d'Arlon 53, B-1040 Brussels

Contractor

S.A. J. Richard, rue de Jemeppe 224, B-4431 Ans (Loncin)

Subcontractors

Steel construction: S.A. Poncin, Grand'rue 72, B-5292 Clavier (Ocquier)

Paint works: Association Namotte S.A. and Appruzzese, rue du Progrès 10, B-4430 Alleur

Working study: Bureau J. Hovelmann, Avenue du Luxembourg 7, B-4020 Liège

REFERENCES

1. Besem, P. H., Dehan, E. and Piraprez, E., Le serrage des boulons à haute résistance dans les assemblages réels (Tightening of high-tensile bolts in actual connections). CRIF—WTCM, July 1987.

2. LEROY, Y. and BESEM, P. H., Mise en oeuvre des boulons à haute résistance. Comparaison des méthodes de serrage (Use of high-tensile bolts. Comparison of the tightening methods). *Annales des Travaux Publics de Belgique*, 4, 1979.
3. BESEM, P. H. and WARNON, C., La remise en état du pont de Daussoulx (The repair of the bridge of Daussoulx). *Annales des Travaux Publics de Belgique*, 1989.
4. Circulaire 576-56 du Ministère des Travaux Publics, 'Protection des métaux ferreux contre la corrosion—Peintures à formule imposée' (Circular No. 576-56 of the Ministry of Public Works, 'Protection of ferrous metals against corrosion—Formula paints').
5. Circulaire 576-63 du Ministère des Travaux Publics, 'Protection des métaux ferreux contre la corrosion—Peintures à performances' (Circular No. 576-63 of the Ministry of Public Works, 'Protection of ferrous metals against corrosion—Performance paints').

66

Hydrodemolition—A Modern Technique of Concrete Removal in Bridge Repair

Renzo Medeot

FIP Industriale SpA, Via Scapacchiò, I-35030 Selvazzano (PD), Italy

ABSTRACT

Given the ever-increasing traffic volume, repair of bridges as key elements of road systems is a problem of growing importance. As is well known, any restoration work first involves the removal of all traces of deteriorated concrete only, while avoiding any damage to sound concrete and restoration materials. The technique of hydrodemolition, developed at the end of the 1970s in Italy, satisfies these requirements, using high-speed water jets without abrasives. This report describes this technology, explaining its theoretical basis and giving many examples of practical applications.

INTRODUCTION

Hydrodemolition is a new word that, borrowing two terms from ancient Greek and Latin, has been created to describe a process as old as the earth itself: the destruction of rock and materials harder than concrete by the relentless force of falling and surging water. The wearing force of water is well known. Over aeons of geological time the Colorado River carved the Grand Canyon, and the Niagara River the famous falls. In practice the hydraulic power of a waterfall of about 10 000 m, 150–300 litres/min, has been harnessed to produce equipment to remove concrete. Hydrodemolition technology, in essence, compresses time from centuries to seconds by speeding water flow to real-time cutting force, to demolish the bonds uniting the concrete aggregate. But let us pause here to give a history of the discovery and development of hydrodemolition.

In the later 1970s it was realised that the problem of removing large areas of deteriorated concrete (for example bridge decks) was becoming increasingly urgent. Apart from the specific repair techniques to be adopted, any restoration work first involves removal of deteriorated concrete. This delicate and often difficult task requires

(a) total removal of all traces of deteriorated concrete;
(b) avoidance of any damage to sound concrete and reinforcing steel; and
(c) good bonding, e.g. a good support surface between existing concrete and restoration materials.

Traditional methods, based essentially on the use of pneumatic hammers, did not guarantee satisfactory results. In particular, the greatest problem was operator difficulty in differentiating between poor quality and good quality concrete, which led to either incomplete removal of poor concrete or excess removal of good concrete. Serious research on possible alternative methods revealed that several studies had been carried out on this subject, thus confirming its importance. However, no practical results had been achieved, although we may quote, for the sake of curiosity, heat treatment methods (all based on producing rapid heating of the damaged area, e.g. by using flamethrowers, plasma beams or even lasers); abrasive processes (based on the use of rotating discs coated with industrial diamonds in a metal matrix or carbonium bound with bakelite); electrical and chemical processes (only applicable in special situations); and—the most curious new method in the group—the use of microwaves.

Research revealed that although pressurised water could demolish concrete no successful attempt had been made. In spite of this, we believed that water jets represented the most promising path and, in spring 1979, we decided to start a research programme aimed at producing equipment for removing concrete by means of high-speed water jets.

The most important discovery made during this research was the following: 'removing a layer of concrete is a process which differs radically from boring and cutting'. To use a familiar example, it is like the difference between sawing and planing a piece of wood: the tools are different, and so are the ways of using them. The failure of attempts all over the world was due essentially to having used techniques which were more suitable for boring or cutting. Strong in this knowledge, we needed a few months to prepare a prototype which was successfully used on the Viadotto del Lago in November 1979 in conjunction with the Italian Road Authorities. The first commercial equipment was ready by spring 1980.

After a series of improvements to perfect the system, hydrodemolition technology was introduced into other countries. The first was Sweden where, in the summer of 1984, the equipment was used on many bridges and was tested by the Swedish Road Authorities (Vägverket), with very flattering results. In spring 1984 it was presented at the World of Concrete in Washington, DC. In the autumn of the same year it began working in Toronto in the Manulife Parking Garage. In 1985 hydrodemolition equipment was successfully used in the USA on the Memorial Bridge Rehabilitation Project. Today hydrodemolition is unanimously accepted as the best process for concrete removal. It has become popular in many countries and the new rehabilitation projects specify this technique at least as an alternative to the traditional methods.

HYDRODEMOLITION

By 'hydrodemolition' we mean the process of selective removal of concrete by means of one or more high-speed water jets. Although the term 'demolition' may recall its synonym, 'destruction', it should be clarified at once that our technique deals with the *selective removal* of deteriorated parts, aiming at static restoration of the structure and not at its total destruction.

Hydrodemolition Mechanism

Concrete is an inhomogeneous material made up of aggregates (sand and gravel) and bonding agent (cement), with gaseous inclusions which make up the so-called porosity. Porosity is generally undesirable, since it alone is an effective weakening agent through which degradation takes place.

The water jet accomplishes its destructive action by means of three separate mechanisms, i.e. direct impact, pressurisation of cracks and cavitation. These three processes reach their maximum efficiency when the water jet strikes the bonding agent. The nozzle is thus played rapidly and continually over the area to be removed and excess water allowed to drain away. However, jet efficiency is a maximum when the jet itself is stable, and stability is influenced by the shape and configuration of the feeding pipe and nozzle, exit speed of water, distance from point of impact, etc.

The conclusion is that an efficient and therefore economic removal process by hydrodemolition may be obtained by carefully combining fluido-dynamic, geometric and kinetic parameters as a function of the existing situation (strength of concrete, presence of reinforcing steel, cracks, etc.) and the type of work required. Obviously satisfactory work requires highly qualified and experienced personnel capable of optimising the equipment and skills necessary for each single case. The equipment must be sufficiently powerful, but sophisticated movements and electronic control systems are also necessary, since without them one of the most important advantages of hydrodemolition is lost, i.e. *selective* removal.

Selective Removal

Clearly, operative conditions being equal, removal involves a greater depth of degraded or generally weaker concrete than it does in the case of sound and resistant concrete, but this *is not* selective removal. Referring to hydrodemolition equipment, selective removal is defined as the capacity to remove completely all and only the deteriorated concrete, independently of the depth to which the damage has penetrated. Deterioration in bridge decks or parking areas may involve thicknesses which vary from point to point—in practice from zero to the whole thickness (Fig. 1).

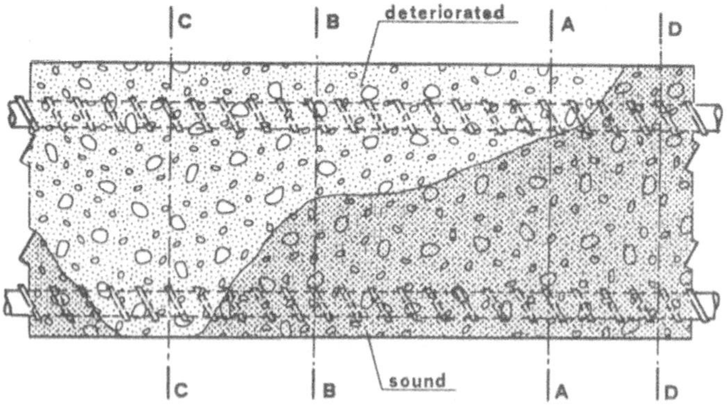

FIG. 1

However, selective removal may also be defined as the capacity to remove only concrete with strength of less than a certain pre-established value, avoiding removal of concrete which has been considered as acceptable by the engineer. As we shall see later, the term 'strength' (commonly understood as compression cubic strength) is incorrect in identifying the type of concrete to be removed. In order to understand the phenomenon of selective removal, we must, albeit briefly, go back to theory.

If gradually increasing force, e.g. compression, is exerted on a material, the latter is deformed according to a curve called the stress–strain characteristic up to breaking point (Fig. 2). By using specific units we have

$$E_r = \int_0^{\varepsilon_1} \sigma \, d\varepsilon$$

where E_r is specific breaking energy (kJ/m^3) and ε_r is breaking strain.

Obviously, if the breaking stress is not reached or if the material receives energy less than the breaking energy, it will remain intact. Clearly materials

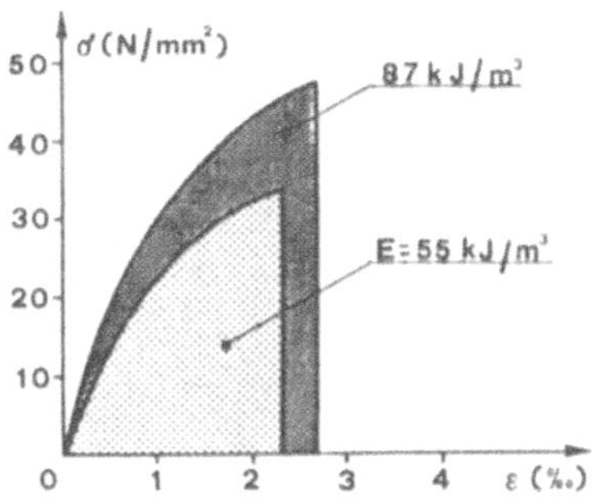

FIG. 2

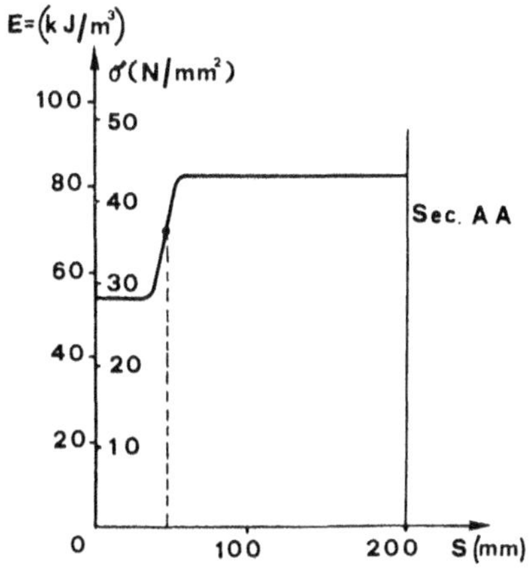

FIG. 3

at higher strength require greater energy, so that there is a law of proportionality between the two parameters. We may thus conclude that

—all materials have a threshold energy value at breaking point and
—there is a law of proportionality between the above threshold energy value and the strength of the same material.

Therefore, if we plot the trend of the strength of concrete as a function of depth, as shown for example in section A–A in Fig. 1, the same diagram may also represent the energy required to break the specimen on a suitable scale (Fig. 3).

If we want to represent the trend of the strength (or breaking energy) in a section in which deterioration extends to a greater depth, as in section B–B, the trend is that shown in Fig. 4.

The power of a water jet of flow rate q (m^3/s) and velocity v (m/s) is given by the equation

$$W = \tfrac{1}{2}\rho q v^2 \quad \text{(W)}$$

where ρ is the specific weight of water.

The energy developed by time interval t is

$$E = Wt \quad \text{(J)}$$

Let us now presume that the operational parameters of the hydrodemolition equipment (pressure or water velocity, flow rate, and the other geometric and kinetic parameters) have been fixed.

We have thus established the amount of energy which may be distributed over one surface unit. Let us now define the trend of energy available in the jet per unit of volume of concrete in increasingly deeper sections (Fig. 5). It

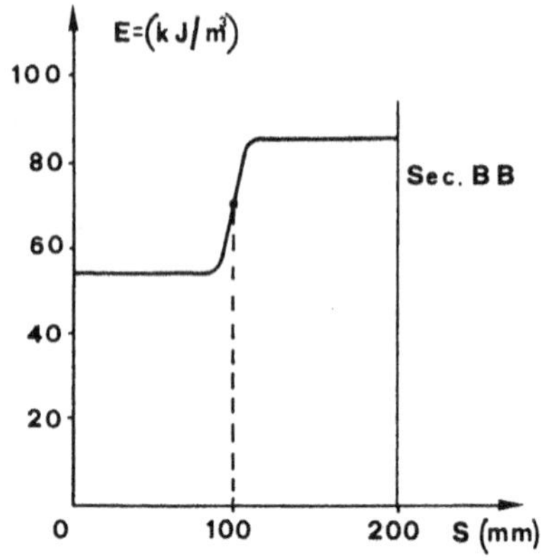

FIG. 4

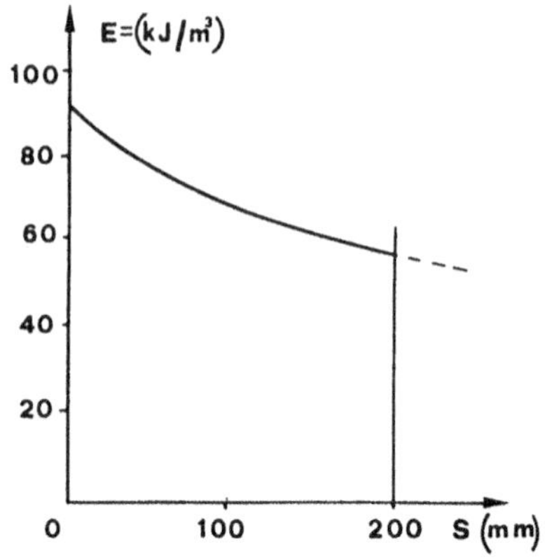

FIG. 5

should be noted that the power of the jet decreases with distance from the nozzle, not only due to dissipation in the concrete and water but also and above all due to the instability of the jet itself, since it produces small drops which rapidly lose their energy even in the air.

If we superimpose the two curves (energy necessary to break the concrete and energy available in the water jet), we see that their cross point defines the thickness of concrete which will be removed (Fig. 6).

If deterioration in another part of the deck had reached a deeper level (for example section B–B of Fig. 1), deeper removal would automatically be obtained (Fig. 7).

Naturally, with sound concrete, a constant thickness may be removed by using a sufficiently powerful jet (Fig. 8).

As already explained, in order to change the curve of available energy at the jet, various parameters may be changed:

$$E = Wt = \tfrac{1}{2}\rho q v t$$

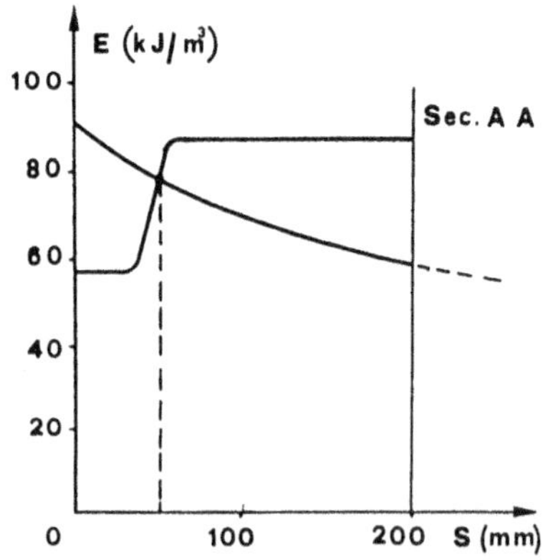

FIG. 6

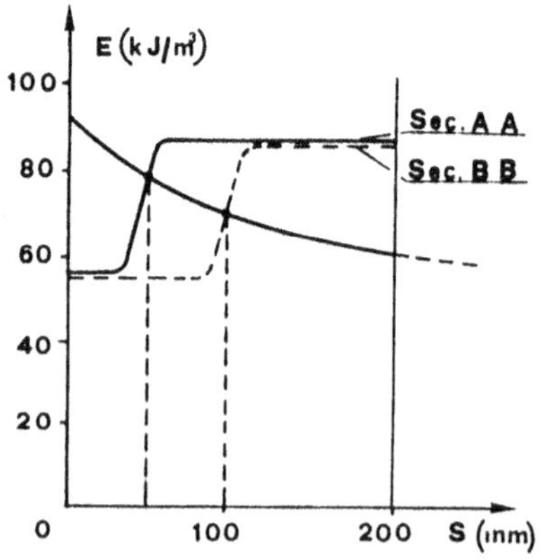

FIG. 7

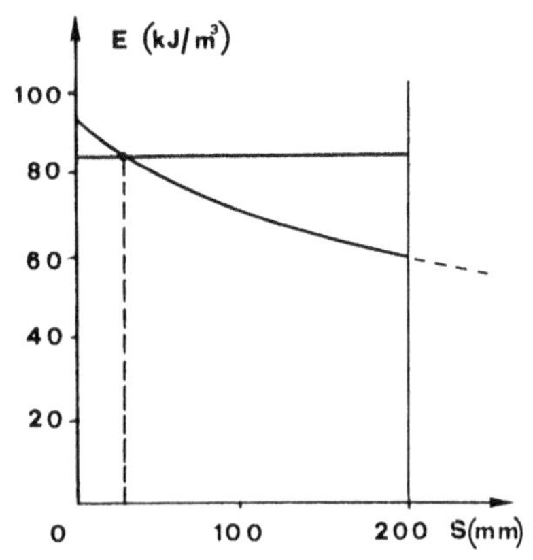

FIG. 8

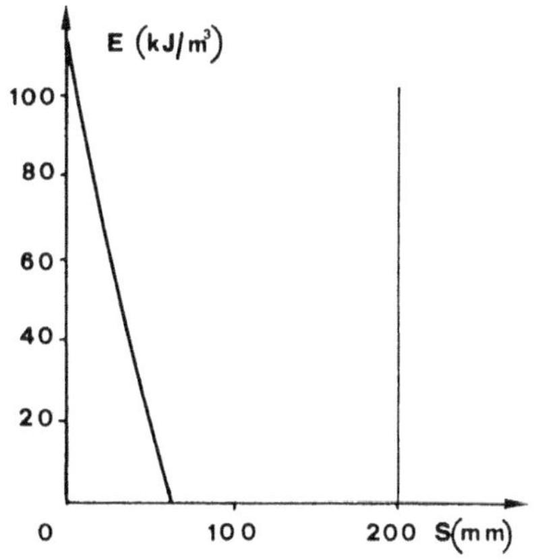

FIG. 9

However, the simplest method is that of varying time *t* by changing kinematic and geometric parameters of the movement. According to the above it seems that selective removal is an intrinsic characteristic of hydrodemolition, and it may be achieved with any equipment capable of controlling a water jet. This is not entirely accurate. In effect, selectivity may be achieved only with equipment supplied with a nozzle-moving system and electronic control for guaranteed constancy of selected parameters in time; moreover, the water jets must be highly stable and powerful. The trend of an unstable jet versus position is shown in Fig. 9.

By overlapping Figs 3, 4 and 9 we see that about the same thickness is removed in each case, and that this thickness does not even change to any great extent if the area in question has *sound* concrete (Fig. 10).

The experimental evidence of the capacity for selective removal of our equipment was shown by two series of tests carried out by the Swedish Road

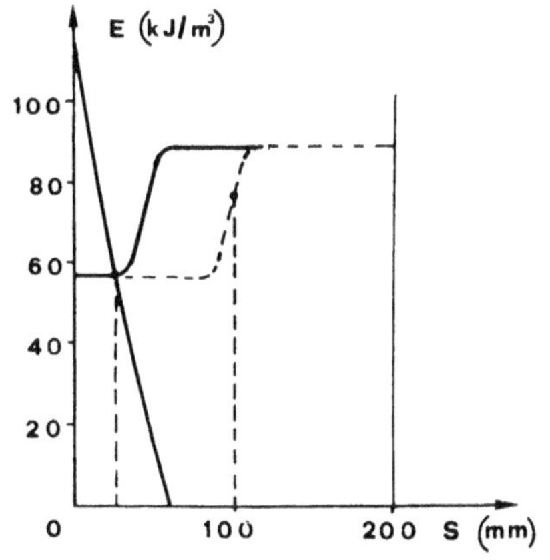

FIG. 10

Administration (Vägverket) in collaboration with the Royal Concrete Institute of Stockholm. Concrete slabs with indentations of regular geometry (squares and rectangles) and varying depths were prepared (Fig. 11). After emplacement of rebars, concrete of lesser strength was poured over the slabs in order to simulate deteriorated concrete.

Hydrodemolition was carried out after curing. It was noted that only the 'deteriorated' concrete was removed, leaving the sound concrete practically intact, both in original geometry and depth of the indentations (Fig. 12).

Modes of Operation

The above shows that hydrodemolition with selective removal does not require detailed testing of the bridge deck in order to identify deteriorated areas and their depth—operations which are expensive and far from precise with existing methods—but it is sufficient to calibrate the equipment carefully and proceed to removal. Within this apparent simplicity there are various modes of operation suiting many different situations. Two of the main ones are described below.

The first case deals with quite widespread deterioration in terms of surface area, with depths varying from zero to the whole thickness and with potential involvement of rebars. Once the minimum thickness to which the repair material can be applied (e.g. 50 mm) has been established, a few square metres of sound concrete are identified. As a first step the strength of the concrete is determined on samples or, more simply, *in situ* using non-destructive methods or pull-out tests. Working parameters are then fixed with the help of diagrams and tables, obtained from previous tests carried out on a slab of predetermined strength.

An initial attempt is made on about 1 m^2 and, if necessary, the parameters are redefined. The equipment is then moved to the area of worst deterioration (the ideal situation would be an area where deterioration involves the entire thickness of the deck). Testing is considered successful if, with the same parameters as before, all the deteriorated concrete is removed.

If there is reason to suppose that the concrete is not homogeneous over the entire deck, the minimum thickness of removed concrete should be checked periodically. Sometimes all the rebars must be exposed and all the deteriorated concrete removed at the same time. The mode of operation is in any case the same.

The second case often encountered is that of decks of relatively good condition but with insufficient cover, leading to delamination. In this case hydroscarification (5–10 mm) of the entire surface is recommended. In this way the upper part, possibly contaminated, is removed and, at the same time, excellent roughness is ensured for good bonding of repair materials.

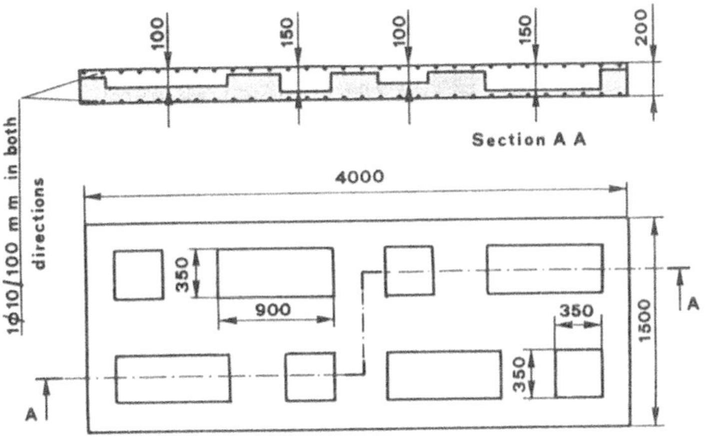
100
150
100
150
200
Section A A
4000
1ϕ10/100 mm in both directions
350
900
350
350
1500
A
A

FIG. 11

FIG. 12

Hydroscarification also shows up possible areas which have undergone some degree of deterioration, but where delamination has not yet occurred. The bound areas of the deteriorated parts are marked with regular geometrical shapes, if possible grouping several adjacent zones into a single patch. Deep removal is then carried out. It should be noted, however, that patching is never recommended. It is done, for reasons of economy, only if the deteriorated areas do not exceed 15–20% of the whole deck.

ADVANTAGES OF HYDRODEMOLITION

Understood as a new process in the field of removing concrete, hydrodemolition generally offers many advantages over traditional methods. Of course additional advantages also derive from the kind of equipment used, its power, manoeuvrability, control system, etc., all of which influence removal speed (or productivity) and quality of work—in other words, the economic result.

From the technical viewpoint, the advantages of hydrodemolition are the following:

—constant, repeatable results, once operating characteristics have been established;
—guaranteed total removal of deteriorated concrete (see paragraph on selective removal);
—no damage caused to sound parts of concrete;
—possibility of working even in the presence of rebars, which are not damaged; on the contrary, they are given a thorough cleaning and any trace of corrosion is removed from even their lower parts, usually not reached by other processes such as sand blasting;
—creation of a very rough surface ensuring excellent bonding to repair materials, much higher than in the case of jacking or chipping hammers;
—no impacts of vibrations, thus on the one hand avoiding damage to reinforcement and on the other ensuring that noise is kept to an acceptable level; moreover, some simultaneous and otherwise impossible operations, such as casting in immediately adjacent areas, may be carried out; and
—no dust or fumes (until now an inevitable accompaniment to concrete removal works).

It should also be noted that work may be done even in poor weather conditions and sub-freezing temperatures.

THE EQUIPMENT

As already noted, historically hydrodemolition came into being to solve the problems of bridge deck restoration. However, it has also been extended to other similar applications, such as repair work on parking garages, airport runways, concrete roads, etc. All these applications refer to horizontal or almost horizontal surfaces.

FIG. 13

FIG. 14

FIG. 15

Even for vertical surfaces and soffits a new demolition unit has been designed and manufactured, using very complex mechanical and oleodynamic systems requiring computerised control. Figure 13 shows a real robot with a demolition head on the end of an articulated arm supported by a 360° swivelling tower. It may be remote controlled for work even in inaccessible areas (e.g. 'hot' areas of nuclear power plants during decommissioning).

For removal of concrete cladding in tunnels, a special truck-mounted piece of equipment has been designed (Fig. 14).

Many accessories have been developed to fulfil special requirements. For areas where access is very difficult, special manual equipment has been designed—the 'Bazooka'. This is a kind of thrust-compensated gun allowing removal of concrete underneath bridge decks and girders (Fig. 15).

CONCLUSIONS

Experience world-wide has shown that hydrodemolition represents a revolutionary but mature technique in concrete removal. It is not only a great improvement over conventional systems but operations which were once impossible may easily be carried out. Structures may be restored even in cases of advanced deterioration. It is a firm belief that to date only a few applications have been explored. Therefore new R&D is to be carried out aimed at improving reliability and efficiency of existing machinery as well as designing new equipment.

67

Aluminium Extrusion Bridge Rehabilitation System

LARS SVENSSON

Lars Svensson JADA AB, Brahegatan 56, S-114 37 Stockholm, Sweden

and

LARS PETTERSSON

The Royal Institute of Technology, Department of Structural Engineering, S-100 44 Stockholm, Sweden

ABSTRACT

The deterioration of road bridges is becoming a serious problem in Sweden. An increasing number of bridge decks are in such poor condition that they must be replaced. This is due mainly to a severe climate, the use of road salt in winter time, and increasing traffic volume and loads. In certain cases major reinforcement of bridge foundations or replacement of bridges has been called for to guarantee safety, often at great cost. A system is described for replacing damaged bridge decks. The system utilises an orthotropic plate that consists of hollow aluminium extrusions fitted together by tongue and groove. Major reduction of dead loads is achieved which allows increased traffic loads without reinforcing foundations or the structural system.

INTRODUCTION

Road bridges are long-term investments with an expected lifetime of at least 50 years. In many cases bridges are in such poor condition after a much shorter time that they need to be repaired or replaced. A severe climate, road salt and increasing traffic loads are three major factors causing deterioration of bridge structures. Deterioration of concrete decks is mainly due to the first two factors. In areas with poor soil conditions increasing traffic loads have caused damage to foundations and supporting structures.

In cases of severe damage to concrete decks the usual course of action is to replace the deck with a new one. There are several methods for undertaking this. If the foundations are damaged, or if the bridge is to carry increased loads, then reinforcement of selected areas is usually necessary, often at great cost.

Bridges are structures in which a large part of the load-bearing capacity is allocated to carrying its own weight. For a steel girder composite bridge most of the dead load lies in the concrete deck. By replacing the concrete deck with a lighter one it is possible to increase the allowable live load without reinforcing the main structure or the foundations.

A lightweight system for replacing damaged concrete bridge decks has been developed and used in Sweden. The basic concept is an orthotropic plate of aluminium. Weight reduction has made it possible to use existing foundations, supports and structural system for increased allowable live loads on bridges that would have to be replaced otherwise.

BASIC CONCEPT AND STRUCTURAL SYSTEM

The basic concept is an orthotropic plate that is built up of hollow aluminium extrusions. The extrusions are fitted together by means of a tongue and groove in the upper flange, as shown in Fig. 1. This type of connection transfers shear force from one extrusion to the other. At the same time it allows each extrusion to rotate independently of the neighbouring ones. The hollow section is used to create a high degree of torsional stiffness in the extrusion.

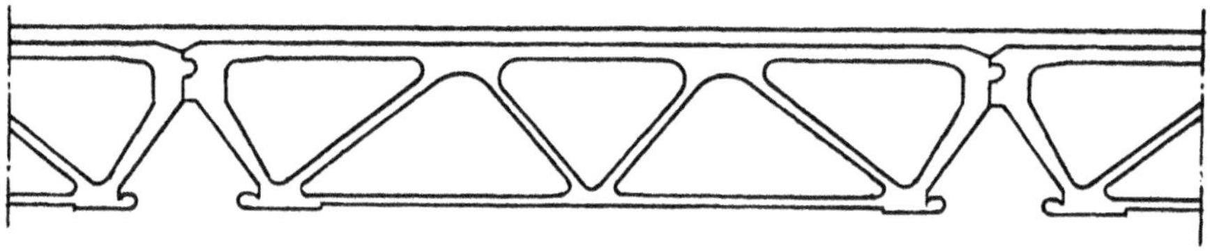

FIG. 1. Section through aluminium bridge deck extrusion.

When subjected to concentrated loads, which is most often the case for bridge decks, the resistance of the deck is due to a combination of the bending and torsional stiffness in the extrusions. In this way the load distribution will be considerable and concentrated loads will be carried by at least seven extrusions at the same time.

The cross-section of the extrusion has the form of a truss. This is to ensure resistance against local deformation due to point loads such as tyre loads caused by vehicles.

The deck is mounted to a secondary structure of steel girders or, as in certain cases, directly to the main structure. The distance between the supporting girders varies between 1·2 and 3·0 m depending on the size and type of extrusions.

Fasteners of extruded aluminium are used to secure the deck to the supporting structure. The fasteners hook on to grooves in the lower flanges of the extrusions and are bolted to the supporting structure.

The surface of the deck is covered with an acrylic-based material called Acrydur. This paving has been applied to bridges for many years and has shown very high resistance to wear in existing bridges.

The weight of the deck system described lies between 50 and 70 kg/m^2. The weight of standard concrete decks is normally between 600 and 700 kg/m^2. The aluminium deck is corrosion resistant even in a marine environment. Construction time is short, which is favourable when repairing bridges in areas with heavy traffic.

APPLICATIONS

The system has been applied to drawbridges, pontoon bridges and stationary bridges with primary structures of steel. It can also be applied to new bridges, often with the added advantage of a reduction in the weight of the primary structures and also in the size of foundations required.

In areas with poor ground conditions the weight reduction caused by incorporating the deck system has allowed the use of existing foundations to support increased loads without the need for additional reinforcement. In piled foundations this often means fewer piles.

Studies by the senior author have shown that the reduction in weight due to the use of the aluminium deck system increases with the size of the bridge. This is largely due to the fact that the relative size of the dead loads increase with increased span.

THEORETICAL AND EXPERIMENTAL INVESTIGATIONS

The system is designed to fulfil requirements given by the Swedish National Road Authority. These include resistance to static and dynamic loads according to the Swedish bridge and building codes.

A first analysis of the deck was undertaken by modelling the structure by the use of finite elements. A model of the deck was formulated using beam finite elements. The extrusions were simulated by beam elements having the flexural and torsional stiffness of the proposed section. The effect of interaction between extrusions was simulated by using connecting beam elements with almost infinite flexural stiffness. The model is shown in Fig. 2.

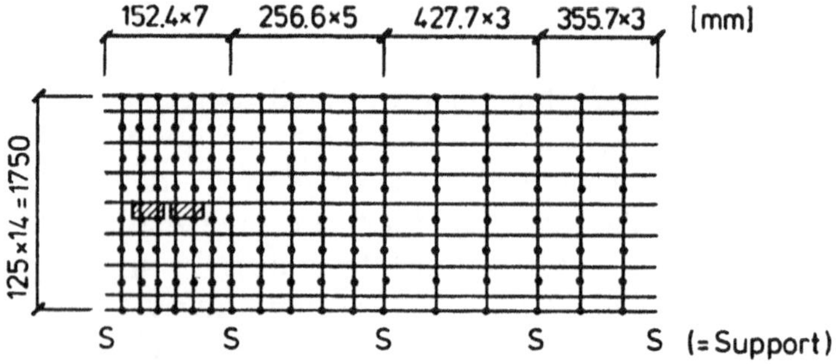

FIG. 2. Finite element model of aluminium bridge deck system.

The applied load represented a pair of truck tyres. The load was placed in the most unfavourable position to cause the largest possible deformation in the structure, as shown in Fig. 3. The deformation reached a maximum of 6·5 mm at the right end of the extrusion subjected to the applied load. The load in this case was 100 kN.

A full-scale model was tested[1] to confirm the results of the theoretical investigations. These tests included static loading as described above and dynamic loading in accordance with the requirements set by the Swedish National Road Authority.

The results of the static loads showed very good agreement with the results from the FEM analysis. A comparison between results is given in Fig. 4. The static load was increased to an ultimate value of 320 kN, which is well beyond the dimensioning load of 100 kN. The maximum deflection of a deck element under a load of 320 kN was 27 mm.

Dynamic loading was applied to investigate resistance to fatigue. The load had an amplitude of 96 kN and a maximum value of 100 kN. The frequency was 1–2 Hz and a total of two million cycles was applied. No cracks or signs of fatigue were visible after the test and the maximum residual deflection was 0·4 mm.

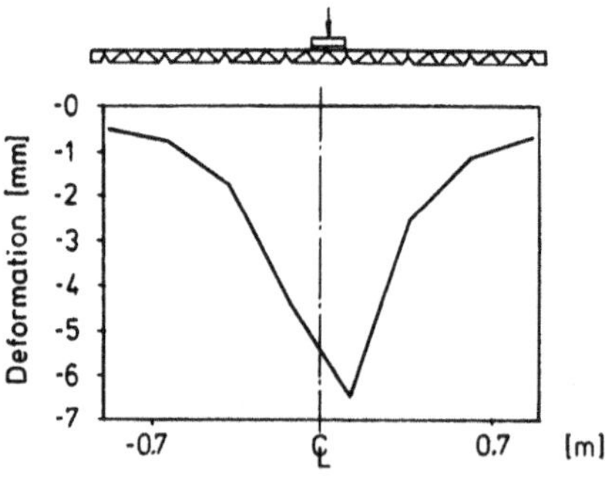

FIG. 3. Deformation of aluminium bridge deck calculated by the finite element method.

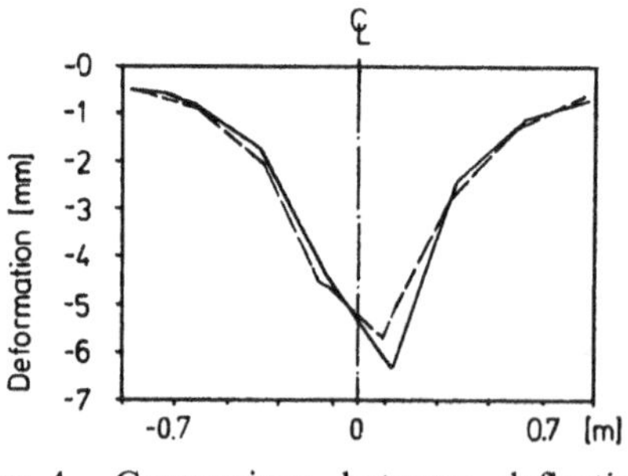

FIG. 4. Comparison between deflections obtained from the finite element analysis and full-scale tests. Load position according to Fig. 3. (———) FEM, (- - - -) full-scale test.

Extruded deck elements were tested in bending and torsion to investigate the influence of the surface material. These showed that the Acrydur surface has negligible influence on the structural behaviour of the deck.

In conclusion, the experimental investigation confirmed the applicability of the FEM analysis. The results were in good agreement with calculated values and the structural resistance of the deck was well within the requirements.

ECONOMIC ASPECTS

In many cases the aluminium bridge deck can be a very competitive technical and economic alternative to the conventional concrete bridge deck. Even though the cost for the aluminium deck is approximately the same as for a corresponding concrete deck, the aluminium deck gives a number of advantages.

—Time for detail design is kept to a minimum.

—Very short time for construction works, which in many cases greatly reduces the total investment costs. At present (June 1989) replacement of a concrete bridge deck by an aluminium deck is under investigation. The plan is to carry out the replacement in just 2 days in order to disturb the traffic as little as possible. In addition, the costs for a temporary bypass road can be avoided.

—The reduction of the dead load can in many cases make it possible to use existing bridge foundations and main girders. This is maybe the greatest advantage and should be studied carefully for each particular case.

—After completion the costs for maintenance are also kept to a minimum because the Acrydur paving is almost impossible to wear down and, in addition, by careful choice of a suitable aluminium alloy the corrosion resistance will also be very good.

CASE STUDY

The new aluminium bridge deck system has been used in several cases in Sweden, with the most recent case being the Tottnaes Bridge south of Stockholm (Fig. 5).

The Tottnaes Bridge, a four-span steel girder bridge with one span designed as a swingspan, was in immediate need of repair, primarily resulting from the deterioration of the concrete bridge deck. In addition, a study of the foundations (wood piles) showed that it would not be possible to

FIG. 5. The Tottnaes Bridge south of Stockholm after renovation with the described aluminium bridge deck system (June 1989).

FIG. 6. Detail of railing and edge beam of aluminium bridge deck system.

increase the allowable live load without reinforcement, which was required if the bridge was to carry modern heavy vehicles.

The only possibility available to keep the old bridge foundations and at the same time increase the allowable live load was to reduce the dead load of the bridge. This was accomplished by the use of the new aluminium bridge deck system. Compared with the old concrete deck a reduction in the dead load of about 550 kg/m^2 was obtained, which was enough to maintain the existing bridge foundations without reinforcement.

Because the Tottnaes Bridge is the only connecting link between the mainland and the island of Toroe, a short time for the restoration of the bridge was of great importance. Using the aluminium bridge deck elements with the Acrydur surface applied to the aluminium extrusions, and by preparing the support details and connections in advance, it was possible to reduce the repair time to a minimum.

All other details such as railings and connections between railing and the bridge deck as well as edge beams (Fig. 6) are readily available products well tested and approved by the Swedish National Road Authority.

SUMMARY

A new bridge deck system is described which is built up from hollow aluminium extrusions. The extrusions fit together by means of a tongue and groove system. In this way the load distribution is considerable, which makes it possible for this very light deck to carry heavy live loads. By replacing old deteriorated concrete bridge decks with an aluminium deck system a considerable reduction of the dead load is achieved, which in many cases means that existing foundations and main girders can be left without additional reinforcement and still carry increased live loads. In addition, the new bridge deck system offers a very short construction time and requires very little maintenance.

REFERENCE

1. SALTIN, M. and McCARTHY, R., Brofarbana av aluminiumprofiler (Aluminium extrusion bridge deck). Diploma work at the Royal Institute of Technology, Department of Structural Engineering, Stockholm, 1988.

Index of Contributors

Abdunur, C., 489
Al-Mandil, M. Y., 549
Azad, A. K., 549

Baluch, M. H., 549
Barr, B., 349
Beales, C., 459
Beckett, D., 3
Berthelsen, F., 29
Besem, P. H., 747
Birnstiel, C., 295
Blyth, M. F., 705
Bouabaz, M., 187
Brookes, C. L., 439
Bucak, Ö., 533
Buchner, S. H., 439
Buckland, P. G., 475, 575
Burke, Jr, M. P., 677

Cairns, J., 619, 643
Camomilla, G., 155
Cesare, M., 319
Cogswell, G., 39
Cope, R. J., 429
Cullimore, S., 205
Cullington, D., 447
Cuninghame, J. R., 459

Dahinter, K., 383
Darby, J. J., 655
Dawe, P. H., 135
Dill, M. J., 655
Dragotti, A., 155
Duchêne, J.-L., 489

Ellick, J. C. A., 585
Ellinas, C., 501
El-Marasy, M., 89

Flesch, R., 327
Fukushi, A., 725

Göhler, B., 287

Halse, W. I., 173
Hammersley, G. P., 655
Harvey, W. J., 515
Hoffman, G. L., 75
Holland, D. A., 135
Horner, R. M. W., 187

Ingvarsson, H., 199
Itzkovitch, M., 607

Jackson, P. A., 429
Johnson, R. A., 501

Kagawa, Y., 725
Kähkönen, A., 101
Keer, J. G., 233

Kernbichler, K., 327
Kirkpatrick, J., 361

Lebek, D., 563
Lee, D., 145
Leeming, M. B., 243
Le Page, B. H., 233
Lichtenstein, A. G., 695
Lindbladh, L., 51
Low, A., 417

Maguire, J. R., 595
Mang, F., 533
Manning, D. G., 305
Marshall, A. R., 63, 101
Masliwec, T., 305
Maxwell, J. W. S., 113
May, P. H., 121
McClure, R. M., 75
Medeot, R., 765
Mehrkar-Asl, S., 439
Mehue, P., 633
Melbourne, C., 523
Middleton, C., 417
Miesseler, H.-J., 395
Mitchell-Baker, D., 205
Montgomery, F. R., 259
Moses, F., 405
Murray, A. McC., 259
Muruganandan, C., 715

Nebbia, G., 155

Palmer, J., 39
Pearson-Kirk, D., 549
Petrangeli, M. P., 373
Pettersson, L., 777
Pretlove, A. J., 585
Pritchard, B., 667

Ramezankhani, M., 349
Read, J. A., 267
Reel, R. S., 715
Reij, A. W. F., 215
Romagnolo, M., 155
Rosenthal, I., 607

Sharif, A. M., 549
Simpson, B., 705
Sriskandan, K., 17
Sloan, T. D., 361
Smith, F. W., 515
Smith, N. J., 223
Söderqvist, M.-K., 63
Sorensen, A. B., 29
Stephens, R. L. C., 173
Svensson, L., 777

Tallin, A. G., 319
Thompson, A., 361

Van der Toorn, A., 215
Vrahimis, S., 121

Waldron, P., 349
Warnon, C., 747
Wenzel, H., 339
Wolff, R., 395
Wood, J. G. M., 501
Woodside, A. R., 737
Woodward, W. D. H., 737
Wouters, M., 747

Subject Index

Ageing mechanism of bridge structures, 219
Al-Darb Bridge, 551–6
Aliasing, 597–8
Alkali aggregate reaction, 506–9
Alkali-silica reaction, 24, 244
Aluminium extrusion bridge rehabilitation system, 777–83
  applications, 779
  basic concept, 778–9
  case study, 781–3
  economic aspects, 781
  FEM analysis, 780–1
  structural system, 778–9
  theoretical and experimental investigations, 779–81
Arch bridges
  analysis of multi-span, 515–22
  assessment of, 523–31
  effect of defects, 523–31
  hinge formation, 526–7
  indeterminacy of, 516–17
  maximum thrust case, 517
  minimum thrust case, 517
  model tests, 524–5
  multiple redundancy, 519–21
  released structure, 517–18
  restoring redundancy, 518–19
  serviceability assessment, 585–94
  six-metre span test, 527–9
  three-metre span tests, 525–7
  vibration tests, 585–94
Arching action, 430
Asphalt cracking, 557
Assessment code, 46, 138–9
Assessment methods, 24–5, 140, 489–99
Averaging, 598
Bascule bridges, 297–8
Belgo Bridge, 483–6
Bending frequency factor (BFF), 591
Berlin–Marienfelde Bridge, 400
Berounka River bridge, 389–90
Blumberg Bridge, 541, 543
Boonton Bridge, 698–700
Box girder bridges, glued segmental, 349–60
Brenner motorway bridges, 327–38
Bridge beams, hammer testing, 600–3
Bridge condition deficiencies, 83–4
Bridge costs. *See* Finance
Bridge data information systems. *See* Computerised database and information systems
Bridge decks
  bending relief by external prestressing, 668–71
  composite, fatigue relief, 674–6
  concrete quality assessment, 552
  condition surveys, 305–17
  construction materials, 40
  damage likelihood chart, 560
  deterioration, 306
  major types of, 306
  structural and material damage, 549–61
Bridge design
  code models, 141
  codes, 25
  criteria, 118
  load concept, 568
  regional influences on, 151
  strategies for improving, 149–51
  vehicles, 553
  see also Load; Loading

Bridge directory, 107–9
  entity attribute relationship, 109
Bridge furniture, 151
Bridge inventory, 20
Bridge management
  basic topics, 216
  computer development, 43–4
  conceptual stage, 17–18
  construction stage, 19–20
  cost-effective strategies in, 715–24
  Cyprus, 121–32
  design stage, 18–19
  DISK system, 89–99
  Europe, 29–38
  Far East, 29–38
  Finland, 63–74, 104–5, 109–11
  highway bridges, 113–20
  implementation of, 130–2
  in-service stage, 20–6
  logic diagram of, 127
  organisation and budget arrangements, 123–4
  overview, 17–27
  Pennsylvania, 75–87
  sequence of, 126–8
  Surrey County Council, 39–50
  Swedish National Road Administration, 51–61
  United States, 63–74
  use of term, 17
  see also DANBRO; Maintenance management
Bridge procurement, 152
Bridge structures
  ageing mechanism of, 219
  deterioration, 23–5
    loading determination in, 504
    reliability analysis, 501–13
    structural analysis in, 504–5
  interventions on, 164–6
BRIDGET, 190
BRIDGIT, 43, 44
Brittle fracture, 23
Bronx–Whitestone Bridge, 486
BS5400, 10, 447–9, 454, 455, 466, 619, 674
BS8110, 645–7

Cable-band bolt tension
  drop in, 725–35
  effects of high tightening torque, 733
  measuring methods and equipment, 729
  safety factors and retightening time, 733–4
Cable-band bolt tension—*contd.*
  tightening torque and tension drop rate, 730–3
Cable stayed bridges, 339–48
  construction and supervision, 343–5
  design, 342
  inspection, 345–8
  maintenance, 345–8
  repair, 345–8
Cable-supported bridges, 150–1
CAN/CSA-S6-88, 575–84
Capital costs, 717
Carbonation depth, 162
Castigliano's theorem, 591
Chloride contamination, 162, 246, 655–66
Chloride profiles, 558, 559
Clifton Suspension Bridge, 205–14
  deflection measurements, 212
  early history, 207–8
  inspection and testing, 211–12
  maintenance strategies, 210–11
  operation, 210–11
  policy of indefinite preservation, 205–7
  post-war period, 208–10
  revised bolt assembly on suspender rod, 209–10
  testing and analysis, 211–12
  traffic statistics, 212
  weight restriction, 208–10
Code of Assessment, 46, 138–9
Code of Practice for Bridges, 619
Code of Practice for Fatigue, 619
Coefficient of thermal expansion, 354
Cogan Viaduct, 351, 352, 356–9
Coherence function, 598–9
Collapse analysis, 419–22
Combined shear and bending failure, 420–1
Composite bridge decks, fatigue relief, 674–6
Composite concrete–steel bridges, repair of, 747–64
Composite steel bridges, 151
Compressive membrane action, 430
Computer-aided sketching of load paths, 515–22
Computer analysis of post-tensioned concrete bridges, 442–3
Computerised database and information systems, 54–6, 58–60, 69, 73, 77–8, 116–18, 128–30, 312–13, 597–600
Concrete
  capillary flow or absorption measurement, 252–3

Concrete—*contd.*
controlling movement of water in, 250–2
corrosion of reinforcement in, 245–7
curing/maturity of, 248
deterioration of, 244–7, 267–71
main faults in, 270–1
remedial measures, 271–4
replacement of members, 274–6
water role in, 243–4
see also FBECR
gas diffusion in, 246
maintenance, 24
moisture content of, 255
quality assessment of bridge decks, 552
removal in bridge repair, 765–76
surface treatments, 233–41, 250–1
water ingress into or through, 248–50
water role in, 247–8
workability/compaction of, 247
Concrete bridges, 24
box girder, 349–60
Czechoslovakia, 383–93
inspection, 140
performance of, 147
rehabilitation programmes, 139–40
strength assessment, 429–38
structural and material damage, 549–61
traffic-induced strain in steel reinforcement, 619–29
see also Prestressed concrete bridges; Reinforced concrete bridges
Conservation viewpoint, 171
Copper wire sensors, 399–400
Corrosion, 559
chloride-induced. *See* Chloride contamination
prestressing tendons, 141
protection of steel structures, 755–62
reinforcement in concrete, 23, 25, 245–7, 267–84, 643
see also FBECR
tendons, 24
COSBEN Program, 721, 724
Cost-effective analysis, 717–21
examples of, 721–4
Cost-effective strategies in bridge management, 715–24
Cost inventory file, 78
Cost model factor (CMF), 190, 192
Cost-significant work packages (CSWPs), 189, 191–2
Costs. *See* Finance
Cover plate terminus, reliability index, 322–5
Crack-bridging coatings, 233–41
need for, 235–6
parameters affecting selection, 235
test methods for, 236–9
Cracks
curvature and stiffness redistribution, effect on, 494
failure patterns, 451
flyover viaducts, 634
mechanical behaviour, effect on, 489–99
reinforced concrete beams, 644
rib-to-deck plate junction, 635–7
rib-to-floor beam junction, 639–40
steel orthotropic decks, 633–43
welded joints, 637–9
Craibstone–Dyce link road, 282
Creep-time relationship, 177–8
Cross Keys Bridge
aspects of repair work, 712–13
future road traffic options, 707–8
history, 706
renovation, 705–13
main contract, 709–12
structural shortfalls, 708
urgent works contract, 708–9
Cyprus, bridge management in, 121–32
Czechoslovakia, concrete bridges, 383–93

DANBRO
bridge management and maintenance system, 29–38
bridge overview, 31
bridge rating, 34
concept of system, 30–7
hierarchic element structure, 31
inspection and bearing capacity module, 32–4
inventory module, 31
main activities covered by system, 30
objectives, 29
outlines for system, 36
ranking and budgeting module, 34–6
special inspection report, 33
system modules, 31–6
Danish State Railways, 37
Darcy's law, 248
DART (deck assessment by radar and thermography), 306–16
costs, 316
data acquisition/processing, 312–13
field operations and experience, 313–14
prototype vehicle, 310–13

Data processing, 166–71
  dynamic testing, 597–600
  see also Computerised database and information systems
Databases. *See* Computerised database and information systems; DISK
Debonding, 307, 309
Deicing problem, 146
Delaminations, 307, 310, 313, 557, 559
Department of Highways, Thailand, 37
Department of Transport, 135–43
Design. *See* Bridge design
Deteriorating bridge structures. *See* Bridge structures
Deterioration curves over time, 164
Deterioration rate modelling, 511
Digital spectral analysis, 597
DIN4150, 587, 592
Discount rate for rehabilitation projects, 718
DISK system, 89–99
  goals of, 90–1
  historical information, 97
  inspection, 92–5
  inventory and administrative information, 91–2
  maintenance, 95–7
  management information, 98–9
  registration of information, 97–8
Dolsan Bridge, 340, 345–7
Dome effect, 430
Dornie Bridge, 282
Dynamic analysis, 163
Dynamic properties, hammer testing, 595–604
Dynamic testing, 163
  analysis of measurements, 334
  data processing, 597–600
  description of method, 332–7
  dynamic calculations, 334
  focus and limitations, 332–3
  highway bridges, 375–6
  history of application, 331–2
  influence of traffic, 335–7
  main aims of, 327–9
  main steps of, 328
  motorway bridges, 327–38
  orthotropic steel bridge decks, 465–6
  outline of method, 329
  sensitivity investigations, 335
  surfaced bridge decks, 469
  systematic system identification, 334–5
  technique, 333

Elastic analyses, 515–16
Elastomeric compression seals, 680
Electrical components, 300–1
Electrical power equipment and controls, 299–302
Elm Street Bridge, 700–4
Europe
  bridge management, 29–38
  maintenance systems, 29–38
Evaluation
  Canadian Standard CAN/CSA-S6-88 Clause 12, 575–84
  guidelines, 407
Expansion joints, 737–45
  failure of, 738–9
  movements experienced at, 739–40
Expert systems, maintenance management, 155–71

Far East
  bridge management, 29–38
  maintenance systems, 29–38
Fast Fourier Transform (FFT), 596, 598
Fatigue, code of practice for, 619
Fatigue assessment of orthotropic steel bridge decks, 459–73
Fatigue costs, 413
Fatigue cracking, 23
Fatigue fractures in railway bridges, 533–5
Fatigue life
  calculation methods, 468–72
  enhancement of, 674–6
Fatigue reliability, 319–26
  definition, 321
Fatigue tests on railway bridges, 539–47
FBECR (fusion-bonded epoxy-coated reinforcement), 267–84
  changes necessary with, 276
  costs incurred using, 276–7
  method of application, 277
  surface preparation prior to coating, 278–81
  UK application, 281–3
Federal Highway Administration (FHWA), 64–5
Federal Sufficiency Rating System (FSRS), 83
Fibre composite materials, 395–402
Figg water permeability test, 254
Finance
  bridge costs, 411–13
  bridge management in Cyprus, 123–4

Finance—*contd.*
capital costs, 717
DART costs, 316
maintenance costs, 13–14, 143, 145–53, 184, 187–97, 199–204, 717
new-build costs, modelling, 188–90
operation costs, 199–204
repair costs, 187–97
Swedish National Road Administration, 52–3
see also Cost
Finite element analysis, 552
Finland
bridge management, 63–74, 104–5, 109–11
bridge statistics, 102
inspection, 73, 105–7
Finnish Roads and Waterways Administration (RWA), 71–4
First-order reliability methods (FORM), 319
Flexible surface mix, 737–45
development of, 740–4
Flexural stiffness, evaluation of, 490–2
Flyover viaducts, cracks in, 634
Forth Rail Bridge
centenary of, 3–15
construction, 8–9
contractors, 14–15
design, 4–8
engineers, 14–15
expansion/contraction lengths, 9
expansion joints, 10
general arrangement of superstructure, 6
historical background, 3–4
logistics, 14
maintenance, 12–14
maintenance costs, 13–14
painting and repainting, 12–13
principal dimensions, 5
provision for movement, 9–10
structural principle of, 5
test loading, 10–11
workforce, 11–12
Foyle Bridge
monitoring
hardware, 364–7
in-service behaviour, 361
results, 368–70
software, 367
system requirements, 364
Fracture mechanics crack growth analysis, 319–20
Freeze/thaw damage, 244
Frequency distribution, 171
Frequency resolution, 597

Gänstorbrücke in Ulm, 331
Gas diffusion in concrete, 246
Germany, prestressed concrete bridges, 287–94
Glued segmental box girder bridges, 349–60
differential temperature effects, 356–9
instrumentation, 352–9
programme objectives, 352
short-term performance, 355
time-dependent effects, 353–5
Golden Gate Bridge, 486
Governing failure mode, 422
Grangetown Viaduct, 351, 352, 355
Gun railway, 537

Haeng Ju Bridge, 341, 342, 344–5, 348
Hammer testing
bridge beams, 600–3
dynamic properties, 595–604
typical setup, 599–600
Health and Safety at Work Act, 12
High alumina cement, 245
Highway bridges, 373–81
design of, 576
dynamic tests, 375–6
inspection, 374–8, 383–93
management flow chart, 115
management systems, 113–20
monitoring systems, 119
policies and objectives, 114–16
protective measures, 259–65
rehabilitation programmes, 135–43
repair, 378–80
serviceability performance, 607–17
steel trusses for, 695–704
strength evaluation, 383–93
structural and material damage, 549–61
technical issues, 118–19
types and dimensions, 419
United States, 405
vibration measurements, 607–17
Hinge formation in arch bridges, 526–7
H-MAC flexible surfacing mix, 738, 744–5
Holland, maintenance strategies, 215–22
Hougomont, 12
Hydrodemolition
advantages of, 774

Hydrodemolition—*contd.*
equipment, 775–6
mechanism of, 767
modes of operation, 772–4
selective removal of concrete by, 768–72
technique, 765–76
Hydrophobic treatments, 239–40
Hydroscarification, 772–4

Impact device design, 588
Incremental benefit/cost analysis, 719–20
Incremental benefit/cost ratio, 716–17, 721
Information systems. *See* Computerised database and information systems
Infrared thermography. *See* Thermography
Initial surface absorption test (ISAT), 253, 256
Innoshima Bridge, 730
Inspection, 20–2
cable stayed bridges, 345–8
Clifton Suspension Bridge, 211–12
complex diagnostic method, 384–5
concrete bridges, 140
cost effectiveness of, 46
Cyprus, 124–6
Finland, 73, 105–7
highway bridges, 374–8, 383–93
movable bridge machinery, 302–4
PennDOT, 78
prestressed concrete bridges, 287–94
reinforced concrete bridges, 261
scheduling system, 67
special methods, 161–4
steel bridges, 319–26
steel trusses, 695–704
Tamar Bridge, 177–8
see also DISK system; DANBRO; Surrey County Council; Swedish National Road Administration
Instrumentation. *See* Monitoring
Integrated bridge construction, 677–93
conversions to continuous span (retrofit), 687–92
cost comparisons, 680
current design trends, 678–9
cycle control joints, 683–7
integral abutment details, 682
passive pressure, 683
performance comparisons, 679–81
pile stresses, 683
structural analysis, 682
structural distress, 681–2

Inventory system in Cyprus, 128–30
ISAT method, 254
ISO4624/78, 762
Italy, highway bridges, 373–81

Jindo Bridge, 340, 343–7

Kanmon Bridge, 730, 733
Kings Bridge, Melbourne, 23
Kishon River bridge, 607–17
Kita Bisan-Seto Bridge, 726
Koblenz/Waldshut railway bridge, 535
Korea, cable stayed bridges, 339–48

Lavant Bridge, 331
Leakage
bridge joints, at, 737
problem of, 598
Level-of-service deficiencies, 83
Life cycles for rehabilitation treatments, 718
Limit state design (LSD), 407
Limit state function, 423
Linear elastic fracture mechanics (LEFM), 319
Lions' Gate Bridge, 478–83, 486
Live load factors, 576, 577, 582
Load and resistance factor design (LRFD), 407, 408
Load capacity
classification, 59–60
evaluation, 408–9
Load factors, 579–81
Load paths
analysis, 516
computer-aided sketching, 515–22
Load relieving techniques, 667–76
Load simulation. *See* Traffic load simulation programme
Loading assessment, 138
Loading determination in deteriorating bridge structures, 504
Loading increases, 26
London Docklands Light Railway, 673–6
LUSAS, 212

Magnesium phosphate concrete, 180
Main-channel at Bamberg, 288–9
Maintenance cost index, 201–3

Maintenance costs, 717
  comparative study of, 199–204
  Forth Rail Bridge, 13–14
  modelling and predicting, 187–97
Maintenance models, 216–22
  application of, 218–21
Maintenance strategies
  activity ranking, 79–80
  activity urgency, 80
  approach to, 182–3
  bridge adequacy, 81
  bridge criticality, 80
  cable stayed bridges, 345–8
  Clifton Suspension Bridge, 210–11
  cost aspects. *See* Maintenance costs
  current research, 228–9
  deficiency point assignment, 81–3
  design criteria, as, 150
  economics of, 143
  Europe, 29–38
  existing situation, 225–6
  expert systems, 155–71
  Far East, 29–38
  Forth Rail Bridge, 12–14
  Holland, 215–22
  implications for new bridges, 228
  management of, 22–3, 47–8, 226–8
  motorway bridges, 155–71
  optimization, 101–11
  past, present and future, 147–9
  PennDOT, 78–83
  prioritization procedure, 79
  qualitative decision tree for, 217
  reinforced concrete bridges, 261–2
  systematics approach to, 215–22
  Tamar Bridge, 176, 178–85
  UK, 223–30
  see also DANBRO; DISK system; Rehabilitation; Repair
Management systems. *See* Bridge management
Marie d'Ivry metro-station, 401
Masonry bridges, 190–6
  see also Arch bridges
Massachusetts Bay Transportation Authority (MBTA), 68–71
Massachusetts Department of Public Works (MDPW), 67
M-beam type bridges, 419–22, 425–6, 430
Method of Measurement for Road and Bridge Works (MMRB), 191
Metropolitan District Commission, Boston, Massachusetts, 66–8
MEXE analysis, 517
Midlands Links Motorway Viaducts, 148
Milton Bridge, 620
Minami Bisan-Seto Bridge, 726
Miner rule, 537
Modal analysis, 596
Monitoring
  glued segmental box girder bridges, 349–60
  highway bridges, 119
  prestressed concrete structures, 395–402
  remote computer-aided bridge performance, 361–71
  traffic-induced strain in steel reinforcement, 621–2
  Wolvercote Viaduct, 659–61
Mortar renderings, 251
Motorway bridges, maintenance strategies, 155–71
Movable bridge machinery, 295–304
  inspection, 302–4
  rehabilitation, 304
Movable bridges, types of, 296–8
Museum Railway Bridge, 539–40, 545

Natural frequencies, 590
Ness Viaduct, 282
Network analysis, 724
New-build costs, modelling, 188–90
Noise, 598
Non-linear analysis, 433–4
Northern Ireland, reinforced concrete bridge protection, 259–65
Northern Ireland Roads Service, 128

Obernberg Bridge, 331
Ohre River bridge, 387–9, 391–2
Olympic Grand Bridge, 340–2, 344, 347–8
Ontario Bridge Code, 616
Operation costs, comparative study of, 199–204
Optical fibre sensors, 399
Orthotropic steel bridge decks
  cracks in, 633–43
  direct measurements under traffic loading, 466
  dynamic testing, 465–6
  effect of surfacing, 471–2
  fatigue assessment of, 459–73
  transverse position of traffic, 471–2
Overloaded vehicle permits, 410–11

Paint degradation, 23–4

Painting and repainting
  Forth Rail Bridge, 12–13
  Tamar Bridge, 178
Pardubice flyover, 390–1
Peel Green underbridges, 447–57
Pelly River Bridge, 581
Pennsylvania Bridge Management System, 75–87
Permits for overloaded vehicles, 410–11
Pig-tail railway, 537
Plastic collapse modes, 418
Poisson's ratio, 354
Portal slab bridges
  collapse analysis, 424
  failure modes, 420
  loadcase governing collapse, 424
  reliability analysis, 424
  sensitivity analysis, 426
Pothole formation, 555–6
Present value analysis, 715–17, 721
Prestressed concrete bridges, 150
  beam assessment, 447–57
  comparison of main defects, 292
  composite I-girder, 389–90
  concrete frame, 392–3
  inspection of, 287–94
  rehabilitation, 293–4
  segmental box girder, 391–2
  shear strength of beams, 448
  stress assessment, 439–46
Prestressed concrete flyover, 390
Prestressed M-beam bridges, 419
  failure modes, 420–2
  probability of failure, 425–6
Prestressing of bridge decks, 668
Probability of failure, 418, 422, 426, 578
PROBAN reliability program, 321
PROINSP computer program, 321
Project level system, 110
Protective measures, 165, 259–63
PRVAL Program, 721
Pull-out tests, 162
Punching shear, 508–9
Punching theory, 432–3

QUADRO, 49
QUADRO4, 23
Quality assurance, 19, 119

Raach Bridge, 331
Radar
  bridge deck condition surveys, 305
  data acquisition/processing, 312–13
  defect detection by, 307–10
  field operations and experience with, 314
  operational characteristics of, 315
Radiography, 24
Railway bridges
  fatigue fractures in, 533–5
  fatigue tests on, 539–47
  residual service life of old structures, 537
  theoretical and experimental investigations on, 533–48
Rakewood Viaduct, 668
Regional influences on bridge design, 151
Rehabilitation programmes
  background studies, 138–40
  concrete bridges, 139–40
  cost-effective strategies, 715–24
  discount rate for, 718
  15-year, 136–8
  highway bridges, 135–43
  management of, 141–2
  movable bridge machinery, 304
  prestressed concrete bridges, 293–4
  research and development, 140–1
  steel trusses, 695–704
  Tamar Bridge, 180–2
  see also Aluminium extrusion bridge rehabilitation system
Rehabilitation/replacement prioritization, 83–6
Reinforced concrete bridges, 190–6, 387–9
  chloride-induced corrosion, 655–66
  condition survey, 263–4
  continuous girder, 386–7
  cracks in beams, 644
  deterioration and repair of beams, 643–53
  deterioration of, 269–70
  expansion joints, 739
  inspection, 261
  laboratory investigations, 263
  maintenance strategies, 261–2
  portal slab, 419
  protective measures, 259–63
  stress redistribution in beams, 643–53
Reinforcing bars,
  corrosion of, 23, 25, 245–7, 267–84, 643
  see also FBECR
Reliability analysis
  advanced level II method, 423–4
  assessment of bridges, in, 417–28
  basic variables, 422–3

Reliability analysis—*contd.*
deteriorating bridge structures, 501–13
FORM, 321
Reliability decrease with time, 564
Remaining bridge life assessment, 409–10
Repair, 165
cable stayed bridges, 345–8
composite concrete-steel bridges, 747–64
concrete removal in, 765–76
costs, modelling and predicting, 187–97
highway bridges, 378–80
materials, 24
Residual life assessment, 718
Residual value assessment, 718
Retractile bridges, 298
Richemont Bridge, 634–5, 639, 640, 642
Rijkswaterstaat, 215
Road and Waterways Administration (RWA), 101–11
Road bridges. *See* Highway bridges
Road data bank, 55
Road Engineering Intelligence and Research, 116
Roads and Waterways Administration (RWA), 65
Rolled sections, reliability index, 325
Royal Border Bridge, 520

Safety aspects, 25, 171, 406–8
Safety index, 578–81
Salt
crystallisation in concrete, 245
scaling in concrete, 245
SAMDA program, 155–71
aims of, 156
flow chart, 158
global assessment, 159
implementation of, 157
record forms, 159
Saudi Arabia, highway bridges, 549–61
Sazava River bridge, 386–7
Scaling, 245, 309–10
Sclerometer tests, 162
SCNALL, 367
Sensors
copper wire, 399–400
optical fibre, 399
Shear failure, 421–2
tests on prestressed concrete beams, 447–57
Shear strength
measured and calculated, 453
prestressed concrete beams, 448
Shock transmission units (STUs), 671–4
Silane for reduction of chloride-induced corrosion, 655–66
Sledgehammer for bridge beam testing, 604
S–N analysis, 319–20
Span drive machinery, 298–300
Special tests, 161–4
Specification criteria, 118
Stabilizing machinery, 298–300
Stahringen Bridge, 538
Standards, 24, 143, 575–84, 587, 592
Steel beams, 151
Steel bridges
inspection, 319–26
see also Orthotropic steel bridges
Steel reinforcement, traffic-induced strain in, 619–29
Steel trusses, inspection and rehabilitation, 695–704
STREG, 43
Strength assessment, 138
bridge decks, 552
concrete bridges, 429–38
conventional approach, 431–2
highway bridges, 383–93
Strengthening methods for existing bridges, 667–76
Stress assessment in post-tensioned concrete bridges, 439–46
Stress intensity functions, 320
Stress redistribution
concrete beams, assessment procedure, 647–53
reinforced concrete beams, 643–53
Stress state correlation with deterioration, 509–11
Structural analysis
bridge decks, 552
deterioration bridge structures, 504–5
Structural assessment, 24–5
transversal cracks, 489–99
Structural dynamics modification (SDM) software, 335
Structural inventory records system (SIRS), 77–81
STRUDL, 552
Stuttgart, Rosensteinbrücke, 291
Stuyvesant Falls Bridge, 696–8
Sulphate attack, 244
Sulphate profiles, 559
Surface treatments
concrete, 233–41, 250–1
tests on coatings, 253–4

Surrey County Council
  assessment strategy, 46–7
  bridge management, 39–50
  capital programme, 42–3
  current allocation, 42–3
  inspection strategy, 44–6
  maintenance management, 47–8
  priorities, 49
Surveillance methods, 161
Suspension bridges
  aerodynamics, 486
  analysis and survey, 477–8
  assessment and rehabilitation, 475–87
  loading, 476–7
  safety factors, 479
  temperature effects, 478
  tension drop in cable-band bolts, 725–35
  see also Clifton Suspension Bridge
Swedish Commission on Maintenance and Costs (DKU), 199, 203
Swedish National Road Administration, 200, 202
  ADP-based system, 54
  bridge age distribution, 52
  bridge management, 51–61
  bridge stock, 51–2
  condition classifications, 59
  finance, 52–3
  inspection system, 57–9
  load-carrying capacity, 52, 59–60
Swing bridges, 296–7
  renovation, 705–13

Tacoma Narrows Bridge, 483
Tamar Bridge, 173–86
  construction details, 175
  inspection, 177–8
  maintenance contract procedures, 183–4
  maintenance strategies, 176, 178–85
  overview, 185–6
  painting and repainting, 178
  pre-construction, 174–5
  rehabilitation, 180–2
  statistics, 173–4
  traffic conditions, 175–7
Target reliability index, 580–1, 583
Tay Bridge, corrosion of reinforcement, 267
Testing. *See* Inspection and under specific test methods
Thermal incompatibility of concrete components (TICC), 559
Thermography
  bridge deck condition surveys, 305

Thermography—*contd.*
  data acquisition/processing, 312–13
  defect detection by, 307
  field operations and experience, 313–14
  operational characteristics, 315
Thrust lines, 515–16
Torpoint Ferry Joint Committee, 173
Torridge Bridge, 351, 353, 356
Total deficiency rating (TDR), 83–5
Tottnaes Bridge, 781–3
Traffic
  bridge repair problems, 751–2
  control devices, 301–2
  delay cost, 49
  disruption, 143
  influence on dynamic testing, 335–7
  loading on orthotropic steel bridge decks, 466
  redistribution analysis, 23
  strain induced in steel reinforcement, 621–2
  transverse position of, 471–2
  volume, mix and weight estimates, 118
Traffic-induced strain in steel reinforcement, 619–29
  test programme, 623–4
  test results, 624–8
  theoretical calculation, 622–3
Traffic load simulation programme, 563–74
  aim of, 567–8
  basis and requirements for development of, 568–70
  description of modules, 570–3
  layout of, 570–3
  principles of, 568
  usage of, 573–4
Traffic loads, Clause 12:, 577–8
Transport and Road Research Laboratory (TRRL), 140–1
Transversal cracks, effect on mechanical behaviour, 489–99
Truck weight regulations, 411–13
Turner proposal, 411

Ulenbergstrasse Bridge, 400
Ulm, Gänstorbrücke, 287–8
Ultrasonic tests, 162
United States
  bridge management, 63–74
  highway bridges, 405
Untermarchtal prestressed-concrete bridge, 289–91

Vapour permeability measurement, 254–5
Vehicle braking efficiency, 119
Vertical lift bridges, 297, 298
Vibration
  bridge decks, 623
  highway bridges, 607–17
  modes, 163
  tests on masonry arch bridges, 585–94
  velocity measurements, 590

Water in concrete. *See* Concrete
Water/cement ratio, 247
Waterproofing problem, 146–7
Welded joints
  cracks in, 637–9
  orthotropic decks, 466–8
Windowing, 598
Windsor probe, 162
Wolvercote Viaduct, 655
  contract details, 665–6
  description of structure, 656
  equipotential contour mapping, 664
  evaluation of patch repairs, 664
  half-cell potential surface mapping, 661
  monitoring, 659–61
  permanent instrumentation, 664
  remedial work, 659
  survey, 657–9

Yellow Mill Pond Bridge, 323
Yield-line analysis, 432
Young's modulus, 354, 600
Yukon Territory, 578–82

GPSR Compliance

The European Union's (EU) General Product Safety Regulation (GPSR) is a set of rules that requires consumer products to be safe and our obligations to ensure this.

If you have any concerns about our products, you can contact us on ProductSafety@springernature.com

In case Publisher is established outside the EU, the EU authorized representative is:

Springer Nature Customer Service Center GmbH
Europaplatz 3
69115 Heidelberg, Germany

Batch number: 10712211

Printed by Printforce, the Netherlands